THE LIBRARY
ST. MARY'S COLLEGE OF MARYLAND
ST. MARY'S CITY, MARYLAND

D1604472

Selected Titles in This Series

182 **I. S. Krasil'shchik and A. M. Vinogradov, Editors,** Symmetries and conservation laws for differential equations of mathematical physics, 1999

181 **Ya. G. Berkovich and E. M. Zhmud',** Characters of finite groups. Part 2, 1999

180 **A. A. Milyutin and N. P. Osmolovskii,** Calculus of variations and optimal control, 1998

179 **V. E. Voskresenskiĭ,** Algebraic groups and their birational invariants, 1998

178 **Mitsuo Morimoto,** Analytic functionals on the sphere, 1998

177 **Satoru Igari,** Real analysis—with an introduction to wavelet theory, 1998

176 **L. M. Lerman and Ya. L. Umanskiy,** Four-dimensional integrable Hamiltonian systems with simple singular points (topological aspects), 1998

175 **S. K. Godunov,** Modern aspects of linear algebra, 1998

174 **Ya-Zhe Chen and Lan-Cheng Wu,** Second order elliptic equations and elliptic systems, 1998

173 **Yu. A. Davydov, M. A. Lifshits, and N. V. Smorodina,** Local properties of distributions of stochastic functionals, 1998

172 **Ya. G. Berkovich and E. M. Zhmud',** Characters of finite groups. Part 1, 1998

171 **E. M. Landis,** Second order equations of elliptic and parabolic type, 1998

170 **Viktor Prasolov and Yuri Solovyev,** Elliptic functions and elliptic integrals, 1997

169 **S. K. Godunov,** Ordinary differential equations with constant coefficient, 1997

168 **Junjiro Noguchi,** Introduction to complex analysis, 1998

167 **Masaya Yamaguti, Masayoshi Hata, and Jun Kigami,** Mathematics of fractals, 1997

166 **Kenji Ueno,** An introduction to algebraic geometry, 1997

165 **V. V. Ishkhanov, B. B. Lur'e, and D. K. Faddeev,** The embedding problem in Galois theory, 1997

164 **E. I. Gordon,** Nonstandard methods in commutative harmonic analysis, 1997

163 **A. Ya. Dorogovtsev, D. S. Silvestrov, A. V. Skorokhod, and M. I. Yadrenko,** Probability theory: Collection of problems, 1997

162 **M. V. Boldin, G. I. Simonova, and Yu. N. Tyurin,** Sign-based methods in linear statistical models, 1997

161 **Michael Blank,** Discreteness and continuity in problems of chaotic dynamics, 1997

160 **V. G. Osmolovskiĭ,** Linear and nonlinear perturbations of the operator div, 1997

159 **S. Ya. Khavinson,** Best approximation by linear superpositions (approximate nomography), 1997

158 **Hideki Omori,** Infinite-dimensional Lie groups, 1997

157 **V. B. Kolmanovskiĭ and L. E. Shaĭkhet,** Control of systems with aftereffect, 1996

156 **V. N. Shevchenko,** Qualitative topics in integer linear programming, 1997

155 **Yu. Safarov and D. Vassiliev,** The asymptotic distribution of eigenvalues of partial differential operators, 1997

154 **V. V. Prasolov and A. B. Sossinsky,** Knots, links, braids and 3-manifolds. An introduction to the new invariants in low-dimensional topology, 1997

153 **S. Kh. Aranson, G. R. Belitsky, and E. V. Zhuzhoma,** Introduction to the qualitative theory of dynamical systems on surfaces, 1996

152 **R. S. Ismagilov,** Representations of infinite-dimensional groups, 1996

151 **S. Yu. Slavyanov,** Asymptotic solutions of the one-dimensional Schrödinger equation, 1996

150 **B. Ya. Levin,** Lectures on entire functions, 1996

149 **Takashi Sakai,** Riemannian geometry, 1996

(Continued in the back of this publication)

Symmetries and Conservation Laws for Differential Equations of Mathematical Physics

Translations of MATHEMATICAL MONOGRAPHS

Volume 182

Symmetries and Conservation Laws for Differential Equations of Mathematical Physics

A. V. Bocharov, V. N. Chetverikov, S. V. Duzhin,
N. G. Khor'kova, I. S. Krasil'shchik (editor),
A. V. Samokhin, Yu. N. Torkhov,
A. M. Verbovetsky, A. M. Vinogradov (editor)

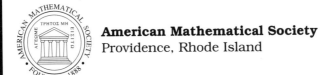

American Mathematical Society
Providence, Rhode Island

EDITORIAL COMMITTEE

AMS Subcommittee
Robert D. MacPherson   Grigorii A. Margulis   James D. Stasheff (Chair)
ASL Subcommittee Steffen Lempp (Chair)
IMS Subcommittee Mark I. Freidlin (Chair)

А. В. Бочаров, А. М. Вербовецкий, А. М. Виноградов (редактор),
С. В. Дужин, И. С. Красильщик (редактор), А. В. Самохин,
Ю. Н. Торхов, Н. В. Хорькова, В. Н. Четвериков

Симметрии и законы сохранения уравнений математической физики

"Факториал", Москва, 1997

The Russian edition was supported in part by the
Russian Foundation for Basic Research under grant #95-01-02825.

Translated from the Russian by A. M. Verbovetsky and I. S. Krasil′shchik

1991 *Mathematics Subject Classification*. Primary 35A30, 58F07;
Secondary 58F05, 58G05.

ABSTRACT. This book is devoted to the geometric theory of differential equations. It covers: ordinary differential equations and their solving by quadratures from the geometric viewpoint; the theory of classical (point) symmetries; contact geometry and its application to first-order partial differential equations; the theory of higher (generalized) symmetries with emphasis on computational techniques and demonstration of their use in solving concrete problems; conservation laws and their computation; Lagrangian formalism; Noether's theorem and relationship between symmetries and conservation laws; Hamiltonian structures on evolutionary equations; nonlocal symmetries; coverings over differential equations; symmetries of integro-differential equations.

The book is rendered as self-contained as possible and includes detailed motivations, extensive examples, and exercises, along with careful proofs of all results.

The book is intended for readers who wish to learn the basics on applications of symmetry methods to differential equations of mathematical physics, but will also be useful for the experts because it assembles a lot of results previously dispersed in numerous journal publications. The text is accessible to advanced graduate students in mathematics, applied mathematics, mathematical and theoretic physics, mechanics, etc.

Library of Congress Cataloging-in-Publication Data

Simmetrii i zakony sokhraneniĭa uravneniĭ matematicheskoĭ fiziki.
English.
Symmetries and conservation laws for differential equations of mathematical physics / A. V. Bocharov ... [et al.], I. S. Krasil′shchik (editor) ... A. M. Vinogradov (editor) ; [translated from the Russian by A. M. Verbovetsky and I. S. Krasil′shchik].
p. cm. — (Translations of mathematical monographs, ISSN 0065-9282 ; v. 182)
Includes bibliographical references and index.
ISBN 0-8218-0958-X (hardcover : alk. paper)
1. Differential equations—Numerical solutions. 2. Symmetry. 3. Conservation laws (Mathematics) 4. Mathematical physics. I. Bocharov, A. V. II. Krasil′shchik, I. S. (Iosif Semenovich) III. Vinogradov, A. M. (Aleksandr Mikhaĭlovich) IV. Title. V. Series.
QC20.7.D5S5613   1999
530.15′535—dc21
                                                               98-53018
                                                               CIP

© 1999 by the American Mathematical Society. All rights reserved.
The American Mathematical Society retains all rights
except those granted to the United States Government.
Printed in the United States of America.

∞ The paper used in this book is acid-free and falls within the guidelines
established to ensure permanence and durability.
Information on copying and reprinting can be found at the back of this volume.
Visit the AMS home page at URL: http://www.ams.org/

Contents

Preface	ix
Chapter 1. Ordinary Differential Equations	1
1. Ordinary differential equations from the geometric viewpoint	1
2. Ordinary differential equations of arbitrary order	6
3. Symmetries of distributions	10
4. Some applications of symmetry theory to integration of distributions	17
4.1. Distributions and characteristic symmetries	17
4.2. Symmetries and dynamical systems	18
4.3. Distributions and noncharacteristic symmetries	20
4.4. Integration of equations by quadratures	21
5. Generating functions	29
6. How to search for equations integrable by quadratures: an example of using symmetries	33
Chapter 2. First-Order Equations	37
1. Contact transformations	37
1.1. Contact elements and the Cartan distribution	37
1.2. Contact transformations	42
1.3. Clairaut equation and its integrals	47
1.4. Contact manifolds in mechanics	49
2. Infinitesimal contact transformations and characteristic fields	50
2.1. Infinitesimal contact transformations	50
2.2. Infinitesimal symmetries of equations	54
2.3. Characteristic vector fields and integration of first-order equations	56
2.4. Symmetries and first integrals	59
3. Complete integrals of first-order differential equations	60
3.1. Complete integrals: a coordinate approach	61
3.2. The construction of complete integrals using symmetry algebras	61
3.3. Complete integrals: an invariant approach	64
3.4. The Lagrange–Charpit method	66
Chapter 3. The Theory of Classical Symmetries	69
1. Equations and the Cartan distribution	69
2. Jet manifolds and the Cartan distribution	72
2.1. Geometric definition of the jet spaces	73
2.2. The Cartan distribution	75
2.3. Integral manifolds of the Cartan distribution	79
3. Lie transformations	83

3.1.	Finite Lie transformations	84
3.2.	Lie fields	89

4. Classical symmetries of equations 92
    4.1. Defining equations 92
    4.2. Invariant solutions and reproduction of solutions 94
5. Examples of computations 96
    5.1. The Burgers equation 96
    5.2. The Korteweg–de Vries equation 98
    5.3. The Khokhlov–Zabolotskaya equation 99
        5.3.1. "Physically meaningful" symmetries 100
        5.3.2. Invariant solutions 101
    5.4. The Kadomtsev–Pogutse equations 102
        5.4.1. Computation of symmetries 103
        5.4.2. Invariant solutions 104
        5.4.3. Reproduction of solutions 106
6. Factorization of equations by symmetries 108
    6.1. Second-order equations in two independent variables 110
7. Extrinsic and intrinsic symmetries 116

Chapter 4. Higher Symmetries 123
1. Spaces of infinite jets and basic differential geometric structures on them 123
    1.1. The manifolds $J^\infty(\pi)$ 124
    1.2. Smooth functions on $J^\infty(\pi)$ 124
    1.3. Prolongations of differential operators 128
    1.4. Vector fields on $J^\infty(\pi)$ 131
    1.5. Differential forms on $J^\infty(\pi)$ 134
    1.6. The horizontal de Rham complex 135
    1.7. Distributions on $J^\infty(\pi)$ and their automorphisms 137
2. The Cartan distribution on $J^\infty(\pi)$ and its infinitesimal automorphisms 138
    2.1. The Cartan distribution 139
    2.2. Integral manifolds 141
    2.3. A computational experiment 142
    2.4. Evolutionary derivations 144
    2.5. Jacobi brackets 148
    2.6. Comparison with Lie fields 148
    2.7. Linearizations 150
3. Infinitely prolonged equations and the theory of higher symmetries 154
    3.1. Prolongations 154
    3.2. Infinitely prolonged equations 156
    3.3. Higher symmetries 158
    3.4. Extrinsic and intrinsic higher symmetries 161
    3.5. Defining equations for higher symmetries 162
4. Examples of computation 164
    4.1. Preparatory remarks 164
    4.2. The Burgers and heat equations 167
    4.3. The plasticity equations 174
    4.4. Transformation of symmetries under change of variables 177

    4.5. Ordinary differential equations   178

Chapter 5. Conservation Laws   185
    1. Introduction: What are conservation laws?   185
    2. The $\mathcal{C}$-spectral sequence   187
        2.1. The definition of the $\mathcal{C}$-spectral sequence   187
        2.2. The term E_0 188
        2.3. The term E_1: preparatory results   189
        2.4. Generalizations   193
        2.5. The term E_1 for $J^\infty(\pi)$ 194
        2.6. The term E_1 in the general case   198
        2.7. Conservation laws and generating functions   201
        2.8. Euler–Lagrange equations   202
        2.9. Hamiltonian formalism on $J^\infty(\pi)$ 203
    3. Computation of conservation laws   206
        3.1. Basic results   206
        3.2. Examples   208
    4. Symmetries and conservation laws   214
        4.1. The Noether theorem   214
        4.2. Hamiltonian equations   216

Chapter 6. Nonlocal Symmetries   221
    1. Coverings   221
        1.1. First examples   221
        1.2. Definition of coverings   224
        1.3. Coverings in the category of differential equations   224
        1.4. Examples of coverings   225
        1.5. Coordinates   226
        1.6. Basic concepts of covering theory   227
        1.7. Coverings and connections   231
        1.8. The horizontal de Rham complex and nonlocal conservation
              laws   231
        1.9. Covering equations   232
        1.10. Horizontal de Rham cohomology and coverings   234
        1.11. Bäcklund transformations   236
    2. Examples of computations: coverings   238
        2.1. Coverings over the Burgers equation   239
        2.2. Coverings over the Korteweg–de Vries equation   242
        2.3. Coverings over the equation $u_t = (B(u)u_x)_x$ 245
        2.4. Covering over the f-Gordon equation   245
        2.5. Coverings of the equation $u_{xx} + u_{yy} = \varphi(u)$ 246
    3. Nonlocal symmetries   249
        3.1. Definition of nonlocal symmetries   249
        3.2. How to find nonlocal symmetries?   249
    4. Examples of computation:
         nonlocal symmetries of the Burgers equation   251
    5. The problem of symmetry reconstruction   257
        5.1. Universal Abelian covering   257
        5.2. Symmetries in the universal Abelian covering   258

	5.3.	Nonlocal symmetries for equations admitting a recursion operator	258
	5.4.	Example: nonlocal symmetries of the Korteweg–de Vries equation	259
	5.5.	Master symmetries	260
	5.6.	Examples	261
	5.7.	General problem of nonlocal symmetry reconstruction	262
	5.8.	Kiso's construction	263
	5.9.	Construction of the covering τ_S	264
	5.10.	The universal property of the symmetry S_τ	265
6.		Symmetries of integro-differential equations	266
	6.1.	Transformation of integro-differential equations to boundary differential form	266
	6.2.	Spaces of $(k, \mathcal{G})$-jets	271
	6.3.	Boundary differential operators	275
	6.4.	The Cartan distribution on $J^\infty(\pi; \mathcal{G})$	279
	6.5.	$\mathcal{G}$-invariant symmetries of the Cartan distribution on $J^\infty(\pi; \mathcal{G})$	284
	6.6.	Higher symmetries of boundary differential equations	287
	6.7.	Examples	290

Appendix.	From Symmetries of Partial Differential Equations Towards Secondary ("Quantized") Calculus	301
Introduction		301
1.	From symmetries to concepts	302
2.	"Troubled times" of quantum field theory	303
3.	"Linguization" of the Bohr correspondence principle	304
4.	Differential equations are diffieties	306
5.	Secondary ("quantized") functions	308
6.	Higher-order scalar secondary ("quantized") differential operators	310
7.	Secondary ("quantized") differential forms	312
8.	Quantization or singularity propagation? Heisenberg or Schrödinger?	314
9.	Geometric singularities of solutions of partial differential equations	316
10.	Wave and geometric optics and other examples	320
	10.1. Σ-characteristic equations	320
	10.2. Maxwell's equations and geometric optics	320
	10.3. On the complementary equations	321
	10.4. Alternative singularities via the homogenization trick	322
	10.5. $\mathbb{R}_{(k)}$-characteristic equations	322
Bibliography		323
Index		329

Preface

The classical symmetry theory for general systems of partial differential equations was created by Sophus Lie more than 100 years ago. The concepts of Lie groups and Lie algebras, so fundamental for modern mathematics, were discovered by Lie [**68**] during these studies. Most of Lie's basic ideas and results on transformation groups were later worked out in numerous papers and books, while his heritage in differential equations remained aside of these developments. The first attempts, after Lie, to apply Lie theory systematically to continuum mechanics were made by L. V. Ovsiannikov and his collaborators in 1950–60 (see [**92**]).

A new, non-Lie epoch in the symmetry theory for partial differential equations began with the discovery of "completely integrable" systems and with subsequent development of the inverse scattering problem method [**1, 22, 28, 89**]. It is well known that any completely integrable equation generates a whole hierarchy consisting of "higher analogs" of the initial equation. Studying these analogs made it possible to understand them as symmetries of a certain equation. Nevertheless, this approach did not comply with Lie theory, and it was the theory of infinite jet spaces which allowed mathematicians to construct the concept of "higher symmetries".

Informally speaking, classical (i.e., Lie type) symmetries are analytically described in terms of independent and dependent variables and first-order derivatives of the latter, whereas non-Lie symmetries may depend on derivatives of arbitrary order. What is more essential is that classical infinitesimal symmetries are vector fields on the submanifold determined in the corresponding jet manifold by the initial equation, whereas "higher" (i.e., nonclassical) ones are cohomology classes of a natural differential complex related to the so-called infinite prolongation of the equation at hand. Therefore, a higher infinitesimal symmetry does not, in general, generate a one-parameter group of (local) diffeomorphisms on the space of its solutions. In other words, the usual relation between Lie groups and Lie algebras ceases to exist in this context. Nevertheless, it still exists "virtually" and always materializes when appropriate additional conditions arise in the problem (e.g., when boundary problems are considered). This nonclassical cohomological approach becomes even more essential when the theory of conservation laws is constructed. Note that *a priori* it was difficult even to assume that the theory of conserved quantities (integrals, fluxes, etc.) admitted by a given system of partial differential equations may be based on homological algebra and uses the theory of spectral sequences as its main technique.

The foundations of the theory of higher symmetries and conservation laws were developed by one of us in 1975–77 [**60, 127, 129**]. Later on, it was tested in particular computations, and sufficiently efficient computational methods were obtained. These methods were applied to compute higher symmetries and conservation laws for particular systems of nonlinear partial differential equations [**137, 142**].

Besides, the most important theoretical details were established [**130**]. They revealed striking parallels between the calculus of vector fields and differential forms on smooth manifolds, on one hand, and the calculus of symmetries and conservation laws, on the other hand. This observation resulted in the conjecture that these parallels are of a much more general nature and can be extended to all concepts of differential calculus. Subsequent analysis led to discovery of *Secondary Differential Calculus*, which is both a powerful tool in the study of general systems of partial differential equations and a natural language for constructing modern quantum field theory on a nonperturbative basis.

Higher symmetries and conservation laws are "local" quantities, i.e., they depend on unknown functions (or "fields", if one uses physical terminology) and on their derivatives. This framework, however, becomes insufficient to describe some important concepts, such as Bäcklund transformations or recursion operators: the reason is that in most interesting cases they operate with "nonlocal" quantities (e.g., integrals of local objects). Note also that integro-differential equations, where nonlocal quantities are present from the very beginning, are of independent interest for the theory of symmetries and conservation laws. The desired extension of the theory is achieved by introducing the notion of coverings [**62**] in the category of differential equations.

The fundamentals of the higher symmetry theory[1] and the corresponding computational algorithms are elaborated sufficiently well now. Concerning the latter, one needs to distinguish between algorithmic techniques oriented to computer applications (see, for example, the collection [**137**]) and analytical methods of studying various model systems. It this context, it is necessary to mention works of the "Ufa school" (see, for example, [**85, 112, 116**] and many other publications). The results of analysis for particular equations and systems are scattered over numerous publications. The most representative (but unfortunately becoming more and more obsolete) digest can be found in [**44, 45, 46**].

There are now a lot of books dealing with geometrical aspects of differential equations and, in particular, with their symmetries (see [**4, 13, 43, 91, 92, 114, 148**]). The most consistent exposition of the geometric and algebraic foundations of the theory of symmetry and conservation laws for partial differential equations is contained in the monograph [**60**] and in the paper [**132**], but these can hardly be considered as "user-friendly" texts. Their deficiencies are partially compensated by [**131**] and by [**61, 62**] (the last two deal with the nonlocal theory). But they do not fill the gap completely.

Therefore the idea of writing a more popular and at the same time mathematically rigorous text on symmetries and conservation laws aroused immediately after the publication of [**60**]. Some chapters of the book you hold now were written then: S. V. Duzhin wrote a chapter on symmetries of ordinary differential equations and a draft version of the chapter on classical symmetries, A. V. Bocharov wrote the chapter on symmetries of first-order equations, and I. S. Krasil'shchik prepared a chapter on higher symmetries. For a number of "historical" reasons, this project could not come to a conclusion then, and we did not return to it until ten years later.

---

[1] In modern literature, the terms *generalized symmetries* (see [**91**]) and *Lie–Bäcklund symmetries* ([**43**]) are also used.

A lot of things changed during this period, including our attitudes towards some aspects of the algebro-geometric foundations of symmetry theory. For this reason, some of the old chapters were considerably updated: using A. V. Bocharov's text, Yu. N. Torkhov wrote the final version of Chapter 2. Chapter 3, in its present form, was written by A. V. Samokhin. I. S. Krasil'shchik prepared the revised version of Chapter 4. We also decided that the book would be incomplete without mentioning conservation law theory, and the corresponding chapter was written by A. M. Verbovetsky. Finally, Chapter 6 was written by N. G. Khor'kova (except V. N. Chetverikov wrote the section on symmetries of integro-differential equations). We also supplied the book with an Appendix, containing the suitably adapted text of A. M. Vinogradov's paper on the Secondary Differential Calculus, describing a deeper perspective on the geometric theory of partial differential equations.

In this book, we expound the basics of the theory of higher and nonlocal symmetries for general systems of partial differential and integro-differential equations. The book concentrates on the following questions:

- What are the higher symmetries and conservation laws admitted by a given system of differential equations?
- What are efficient methods for computing them?
- If we find a symmetry or a conservation law, how can we use it?

Concerning the last question, we had to restrict ourselves to the simplest and most straightforward applications of the theory. A more detailed exposition would need another couple of volumes (which we hope to write in the future).

We have tried to take into account the interests of two groups of readers of this book: those who are mainly interested in theoretical aspects, and those who are interested in applications. To the latter (and we believe that people who work in theoretical and mathematical physics, field theory, and continuum mechanics belong to this group) we advise skipping the proofs and general conceptual discussions and mostly paying attention to the algorithms and techniques needed in particular computations. We hope that our exposition of these matters is sufficiently clear. On the other hand, we saw no way to make the exposition more popular by passing to the language of local coordinates, which is standard in mathematical physics. For understanding the conceptual part of the theory, as well as for efficient use of the algorithms, such an approach would be self-defeating. We can claim here that a lot of papers on the topics of this book are fighting with coordinate difficulties rather than with real problems.

We hope that after reading this book you will not only be able to look from a different point of view at various problems related to nonlinear differential equations, but will take pen and paper (or switch on your computer and load a package for symbolic computations) to find something new and interesting in your own professional problems. Formally speaking, to start looking for symmetries and conservation laws, it suffices to know mathematics as taught in the standard university courses, and to use the formulas presented in this book. To understand the material more deeply, one needs knowledge on:

- geometry of smooth manifolds and vector bundles over these manifolds [**3, 25, 95, 115, 146**],
- symplectic geometry [**7, 143**],
- Lie groups and Lie algebras [**94, 105, 122**],
- commutative algebra [**9**],

- homological algebra [**15, 31, 35, 79, 84**].

In addition, for "philosophical reasons", we also recommend at least a preliminary knowledge of algebraic geometry [**108**] and category theory [**31, 79**].

Since the late sixties, the authors of this book have been participants in the seminar on the algebraic and geometric foundations of the theory of nonlinear differential equations at the Faculty of Mathematics and Mechanics of Moscow State University. On behalf of all the authors, we would like to express our gratitude to all the participants who helped us both in forming a general approach and in clarifying particular problems. We are especially grateful to: D. M. Gessler, V. V. Lychagin, V. E. Shemarulin, M. M. Vinogradov, and V. A. Yumaguzhin. We are also grateful to one of our first Russian readers, A. V. Shepetilov, for his useful remarks on some errors in the book.

Our special thanks are due to the Russian Foundation for Basic Research (RFFI), whose financial support was crucial for the publication of the Russian edition. While writing the book, I. S. Krasil'shchik, A. V. Samokhin, and A. M. Verbovetsky were also partially supported by RFFI Grant 97-01-00462, and Yu. N. Torkhov was supported by RFFI Grant 96-01-01360.

It is our pleasure to acknowledge the use of Paul Taylor's Commutative Diagrams package for typesetting commutative diagrams.

Finally, we would like to point out that this book can be considered as an introduction to the monograph series on the foundations of the secondary calculus launched by the Diffiety Institute of the Russian Academy of Natural Sciences. Hot information on the Diffiety Institute's activities can be found on the World Wide Web starting at URL:

```
http://www.botik.ru/~diffiety/
```

or

```
http://ecfor.rssi.ru/~diffiety/
```

A. M. Vinogradov
I. S. Krasil'shchik

CHAPTER 1

Ordinary Differential Equations

Of all differential equations, two classes, ordinary differential equations and scalar partial differential equations of order one, stand out because of a number of features facilitating their study and making the theory more readily applicable. This chapter discusses the geometric approach to ordinary differential equations. Here we first introduce, in the simplest situation, the concepts that will be extensively used throughout the book. We also show that the geometric theory of symmetries makes it possible to understand and generalize the standard procedures of explicit integration of ordinary differential equations, and to obtain new results in this direction as well.

1. Ordinary differential equations from the geometric viewpoint

It is well known (see, e.g., [6]) that a first-order ordinary differential equation solved for the derivative

$$u' = f(x, u) \tag{1.1}$$

can be geometrically interpreted as a vector field on the (x, u)-plane. To this end, at each point (x_0, u_0) one has to consider the vector $(1, f(x_0, u_0))$ and the operator $\partial/\partial x + f(x_0, u_0)\partial/\partial u$ of derivation in the direction of this vector. The trajectories of the field $\partial/\partial x + f(x, u)\partial/\partial u$ are called the *integral curves* of equation (1.1). They are graphics of solutions of the equation under consideration (see Figure 1.1).

To interpret the equation

$$F(x, u, u') = 0 \tag{1.2}$$

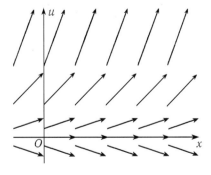

FIGURE 1.1. The vector field $X = \partial/\partial x + u\partial/\partial u$ corresponding to the differential equation $u' = u$

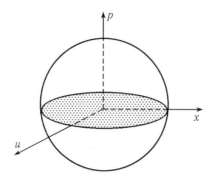

FIGURE 1.2

in the same fashion, one needs to solve it for the derivative u'. When doing so, the following difficulty is encountered: several values of u' (or no values at all) may correspond to some values of x and u by (1.2). Moreover, in a neighborhood of a singular point (that is, of a point where $\partial F/\partial u' = 0$ and, hence, the implicit function theorem is not valid), the function expressing u' in terms of x and u, even though it is well-defined, need not be smooth.

EXAMPLE 1.1. Let
$$F(x, u, u') = u'^2 + u^2 + x^2 - 1.$$
Then $u' = \pm\sqrt{1 - x^2 - u^2}$. This function is defined on the disk $x^2 + u^2 \leq 1$. It is two-valued inside the disk and its derivatives become infinite on the boundary (see Figure 1.2).

These difficulties can be overcome in the following way. Consider the space $\mathbb{R}^3$ with the coordinates x, y, p and the surface $\mathcal{E}$ given by $F(x, u, p) = 0$. To every solution $u = f(x)$ of equation (1.2) there corresponds the curve on this surface defined by the equations
$$u = f(x), \qquad p = f'(x). \tag{1.3}$$

The coordinate x can be taken as a parameter, i.e., the projection of this curve to the x-axis is a diffeomorphism. Furthermore, the functions expressing the coordinates u and p in terms of x are not arbitrary: the latter is the derivative of the former. Consequently, not every curve in $\mathbb{R}^3$ can be written in the form (1.3).

Since $p_0 = f'(x_0)$, the tangent vector to the curve (1.3) at a point $a = (x_0, u_0, p_0)$ takes the form
$$\frac{\partial}{\partial x} + p_0 \frac{\partial}{\partial u} + f''(x_0) \frac{\partial}{\partial p}. \tag{1.4}$$
Therefore this vector lies in the plane given by the equation
$$u - u_0 = p_0(x - x_0), \tag{1.5}$$
i.e., it belongs to the kernel of the 1-form
$$\omega = du - p\, dx \tag{1.6}$$
at the point a. It is also obvious that the span of all vectors of the form (1.4) coincides with the entire plane (1.5).

The converse statement holds as well: a curve in $\mathbb{R}^3$ projecting diffeomorphically to the x-axis and integral for the 1-form ω (or, which is the same, for the distribution $\mathcal{C}$ of codimension 1 given by the form ω) has the form (1.3).

Thus, the two-dimensional distribution $\mathcal{C}$ given by the 1-form (1.6) is the geometric structure which distinguishes in a natural way the class of curves corresponding to solutions of ordinary differential equations of the first order. This distribution $\mathcal{C}$ is called the *Cartan distribution*.

By means of the Cartan distribution, solutions of a differential equation $\mathcal{E} \subset \mathbb{R}^3$ can be interpreted as integral curves of the distribution $\mathcal{C}$ belonging to the surface $\mathcal{E}$ and projecting to the x axis without degeneration.

Note that the 2-form $d\omega = dx \wedge dp$ cannot be written as $\gamma \wedge \omega$, where γ is a 1-form. Hence, by the Frobenius theorem (see, e.g., [**115, 146**]), the Cartan distribution is not completely integrable. Therefore its maximal integral manifolds are one-dimensional, and the set of points where the plane of the Cartan distribution is tangent to the surface $\mathcal{E}$ is a closed nowhere dense subset of $\mathcal{E}$. We call such points *singular*. Thus, the set of nonsingular points of the surface $\mathcal{E}$ is open and everywhere dense.

Singular points of equation (1.2) can be found from the condition that the differential dF and the form ω are collinear at these points. This condition can be written as two relations:

$$\frac{\partial F}{\partial p} = 0, \qquad \frac{\partial F}{\partial x} + p\frac{\partial F}{\partial u} = 0.$$

If $a \in \mathcal{E}$ is a nonsingular point, then the intersection of the plane tangent to the surface $\mathcal{E}$ at the point a with the plane $\mathcal{C}_a$ of the distribution $\mathcal{C}$ is a straight line l_a tangent to $\mathcal{E}$ at a. Thus on $\mathcal{E}$ there arises the direction field (one-dimensional distribution) $a \mapsto l_a$, which will be denoted by $\mathcal{C}(\mathcal{E})$. A curve Γ is integral for the direction field $\mathcal{C}(\mathcal{E})$ if and only if Γ is an integral curve of the distribution $\mathcal{C}$ and lies on $\mathcal{E}$ (with the exception of the set of singular points). Therefore, it can be concluded that solutions of the equation $\mathcal{E}$ are integral curves of the direction field $\mathcal{C}(\mathcal{E})$ projecting to the x-axis without degeneration.

Note that arbitrary integral curves of the direction field $\mathcal{C}(\mathcal{E})$, which do not necessarily satisfy the condition of nonsingularity of projection to the x-axis, can be interpreted as multi-valued solutions of the equation $\mathcal{E}$.

EXAMPLE 1.2. The curve given by the equations

$$\begin{cases} u^3 - u + x = 0, \\ 3u^2 p - p + 1 = 0, \end{cases}$$

lies on the surface

$$(3x - 2u)p = u$$

(see Figure 1.3) and is integral for the Cartan distribution, but it is not of the form (1.3). Hence this curve is a multi-valued solution to the equation $(3x - 2u)u' = u$.

EXAMPLE 1.3. From the geometric point of view, the equation

$$\left(\frac{du}{dx}\right)^2 + u^2 = 1$$

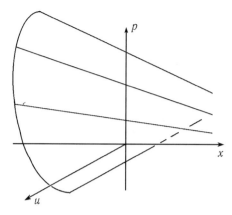

FIGURE 1.3. The graph of the surface $(3x - 2u)p = u$ for $p \geq -1/2$

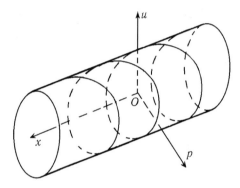

FIGURE 1.4

is the cylinder $\mathcal{E}$ (see Figure 1.4) in the space $\mathbb{R}^3$ with coordinates x, u, p defined by the relation $p^2 + u^2 = 1$. Consider the coordinate system (x, φ) on this cylinder, with φ being defined by

$$u = \sin \varphi, \qquad p = \cos \varphi.$$

Then the restriction of $\omega = du - p\, dx$ to $\mathcal{E}$ is

$$\omega_{\mathcal{E}} = \cos \varphi\, d(\varphi - x).$$

The points of tangency of the distribution $\mathcal{C}$ and the cylinder $\mathcal{E}$ are those where $\omega_{\mathcal{E}}$ vanishes. It is obvious that these points are situated on the two straight lines obtained by intersection of $\mathcal{E}$ with the plane $p = 0$. These are singular solutions $u = \pm 1$ to the equation at hand. Further, the equation $\omega_{\mathcal{E}} = 0$ reduces to the equation $d(\varphi - x) = 0$ and implies that the integral curves of the distribution $\mathcal{C}(\mathcal{E})$ are the circular helices

$$\varphi = x + c, \qquad c \in \mathbb{R}.$$

All these curves project to the x-axis without degeneration and give rise to usual solutions. Expressing φ in terms of u, these solutions can be written in the following explicit form:

$$u = \sin(x + c).$$

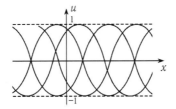

FIGURE 1.5

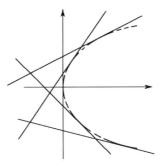

FIGURE 1.6

Projecting to the (x, u)-plane, the solutions fill the region $|u| \leq 1$ (see Figure 1.5).

The singular solutions (the straight lines $u = \pm 1$) are envelopes of the sine curves.

EXAMPLE 1.4. Consider the Clairaut equation
$$u - x\frac{du}{dx} = f\left(\frac{du}{dx}\right),$$
where f is a smooth function. The form of the surface $\mathcal{E} = \{u - xp = f(p)\}$ depends on the choice of f. As coordinates on $\mathcal{E}$, we can take the variables x, p. The form ω in these coordinates is written as
$$\omega_{\mathcal{E}} = d(xp + f(p)) - p\,dx = (x + f'(p))\,dp.$$
This form vanishes at the points of the surface $\mathcal{E}$ belonging to the curve $x = -f'(p)$. The projection of this curve to the (x, u)-plane will be the graph of a singular solution of the Clairaut equation. This solution can be obtained in analytic form by solving the equation $x = -f'(p)$ for p and substituting the result to the equation $u - xp = f(p)$. For example, if $f(p) = p^\alpha$, $\alpha < 0$, then
$$u = (1 - \alpha)\left(-\frac{x}{\alpha}\right)^{\frac{\alpha}{\alpha-1}}.$$
Note that a singular solution can be multi-valued (e.g., $u = \pm 2\sqrt{x}$ for $\alpha = -1$; see Figure 1.6).

The integral curves of the distribution $\mathcal{C}(\mathcal{E})$ at nonsingular points can be found from the equation $\omega_{\mathcal{E}} = 0$ subject to the condition $x + f'(p) \neq 0$. This yields $dp = 0$, $p = c$, and $u = cx + f(c)$, where $c \in \mathbb{R}$.

As in Example 1.3, we see that singular solutions are envelopes of one-parameter families of nonsingular solutions.

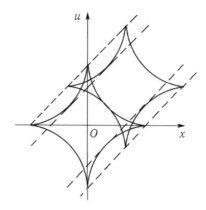

FIGURE 1.7

Let us remark that the formal procedure for searching the envelope, i.e., elimination of p from
$$F(x, u, p) = 0,$$
$$\frac{\partial F(x, u, p)}{\partial p} = 0, \qquad (1.7)$$
can yield a curve with cusps and points of tangency to integral curves[1].

Consider, for instance, the equation [**100**]
$$(x-u)^2 \left(1 + \left(\frac{du}{dx}\right)^2\right)^3 = a^2 \left(1 + \left(\frac{du}{dx}\right)^3\right)^2, \qquad a > 0.$$
It can easily be checked that the functions defined by
$$(x-c)^{2/3} + (y-c)^{2/3} = a^{2/3},$$
are solutions to this equation, the solutions of (1.7) being of the form
$$x - u = \pm a \quad \text{or} \quad x - u = \pm a/\sqrt{2}.$$
This family of curves contains envelopes of solutions as well as cusps and points of tangency to integral curves (see Figure 1.7).

2. Ordinary differential equations of arbitrary order

By analogy with §1, we can develop the geometric theory of ordinary differential equations of arbitrary order.

The equation
$$F\left(x, u, \frac{du}{dx}, \ldots, \frac{d^k u}{dx^k}\right) = 0$$
can be interpreted as a hypersurface in the $(k+2)$-dimensional space $\mathbb{R}^{k+2}$ with coordinates $x, u, p_1, \ldots, p_k$ defined by $F(x, u, p_1, \ldots, p_k) = 0$.

---

[1] For more details on finding envelopes, see [**27, 97**]. A detailed discussion of singular solutions to differential equations can be found in [**47, 100, 113**].

2. ORDINARY DIFFERENTIAL EQUATIONS OF ARBITRARY ORDER

Consider the 2-dimensional distribution in $\mathbb{R}^{k+2}$ given by

$$\omega_0 = du - p_1 \, dx,$$
$$\omega_1 = dp_1 - p_2 \, dx,$$
$$\dots\dots\dots\dots\dots\dots$$
$$\omega_{k-1} = dp_{k-1} - p_k \, dx.$$

For $k = 1$, this distribution coincides with the Cartan distribution discussed in §1, so we shall likewise call it the *Cartan distribution* and denote it by $\mathcal{C}$.

Note that a curve in $\mathbb{R}^{k+2}$ is the graph of a function $u = f(x)$ and of its derivatives

$$L_f = \{\, u = f(x),\ p_1 = f'(x), \dots, p_k = f^{(k)}(x) \,\}$$

if and only if it projects to the x-axis without degeneration and is an integral curve for the distribution $\mathcal{C}$ (in §1 we proved this for $k = 1$).

Using the Cartan distribution $\mathcal{C}$, one can give a geometric interpretation to the well-known method for integrating ordinary differential equations not containing the independent variable in explicit form.

EXAMPLE 2.1 (see also [91]). Let us consider a second-order equation of the form $F(u, du/dx, d^2u/dx^2) = 0$ or, in other words, a hypersurface $\mathcal{E}$ in $\mathbb{R}^4$ defined by

$$F(u, p_1, p_2) = 0.$$

Restrict the distribution given by 1-forms

$$\omega_0 = du - p_1 \, dx, \quad \omega_1 = dp_1 - p_2 \, dx \tag{2.1}$$

to this hypersurface. The hypersurface $\mathcal{E}$ and the forms ω_0, ω_1 (and, hence, the distribution $\mathcal{C}$) are invariant under translations along the x-axis, i.e., transformations $t_c\colon (x, u, p_1, p_2) \mapsto (x + c, u, p_1, p_2)$, where c is a constant. This makes it possible to construct the quotient equation by the variable x of which the equation is independent (more precisely, by the action of one-parameter group T of parallel translations t_c).

In other words, let us consider the quotient mapping

$$\pi\colon (x, u, p_1, p_2) \longmapsto (u, p_1, p_2)$$

from $\mathbb{R}^4$ to the 3-dimensional space $\mathbb{R}^4/T$. The image of the hypersurface $\mathcal{E}$ under this mapping is the hypersurface $\mathcal{E}/T$ in the space $\mathbb{R}^4/T$ given by the same equation $F(u, p_1, p_2) = 0$. The distribution $\mathcal{C}$ projects without degeneration to the space $\mathbb{R}^4/T$ and, thus, gives rise to the two-dimensional quotient distribution $\mathcal{C}/T$. This distribution can be defined by a 1-form on $\mathbb{R}^4/T$: for example, one can take the following linear combination of the forms (2.1):

$$\widetilde{\omega} = \omega_1 - \frac{p_2}{p_1}\omega_0 = dp_1 - \frac{p_2}{p_1}\, du.$$

Thus, as a result of the factorization, we obtain the 3-dimensional space $\mathbb{R}^4/T$, endowed with the 2-dimensional distribution given by $\widetilde{\omega} = 0$, and the 2-dimensional surface $\mathcal{E}/T \subset \mathbb{R}^4/T$, i.e., an ordinary first-order differential equation. This will be in complete agreement with §1 if we identify the form $\widetilde{\omega}$ with the canonical form ω. To this end, let us introduce the coordinates

$$x' = u, \quad u' = p_1, \quad p' = \frac{p_2}{p_1} \tag{2.2}$$

on $\mathbb{R}^4/T$. Then we have $\widetilde{\omega} = du' - p'dx'$. In the new coordinates, the equation of $\mathcal{E}/T$ takes the form

$$F'(x', u', p') = 0,$$

where $F'(x', u', p') = F(x', u', p'u')$.

Thus, the quotient mapping π takes the initial equation of order two to a first-order equation

$$F'\left(x', u', \frac{du'}{dx'}\right) = 0,$$

which is called the *quotient equation*. Therein lies the geometric meaning of the change of coordinates (2.2).

REMARK 2.1. The distribution $\mathcal{C}/T$ can be defined by another form. Consider, for example, the form

$$\widetilde{\omega} = \omega_0 - \frac{p_1}{p_2}\omega_1 = du - \frac{p_1}{p_2}dp_1 = du - \frac{1}{p_2}d\left(\frac{p_1^2}{2}\right).$$

To identify it with the canonical form ω, we make the change of coordinates

$$x' = \frac{p_1^2}{2}, \quad u' = u, \quad p' = \frac{1}{p_2},$$

which reduces the initial equation to the first-order equation

$$F\left(u', \pm\sqrt{2x'}, \frac{dx'}{du'}\right) = 0.$$

In a similar manner, one can reduce by 1 the order of any equation not explicitly depending on x. Factoring by x the space $\mathbb{R}^{k+2}$ with the coordinates $x, u, p_1, \ldots, p_k$, we obtain the space $\mathbb{R}^{k+2}/T$ with the coordinates $u, p_1, \ldots, p_k$. The image of the Cartan distribution $\mathcal{C}$ on $\mathbb{R}^{k+2}$ under this factorization is the two-dimensional distribution $\mathcal{C}/T$ on $\mathbb{R}^{k+2}/T$ given by

$$\widetilde{\omega}_0 = \omega_1 - \frac{p_2}{p_1}\omega_0 = dp_1 - \frac{p_2}{p_1}du,$$

$$\cdots\cdots\cdots\cdots\cdots\cdots\cdots\cdots\cdots\cdots$$

$$\widetilde{\omega}_{k-2} = \omega_{k-1} - \frac{p_k}{p_1}\omega_0 = dp_{k-1} - \frac{p_k}{p_1}du.$$

We claim that the coordinates $u, p_1, \ldots, p_k$ in the space $\mathbb{R}^{k+2}/T$ can be changed for new coordinates $x', u', p'_1, \ldots, p'_{k-1}$ such that the quotient distribution $\mathcal{C}/T$ is given by

$$\widetilde{\omega}'_0 = du' - p'_1\,dx', \ldots, \widetilde{\omega}'_{k-2} = dp'_{k-2} - p'_{k-1}\,dx'.$$

Indeed, define x', u', and $p' = p'_1$ by formulas (2.2). Since

$$dp'_1 = d\left(\frac{p_2}{p_1}\right) = \frac{1}{p_1}dp_2 - \frac{p_2}{p_1^2}dp_1,$$

we can put
$$\widetilde{\omega}_1' = \frac{1}{p_1}\widetilde{\omega}_1 - \frac{p_2}{p_1^2}\widetilde{\omega}_0 = dp_1' - \left(\frac{p_3}{p_1^2} - \frac{p_2^2}{p_1^3}\right)du.$$
Therefore
$$p_2' = \frac{p_3}{p_1^2} - \frac{p_2^2}{p_1^3}.$$
Continuing this line of reasoning, we obtain the desired change of variables. If we replace the variables $u, \ldots, p_k$ in the initial k-order equation
$$F(u, p_1, \ldots, p_k) = 0$$
by $x', u', \ldots, p_{k-1}'$, we get the quotient equation of order $k-1$.

As it was noted above, the restriction $\mathcal{C}(\mathcal{E})$ of the two-dimensional Cartan distribution on $\mathbb{R}^{k+2}$ to the hypersurface $\mathcal{E}$ is one-dimensional at all points except for a nowhere dense set of singular points where the tangent plane to $\mathcal{E}$ contains the plane of $\mathcal{C}$. Therefore, the distribution $\mathcal{C}(\mathcal{E})$ can be defined by one vector field. Let us explicitly describe such a field and establish the conditions distinguishing between singular and nonsingular points.

Let
$$Y = \alpha\frac{\partial}{\partial x} + \beta_0\frac{\partial}{\partial u} + \beta_1\frac{\partial}{\partial p_1} + \cdots + \beta_k\frac{\partial}{\partial p_k}$$
be a field lying in the distribution $\mathcal{C}$ and tangent to the surface $\mathcal{E}$. Since the field Y belongs to the Cartan distribution, we have
$$\omega_i(Y) = 0, \quad i = 0, 1, \ldots, k-1,$$
i.e., for all i except for $i = k$ we have $\beta_i = p_{i+1}\alpha$. The requirement that the field Y is tangent to $\mathcal{E}$ is equivalent to the condition $Y(F)|_\mathcal{E} = 0$, i.e., to existence of a function λ such that
$$\alpha\left(\frac{\partial F}{\partial x} + p_1\frac{\partial F}{\partial u} + p_2\frac{\partial F}{\partial p_1} + \cdots + p_k\frac{\partial F}{\partial p_{k-1}}\right) + \beta_k\frac{\partial F}{\partial p_k} = \lambda F. \tag{2.3}$$
Obviously, we can satisfy this condition by putting
$$\lambda = 0, \quad \alpha = -\frac{\partial F}{\partial p_k}, \quad \beta_k = \frac{\partial F}{\partial x} + p_1\frac{\partial F}{\partial u} + \cdots + p_k\frac{\partial F}{\partial p_{k-1}}.$$
Thus, the desired vector field can be written as
$$Y_F = -\frac{\partial F}{\partial p_k}\left(\frac{\partial}{\partial x} + p_1\frac{\partial}{\partial u} + \cdots + p_k\frac{\partial}{\partial p_{k-1}}\right)$$
$$+ \left(\frac{\partial F}{\partial x} + p_1\frac{\partial F}{\partial u} + \cdots + p_k\frac{\partial F}{\partial p_{k-1}}\right)\frac{\partial}{\partial p_k}.$$

In particular, for a first-order equation $F(x, u, p) = 0$, the field Y_F takes the form
$$Y_F = -\frac{\partial F}{\partial p}\left(\frac{\partial}{\partial x} + p\frac{\partial}{\partial u}\right) + \left(\frac{\partial F}{\partial x} + p\frac{\partial F}{\partial u}\right)\frac{\partial}{\partial p}. \tag{2.4}$$

If the equation is solved with respect to the derivative, i.e., $F = -p + f(x, u)$, then $\mathcal{E} = \{p = f(x, u)\}$ and
$$Y_F = \frac{\partial}{\partial x} + p\frac{\partial}{\partial u} + (f_x + pf_u)\frac{\partial}{\partial p}.$$

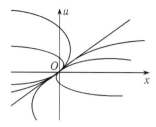

FIGURE 1.8

The projection of this field to the (x, u)-plane has the form
$$\frac{\partial}{\partial x} + f(x, u)\frac{\partial}{\partial u}.$$
The trajectories of the field obtained are, as we know, integral curves for the equation at hand.

The point P of the equation $\mathcal{E}$ is singular if relation (2.3) at this point holds for all α and β_k. Since the right-hand side of this relation vanishes on $\mathcal{E}$, the equalities
$$F = \frac{\partial F}{\partial x} + p_1\frac{\partial F}{\partial u} + \cdots + p_k\frac{\partial F}{\partial p_{k-1}} = \frac{\partial F}{\partial p_k} = 0$$
are necessary and sufficient condition for the point under consideration to be singular. Thus, the singular points of the hypersurface $\mathcal{E}$ are those where the field Y_F vanishes.

The field Y_F is called the *characteristic field* of the equation $\mathcal{E}$. The projections of its trajectories to the (x, u)-plane are the graphs of solutions of this equation.

EXAMPLE 2.2. Consider the equation $(3x - 2u)u' = u$. Then $F = (3x - 2u)u' - u$, and by formula (2.4) we have
$$Y_F = (2u - 3x)\frac{\partial}{\partial x} + p(2u - 3x)\frac{\partial}{\partial u} + (2p - 2p^2)\frac{\partial}{\partial p}$$
$$= (2u - 3x)\frac{\partial}{\partial x} - u\frac{\partial}{\partial u} + (2p - 2p^2)\frac{\partial}{\partial p}$$
(the last equality holds on the equation, i.e., on the surface $\mathcal{E} = \{(3x - 2u)p = u\}$). Let us take x and u as coordinates on $\mathcal{E}$ and find the trajectories of the field Y_F from the system of equations
$$\begin{cases} \dot{x} = 2u - 3x, \\ \dot{u} = -u. \end{cases}$$

Solving this system, we get $x = C_1 e^{-t} + C_2 e^{-3t}$, $u = C_1 e^{-t}$. Hence $x = u + au^3$, $a = \text{const}$. The family of trajectories of Y_F is shown in Figure 1.8. Note that at the point $(0, 0, 0)$, which is a singular point of the equation under consideration, the uniqueness theorem for solutions of ordinary differential equations is not valid.

3. Symmetries of distributions

Since solutions of ordinary differential equations (including the multi-valued ones) are integral curves for the corresponding Cartan distributions, we begin the

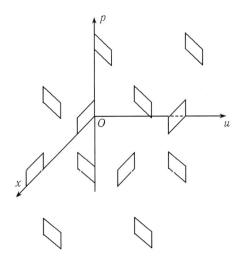

FIGURE 1.9. The Cartan distribution $\mathcal{C}$ on $\mathbb{R}^3$

study of symmetries of differential equations by discussing symmetries in the more general context of distributions.

Recall (see [**115, 146**]) that a p-dimensional *distribution* P on a manifold M of dimension n is a correspondence $a \mapsto P_a$ that takes each point $a \in M$ to a p-dimensional subspace P_a of the tangent space $T_a M$. The distribution P is called *smooth* if for any point $a \in M$ there exist a neighborhood $U \ni a$ and p smooth vector fields $X_1, \ldots, X_p$ generating the distribution at every point of the neighborhood.

Let P be a smooth distribution on a smooth manifold M.

DEFINITION 3.1. A diffeomorphism $\varphi \colon M \to M$ is called a (finite) *symmetry* of the distribution P if it preserves this distribution, i.e., $\varphi_*(P_a) \subset P_a$ for all $a \in M$.

EXAMPLE 3.1. The Cartan distribution on the space $\mathbb{R}^3$ (see Figure 1.9) with coordinates x, u, p is given by the 1-form $\omega = du - p\,dx$. It is obviously invariant under translations along x and along u, i.e., the diffeomorphism $\varphi_1(x, u, p) = (x + a, u + b, p)$ is a symmetry of $\mathcal{C}$ for all a and b. Note that $\varphi_1^*(\omega) = \omega$. Translations $\varphi_2(x, u, p) = (x, u, p + c)$ along p are not symmetries. Indeed, $\varphi_2^*(\omega) = du - (p + c)\,dx$, and this form is not proportional to ω.

The distribution under study has less trivial symmetries as well.

For example, it can easily be checked that the so-called Legendre transformation $\varphi(x, u, p) = (p, u - xp, -x)$ is a symmetry, since it preserves the form ω.

The transformation
$$\psi \colon \mathbb{R}^3 \to \mathbb{R}^3, \qquad \psi(x, u, p) = (p, xp - u, x),$$
resembles the previous one, but does not preserve the form ω, since $\psi^*(\omega) = -\omega$. Nevertheless, the transformation ψ is a symmetry of the distribution $\mathcal{C}$.

The set of all symmetries will be denoted by $\operatorname{Sym} P$. It is obvious that the composition of two symmetries is again a symmetry. The map inverse to a symmetry is also a symmetry. Thus, $\operatorname{Sym} P$ is the group with respect to the composition.

Let $P\mathcal{D}$ be the $C^\infty(M)$-module of vector fields X such that the vector X_a belongs to P_a for all $a \in M$.

Suppose that the distribution P is generated by fields $X_1, \ldots, X_l \in PD$. Definition 3.1 is equivalent to $\varphi_*(X_i) \in PD$ for all $i = 1, \ldots, l$ or, which is the same,

$$\varphi_*(X_i) = \sum_j \mu_{ij} X_j, \qquad i = 1, \ldots, l, \tag{3.1}$$

for some functions μ_{ij} on M.

Assume that the differential forms $\omega_1, \ldots, \omega_k$, where $\omega_i = \sum_j \omega_{ij}\, dx_j$, $\omega_{ij} \in C^\infty(M)$, generate the distribution P. The condition $\varphi \in \operatorname{Sym} P$ means that

$$\varphi^*(\omega_i) = \sum_j \lambda_{ij} \omega_j, \qquad i = 1, \ldots, k, \tag{3.2}$$

for some functions λ_{ij} on M or, equivalently,

$$\sum_{j,s} \omega_{ij}(\varphi(x)) \frac{\partial \varphi_j}{\partial x_s}\, dx_s = \sum_{j,s} \lambda_{ij}(x) \omega_{js}(x)\, dx_s, \qquad i = 1, \ldots, k,$$

where $s = 1, \ldots, n = \dim M$. Hence the symmetries φ of the distribution at hand can be found from the following system of equations:

$$\sum_j \omega_{ij}(\varphi(x)) \frac{\partial \varphi_j}{\partial x_s} = \sum_j \lambda_{ij}(x) \omega_{js}(x), \tag{3.3}$$

where λ_{ij} are arbitrary smooth functions, $i = 1, \ldots, k$, $j = 1, \ldots, n$.

The problem of solving these equation is not easier than the problem of finding the integral manifolds of the distribution under consideration. Moreover, the direct attempt to describe the set $\operatorname{Sym} P$ of all symmetries of a given distribution P is doomed to failure. However, passing to the infinitesimal point of view, i.e., passing from the above-defined finite symmetries to infinitesimal ones, considerably simplifies the situation.

DEFINITION 3.2. A vector field $X \in \mathrm{D}(M)$ is said to be an *infinitesimal symmetry* of the distribution P, if the flow A_t generated by the field X consists of finite symmetries.

The set of all infinitesimal symmetries will be denoted by D_P.

Definition 3.2 is generally inconvenient to use in concrete computations. But it can be restated in a more constructive way:

THEOREM 3.1. *Let P be a distribution on M. The following conditions are equivalent*:

1. $X \in \mathrm{D}_P$.
2. *If $X_1, \ldots, X_l$ are vector fields generating P, then there exist smooth functions μ_{ij} such that $[X, X_i] = \sum_j \mu_{ij} X_j$ for all $i = 1, \ldots, l$.*
3. *If $\omega_1, \ldots, \omega_k$ are 1-forms defining P, then there exist smooth functions ν_{ij} such that $X(\omega_i) = \sum_j \nu_{ij} \omega_j$ for all $i = 1, \ldots, k$.*

PROOF. $1 \implies 2$. Let X be a symmetry of the distribution P and A_t be the one-parameter transformation group corresponding to the vector field X. Since the diffeomorphism A_t preserves the distribution P for every $t \in \mathbb{R}$, the image of the vector field $X_i \in PD$ under this diffeomorphism belongs to PD:

$$(A_t)_*(X_i) = \sum_j \alpha_{ij}(t) X_j,$$

where $\alpha_{ij}(t)$ is a family of smooth functions on M smoothly depending on the parameter $t \in \mathbb{R}$. Differentiating this expression with respect to t at $t = 0$, we get

$$[X, X_i] = \sum_j \mu_{ij} X_j,$$

with $\mu_{ij} = -d\alpha_{ij}(t)/dt|_{t=0}$.

$2 \Longrightarrow 3$. Suppose that a vector field X satisfies Condition 2. We claim that if a 1-form ω vanishes on all vector fields $X_1, \ldots, X_l$, then the form $X(\omega)$ has the same property (and, hence, can be represented as a linear combination of $\omega_1, \ldots, \omega_k$). Indeed, for all i we have

$$X(\omega)(X_i) = -\omega([X, X_i]) = 0.$$

$3 \Longrightarrow 1$. First note that the equality $A^*_{t+s} = A^*_t \circ A^*_s$ implies

$$\left.\frac{d}{dt}\right|_{t=s} A^*_t(\omega) = A^*_s(X(\omega)).$$

Further, consider the following $(k+1)$-form dependent upon the parameter t:

$$\Omega_i(t) = A^*_t(\omega_i) \wedge \omega_1 \wedge \cdots \wedge \omega_k.$$

Since $A^*_0(\omega_i) = \omega_i$, we have $\Omega_i(0) = 0$. We claim that $\Omega_i(t) \equiv 0$. Indeed,

$$\frac{d}{dt}\Omega_i(t) = A^*_t(X(\omega_i)) \wedge \omega_1 \wedge \cdots \wedge \omega_k = \sum A^*_t(\nu_{ij})\Omega_j(t).$$

Hence, the vector consisting of $(k+1)$-forms $(\Omega_1(t), \ldots, \Omega_k(t))$ is a solution to a linear homogeneous system of ordinary differential equations with zero initial conditions. Therefore, $\Omega_i(t) \equiv 0$ for all i.

Thus, $A^*_t(\omega_i)$ is a linear combination of $\omega_1, \ldots, \omega_k$ for all t, i.e., A_t is a symmetry of the distribution P. $\square$

COROLLARY 3.2. *If $X, Y \in \mathrm{D}_P$, then $\alpha X + \beta Y \in \mathrm{D}_P$ ($\alpha, \beta \in \mathbb{R}$) and, moreover, $[X, Y] \in \mathrm{D}_P$. In other words, D_P is a Lie algebra over $\mathbb{R}$ with respect to the commutation operation.*

PROOF. By condition 2 of Theorem 3.1, we can write $[X, X_i] = \sum_j \mu_{ij} X_j$ and $[Y, X_i] = \sum_j \lambda_{ij} X_j$. Then $[\alpha X + \beta Y, X_i] = \sum_j (\alpha \mu_{ij} + \beta \lambda_{ij}) X_j$. This yields $\alpha X + \beta Y \in \mathrm{D}_P$.

Further, by virtue of the Jacobi identity

$$[[X, Y], X_i] = [[X, X_i], Y] - [[Y, X_i], X]$$

$$= \sum_j ([\mu_{ij} X_j, Y] - [\lambda_{ij} X_j, X])$$

$$= \sum_{j,k} (\lambda_{ij}\mu_{ik} - \mu_{ij}\lambda_{jk})X_k + \sum_j (X(\lambda_{ij}) - Y(\mu_{ij}))X_j.$$

This shows that $[X, Y] \in \mathrm{D}_P$. $\square$

Using Theorem 3.1, we can write down coordinate conditions for a field $X = \sum_i X^i \partial/\partial x_i$ to be a symmetry of the distribution given by the system of 1-forms $\omega_1, \ldots, \omega_k$, where $\omega_j = \sum_s \omega_{js} dx_s$.

We have

$$X(\omega_j) = \sum_{i,s} \left(X^i \frac{\partial \omega_{js}}{\partial x_i} + \frac{\partial X^i}{\partial x_s} \omega_{ji} \right) dx_s.$$

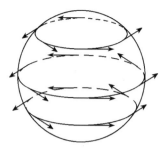

Figure 1.10

By Condition 3 of Theorem 3.1, the 1-form $X(\omega_j)$ is represented as $\sum_i \nu_{ji}\omega_i$, i.e., there exist smooth functions ν_{ji} such that

$$\sum_i \left(X^i \frac{\partial \omega_{js}}{\partial x_i} + \frac{\partial X^i}{\partial x_s} \omega_{ji} \right) = \sum_i \nu_{ji} \omega_{is}, \tag{3.4}$$

where $j = 1, \ldots, k$, $s = 1, \ldots, n$.

Note that in contrast to (3.3), the system of equations (3.4) is linear in the fields $X^1, \ldots, X^k$. It is called the *system of linear Lie equations* corresponding to the *system of nonlinear Lie equations* (3.3). Just as the nonlinear Lie equations serve to find finite symmetries of a given distribution, the linear Lie equations serve to find its infinitesimal symmetries.

Further on we primarily study infinitesimal symmetries, so the word "symmetry" will always mean "infinitesimal symmetry" unless otherwise explicitly specified.

EXAMPLE 3.2. Symmetries of a one-dimensional distribution generated by a vector field Y are vector fields X such that $[X, Y] = \lambda Y$ for some function λ.

Consider, as an example, the vector field Y on the sphere that in the "geographical" coordinates φ, θ has the form $Y = \partial/\partial \varphi$ (see Figure 1.10). Take a field $X = \alpha \partial/\partial \theta + \beta \partial/\partial \varphi$. We have $[X, Y] = -\alpha_\varphi \partial/\partial \theta - \beta_\varphi \partial/\partial \varphi$, so the field X is a symmetry, if $\alpha_\varphi = 0$. Thus, the symmetries of the distribution under study are the fields of the form $\alpha \partial/\partial \theta + \beta \partial/\partial \varphi$, where β is an arbitrary function on the sphere and α is a function constant on the parallels.

The finite symmetries of this distribution are given by pairs of functions $\bar\theta = f(\theta)$, $\bar\varphi = g(\theta, \varphi)$.

EXAMPLE 3.3. Let us discuss the local structure of the symmetry algebra of a completely integrable distribution.

It is clear that for any completely integrable distribution P at a generic point there exists a coordinate system such that P is given by the basis

$$X_1 = \frac{\partial}{\partial x_1}, \ldots, X_l = \frac{\partial}{\partial x_l}.$$

Let $X = \sum_i X^i \partial/\partial x_i$. Then $[X_j, X] = \sum_i (\partial X^i/\partial x_j)\partial/\partial x_i$, so that the condition $[X, X_j] \in PD$, $j = 1, \ldots, l$, is equivalent to the equalities $\partial X^i/\partial x_j = 0$, $j \le l < i$.

The field X splits into the longitudinal component $\sum_{i \le l} X^i \partial/\partial x_i$ and the transversal component $\sum_{i > l} X^i \partial/\partial x_i$. Using this decomposition, we can say that X is a symmetry of the distribution P if and only if the coefficients of its transversal component are constant on the leaves of P, i.e., on the maximal integral manifolds

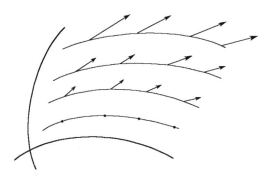

FIGURE 1.11

of the distribution P which are of the form $x_{l+1} = C_{l+1}, \ldots, x_n = C_n$, $C_i = \text{const}$ (see Figure 1.11).

EXAMPLE 3.4. Let us find all (infinitesimal) symmetries of the Cartan distribution $\mathcal{C}$ on $\mathbb{R}^3$. Recall that it is given by the form $\omega = du - p\,dx$. Let

$$X = \alpha \frac{\partial}{\partial x} + \beta \frac{\partial}{\partial u} + \gamma \frac{\partial}{\partial p} \qquad (3.5)$$

be a symmetry of $\mathcal{C}$. We have

$$X(\omega) = (\beta_x - p\alpha_x - \gamma)\,dx + (\beta_u - p\alpha_u)\,du + (\beta_p - p\alpha_p)\,dp.$$

The condition that $X(\omega)$ is proportional to ω reads

$$\begin{cases} \beta_p - p\alpha_p = 0, \\ \beta_x - p\alpha_x - \gamma = -p(\beta_u - p\alpha_u). \end{cases}$$

This system implies that the symmetries of the distribution $\mathcal{C}$ are vector fields of the form (3.5) such that α and β are arbitrary functions related by the equality $\beta_p = p\alpha_p$, and γ is expressed in terms of these functions by the formula $\gamma = \beta_x + p\beta_u - p\alpha_x - p^2\alpha_u$. Consider the function $f = \beta - p\alpha$. It is clear that $\alpha = -f_p$, $\beta = f - pf_p$, $\gamma = f_x + pf_u$, i.e.,

$$X = -f_p \frac{\partial}{\partial x} + (f - pf_p) \frac{\partial}{\partial u} + (f_x + pf_u) \frac{\partial}{\partial p}.$$

Thus, a symmetry X of the Cartan distribution is uniquely determined by the function $f = f(x, u, p)$, which can be chosen arbitrarily[2].

Among symmetries of a given distribution P there is a distinguished class of symmetries that lie in P, i.e., belong to $\mathrm{D}_P \cap PD$. These are called *characteristic* (or *trivial*) symmetries. Symmetries that do not belong to PD will be referred to as *nontrivial*.

PROPOSITION 3.3. *Let $X, Y \in \mathrm{D}_P$, the symmetry Y being trivial. Then the symmetry $[X, Y]$ is also trivial. In other words, the set of all characteristic symmetries is an ideal of the Lie algebra D_P.*

---

[2]This result is a particular case of the general theory of contact fields (see Chapter 2).

PROOF. Let $X_1, \ldots, X_l$ be vector fields generating the distribution P. Suppose that $Y = \sum f_i X_i$. By Theorem 3.1, we have $[X, X_i] = \sum_j \mu_{ij} X_j$. Therefore,

$$[X, Y] = \left[X, \sum_i f_i X_i\right] = \sum_i X(f_i) X_i + \sum_{i,j} f_i \mu_{ij} X_j.$$

Thus, the field $[X, Y]$ lies in P. By virtue of Corollary 3.2, it is a symmetry of the distribution P. □

Let $\operatorname{char} P \subset D_P$ denote the ideal of trivial (characteristic) symmetries. By definition, $\operatorname{char} P = D_P \cap PD$.

In view of Proposition 3.3, we can consider the quotient Lie algebra

$$\operatorname{sym} P = D_P / \operatorname{char} P,$$

which is said to be the *Lie algebra of nontrivial symmetries* of the distribution P. Note that for completely integrable distributions one has $\operatorname{char} P = PD$. Consequently $\operatorname{sym} P = D_P / PD$ in this case.

Let us consider symmetries of a completely integrable distribution P. Clearly, the one-parameter transformation group generated by a symmetry preserves the leaves of P. By the Frobenius theorem, transformations corresponding to a characteristic symmetry move every leaf of P along itself. By contrast, nontrivial symmetries shuffle the leaves.

These observations can be generalized to nonintegrable distributions.

PROPOSITION 3.4. *A characteristic symmetry of the distribution P is tangent to every maximal integral manifold of P.*

PROOF. Assume that there exists a characteristic symmetry X which is not tangent to an integral manifold L. Consider the flow A_t of X and the manifold $\bar{L} = \bigcup_t A_t(L)$. Clearly, L is strictly embedded to $\bar{L}$.

We claim that the manifold $\bar{L}$ is integral for P. Indeed, the tangent space to $\bar{L}$ at a point x spans the vector X_x and the tangent space to the submanifold $A_t(L)$ passing through x. Since both lie in P_x, the same is true for $T_x(\bar{L})$. □

REMARK 3.1. Thus, every maximal integral manifold of the distribution P is composed of its characteristics, i.e., of the trajectories of characteristic symmetries of P. A similar construction occurs in the theory of first-order partial differential equations, where our terminology came from.

Similar to the case of completely integrable distributions, transformations corresponding to nontrivial symmetries of an arbitrary distribution preserve the set of maximal integral manifolds, but, unlike trivial symmetries, shuffle them. Note that the action of elements of $\operatorname{sym} P$ on the set of maximal integral manifolds is well-defined, since by Proposition 3.4 such a manifold is invariant under the action of a trivial symmetry.

PROPOSITION 3.5. *The set of characteristic fields is a module over the function ring, i.e., if $X \in \operatorname{char} P$ and $f \in C^\infty(M)$, then $fX \in \operatorname{char} P$.*

PROOF. For an arbitrary vector field $Y \in PD$, we have

$$[fX, Y] = f[X, Y] - Y(f)X \in PD.$$

Hence, $fX \in D_P$. In addition, $fX \in PD$; therefore $fX \in \operatorname{char} P$. □

COROLLARY 3.6. *The set* char P *consists of vector fields lying in the distribution.*

The distribution char P is called *characteristic*. The dimension of this distribution may decrease at some points.

THEOREM 3.7. *The characteristic distribution is completely integrable.*

PROOF. It follows from Proposition 3.3 that char P is a Lie algebra. The theorem is proved. □

Now, let us use Examples 3.2–3.4 to illustrate the above theory.

In Example 3.2, we dealt with a one-dimensional distribution on the sphere. Any one-dimensional distribution is completely integrable, so that in this case the characteristic distribution coincides with the one considered.

The same obviously holds for any completely integrable distribution.

Further, let X be the characteristic symmetry of the Cartan distribution discussed in Example 3.4. Since $X \in D_P$, we have

$$X = -f_p \frac{\partial}{\partial x} + (f - p f_p) \frac{\partial}{\partial u} + (f_x + p f_u) \frac{\partial}{\partial p}.$$

The fact that $X \in$ char P implies $\omega(X) = 0$. But $\omega(X) = X \lrcorner (du - p dx) = f$. Therefore $f = 0$ and, hence, $X = 0$. Thus, in this case the characteristic distribution is zero-dimensional.

4. Some applications of symmetry theory to integration of distributions

In this section, we show how to describe maximal integral manifolds of a distribution using its symmetries.

4.1. Distributions and characteristic symmetries. Consider a typical situation. Let P be a distribution and X be a characteristic symmetry of P. Given an integral manifold L transversal to the trajectories of X, we can construct the manifold $N = \bigcup_t A_t(L)$, where A_t is the flow of X. Since maximal integral manifolds of a distribution are composed of its characteristics, the manifold N is integral for P. In particular, taking a point as the manifold L, we get one trajectory of the field X, which is a one-dimensional integral manifold.

EXAMPLE 4.1. In Example 2.2, we discussed the differential equation

$$(3x - 2u)u' = u.$$

For the distribution $\mathcal{C}_\mathcal{E}$ given by the form $du - p\, dx$ the characteristic symmetry Y_F was constructed. The trajectories of this field,

$$\begin{cases} x = (x_0 - u_0)e^{-3t} + u_0 e^{-t}, \\ u = u_0 e^{-t}, \end{cases}$$

are solutions of the equation at hand passing through the point (x_0, u_0, p_0).

EXAMPLE 4.2. Consider the upper half-plane

$$H = \{x, y \mid y > 0\} \subset \mathbb{R}^2$$

and the manifold M of all unit tangent vectors to H. As coordinates on M take x, y, u, with $u \in \mathbb{R} \mod 2\pi$ being the angle between the upward vertical direction and the given vector. Further, consider the distribution P defined by the 1-form

$$\omega = du + \frac{1 - \cos u}{y} dx - \frac{\sin u}{y} dy.$$

The reader can readily verify that this distribution satisfies the Frobenius complete integrability condition.

This distribution defines the so-called *oricycle foliation*. Its geometric meaning can be explained as follows.

Recall that the upper half-plane with the metric $ds^2 = (dx^2 + dy^2)/y^2$ is a model of the Lobachevskian plane. The geodesics ("straight lines") in this model are semicircles with centers on the x-axis and vertical rays.

A leaf of the oricycle foliation consists of all vectors tangent to the geodesics passing through the same point of the x-axis.

The lines $L = \{u = \pi, x = c\}$, where c is a constant, are integral manifolds of P, since ω vanishes on them. The field

$$D_x = \frac{\partial}{\partial x} - \frac{1 - \cos u}{y} \frac{\partial}{\partial u}$$

is a characteristic symmetry of P, because the distribution P is completely integrable and the form ω vanishes on the field D_x. Moreover, it is obvious that the field D_x is transversal to the field $\partial/\partial y$, which is tangent to the line L.

Now, we consider all trajectories of the field D_x passing through L to obtain a two-dimensional integral surface of the distribution P.

Solving the system

$$\begin{cases} \dot{x} = 1, \\ \dot{y} = 0, \\ \dot{u} = -\frac{1 - \cos u}{y}, \end{cases}$$

yields the trajectories of the field D_x:

$$y = C_1, \qquad x - y \cot \frac{u}{2} = C_2,$$

where C_1 and C_2 are constants.

If points (x, y, u) and $(c, y_0, \pi) \in L$ lie on the same trajectory, then $y = y_0$, $x - y \cot u/2 = c - y_0 \cot \pi/2$. It follows that $x - y \cot u/2 = c$. This is exactly the surface $\bigcup_t A_t(L)$ (see Figure 1.12).

Geometrically, this surface is the set of all unit vectors tangent to the semicircles such that their centers lie on the x-axis and they pass through the same point of this axis (Figure 1.13).

4.2. Symmetries and dynamical systems. Many standard methods for solving differential equations make use of symmetry theory. Let us discuss an example.

A system of first-order ordinary differential equations solved with respect to derivatives is interpreted as a vector field X (or, in other words, as a dynamical system). A vector field Y is called a *symmetry* of the field X if $[X, Y] = 0$. It is readily seen that a shift along the trajectories of Y shuffles the trajectories of X. Given a symmetry, one usually changes coordinates to "straighten" the field Y (i.e., chooses new coordinates, $y_1, \ldots, y_n$ in such a way that all except one of them,

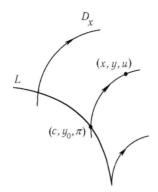

FIGURE 1.12

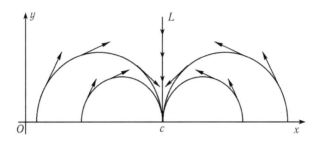

FIGURE 1.13

say y_n, are constant on the trajectories of the field Y). In these new coordinates, the field Y takes the form $\partial/\partial y_n$, so that the condition $[X, Y] = 0$ implies that the coefficients of the field X do not depend on y_n. Thus, finding trajectories of the field X reduces to integration of an $(n-1)$-dimensional dynamical system and one quadrature.

EXAMPLE 4.3. Consider the 3-dimensional dynamical system

$$\begin{cases} \dot{x}_1 = t\left(\dfrac{x_2^2}{x_3} + \dfrac{x_1^2}{x_2^2}x_3\right) + \dfrac{x_1^2}{x_2}, \\ \dot{x}_2 = t\dfrac{x_1 x_3}{x_2} + x_1, \\ \dot{x}_3 = \dfrac{x_1 x_3}{x_2}. \end{cases}$$

The vector field

$$Y = x_1 \frac{\partial}{\partial x_1} + x_2 \frac{\partial}{\partial x_2} + x_3 \frac{\partial}{\partial x_3}$$

is a symmetry of this system. Indeed, it can easily be checked that the commutator of the field

$$X = \left(t\left(\frac{x_2^2}{x_3} + \frac{x_1^2}{x_2^2}x_3\right) + \frac{x_1^2}{x_2}\right)\frac{\partial}{\partial x_1} + \left(t\frac{x_1 x_3}{x_2} + x_1\right)\frac{\partial}{\partial x_2} + \frac{x_1 x_3}{x_2}\frac{\partial}{\partial x_3}$$

and the field Y is equal to zero. Now choose the new coordinates

$$\begin{cases} x'_1 = x_1/x_2, \\ x'_2 = x_2/x_3, \\ x'_3 = \ln x_3. \end{cases}$$

In these coordinates, the field Y takes the form $\partial/\partial x'_3$, while the dynamical system X is

$$\begin{cases} \dot{x}'_1 = tx'_2, \\ \dot{x}'_2 = tx'_1, \\ \dot{x}'_3 = x'_1. \end{cases}$$

It follows that finding the trajectories of the dynamical system X reduces to the integration of the two-dimensional dynamical system

$$\begin{cases} \dot{x}'_1 = tx'_2, \\ \dot{x}'_2 = tx'_1 \end{cases}$$

and one quadrature $x'_3 = \int x'_1(t)\,dt$.

We now turn to discuss applications of nontrivial (i.e., noncharacteristic) symmetries to the problem of integrating distributions.

4.3. Distributions and noncharacteristic symmetries. Let X be a symmetry of a distribution P, and let A_t be the corresponding flow. Given an integral manifold L of the distribution P, we can construct the whole family $A_t(L)$ of such manifolds.

EXAMPLE 4.4. Consider the equation $(3x - 2u)u' = u$ from Example 2.2. The field $X = u^3 \partial/\partial x$, written in the local coordinates (x, u) on $\mathcal{E}$, obviously commutes with the field Y_F (see Example 2.2) restricted to $\mathcal{E}$ (check that this restriction has the form $(2u-3x)\partial/\partial x - u\partial/\partial u$). Therefore X is a symmetry of the one-dimensional distribution $\mathcal{C}(\mathcal{E})$ on $\mathcal{E}$. Take the solution $L = \{u = x\}$ of the equation under study. Shifts of this solution along the trajectories of the field X yield all solutions of our equation (except for $u = 0$). In fact, the shift along the trajectories of the field X at time t has the form

$$\begin{cases} x = u_0^3 t + x_0, \\ u = u_0. \end{cases}$$

From these formulas it easily follows that the image of the straight line $L = \{u = x\}$ is the curve $A_t(L) = \{tu^3 + u = x\}$ (see Figure 1.14).

EXERCISE 4.1. Find a symmetry of this equation that yields every (or almost every) solution from the zero solution.

REMARK 4.1. Under shifts along the trajectories of the field X from Example 4.4, the solution $u = 0$ moves along itself. Such a solution is said to be *invariant*.

EXERCISE 4.2. Find the solution of the equation $(3x - 2u)u' = u$ invariant under the infinitesimal homothety $x\partial/\partial x + u\partial/\partial u$.

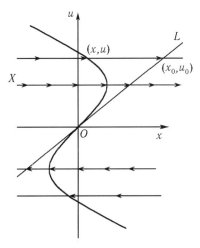

FIGURE 1.14

4.4. Integration of equations by quadratures. The second aspect of applications of nontrivial symmetries to which we would like to call attention, is integration of equations by quadratures. The point is that if for a completely integrable distribution one knows a solvable Lie algebra of nontrivial symmetries and the dimension of this algebra equals the codimension of the distribution, then the integral manifolds of this distribution can be found by quadratures. Before describing this procedure, let us give the following definition.

DEFINITION 4.1. A function $f \in C^\infty(M)$ is called a *first integral of a distribution* P, if $X(f) = 0$ for any $X \in PD$.

It is obvious that if f is a first integral of the distribution P, then it is constant on every integral manifold of this distribution, or, in other words, every integral manifold of P lies entirely in a level surface $\{f = c\}$ of the first integral.

Let us find first integrals for distributions from Examples 3.2–3.4.

Clearly, first integrals of the field $X = \partial/\partial\varphi$ on the sphere are functions constant on the parallels of the sphere. This example is easily generalized to the case of an arbitrary one-dimensional distribution[3].

First integrals of a completely integrable distribution P locally given by the fields
$$X_1 = \frac{\partial}{\partial x_1}, \ldots, X_l = \frac{\partial}{\partial x_l},$$
are of the form $f(x_{l+1}, \ldots, x_n)$, where f is an arbitrary function.

Finally, let f be a first integral of the Cartan distribution given by the form $\omega = du - p\,dx$. Since the fields $\partial/\partial p$ and $\partial/\partial x + p\partial/\partial u$ lie in the Cartan distribution, we have $\partial f/\partial p = \partial f/\partial x + p\partial f/\partial u = 0$; therefore $f = $ const. Thus, the Cartan distribution does not possess nontrivial first integrals.

---

[3]Note that the terminology used here is consistent with that used in the theory of first-order partial differential equations: if the distribution P is given by a vector field X, then its first integrals are the first integrals of the equation $X(\varphi) = 0$, i.e., functions constant on the trajectories of the field X. See also Chapter 2.

Assume that we know a first integral of a distribution P on M. Then the problem of finding all maximal integral manifolds reduces to integration of a distribution on a manifold whose dimension is diminished by one with respect to M. Namely, the problem amounts to integrating the restriction of the distribution P to a level surface of the first integral. Each new independent first integral reduces the dimension by one. In the case when functionally independent first integrals $f_1, \ldots, f_k$, where $k = \operatorname{codim} P$, are known, the distribution under consideration can be given by the set of exact 1-forms $df_1, \ldots, df_k$. Therefore this distribution is completely integrable, and its leaves are the mutual level surfaces $\{\, x \in M \mid f_i(x) = c_i \,\}$ of the functions $f_1, \ldots, f_k$, where $c_1, \ldots, c_k$ are constants.

The problem of finding a first integral can be stated as the problem of finding an exact 1-form $\omega = df$ vanishing on vectors belonging to the distribution P, since in this case $\omega(X) = X(f) = 0$ for all $X \in PD$. Thus, complete integrability of P is equivalent to existence of $k = \operatorname{codim} P$ such forms, provided that they are independent.

DEFINITION 4.2. A linear subspace $\mathcal{Y} \subset \operatorname{sym} P$ is called *nondegenerate*, if for any point $x \in M$ and any field $X \in \mathcal{Y}$ the condition $X_x \in P_x$ is equivalent to the condition $X = 0$.

Let $x \in M$ and $\mathcal{Y}_x = \{\, X_x \mid X \in \mathcal{Y} \,\}$. If the subspace $\mathcal{Y}$ is nondegenerate, then from Definition 4.2 it follows that

$$\dim \mathcal{Y} = \dim \mathcal{Y}_x \leq \operatorname{codim} P.$$

Let $\omega_1, \ldots, \omega_k$ be a set of forms defining the distribution P and $X_1, \ldots, X_l$ be a basis of a nondegenerate subspace $\mathcal{Y} \subset \operatorname{sym} P$. Clearly, $l \leq k$ and the matrix $\|\omega_i(X_j)\|$ has the rank l at any point of the manifold M. If $l = k$, then $\det \|\omega_i(X_j)\| \neq 0$ and one can choose another set of forms defining P, say $\omega'_1, \ldots, \omega'_k$, such that $\omega'_i(X_j) = \delta_{ij}$. To do this, it suffices to multiply the column $(\omega_1, \ldots, \omega_k)^t$ by the matrix inverse to $\|\omega_i(X_j)\|$. Assume that the space $\mathcal{Y}$ is closed with respect to the commutation; then we have the following.

THEOREM 4.1. *Let P be a completely integrable distribution defined by the set of 1-forms $\omega_1, \ldots, \omega_k$. Let $X_1, \ldots, X_k$ be a basis of a nondegenerate Lie algebra $\mathcal{Y}$ of symmetries of this distribution. Suppose that $\omega_i(X_j) = \delta_{ij}$ and*

$$[X_i, X_j] = \sum_s c_{ij}^s X_s,$$

where the c_{ij}^s are constants. Then

$$d\omega_s = -\frac{1}{2} \sum_{i,j} c_{ij}^s \omega_i \wedge \omega_j.$$

PROOF. By the Frobenius theorem, there exist 1-forms γ_{ij} such that

$$d\omega_s = \sum_j \gamma_{sj} \wedge \omega_j.$$

Since $X_i \in \operatorname{sym} P$, the 1-forms

$$X_i(\omega_s) = X_i \,\lrcorner\, d\omega_s + d(X_i \,\lrcorner\, \omega_s) = X_i \,\lrcorner\, d\omega_s$$

vanish on vectors from the distribution P. The equality

$$X_i \lrcorner\, d\omega_s = \sum_j \gamma_{sj}(X_i)\omega_j - \gamma_{si}$$

implies that the forms γ_{sj} are also zero on vectors from P and, therefore, $\gamma_{ij} = \sum_s a_{ij}^s \omega_s$ for some appropriate functions $a_{ij}^s \in C^\infty(M)$. Hence,

$$d\omega_s = \sum_{i<j} a_{ij}^s \omega_i \wedge \omega_j.$$

It follows that

$$d\omega_s(X_i, X_j) = a_{ij}^s, \qquad \text{for } i < j.$$

On the other hand,

$$d\omega_s(X_i, X_j) = X_i(\omega_s(X_j)) - X_j(\omega_s(X_i)) - \omega_s([X_i, X_j])$$
$$= -\omega_s\left(\sum_t c_{ij}^t X_t\right) = -c_{ij}^s.$$

Therefore, $a_{ij}^s = -c_{ij}^s$ and

$$d\omega_s = -\sum_{i<j} c_{ij}^s \omega_i \wedge \omega_j = -\frac{1}{2}\sum_{i,j} c_{ij}^s \omega_i \wedge \omega_j. \qquad \square$$

COROLLARY 4.2. *Under the conditions of Theorem 4.1, assume that the algebra $\mathcal{Y}$ is commutative. Then all the forms ω_i are closed and, thus, locally exact: $\omega_i = dh_i$ for some smooth functions h_i. These first integrals can be locally found by computing the integrals*

$$h_i(a) = \int_{a_0}^{a} \omega_i,$$

where a_0 is a fixed point of the manifold M.

In particular, since ordinary differential equations of order k are identified with one-dimensional distributions on $(k+1)$-dimensional manifolds, Corollary 4.2 immediately implies

COROLLARY 4.3. *If the distribution corresponding to an ordinary differential equation of order k possesses a k-dimensional commutative nondegenerate Lie algebra of symmetries, then this equation is integrable by quadratures.*

EXAMPLE 4.5. Consider a first-order equation solved with respect to the derivative. Assume that the right-hand side does not depend on x:

$$\frac{du}{dx} = f(u).$$

The corresponding distribution is defined on the surface $p = f(u)$ by the 1-form $\omega = du - f\,dx$. The vector field

$$X = \frac{\partial}{\partial x}$$

is tangent to this surface and is a nontrivial symmetry of the distribution $\mathcal{C}(\mathcal{E})$.

By virtue of Theorem 4.1 and Corollary 4.2, the form $\omega_\mathcal{E}/f$ is exact. Indeed,

$$\frac{\omega_\mathcal{E}}{f} = \frac{1}{f}(du - f\,dx) = d\left(\int \frac{du}{f(u)} - x\right).$$

In multiplication by $1/f$, one easily recognizes the well-known method of "separation of variables" for solving equations of the type considered. Thus, this method essentially amounts to transforming the 1-form defining the equation to an exact one.

EXERCISE 4.3. Show that on the equation $\mathcal{E}$ the field X coincides with the field
$$f\frac{\partial}{\partial u} + f'p\frac{\partial}{\partial p}.$$

The previous example serves to illustrate the following general fact: with a knowledge of one nontrivial symmetry of a completely integrable distribution of codimension 1 (in particular, of a first-order ordinary differential equation), it is possible to describe leaves of this distribution (in particular, solutions to this differential equation) by means of quadratures. If X is a nontrivial symmetry of a distribution P given by 1-form ω, then the form $f\omega$, with $f = 1/\omega(X)$, which also defines the distribution P, is closed[4]. In this case, the leaves of P coincide with the level surfaces of the integral of this form.

EXAMPLE 4.6. Consider the distribution P from Example 4.2. Since the coefficients of the form
$$\omega = du + \frac{1-\cos u}{y}dx - \frac{\sin u}{y}dy$$
do not depend on x, the field $X = \partial/\partial x$ is a symmetry of P. We have $\omega(X) = (1-\cos u)/y \neq 0$, so that X is a nontrivial symmetry. Thus, the integrating factor equals $f = 1/\omega(X) = y/(1-\cos u)$, and
$$f\omega = dx - \frac{\sin u}{1-\cos u}dy + \frac{y}{1-\cos u}du = d\left(x - y\cot\frac{u}{2}\right).$$
The equation $x - y\cot(u/2) = c$, $c = \text{const}$, defines all the leaves of the distribution at hand, except for the leaf $\{u = 0\}$ (the function f is undefined for $u = 0$).

EXAMPLE 4.7. Any homogeneous first-order equation $\mathcal{E}$
$$u' = \varphi\left(\frac{u}{x}\right)$$
has the symmetry $X = x\partial/\partial x + u\partial/\partial u$. Indeed, in the coordinates x and u the Cartan distribution on $\mathcal{E}$ is given by the form
$$\omega_\mathcal{E} = du - \varphi\left(\frac{u}{x}\right)dx.$$
It can easily be checked that $X(\varphi) = 0$ and $X(\omega_\mathcal{E}) = \omega_\mathcal{E}$; hence X is a symmetry. For an integrating factor one can take the function
$$f = \frac{1}{\omega_\mathcal{E}(X)} = \frac{1}{u - x\varphi\left(\frac{u}{x}\right)}.$$
In particular, for the equation $(3x - 2u)u' = u$ (see Example 2.2), which is equivalent to the equation
$$u' = \frac{u}{3x - 2u},$$

---

[4]Recall that the function f is said to be an *integrating factor*.

everywhere except for the line $3x - 2u = 0$, we have
$$\varphi(\xi) = \frac{\xi}{3 - 2\xi}, \qquad f = \frac{3x - 2u}{2u(x - u)},$$
and, therefore,
$$\omega'_{\mathcal{E}} = \frac{3x - 2u}{2u(x - u)} \omega_{\mathcal{E}} = d\left(\frac{1}{2} \ln \frac{u^3}{x - u}\right).$$

As above, this yields the solutions of equation under study: $u^3/(x - u) = c$, $c = $ const.

EXAMPLE 4.8. Consider a linear nonhomogeneous equation
$$a_k u^{(k)} + \cdots + a_1 u' + a_0 u = f,$$
where $a_0, a_1, \ldots, a_k$ and f are given functions of x. It is easy to prove that the field
$$X_\varphi = \varphi \frac{\partial}{\partial u} + \varphi' \frac{\partial}{\partial p_1} + \cdots + \varphi^{(k)} \frac{\partial}{\partial p_k},$$
where $\varphi(x)$ is an arbitrary solution of the corresponding homogeneous equation
$$a_k u^{(k)} + \cdots + a_1 u' + a_0 u = 0,$$
is a symmetry of the initial equation. Such symmetries make a k-dimensional commutative Lie algebra, and if one knows its basis, i.e., a fundamental system $\varphi_1, \ldots, \varphi_k$ of solutions of the homogeneous equation, then, by Corollary 4.3, the nonhomogeneous equation can be integrated by quadratures. Note that $\omega_i(X_\varphi) = \varphi^{(i)}$, where, as before, $\omega_i = dp_i - p_{i+1}\, dx$, Therefore, the matrix $\|\omega_i(X_{\varphi_j})\|$, which in the course of solution must be inverted to find an integrating factor, is nothing but the usual Wronskian matrix of the fundamental system involved.

Now, suppose that we know a nondegenerate noncommutative Lie algebra $\mathcal{Y}$ of symmetries of a completely integrable distribution P. Consider its commutator subalgebra
$$\mathcal{Y}^{(1)} = [\mathcal{Y}, \mathcal{Y}] = \left\{ \sum [X, Y] \mid X, Y \in \mathcal{Y} \right\}$$
and assume that $\mathcal{Y}^{(1)} \neq \mathcal{Y}$. Then one can choose a basis $X_1, \ldots, X_k$ of $\mathcal{Y}$ in such a way that $X_1, \ldots, X_r \notin \mathcal{Y}^{(1)}$, while $X_{r+1}, \ldots, X_k \in \mathcal{Y}^{(1)}$. In this case, for any two fields X_i and X_j we have $[X_i, X_j] \in \mathcal{Y}^{(1)}$, so that $c_{ij}^s = 0$ for all i, j, if $s \leq r$. Hence, the forms $\omega_1, \ldots, \omega_r$ are closed and, therefore, $\omega_1 = dh_1, \ldots, \omega_r = dh_r$ in some open domain. The level surfaces
$$H_c = \{ h_1 = c_1, \ldots, h_r = c_r \}, \qquad c = (c_1, \ldots, c_r), \qquad c_i \in \mathbb{R},$$
are invariant under the commutator subalgebra $\mathcal{Y}^{(1)}$, since $X_j(h_i) = \omega_i(X_j) = 0$ if $i \leq r$ and $j \geq r + 1$.

Let P_c be the restriction of the distribution P to the surface H_c. The distribution P_c is completely integrable. Indeed, the foliation of M whose leaves are maximal integral manifolds of P cuts a foliation on the surface H_c. The same fact can be proved analytically. By Theorem 4.1,
$$d\omega_s = -\frac{1}{2} \sum_{i,j > r} c_{ij}^s \omega_i \wedge \omega_j, \qquad s \geq r + 1.$$

The forms ω_i vanish on H_c for $i \leq r$. Hence, by the Frobenius theorem, the distribution P_c is completely integrable.

Observe now that $\mathcal{Y}^{(1)}$ is a nondegenerate Lie algebra of symmetries of the distribution P_c. Actually, a shift along the trajectories of any field $X \in \mathcal{Y}^{(1)}$ shuffles the leaves of P and preserves the manifold H_c; hence it must also shuffle the leaves of P_c, which is the restriction of P to H_c. Nondegeneracy of $\mathcal{Y}^{(1)}$ follows from the fact that $\|\omega_i(X_j)\|$, $r < i, j \leq k$, is the unit matrix and, therefore, is nondegenerate.

Because of this, we can subject $\mathcal{Y}^{(1)}$ to the same procedure that was formerly applied to $\mathcal{Y}$. Namely, let

$$\mathcal{Y}^{(2)} = [\mathcal{Y}^{(1)}, \mathcal{Y}^{(1)}]$$

be the commutator subalgebra of $\mathcal{Y}^{(1)}$. If $\mathcal{Y}^{(2)} \neq \mathcal{Y}^{(1)}$, then some of the 1-forms $\omega_{r+1}|_{dH_c}, \ldots, \omega_k|_{dH_c}$ are closed and give rise to local first integrals of the distribution P_c. The distribution P_c can be restricted to the mutual level surface of these integrals, etc.

We continue in this fashion, obtaining the sequence of subalgebras

$$\mathcal{Y} \supset \mathcal{Y}^{(1)} \supset \mathcal{Y}^{(2)} \supset \cdots,$$

where $\mathcal{Y}^{(i+1)} = [\mathcal{Y}^{(i)}, \mathcal{Y}^{(i)}]$. If the algebra $\mathcal{Y}$ is *solvable*, i.e., $\mathcal{Y}^{(l)}$ is commutative for some l (which is equivalent to $\mathcal{Y}^{(l+1)} = 0$), then after l steps we shall be able to apply Corollary 4.3. All this work yields the following theorems (which, by the way, explain the term "solvable"):

THEOREM 4.4. *If a completely integrable distribution P of codimension k possesses a solvable k-dimensional nondegenerate Lie algebra $\mathcal{Y} \subset \operatorname{sym} P$ of symmetries, then P is integrable by quadratures, i.e., one can find a complete set of first integrals for P by integrating closed 1-forms and solving functional equations.*

THEOREM 4.5 (Bianchi–Lie). *If an ordinary differential equation of order k possesses a solvable k-dimensional nondegenerate Lie algebra of symmetries, then it is integrable by quadratures.*

It is these results which caused the concept of solvability, which had originated in the Galois theory of algebraic equations, to be extended to the theory of Lie groups and algebras. Note also that the Lie groups of finite symmetries can be used instead of the algebras of symmetries, since one can always pass to the corresponding algebra by considering infinitesimal generators of the group.

EXAMPLE 4.9. Consider the equation

$$au^2 u''' + buu'u'' + cu'^3 = 0.$$

It possesses a solvable three-dimensional nondegenerate algebra of symmetries consisting of the translation along x (i.e., the field $\partial/\partial x$) and of two independent scale transformations (i.e., the fields $x\partial/\partial x$ and $u\partial/\partial u$). It is convenient to take the fields

$$X_1 = \frac{\partial}{\partial x},$$
$$X_2 = -x\frac{\partial}{\partial x} + u\frac{\partial}{\partial u} + 2p_1\frac{\partial}{\partial p_1} + 3p_2\frac{\partial}{\partial p_2} + 4p_3\frac{\partial}{\partial p_3},$$
$$X_3 = u\frac{\partial}{\partial u} + p_1\frac{\partial}{\partial p_1} + p_2\frac{\partial}{\partial p_2} + p_3\frac{\partial}{\partial p_3}$$

for a basis of the symmetry algebra, and follow the above-described scheme to accomplish the integration.

REMARK 4.2. Interestingly, in Kamke's well-known reference book [49] on ordinary differential equations, for each of the following two differential equations (numbers 7.8 and 7.9 in Kamke's book):

$$4u^2 u''' - 18uu'u'' + 15u'^3 = 0,$$
$$9u^2 u''' - 45uu'u'' + 40u'^3 = 0,$$

which are both of the type considered in Example 4.9, a different solution procedure is given.

EXERCISE 4.4. Prove that any Lie algebra consisting of infinitesimal translations and infinitesimal scale transformations is solvable. To this end, check that if $\mathcal{Y}$ is a Lie algebra of the type considered, then its commutator subalgebra $\mathcal{Y}^{(1)}$ consists of translations only and, hence, is commutative.

Note that any two-dimensional Lie algebra is solvable. Indeed, if X_1 and X_2 constitute a basis of $\mathcal{Y}$, then the subalgebra $\mathcal{Y}^{(1)}$ is generated by $[X_1, X_2]$ and, thus, $\mathcal{Y}^{(2)} = 0$. Therefore, any completely integrable distribution of codimension two, in particular, any second-order ordinary differential equation, can be integrated by quadratures, if a two-dimensional Lie algebra of its symmetries is known.

EXAMPLE 4.10. Consider the second-order equation

$$u'' = u' + u^n - \frac{2n+2}{(n+3)^2} u, \qquad (4.1)$$

where $n \neq -3$, $n \in \mathbb{R}$.

The independent variable x does not enter into the equation explicitly; hence the field

$$X_1 = \frac{\partial}{\partial x}$$

is a symmetry. It is not hard to prove that the field

$$X_2 = e^{kx} \left[\frac{\partial}{\partial x} + \frac{k+1}{2} u \frac{\partial}{\partial u} + \left(\frac{k(k+1)}{2} u + \frac{1-k}{2} p \right) \frac{\partial}{\partial p} \right],$$

where $k = (1-n)/(n+3)$, is also a symmetry of the equation under consideration[5].

The geometric image associated with the given equation is the hypersurface $\mathcal{E}$ in the four-dimensional space with coordinates x, u, p, q, where p and q stand for u' and u'' respectively, defined by the equation

$$q = p + u^n - \frac{2n+2}{(n+3)^2} u,$$

together with the one-dimensional distribution on $\mathcal{E}$ given by the forms $\omega_1 = du - p\,dx$ and $\omega_2 = dp - q\,dx$.

---

[5]Contrary to the previous example, where all symmetries were seen with "the naked eye", finding X_2 is not as straightforward. To do this, one needs the method of generating functions for symmetries of the Cartan distribution. It is discussed in §5.

The functions x, u, p can be taken for coordinates on $\mathcal{E}$. Consider the matrix

$$M = \|\omega_i(X_j)\| = \begin{pmatrix} -p & e^{kx}\left(\dfrac{k+1}{2}u - p\right) \\ -q & e^{kx}\left(\dfrac{k(k+1)}{2}u + \dfrac{1-k}{2}p - q\right) \end{pmatrix}.$$

The inverse matrix has the form

$$M^{-1} = \frac{1}{\Delta} \begin{pmatrix} e^{kx}\left(\dfrac{k(k+1)}{2}u + \dfrac{1-k}{2}p - q\right) & -e^{kx}\left(\dfrac{k+1}{2}u - p\right) \\ q & -p \end{pmatrix},$$

where $\Delta = \det M = e^{kx} T$,

$$T = -\frac{k(k+1)}{2} up - \frac{1-k}{2} p^2 + \frac{k+1}{2} uq.$$

Multiplying the matrix M^{-1} by the column consisting of ω_1 and ω_2, we obtain new basis forms defining $\mathcal{C}(\mathcal{E})$:

$$\omega_1' = \frac{1}{T}\left[\left(\frac{k(k+1)}{2}u + \frac{1-k}{2}p - q\right) du - \left(\frac{k+1}{2}u - p\right) dp\right] + dx,$$

$$\omega_2' = \frac{1}{\Delta}(q\, du - p\, dp).$$

By Theorem 4.1 the form ω_1' is closed, since $[X_1, X_2] = kX_2$. To find its integral, note that

$$\frac{\partial T}{\partial u} = \frac{1-k^2}{2}p + (1-k)u^n + \frac{(k+1)^2(k-1)}{4}u,$$

$$\frac{\partial T}{\partial p} = \frac{1-k^2}{2}u - (1-k)p.$$

Therefore, if $k \neq 1$, then the form ω_1' can be written as

$$\omega_1' = \frac{1}{(k-1)T}\left(\frac{\partial T}{\partial u} du + \frac{\partial T}{\partial p} dp\right) + dx = \frac{1}{k-1} d\ln(Te^{(k-1)x}).$$

Thus, $f = Te^{(k-1)x}$ is a first integral of the distribution. Note that we have reduced the order of equation (4.1), because it is equivalent to the family of first-order equations of the form

$$\frac{k-1}{2}p^2 + \frac{1-k^2}{2}up + \frac{k+1}{2}u^{n+1} + \frac{(k+1)^2(k-1)}{8}u^2 - ce^{(1-k)x} = 0, \quad (4.2)$$

where c is an arbitrary constant.

Let us now restrict the distribution $\mathcal{C}(\mathcal{E})$ to the surface H_c defined by (4.2). On this surface, p can be expressed in terms of x and u as follows:

$$p = \frac{k+1}{2}u \pm \sqrt{\frac{1+k}{1-k}u^{n+1} + \frac{2c}{k-1}e^{(1-k)x}}. \quad (4.3)$$

Furthermore, note that on H_c we have $\Delta = ce^x$. Therefore,

$$\omega_2'|_{H_c} = -d\left(\frac{e^{-x}p^2}{2c}\right) + \frac{e^{-x}}{c}\left(p + u^n + \frac{k^2-1}{4}u\right) du - \frac{e^{-x}}{2c}p^2\, dx,$$

where p is given by (4.3). From the proof of Theorem 4.4 it follows that this form is closed; its integral is

$$g = -\frac{e^{-x}}{2c}\left(\frac{k+1}{2}u \pm \sqrt{\frac{1+k}{1-k}u^{n+1} + \frac{2c}{k-1}e^{(1-k)x}}\right)^2 + \frac{e^{-kx}}{k(k-1)}$$
$$+ \frac{e^{-x}}{c}\left(\frac{u^{n+1}}{n+1} + \frac{(k+1)^2}{8}u^2\right) \pm \frac{e^{-x}}{c}\int_0^u \sqrt{\frac{1+k}{1-k}\eta^{n+1} + \frac{2c}{k-1}e^{(1-k)x}}\, d\eta.$$

All solutions of the equation at hand are given by the implicit relation $g = c_1$, $c_1 = \text{const}$. The Chebyshev theorem implies that the integral g is an elementary function whenever $2/(n+1)$ is integer.

EXERCISE 4.5. 1. Complete the integration of 4.1 in the case $n = -1$.

2. Integrate the equation $u'' = u' + ae^{bu} - 2/b$ by means of the symmetry algebra with the generators

$$X_1 = \frac{\partial}{\partial x}, \qquad X_2 = e^{-x}\left(\frac{\partial}{\partial x} + \frac{2}{b}\frac{\partial}{\partial u} + \left(p - \frac{2}{b}\right)\frac{\partial}{\partial p}\right).$$

5. Generating functions

In this section, we describe the generating function method for finding symmetries of the Cartan distribution.

Let $\mathcal{E}$ be an ordinary differential equation of order k resolved with respect to the highest-order derivative:

$$u^{(k)} = f(x, u, u^{(1)}, \ldots, u^{(k-1)}). \tag{5.1}$$

Let us view x, $u = p_0, p_1, \ldots, p_{k-1}$ as coordinates on the hypersurface $\mathcal{E} = \{p_k = f(x, u, p_1, \ldots, p_{k-1})\}$. In these coordinates, the restriction of any vector field X to $\mathcal{E}$ is written as

$$X = \alpha\frac{\partial}{\partial x} + \beta_0\frac{\partial}{\partial p_0} + \cdots + \beta_{k-1}\frac{\partial}{\partial p_{k-1}}.$$

Note that the algebra of characteristic symmetries of $\mathcal{C}(\mathcal{E})$ consists of fields of the form $\lambda\bar{D}$, where λ is an arbitrary function on $\mathcal{E}$ and

$$\bar{D} = \frac{\partial}{\partial x} + p_1\frac{\partial}{\partial p_0} + \cdots + p_{k-1}\frac{\partial}{\partial p_{k-2}} + f\frac{\partial}{\partial p_{k-1}}. \tag{5.2}$$

is the *total derivative operator* with respect to x on the equation $\mathcal{E}$.

Let us remark that $\bar{D}(p_i) = p_{i+1}$, $i < k-1$, and, by (5.1), $\bar{D}(p_{k-1}) = f$.

In the quotient space of all fields by $\operatorname{char}\mathcal{C}(\mathcal{E})$, one has

$$\frac{\partial}{\partial x} = -p_1\frac{\partial}{\partial p_0} - \cdots - p_{k-1}\frac{\partial}{\partial p_{k-2}} - f\frac{\partial}{\partial p_{k-1}},$$

so that it is sufficient to search for symmetries only among vector fields of the form

$$X = \beta_0\frac{\partial}{\partial p_0} + \cdots + \beta_{k-1}\frac{\partial}{\partial p_{k-1}}. \tag{5.3}$$

For further calculations, it is convenient to consider the operation of *horizontalization* on the space of 1-forms on $\mathcal{E}$, which is defined by

$$\urcorner dx = dx,$$
$$\urcorner dp_i = p_{i+1}\, dx, \quad i < k-1,$$
$$\urcorner dp_{k-1} = f\, dx,$$

and extended to all of the space of 1-form as a linear map over the ring of smooth functions.

LEMMA 5.1. *The following statements hold*:
1. *The operation of horizontalization takes a 1-form ω to zero if and only if ω vanishes on the Cartan distribution.*
2. *For any function $g \in C^\infty(\mathcal{E})$, one has*

$$\urcorner dg = \bar{D}(g)\, dx.$$

PROOF. 1. Let $\omega = \gamma\, dx + \sum_i \gamma_i\, dp_i$. The equality $\urcorner \omega = 0$ means that $\gamma + \sum_i \gamma_i p_{i+1} + \gamma_{k-1} f = 0$. Therefore,

$$\omega = -\left(\sum_i \gamma_i p_{i+1} + \gamma_{k-1} f\right) dx + \sum_i \gamma_i\, dp_i = \sum_i \gamma_i \omega_i.$$

2. It is straightforward to compute that

$$\urcorner dg = \urcorner\left(\frac{\partial g}{\partial x}\, dx + \sum_i \frac{\partial g}{\partial p_i}\, dp_i\right)$$
$$= \left(\frac{\partial g}{\partial x} + \sum_i p_{i+1}\frac{\partial g}{\partial p_i} + f\frac{\partial g}{\partial p_{k-1}}\right) dx = \bar{D}(g)dx. \quad \square$$

Suppose now that a vector field X of the form (5.3) is a symmetry of the Cartan distribution. By virtue of Theorem 3.1 and Lemma 5.1, this is equivalent to

$$\urcorner X(dp_i - p_{i+1}\, dx) = 0, \quad i < k-1, \tag{5.4}$$
$$\urcorner X(dp_{k-1} - f\, dx) = 0. \tag{5.5}$$

From (5.4) it follows that

$$(\bar{D}(\beta_i) - \beta_{i+1})\, dx = 0.$$

Hence, $\beta_{i+1} = \bar{D}(\beta_i)$ for all $i < k-1$. Denoting β_0 by φ, we arrive at the following expression for X:

$$X = \sum_{i=0}^{k-1} \bar{D}^i(\varphi)\frac{\partial}{\partial p_i}. \tag{5.6}$$

In particular, we see that the field X, which is a symmetry of the Cartan distribution, is uniquely defined by one function $\varphi = X(u)$. The function φ will be referred to as the *generating function* of the field X given by (5.6). The symmetry with the generating function φ will be denoted by X_φ.

Clearly, the generating function of a symmetry for a given differential equation cannot be arbitrary. From (5.5) it follows that

$$\bar{D}^k(\varphi) - \sum_{i=0}^{k-1} \frac{\partial f}{\partial p_i}\bar{D}^i(\varphi) = 0. \tag{5.7}$$

The linear differential operator on the left-hand side of (5.7) is denoted by $\ell_F^{\mathcal{E}}$ and is called the *universal linearization*[6] of the function $F = p_k - f$ restricted to the equation $\mathcal{E}$.

The above discussion can be summarized as follows.

THEOREM 5.2. *Suppose that the equation $\mathcal{E}$ has the form $F = 0$, where $F = p_k - f(x, u, \ldots, p_{k-1})$. Then the correspondence $\varphi \mapsto X_\varphi$ defines the isomorphism*

$$\ker \ell_F^{\mathcal{E}} \cong \operatorname{sym} \mathcal{E}.$$

REMARK 5.1. The assumption that the equation is solved with respect to the highest-order derivative is not in fact essential. In Chapter 4, we shall prove that this theorem holds for "almost all" functions F. In the general case, the operator $\ell_F^{\mathcal{E}}$ is defined by

$$\ell_F^{\mathcal{E}} = \sum_i \frac{\partial F}{\partial p_i} \bar{D}^i.$$

Obviously, this formula gives the above expression for $\ell_F^{\mathcal{E}}$ in the case when $F = p_k - f(x, u, \ldots, p_{k-1})$.

Since the fields of the form (5.6) are in one-to-one correspondence with the generating functions, the generating function φ of the field X_φ is recovered by means of the formula

$$\varphi = X_\varphi \,\lrcorner\, \omega_0 = \omega_0(X_\varphi), \tag{5.8}$$

where $\omega_0 = dp_0 - p_1\, dx$. Note that the form ω_0 vanishes on the characteristic symmetries, because $\omega_0(\bar{D}) = 0$, i.e., the right-hand side of (5.8) is defined by the coset of X_φ modulo the characteristic symmetries. So, to find the generating function, it is not necessary for the symmetry to be reduced to the form (5.6).

Note also that, since no differential other than dx and du appears in the expression of ω_0, the generating function of a symmetry X is uniquely determined by the coefficients at $\partial/\partial x$ and $\partial/\partial u$ in the coordinate expression for this symmetry:

$$\omega_0 \left(\alpha \frac{\partial}{\partial x} + \beta \frac{\partial}{\partial u} + \cdots \right) = \beta - \alpha p_1.$$

Consider now a symmetry X such that its coefficients α and β depend on x and u only. Geometrically, this means that the image X_0 of X under the projection $\pi \colon \mathbb{R}^{k+2} \to \mathbb{R}^2$, $\pi(x, u, p_1, \ldots, p_k) = (x, u)$, is well defined. The field X_0 has the form

$$X_0 = \alpha \frac{\partial}{\partial x} + \beta \frac{\partial}{\partial u}.$$

In the case considered, the symmetry X is uniquely determined by the field X_0 and is called the *lifting* of X_0. Vector fields on the plane $\mathbb{R}^2$, i.e., infinitesimal changes of independent and dependent variables, are said to be infinitesimal *point transformations*[7].

Thus, point transformations (or, more precisely, the liftings of point transformations) are those symmetries of the Cartan distribution whose generating functions

---

[6]The general theory of such operators is discussed in Chapter 4.

[7]The meaning of this term will be clarified in Chapters 2 and 3 in studies of more general contact transformations.

depend on x, u and p_1 only, with linear dependence on p_1. If the generating function is an arbitrary function of x, u and p_1, then the corresponding vector field is called *contact*. Symmetries whose generating functions depend on all coordinates $x, u, p_1, \ldots, p_{k-1}$ are called *higher*. The general theory of higher symmetries will be developed in Chapter 4.

Here are some point transformations which are most frequently encountered in applications:

1. Translation along x: $X_0 = \partial/\partial x$. The generating function is $-p_1$.
2. Translation along u: $X_0 = \partial/\partial u$. The generating function is 1.
3. Scale transformation: $X_0 = ax\partial/\partial x + bu\partial/\partial u$, where a and b are constant. The generating function is $bu - axp_1$.

Consider now a system of ordinary differential equations. For the sake of simplicity, we restrict ourselves to first-order systems solved with respect to the highest-order derivatives:

$$\begin{cases} \dot{u}_1 = f_1(x, u_1, \ldots, u_n), \\ \cdots\cdots\cdots\cdots\cdots\cdots \\ \dot{u}_n = f_n(x, u_1, \ldots, u_n). \end{cases} \tag{5.9}$$

As before, such a system is geometrized by considering the manifold $\mathcal{E}$ with the coordinates $x, u_1, \ldots, u_n$ equipped with the one-dimensional completely integrable Cartan distribution $\mathcal{C}(\mathcal{E})$. This distribution is defined by the forms

$$\begin{cases} \omega_1 = du_1 - f_1\, dx, \\ \cdots\cdots\cdots\cdots\cdots \\ \omega_n = du_n - f_n\, dx. \end{cases}$$

Its integral curves are graphs of solutions of the system under consideration. A counterpart of Theorem 5.2 in the case of systems of ordinary equations can be formulated as follows:

THEOREM 5.3. *The space of nontrivial infinitesimal symmetries of system* (5.9) *is isomorphic to the kernel of the matrix operator*

$$\ell_F^{\mathcal{E}} = \begin{pmatrix} \bar{D} - \partial f_1/\partial u_1 & -\partial f_1/\partial u_2 & \cdots & -\partial f_1/\partial u_n \\ -\partial f_2/\partial u_1 & \bar{D} - \partial f_2/\partial u_2 & \cdots & -\partial f_2/\partial u_n \\ \cdots\cdots\cdots\cdots\cdots\cdots\cdots\cdots\cdots\cdots\cdots\cdots\cdots\cdots\cdots\cdots \\ -\partial f_n/\partial u_1 & -\partial f_n/\partial u_2 & \cdots & \bar{D} - \partial f_n/\partial u_n \end{pmatrix},$$

where $\bar{D} = \partial/\partial x + f_1 \partial/\partial u_1 + \cdots + f_n \partial/\partial u_n$, *acting on the space of n-vector functions of variables* $x, u_1, \ldots, u_n$.

The symmetry corresponding to a vector function $\varphi = (\varphi_1, \ldots, \varphi_n) \in \ker \ell_F^{\mathcal{E}}$ *under this isomorphism has the form*

$$X_\varphi = \varphi_1 \frac{\partial}{\partial u_1} + \cdots + \varphi_n \frac{\partial}{\partial u_n}.$$

The vector function φ is called the *generating section* of the symmetry X_φ.

Let us now discuss the Lie algebra structure on the space of generating functions and sections induced by the commutator operation of symmetries. Computing the commutator of symmetries $X_\varphi = \sum_i \bar{D}^i(\varphi) \partial/\partial p_i$ and $X_\psi = \sum_i \bar{D}^i(\psi) \partial/\partial p_i$ yields

$$[X_\varphi, X_\psi] = \sum_j \sum_i \left(\bar{D}^i(\varphi) \frac{\partial \bar{D}^j(\psi)}{\partial p_i} - \bar{D}^i(\psi) \frac{\partial \bar{D}^j(\varphi)}{\partial p_i} \right) \frac{\partial}{\partial p_j}.$$

Denote the generating function of the field obtained by $\{\varphi, \psi\}$. Contracting this field with $\omega_0 = dp_0 - p_1\, du$, we get

$$\{\varphi, \psi\} = \sum_{i=0}^{k-1}\left(\bar{D}^i(\varphi)\frac{\partial \psi}{\partial p_i} - \bar{D}^i(\psi)\frac{\partial \varphi}{\partial p_i}\right).$$

This bracket is called the *higher Jacobi bracket* (on the equation $\mathcal{E}$).

For a system of the form (5.9) the Jacobi bracket $\chi = \{\varphi, \psi\}$ of generating sections $\varphi = (\varphi_1, \ldots, \varphi_n)$ and $\psi = (\psi_1, \ldots, \psi_n)$ is given by the formula

$$\chi_i = \sum_{j=1}^{n}\left(\varphi_j\frac{\partial \psi_i}{\partial u_j} - \psi_j\frac{\partial \varphi_i}{\partial u_j}\right), \qquad j = 1, \ldots, n,$$

which immediately follows from the fact that the symmetry with the generating section φ has the form $X_\varphi = \varphi_1 \partial/\partial u_1 + \cdots + \varphi_n \partial/\partial u_n$ and that the Jacobi bracket corresponds to the commutator of symmetries.

6. How to search for equations integrable by quadratures: an example of using symmetries

Let us find all equations of the form

$$u'' = u' + f(u) \tag{6.1}$$

possessing a two-dimensional Lie algebra of point symmetries [**26, 77**]. As mentioned above, all two-dimensional algebras are solvable, and therefore all such equations are integrable by quadratures.

Since the independent variable x does not enter explicitly into the equation, the translation along x is a symmetry. So, the problem is to determine when this equation has one more symmetry with a generating function of the form

$$\varphi = \alpha p + \beta, \tag{6.2}$$

where $p = p_1$, α and β are functions of x, u, such that $\{p, \varphi\}$ is a linear combination of p and φ. In what follows, we exclude the trivial case when the function $f(u)$ is linear.

We shall use variables x, u, p as coordinates on the surface $\mathcal{E}$ corresponding to equation (6.1). In these coordinates, the operator of total derivative has the form

$$\bar{D} = \frac{\partial}{\partial x} + p\frac{\partial}{\partial u} + (p + f)\frac{\partial}{\partial p}. \tag{6.3}$$

The generating function of a symmetry $\varphi(x, u, p)$ must satisfy the equation

$$\bar{D}^2(\varphi) - \bar{D}(\varphi) - f'\varphi = 0.$$

Taking (6.2) and (6.3) into account, we can rewrite this equation as

$$\alpha_{uu}p^3 + (2\alpha_u + 2\alpha_{xu} + \beta_{uu})p^2 + (3\alpha_u f + \alpha_x + \alpha_{xx} + 2\beta_{xu})p \\ + (2\alpha_x f + \beta_u f + \beta_{xx} - \beta_x - \beta f') = 0, \tag{6.4}$$

where the indices x and u label the derivative with respect to these variables. The left-hand side of this expression is a polynomial in p, so that the coefficients of this polynomial vanish. Thus, equation (6.4) is equivalent to the system

$$\begin{cases} \alpha_{uu} = 0, \\ 2\alpha_u + 2\alpha_{xu} + \beta_{uu} = 0, \\ 3f\alpha_u + \alpha_x + \alpha_{xx} + 2\beta_{xu} = 0, \\ 2f\alpha_x + f\beta_u + \beta_{xx} - \beta_x - f'\beta = 0. \end{cases} \quad (6.5)$$

The first equation yields

$$\alpha = \gamma u + \delta, \qquad \gamma, \delta \in C^\infty(x),$$

where $C^\infty(x)$ stands for the ring of smooth functions of x. Substituting this expression into the second equation, we obtain

$$\beta = -(\gamma' + \gamma)u^2 + \varepsilon u + \zeta, \qquad \varepsilon, \zeta \in C^\infty(x).$$

Then the third equation is reduced to

$$3f\gamma = 3(\gamma' + \gamma'')u - \delta' - \delta'' - 2\varepsilon'.$$

Since the function $f(u)$ was supposed to be nonlinear, it follows that $\gamma = 0$ and $\varepsilon = (\varkappa - \delta - \delta')/2$, where $\varkappa = \text{const}$.

Now, the last equation of (6.5) takes the form

$$(\varepsilon u + \zeta)f' - \eta f = \theta u + \lambda, \quad (6.6)$$

where $\eta = 2\delta' + \varepsilon$, $\theta = \varepsilon'' - \varepsilon'$, $\lambda = \zeta'' - \zeta'$ are functions belonging to $C^\infty(x)$.

Equation (6.6) is an ordinary differential equation for $f(u)$, involving the variable x as a parameter. If $\varepsilon \neq 0$, $\eta \neq 0$, and $\varepsilon \neq \eta$, its general solution has the form

$$f(u) = \mu \left(u + \frac{\zeta}{\varepsilon} \right)^{\frac{\eta}{\varepsilon}} + \frac{\theta}{\varepsilon - \eta}\left(u + \frac{\zeta}{\varepsilon} \right) + \frac{\theta\zeta - \varepsilon\lambda}{\varepsilon\eta}, \quad (6.7)$$

where $\mu \in C^\infty(x)$.

Of all the functions (6.7), we have to choose those which do not depend on x and are nonlinear in u. If $\eta/\varepsilon \neq 2$, then these requirements hold if and only if

$$\mu = a, \qquad \frac{\zeta}{\varepsilon} = b, \qquad \frac{\eta}{\varepsilon} = c,$$
$$\frac{\theta}{\varepsilon - \eta} = d, \qquad \frac{\theta\zeta - \varepsilon\lambda}{\varepsilon\eta} = e, \quad (6.8)$$

where a, b, c, d, and e are constants.

Taking into consideration all the relations on the functions δ, ε, η, ζ, θ, λ and using (6.8), we easily obtain the following expressions:

$$\varepsilon = he^{kx}, \quad \zeta = b\varepsilon, \quad \eta = c\varepsilon, \quad \theta = (k^2 - k)\varepsilon, \quad \lambda = b\theta,$$

where $k = (1 - c)/(c + 3)$ and h is a constant. Hence,

$$f(u) = a(u + b)^c - \frac{2c + 2}{(c + 3)^2}(u + b). \quad (6.9)$$

From the relations on δ, ε, η, ζ, θ, and λ it follows that if $\eta = 2\varepsilon$ and the function (6.7) does not depend on x, then it has the form (6.9) for $c = 2$.

Now consider the possibilities previously excluded. Either of the assumptions $\eta = 0$ and $\eta = \varepsilon$ results in the linearity of $f(u)$. In the case $\varepsilon = 0$, we have a new series of solutions to equation (6.6):
$$f(u) = ae^{bu} - \frac{2}{b}, \qquad a, b = \text{const.}$$

From the above computation, we obtain the following assertion.

THEOREM 6.1. *Among all nonlinear equations of the form*
$$u'' = u' + f(u)$$
only equations of the two series

1. $u'' = u' + a(u+b)^c - \dfrac{2c+2}{(c+3)^2}(u+b) \qquad a, b, c \in \mathbb{R}, \quad c \neq 1, -3,$

2. $u'' = u' + ae^{bu} - \dfrac{2}{b}, \qquad a, b \in \mathbb{R}, \quad b \neq 0,$

possess a two-dimensional algebra of point symmetries. In the first case, this algebra spans the generating functions
$$\begin{cases} \varphi_1 = p, \\ \varphi_2 = e^{kx}\left(p - \dfrac{k+1}{2}(u+b)\right), \end{cases} \quad k = \dfrac{1-c}{c+3}.$$

In the second case, the algebra is generated by
$$\begin{cases} \varphi_1 = p, \\ \varphi_2 = e^{-x}\left(p - \dfrac{2}{b}\right). \end{cases}$$

CHAPTER 2

First-Order Equations

In this chapter, the geometric approach to differential equations and their solutions, discussed above in the case of ordinary differential equations, is extended to scalar differential equations of first order.

As in Chapter 1, we interpret such differential equations as manifolds endowed with the Cartan distribution and, in this language, study their symmetries, integrability, etc.

The geometric theory of first-order equations is closely related to symplectic geometry and, in particular, Hamiltonian mechanics. These questions are also the concern of the present chapter.

1. Contact transformations

1.1. Contact elements and the Cartan distribution. Let $x_1, \ldots, x_n$ denote coordinates on $\mathbb{R}^n$. In these coordinates, a first-order differential equation for one unknown function u has the form

$$F\left(x_1, \ldots, x_n, u, \frac{\partial u}{\partial x_1}, \ldots, \frac{\partial u}{\partial x_n}\right) = 0,$$

where F is a smooth function of $2n+1$ variables.

Denote the partial derivatives of u by $p_1 = \partial u/\partial x_1, \ldots, p_n = \partial u/\partial x_n$, and consider the $(2n+1)$-dimensional vector space $J^1(\mathbb{R}^n)$ with coordinates $x_1, \ldots, x_n$, $u, p_1, \ldots, p_n$. A differential equation gives rise to a locus

$$\mathcal{E} = \{F(x_1, \ldots, x_n, u, p_1, \ldots, p_n) = 0\}$$

in this space.

As a rule, we shall assume that the differential

$$dF = \sum_{i=1}^{n} \frac{\partial F}{\partial x_i} dx_i + \frac{\partial F}{\partial u} du + \sum_{i=1}^{n} \frac{\partial F}{\partial p_i} dp_i$$

is nonzero on some everywhere dense subset of the hypersurface $\mathcal{E}$.

Under this assumption, any function G vanishing on $\mathcal{E}$ has the form $G = \mu F$ near $\mathcal{E}$, where μ is a smooth function of $x_1, \ldots, x_n, u, p_1, \ldots, p_n$.

Differential equations on manifolds other than $\mathbb{R}^n$ (e.g., on a torus) are not infrequent in applications. Thus, let us define, in a coordinate-free manner, the space $J^1(M)$ for an arbitrary smooth manifold M.

Consider the $(n+1)$-dimensional manifold $J^0(M) = M \times \mathbb{R}$; a point of $J^0(M)$ is a pair (x, u), where $x \in M$, $u \in \mathbb{R}$. A *contact element* at a point $\theta \in J^0(M)$ is a pair (θ, L), where $L \subset T_\theta(J^0(M))$ is an n-dimensional plane. A contact element (θ, L) is called *nonsingular*, if the plane L does not contain the vertical vector $\partial/\partial u$.

EXERCISE 1.1. Prove that the set of all nonsingular contact elements on $J^0(M)$ is a smooth manifold.

DEFINITION 1.1. The manifold of all nonsingular contact elements on $J^0(M)$ is called the *manifold of 1-jets* of smooth functions on M and is denoted by $J^1(M)$.

Note that the projections

$$\pi_{1,0}\colon J^1(M) \longrightarrow J^0(M), \quad \pi_{1,0}(\theta, L) = \theta \tag{1.1}$$

$$\pi_1\colon J^1(M) \longrightarrow M, \quad \pi_1((x,u), L) = x, \quad (x,u) \in J^0(M), \tag{1.2}$$

are well-defined and are smooth vector bundles. The $C^\infty(M)$-module of sections of the bundle (1.2) will be denoted by $\mathcal{J}^1(M)$.

The manifold $J^1(M)$ is basic to the geometric formulation of first-order differential equations.

Let $\Gamma(f) \subset J^0(M)$ be the graph of a function $f \in C^\infty(M)$, i.e., $\Gamma(f)$ consists of the points $b = (a, f(a))$, where $a \in M$. At every point $b = (a, f(a))$ of this graph, consider the nonsingular contact element (b, L), where $L = T_a(\Gamma(f))$ is the tangent plane to the graph. Clearly, every nonsingular contact element $\theta = (b, L_\theta) \in J^1(M)$ is tangent to the graph of an appropriate function.

Let us give a coordinate description of contact elements. Take a function $f \in C^\infty(M)$ and suppose that in local coordinates $x_1, \ldots, x_n, u$ it has the form $u = f(x_1, \ldots, x_n)$. Let $\xi_1, \ldots, \xi_n, \eta$ be coordinates on $T_b(J^0(M))$, where $b = (a, f(a))$, $a \in M$, with respect to the basis $\partial/\partial x_1, \ldots, \partial/\partial x_n, \partial/\partial u$. Clearly, the plane tangent to the graph of f is given by the equation

$$\eta = p_1 \xi_1 + \cdots + p_n \xi_n, \tag{1.3}$$

where $p_1 = \partial f/\partial x_1(a), \ldots, p_n = \partial f/\partial x_n(a)$.

Thus, on the manifold of 1-jets there exist *special* (or *canonical*) *local coordinates* $x_1, \ldots, x_n, u, p_1, \ldots, p_n$. Their geometric meaning is as follows: if $\theta = (b, L_\theta) \in J^1(M)$, where L_θ is the plane tangent to the graph of the function $u = u(x_1, \ldots, x_n)$ at $b = (a, f(a))$, then the contact element (b, L_θ) has the coordinates

$$\left(a_1, \ldots, a_n, f(a), \frac{\partial f}{\partial x_1}(a), \ldots, \frac{\partial f}{\partial x_n}(a)\right).$$

EXERCISE 1.2. Show that $J^1(M) = T^*(M) \times \mathbb{R}$.

By this exercise, one can define the projection $\pi\colon J^1(M) \to T^*(M)$. In the special coordinates it takes the form

$$\pi(x_1, \ldots, x_n, u, p_1, \ldots, p_n) = (x_1, \ldots, x_n, p_1, \ldots, p_n). \tag{1.4}$$

Let us now discuss a geometric interpretation of solutions to a given differential equation. For any smooth function $f \in C^\infty(M)$, define the map

$$j_1(f)\colon M \longrightarrow J^1(M), \quad j_1(f)\colon a \longmapsto (b, L_{f(a)}), \tag{1.5}$$

which takes each point $a \in M$ to the nonsingular contact element consisting of the point $b = (a, f(a))$ and the plane tangent to the graph of f at this point.

The map $j_1(f)$ is said to be the *1-jet* of the function f. Its graph $N_f = j_1(f)(M) \subset J^1(M)$ is an n-dimensional manifold. In the special local coordinates

1. CONTACT TRANSFORMATIONS

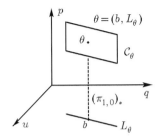

FIGURE 2.1

$x_1, \ldots, x_n$, u, $p_1, \ldots, p_n$ this manifold is obviously given by the equations

$$\begin{cases} u = f(x_1, \ldots, x_n), \\ p_1 = \dfrac{\partial f}{\partial x_1}(x_1, \ldots, x_n), \\ \cdots\cdots\cdots\cdots\cdots \\ p_n = \dfrac{\partial f}{\partial x_n}(x_1, \ldots, x_n). \end{cases} \qquad (1.6)$$

Note that, similarly to the case of one independent variable, not every n-dimensional submanifold in $J^1(M)$ has the form N_f. To describe submanifolds of the form N_f, we supply the manifold $J^1(M)$ with the *Cartan distribution*, which is defined below.

Let $\theta = (b, L_\theta)$ be a point of $J^1(M)$, $L_\theta \subset T_b(J^0(M))$. Denote by $\mathcal{C}_\theta$ the set of vectors in the tangent space $T_\theta(J^1(M))$ whose images under the projection $(\pi_{1,0})_*$ belong to the plane L_θ:

$$\mathcal{C}_\theta = \{\, \xi \in T_\theta(J^1(M)) \mid (\pi_{1,0})_*(\xi) \in L_\theta \,\}.$$

$\mathcal{C}_\theta$ is a hyperplane called the *Cartan plane* at $\theta \in J^1(M)$ (see Figure 2.1).

DEFINITION 1.2. The distribution

$$\theta \longmapsto \mathcal{C}_\theta, \qquad \theta \in J^1(M),$$

on the manifolds of 1-jets $J^1(M)$ is called the *Cartan distribution*.

PROPOSITION 1.1. *Let $f \in C^\infty(M)$ be a function, $a \in M$, $\theta = j_1(f)(a) \in N_f$. Then the manifold N_f is tangent to $\mathcal{C}_\theta$.*

PROOF. The map $\pi_{1,0}$ takes the manifold N_f to the graph of f. Therefore the tangent plane $T_\theta(N_f)$ is projected to the tangent plane to the graph of f at the point $b = (a, f(a))$, i.e., on L_θ. Hence, $T_\theta(N_f) \subset \mathcal{C}_\theta$. $\square$

This proposition can be restated as follows:

COROLLARY 1.2. *The image of the map $j_1(f) \colon M \to J^1(M)$ is an integral manifold of the Cartan distribution.*

EXERCISE 1.3. a. Show that the fibers of the projection (1.1) are integral manifolds of the Cartan distribution.

b. Let V be an arbitrary k-dimensional submanifold of $J^0(M)$, $1 \leq k \leq n = \dim M$. Consider the manifold $P \subset J^1(M)$ consisting of contact elements (b, L)

such that $b \in V$ and $L \supset T_b(V)$. Prove that P is an integral manifold of the Cartan distribution.

Let us now define a differential form U_1 which determines the Cartan distribution on $J^1(M)$. Take a contact element $\theta = (b, L_\theta)$, where $b \in J^0(M)$ and $L_\theta \subset T_b(J^0(M))$. The image of a tangent vector $\xi \in T_\theta(J^1(M))$ under the projection $\pi_{1,0}\colon J^1(M) \to J^0(M)$, $\pi_{1,0}(\theta) = b$, can be uniquely decomposed into the sum $\xi = \xi_0 + \xi_0'$, where $\xi_0 = (\pi_{1,0})_*(\xi) = \rho_\xi\, \partial/\partial u|_b$, $\xi_0' \in L_\theta$. Set $U_1(\xi) = \rho_\xi$. The following proposition is obvious.

PROPOSITION 1.3. $\xi \in \mathcal{C}_\theta$ if and only if $U_1(\xi) = 0$.

Thus, the 1-form U_1 defines the Cartan distribution. Let us describe this form in local coordinates.

Let $x_1, \ldots, x_n$, u, $p_1, \ldots, p_n$ be special local coordinates on $J^1(M)$ in the vicinity of a point θ. Since the plane $L_\theta \subset T_b(J^0(M))$ is given by the equation $\eta = p_1\xi_1 + \cdots + p_n\xi_n$, where $\xi_1, \ldots, \xi_n, \eta$ are the coordinates in $T_b(J^0(M))$, we have

$$\sum_{i=1}^n \xi_i \frac{\partial}{\partial x_i} + \left(\sum_{i=1}^n p_i\xi_i\right)\frac{\partial}{\partial u} \in L_\theta. \tag{1.7}$$

Suppose that

$$\xi = \sum_{i=1}^n \xi_i \frac{\partial}{\partial x_i} + \eta\frac{\partial}{\partial u} + \sum_{i=1}^n \zeta_i \frac{\partial}{\partial p_i} \in T_\theta(J^1(M));$$

then

$$(\pi_{1,0})_*(\xi) = \sum_{i=1}^n \xi_i \frac{\partial}{\partial x_i} + \eta\frac{\partial}{\partial u}.$$

From (1.7) it follows that

$$(\pi_{1,0})_*(\xi) = \left(\eta - \sum_{i=1}^n p_i\xi_i\right)\frac{\partial}{\partial u} + \chi,$$

with $\chi \in L_\theta$. Hence,

$$U_1 = du - \sum_{i=1}^n p_i\, dx_i. \tag{1.8}$$

REMARK 1.1. Contact geometry closely resembles symplectic geometry. This is essentially due to the isomorphism $J^1(M) = T^*M \times \mathbb{R}$. For a detailed discussion of symplectic geometry see, e.g., [**33, 3, 7**].

In view of this remark, it is not surprising that there exists a universal element on the manifold $J^1(M)$ similar to the universal form $p\, dq$ in symplectic geometry [**72, 77**].

PROPOSITION 1.4. There exists a unique element $\rho \in \mathcal{J}^1(J^1(M))$ on the manifold $J^1(M)$ such that for any section $\theta \in \mathcal{J}^1(M)$ one has $\theta^*(\rho) = \theta$.

PROOF. Consider an arbitrary point $x \in J^1(M)$. It is clear that there exists a function $f \in C^\infty(M)$ such that $x = j_1(f)|_a$, $a = \pi_1(x)$. By definition, put $\rho|_x = j_1(\pi_1^*(f))|_x$. Let us prove the equality $\theta^*(\rho) = \theta$. Take $x = \theta(a)$. We have

$$\theta^*(\rho)|_a = \theta^*(j_1(\pi_1^*(f))|_x) = j_1(\theta^*\pi_1^*(f))|_a = j_1(f)|_a = x = \theta(a).$$

To prove uniqueness of ρ, observe that if one has $(\theta)^*(\rho') = 0$ for some element ρ' and all $\theta \in \mathcal{J}^1(M)$, then $\rho' = 0$. $\square$

EXERCISE 1.4. Prove that in the local coordinates the element ρ has the form $\sum_{i=1}^n p_i\, dx_i$.

The decomposition $J^1(M) = T^*(M) \times \mathbb{R}$ induces the decomposition $\mathcal{J}^1(M) = \Lambda^1(M) \oplus C^\infty(M)$. Define an operator $S\colon \mathcal{J}^1(M) \to \Lambda^1(M)$ by the formula[1] $S(f, \omega) = df - \omega$.

EXERCISE 1.5. Prove that the image of the operator $j_1\colon C^\infty(M) \to \mathcal{J}^1(M)$ coincides with the kernel of the operator[2] S.

It is readily seen that $U_1 = S(\rho)$.

The following proposition states that the converse of Corollary 1.2 is also true:

PROPOSITION 1.5. *A submanifold $N \subset J^1(M)$ that projects to M without degeneration is a maximal integral manifold of the Cartan distribution if and only if it is the graph of a function $f \in C^\infty(M)$, i.e., $N = j_1(f)(M)$.*

PROOF. An n-dimensional manifold $N \subset J^1(M)$ that projects to M without degeneration has the form
$$\begin{cases} u = f(x_1, \ldots, x_n), \\ p_1 = g_1(x_1, \ldots, x_n), \\ \cdots\cdots\cdots\cdots\cdots\cdots \\ p_n = g_n(x_1, \ldots, x_n). \end{cases}$$
The manifold N is an integral manifold of the Cartan distribution if and only if $U_1|_N = 0$. In the coordinates $x_1, \ldots, x_n$ on N, the restriction of U_1 to N can be written as $\sum_{i=1}^n (\partial f/\partial x_i - g_i)\, dx_i$. Thus, N will be an integral manifold if and only if $g_i(x_1, \ldots, x_n) = \partial f/\partial x_i$, $i = 1, \ldots, n$, and this is precisely the assertion of the proposition. $\square$

Thus, the Cartan distribution on $J^1(M)$ is the geometric structure which distinguishes the graphs of 1-jets from all n-dimensional submanifolds of $J^1(M)$.

Consider a submanifold $\mathcal{E} \subset J^1(M)$. The Cartan distribution on $J^1(M)$ induces the *Cartan distribution* $\mathcal{C}(\mathcal{E})$ on $\mathcal{E}$, $\mathcal{C}(\mathcal{E})_\theta = \mathcal{C}_\theta \cap T_\theta(\mathcal{E})$.

DEFINITION 1.3.
1. A submanifold $\mathcal{E} \subset J^1(M)$ of codimension one supplied with the induced Cartan distribution is called a *first-order differential equation*.
2. A *generalized solution* $L \subset \mathcal{E}$ of an equation $\mathcal{E}$ is an n-dimensional maximal integral manifold of the Cartan distribution on $\mathcal{E}$.

EXAMPLE 1.1. Consider the stationary Hamilton–Jacobi equation
$$\left(\frac{\partial S}{\partial q}\right)^2 + q^2 = C, \qquad C = \text{const},$$

---

[1] The operator S is the simplest operator in the series of the Spencer operators
$$S_{k,l}\colon \mathcal{J}^k \otimes \Lambda^l \to \mathcal{J}^{k-1} \otimes \Lambda^{l+1}, \qquad S_{k,l}(aj_k(f) \otimes \omega) = j_{k-1}(f) \otimes (da \wedge \omega).$$
For discussion of the Spencer operators see [60].

[2] The exact sequence
$$0 \longrightarrow C^\infty(M) \xrightarrow{j_1} \mathcal{J}^1(M) \xrightarrow{S} \Lambda^1(M) \longrightarrow 0$$
is the simplest Jet-Spencer complex (see [60]).

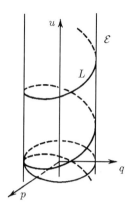

FIGURE 2.2

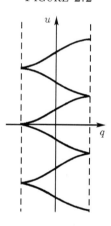

FIGURE 2.3

describing a one-dimensional harmonic oscillator. The corresponding locus $p^2+q^2 = C$ is the cylinder in the (q, u, p)-space. The Cartan distribution $\mathcal{C}(\mathcal{E})$ is a direction field on this cylinder. Integral curves of $\mathcal{C}(\mathcal{E})$ (see Figure 2.2) are generalized solutions of the equation. It is obvious that these curves project to the q-axis with degeneration. Figure 2.3 shows the projections of these curves to the (q, u)-plane.

If the tangent spaces $T_\theta(\mathcal{E})$ nowhere coincide with the Cartan planes $\mathcal{C}_\theta$, then the induced Cartan distribution $\mathcal{C}(\mathcal{E})$ is of codimension 1.

DEFINITION 1.4. A point $\theta \in \mathcal{E}$ is called a *singular point of the equation* $\mathcal{E}$ if $T_\theta(\mathcal{E}) = \mathcal{C}_\theta$.

EXERCISE 1.6. Show that the origin of coordinates is a singular point of the equations

$$x_1 \frac{\partial f}{\partial x_1} + x_2 \frac{\partial f}{\partial x_2} = f \quad \text{and} \quad \left(\frac{\partial f}{\partial x_1}\right)^2 + \left(\frac{\partial f}{\partial x_2}\right)^2 = f.$$

Below, we work, as a rule, in the vicinity of nonsingular points of $\mathcal{E}$.

1.2. Contact transformations. A smooth transformation $J^1(M) \to J^1(M)$ is called a *contact transformation*, if it takes the Cartan distribution to itself.

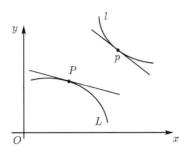

FIGURE 2.4

Thus, $F\colon J^1(M) \to J^1(M)$ is a contact transformation if either of the following two equivalent conditions holds:
1. $F_*(\mathcal{C}_\theta) \subset \mathcal{C}_{F(\theta)}$ for all $\theta \in J^1(M)$.
2. $F^*(U_1) = \lambda U_1$, where $\lambda \in C^\infty(J^1(M))$.

In coordinates the second condition takes the form

$$dU - \sum_{i=1}^n P_i\, dX_i = \lambda\left(du - \sum_{i=1}^n p_i\, dx_i\right), \tag{1.9}$$

where $X_i = F^*(x_i)$, $U = F^*(u)$, $P_i = F^*(p_i)$, $\lambda \in C^\infty(J^1(M))$.

Historically, contact transformations were understood as "transformations of curves on the (x,y)-plane" of the form

$$\begin{aligned}X &= f(x,y,y'),\\ Y &= \varphi(x,y,y'),\end{aligned} \tag{1.10}$$

satisfying the condition that the images of any two curves tangent to each other are also tangent. Consider a curve l given by an equation $y = y(x)$ and a point $p \in l$. Suppose that a transformation F of the form (1.10) takes l to a curve L given by an equation $Y = Y(X)$ and $P = F(p)$ (see Figure 2.4). It is clear that

$$\frac{dY}{dX} = \frac{\dfrac{\partial \varphi}{\partial x} + \dfrac{\partial \varphi}{\partial y}y' + \dfrac{\partial \varphi}{\partial y'}y''}{\dfrac{\partial f}{\partial x} + \dfrac{\partial f}{\partial y}y' + \dfrac{\partial f}{\partial y'}y''} = \frac{\left(\dfrac{\partial \varphi}{\partial x} + \dfrac{\partial \varphi}{\partial y}y'\right)\left(1 + \dfrac{\dfrac{\partial \varphi}{\partial y'}y''}{\dfrac{\partial \varphi}{\partial x} + \dfrac{\partial \varphi}{\partial y}y'}\right)}{\left(\dfrac{\partial f}{\partial x} + \dfrac{\partial f}{\partial y}y'\right)\left(1 + \dfrac{\dfrac{\partial f}{\partial y'}y''}{\dfrac{\partial f}{\partial x} + \dfrac{\partial f}{\partial y}y'}\right)}.$$

Obviously, dY/dX does not depend on y'' (this is just the condition that the transformation F preserves the tangency of curves), if

$$\frac{\partial \varphi}{\partial y'}\left(\frac{\partial f}{\partial x} + \frac{\partial f}{\partial y}y'\right) = \frac{\partial f}{\partial y'}\left(\frac{\partial \varphi}{\partial x} + \frac{\partial \varphi}{\partial y}y'\right). \tag{1.11}$$

It can easily be checked that (1.11) exactly means that the transformation $J^1(\mathbb{R}^1) \to J^1(\mathbb{R}^1)$, $j_1(y(x))|_p \mapsto j_1(Y(X))|_P$ preserves the differential form $dy - p\,dx$, up to a factor.

Contact transformations can be obtained by lifting smooth transformations of the space $J^0(M)$. Namely, let $F\colon J^0(M) \to J^0(M)$ be such a transformation. In the coordinates $x_1, \ldots, x_n, u$, it has the form

$$(x_1, \ldots, x_n, u) \longmapsto (a_1(x, u), \ldots, a_n(x, u), b(x, u)). \tag{1.12}$$

This transformation induces the transformation of $J^1(M)$. The coordinate description of the latter is as follows.

EXERCISE 1.7. Show that under the transformation (1.12)

$$\begin{pmatrix} p_1 \\ \ldots \\ p_n \end{pmatrix} \longmapsto \Delta^{-1} \times \begin{pmatrix} \partial b/\partial x_1 + p_1 \partial b/\partial u \\ \ldots \\ \partial b/\partial x_n + p_n \partial b/\partial u \end{pmatrix}, \tag{1.13}$$

where Δ is the matrix with the elements $\Delta_{ik} = \partial a_k/\partial x_i + p_i \partial a_k/\partial u$.

EXERCISE 1.8. Prove that the transformation given by (1.12) and (1.13) is contact.

Formulas (1.12) and (1.13) define the transformation of $J^1(M)$, which is called the *lifting* of a given transformation of $J^0(M)$ to $J^1(M)$. The lifting of F takes each contact element $\theta = (b, L_\theta)$ to the element $\bar{F}(\theta) = (F(b), (F_*)_b(L_\theta))$.

Since $\pi_{1,0} \circ \bar{F} = F \circ \pi_{1,0}$ and $(F_*)_b(L_\theta) = L_{\bar{F}(\theta)}$, we have $(\bar{F}_*)_b(\mathcal{C}_\theta) \subset \mathcal{C}_{\bar{F}(\theta)}$. Thus, the transformations of the form $\bar{F}$ preserve the Cartan distribution and, therefore, take the integral manifolds of this distribution into themselves[3].

Contact transformations that are liftings of transformations of $J^0(M)$ are called *point transformations.*.

Point transformations are not the only contact transformations.

EXAMPLE 1.2. Consider the *Legendre transformation*

$$L\colon \begin{cases} x_i \longmapsto p_i, \\ u \longmapsto \sum_{k=1}^n x_k p_k - u, \\ p_i \longmapsto x_i, \quad i = 1, \ldots, n, \end{cases} \tag{1.14}$$

where $x_1, \ldots, x_n, u, p_1, \ldots, p_n$ are special local coordinates.

EXERCISE 1.9. Show that the Legendre transformation preserves the Cartan distribution.

EXERCISE 1.10. Let $F\colon J^0(M) \to J^0(M)$ be a transformation of the form $F(x, u) = (f(x), u)$, where $f\colon M \to M$. Prove that the lifting $\bar{F}$ of F preserves the form U_1, i.e., $\bar{F}^*(U_1) = U_1$.

---

[3]Strictly speaking, the lifting fails to exist at the points θ for which the element $\bar{F}(\theta)$ is singular. Analytically, this means that the matrix Δ is not invertible at these points. In what follows we do not consider such points.

1. CONTACT TRANSFORMATIONS

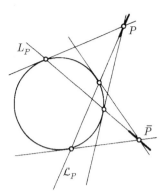

FIGURE 2.5

EXAMPLE 1.3. The following transformation is a generalization of the Legendre transformation and is called the *Euler transformation*:

$$E: \begin{cases} x_i \longmapsto p_i, \\ x_l \longmapsto x_l, \\ u \longmapsto \sum_{s=1}^{k} x_s p_s - u, \\ p_i \longmapsto x_i, \\ p_l \longmapsto -p_l, \end{cases} \quad i = 1, \ldots, k, \quad l = k+1, \ldots, n.$$

It can easily be checked that $E^*(U_1) = -U_1$, so the Euler transformation is contact.

Here are more examples of contact transformations (see [**33**]).

EXAMPLE 1.4. Consider the transformation given by

$$\bar{x}_i = x_i + \frac{\theta p_i}{\sqrt{1 + \sum_{k=1}^{n} p_k^2}}, \quad \bar{u} = u - \frac{\theta}{\sqrt{1 + \sum_{k=1}^{n} p_k^2}}, \quad \bar{p}_i = p_i. \tag{1.15}$$

EXERCISE 1.11. Prove that transformation (1.15) is contact. Is it a point transformation?

EXAMPLE 1.5. The transformation of $\mathbb{R}^3$ given by the formulas

$$\bar{q} = \frac{p}{pq - u}, \quad \bar{u} = -\frac{1}{pq - u}, \quad \bar{p} = -\frac{q}{u}, \tag{1.16}$$

is contact. Its geometric meaning is as follows. Consider the unit circle

$$q^2 + u^2 = 1.$$

(see Figure 2.5). Recall that the polar line of a point $P = (a, b)$ with respect to this circle is the straight line L_P given by the equation

$$aq + bu - 1 = 0.$$

If P lies outside the circle, then its polar line passes through the intersection points of the tangents to the circles passing through P with the circle.

Consider the straight line $\mathcal{L}_P$ passing through P with a slope equal to p_0. There exists a unique point on L_P such that $\mathcal{L}_P$ is its polar line with respect to the circle.

It is not hard to prove that the coordinates of $\bar{P}$ are given by (1.16).

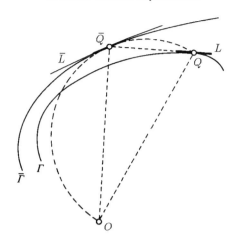

FIGURE 2.6

By taking an arbitrary second-order curve instead of the circle, one can define a contact transformation in a similar way.

EXAMPLE 1.6. Consider the contact transformation of $\mathbb{R}^3$ given by the formulas

$$\bar{x} = \frac{(xp-u)p}{1+p^2}, \quad \bar{u} = -\frac{xp-u}{1+p^2}, \quad \bar{p} = \frac{xp^2 - x - 2up}{up^2 - u + 2xp}. \tag{1.17}$$

and called the *pedal transformation*. Let us give its geometric interpretation (see Figure 2.6).

Let O be the origin of coordinates. Consider a point $Q = (x_0, u_0)$ and the straight line L passing through Q with a slope equal to p. Drop a perpendicular to L from O. Through the foot of the perpendicular let us pass a tangent to the circle with the diameter OQ. It can easily be checked that formulas (1.17) give the coordinates of $\bar{Q}$ and the slope of the tangent $\bar{L}$.

Let us discuss the action of contact transformations on differential equations.

A contact transformation $F \colon J^1(M) \to J^1(M)$ takes a hypersurface $\mathcal{E}$, corresponding to an equation, to a hypersurface $\mathcal{E}'$ (which may coincide with $\mathcal{E}$). Since contact transformations preserve the Cartan distribution, $\mathcal{E}'$ is also "almost everywhere" a differential equation. Note that the transformation $\bar{F}$ establishes a one-to-one correspondence between (generalized) solutions of the initial equation and the transformed one.

DEFINITION 1.5. Equations $\mathcal{E}$ and $\mathcal{E}'$ in $J^1(M)$ are called (*locally*) *equivalent*, if there exists a contact transformation of $J^1(M)$ such that the image of $\mathcal{E}$ under this transformation is $\mathcal{E}'$.

EXAMPLE 1.7. Consider the action of the lifting of the transformation $J^0(M) \to J^0(M)$, $(x_1, x_2, u) \mapsto (u, x_2, x_1)$ on the equation $\mathcal{E}_1 = \{u = \sqrt{p_2^2/p_1 - x_1^2}\}$. From formula (1.13) it follows that

$$(x_1, x_2, u, p_1, p_2) \longmapsto (u, x_2, x_1, 1/p_1, -p_2/p_1).$$

The transformed equation $\mathcal{E}'_1 = \{x_1 = \sqrt{p_2^2/p_1 - u^2}\}$ coincides with the initial one for $x_1 > 0$, $u > 0$.

The same transformation takes the equation $\mathcal{E}_2 = \{4u - p_1^2 - p_2^2 = 0\}$ to $\mathcal{E}_2' = \{4x_1 p_1^2 - 1 - p_2^2 = 0\}$. The image of the graph of the solution to $\mathcal{E}_2$ given by $u = x_1^2 + x_2^2$ is the surface $u^2 = x_1 - x_2^2$, which is a multi-valued solution to $\mathcal{E}_2'$. The points $(x_1 = x_2^2, u = 0)$ on this surface are singular, since the surface is tangent to the vector $\partial/\partial u$ at these points.

For $x_1 > x_2^2$ the solutions $u = \pm\sqrt{x_1 - x_2^2}$ are single-valued and smooth.

DEFINITION 1.6. Let $\mathcal{E} \subset J^1(M)$ be a first-order differential equation. A contact transformation $F \colon J^1(M) \to J^1(M)$ is called a *finite contact symmetry* of $\mathcal{E}$ if $F(\mathcal{E}) = \mathcal{E}$.

Consider an equation $\mathcal{E}$ given by a hypersurface $H = 0$, where $H \in C^\infty(J^1(M))$ is a smooth function with nonzero differential. It is obvious that a contact transformation F is a symmetry of $\mathcal{E}$ if and only if $F^*(H) = \lambda H$ for some function $\lambda \in C^\infty(J^1(M))$.

Let us also stress that symmetries of $\mathcal{E}$ take (generalized) solutions of this equation to solutions of the same equation.

EXAMPLE 1.8. The Legendre transformation is a finite contact symmetry of the equation

$$\mathcal{E} = \left\{ u - \frac{1}{2} \sum_{i=1}^n x_i p_i = 0 \right\}. \tag{1.18}$$

Equation (1.18) has the n-parameter family of solutions $u = \sum_{i=1}^n b_i x_i^2$, where $b_i \in \mathbb{R}$. This family is invariant under the Legendre transformation:

$$L\left(j_1\left(\sum_{i=1}^n b_i x_i^2 \right) \right) = j_1\left(\sum_{i=1}^n \frac{x_i^2}{4 b_i} \right).$$

The graph of the constant function $u = c$ transformed by the Legendre transformation,

$$\begin{cases} u = c, \\ p_1 = 0, \\ \cdots \cdots \\ p_n = 0, \end{cases} \longmapsto \begin{cases} u = -c, \\ x_1 = 0, \\ \cdots \cdots \\ x_n = 0, \end{cases}$$

is an integral manifold whose projection to the manifold of independent variables is the point $(x_1 = \cdots = x_n = 0)$.

Let us now use the Clairaut equation to illustrate applications of such degenerate integral manifolds.

1.3. Clairaut equation and its integrals. The general Clairaut equation has the form

$$F\left(\sum_{i=1}^n x_i p_i - u, p_1, \ldots, p_n \right) = 0, \tag{1.19}$$

where F is a smooth function of $n + 1$ variables. The image of this equation under the Legendre transformation is the equation

$$F(u, x_1, \ldots, x_n) = 0, \tag{1.20}$$

which contains no derivatives. The locus associated to (1.20) is an n-dimensional surface in the $(n+1)$-dimensional space $J^0(\mathbb{R}^n)$, the corresponding first-order differential equation being its inverse image under the projection $\pi_{1,0}\colon J^1(\mathbb{R}^n) \to J^0(\mathbb{R}^n)$.

Take a point $(a_1,\ldots,a_n,b) \in J^0(\mathbb{R}^n)$ such that $F(b,a_1,\ldots,a_n) = 0$. The fiber of $\pi_{1,0}\colon J^1(\mathbb{R}^n) \to J^0(\mathbb{R}^n)$ over this point is a generalized solution of (1.20).

Applying the Legendre transformation to these generalized solutions yields an n-parameter family of single-valued solutions to (1.19), which is said to be a *complete integral of the Clairaut equation*.

On the other hand, solving (1.20) for u in the vicinity of a nonsingular point (i.e., where $\partial F/\partial u \neq 0$) yields

$$u = f(x_1,\ldots,x_n). \tag{1.21}$$

To find the image of this solution under the Legendre transformation, observe that the graph of the 1-jet of (1.21) has the form

$$\begin{cases} p_1 = \dfrac{\partial f}{\partial x_1}(x_1,\ldots,x_n), \\ \ldots\ldots\ldots\ldots\ldots\ldots \\ p_n = \dfrac{\partial f}{\partial x_n}(x_1,\ldots,x_n), \\ F(u,x_1,\ldots,x_n) = 0. \end{cases} \tag{1.22}$$

Under the Legendre transformation, (1.22) maps to the system

$$\begin{cases} x_1 = \dfrac{\partial f_l}{\partial x_1}(p_1,\ldots,p_n), \\ \ldots\ldots\ldots\ldots\ldots\ldots \\ x_n = \dfrac{\partial f_l}{\partial x_n}(p_1,\ldots,p_n), \\ F\left(\sum_{i=1}^n x_i p_i - u, p_1,\ldots,p_n\right) = 0, \end{cases} \tag{1.23}$$

which is compatible.

Express $p_1,\ldots,p_n$ in terms of $x_1,\ldots,x_n$ by the first n equations. Substituting them into the last equation and solving it for u in terms of $x_1,\ldots,x_n$ yields a finite number of solutions to the Clairaut equation, which are called its *exceptional integrals*.

EXAMPLE 1.9. Consider the following Clairaut equation with two independent variables:

$$x_1 p_1 + x_2 p_2 - u = \tfrac{1}{3}(p_1^3 + p_2^3). \tag{1.24}$$

Using the Legendre transformation, we obtain the equation

$$u = \tfrac{1}{3}(x_1^3 + x_2^3),$$

which gives the following complete integral of the initial equation:

$$u = a_1 x_1 + a_2 x_2 - \tfrac{1}{3}(a_1^3 + a_2^3), \qquad a_1, a_2 = \text{const}.$$

Exceptional integrals are found from the system
$$\begin{cases} u = x_1 p_1 + x_2 p_2 - \tfrac{1}{3}(p_1^3 + p_2^3), \\ x_1 = p_1^2, \\ x_2 = p_2^2. \end{cases} \quad (1.25)$$

It follows that exceptional integrals exist in the quadrant $x_1 > 0$, $x_2 > 0$ and have the form
$$p_1 = \alpha_1 \sqrt{x_1}, \qquad p_2 = \alpha_2 \sqrt{x_2},$$
$$u = \tfrac{2}{3}(\alpha_1 x_1^{3/2} + \alpha_2 x_2^{3/2}), \quad (1.26)$$
with $\alpha_1 = \pm 1$ and $\alpha_2 = \pm 1$. Thus, formula (1.26) defines four exceptional integrals. However, from the geometric point of view, the system (1.25) defines only one generalized solution, which is a surface of the fourth order and degenerately projects to the (x_1, x_2)-plane. Formula (1.26) defines the four branches of this generalized solution.

1.4. Contact manifolds in mechanics.

DEFINITION 1.7. A $(2n+1)$-dimensional manifold M equipped with a 1-form θ is called a *contact manifold* if $\theta \wedge (d\theta)^n$ is a volume form.

An odd-dimensional manifold equipped with a closed 2-form of maximal rank will be called an *almost contact manifold*.

It immediately follows from §1.1 that the manifold of 1-jets $J^1(M)$ is a contact manifold.

Note that at any point of a contact manifold the restriction of $d\theta$ to the kernel of θ is nondegenerate. In particular, any contact manifold is almost contact (with the 2-form $d\theta$).

Given a 2-form ω on a manifold M, the set of vectors ξ such that $\xi \lrcorner \omega = 0$ is called the *characteristic distribution* of the form ω. A vector field X on an almost contact manifold (M, ω) is called *characteristic* if $X \lrcorner \omega = 0$.

Let us give some examples of contact manifolds ([**2**]).

EXAMPLE 1.10. Let (M, ω, H) be a Hamiltonian system (i.e., (M, ω) is a symplectic manifold and H is a smooth function on M), and let H_e be a regular energy surface. We claim that $(H_e, \omega|_{H_e})$ is an almost contact manifold and the restriction of the Hamiltonian field X_H to H_e is a characteristic vector field on this manifold.

Indeed, $d(\omega|_{H_e}) = d\omega|_{H_e} = 0$, so that the form $\omega|_{H_e}$ is closed. Since the form ω is nondegenerate and codimension of H_e is equal to 1, ω is of maximal rank on H_e. Further, the field X_H is tangent to H_e; therefore $X_H|_{H_e} \lrcorner \omega|_{H_e} = 0$.

EXAMPLE 1.11. Let (M, ω) be a symplectic manifold and $\pi_2 \colon \mathbb{R} \times M \to M$ the natural projection $\pi_2(t, x) = x$. We claim that $(\mathbb{R} \times M, \pi_2^*(\omega))$ is almost contact.

Indeed, $d\pi_2^*(\omega) = 0$, and so the form $\widetilde{\omega} = \pi_2^*(\omega)$ is closed. To check the maximal rank condition, it is sufficient to show that the characteristic distribution of this form is one-dimensional. If the vector $((t, r), v_p) \in T_{(t,p)}(\mathbb{R} \times M)$ lies in the characteristic space, then
$$\pi_2^*(\omega)_{(t,p)}(((t, r), v_p), ((t, s), w_p)) = 0,$$
where $w_p \in T_p M$ and $s = \partial/\partial t$. By definition of $\pi_2^*(\omega)$, we have $\omega_p(v_p, w_p) = 0$ for any w_p. Thus, $v_p = 0$ and the characteristic distribution is generated by the vector

field

$$\left.\overline{\frac{\partial}{\partial t}}\right|_{(t,p)} = ((t,1),0) \in T_{(t,p)}(\mathbb{R} \times M) = T_t\mathbb{R} \times T_pM.$$

Note that if $\omega = d\theta$ and $\widetilde{\theta} = dt + \pi_2^*(\theta)$, where $t\colon \mathbb{R} \times M \to \mathbb{R}$, $t(t,x) = t$, then $\widetilde{\omega} = d\widetilde{\theta}$ and $(\mathbb{R} \times M, \widetilde{\theta})$ is a contact manifold.

A smooth mapping $X\colon \mathbb{R} \times M \to TM$ is called a *vector field depending on time*, if for any $t \in \mathbb{R}$ the mapping $X\colon \{t\} \times M \to TM$ is a vector field on M. To any vector field depending on time, one associates the vector field $\widetilde{X}$ on $\mathbb{R} \times M$ by $\widetilde{X} = \overline{\partial/\partial t} + X$.

Take a function $H \in C^\infty(\mathbb{R} \times M)$. For each $t \in \mathbb{R}$, there is the Hamiltonian vector field X_{H_t}, where $H_t\colon M \to \mathbb{R}$, $H_t(x) = H(t,x)$. Define the vector field depending on time $X_H\colon \mathbb{R} \times M \to TM$ by the formula $X_H(t,x) = X_{H_t}(x)$. Let $\widetilde{X}_H$ be the corresponding field on $\mathbb{R} \times M$.

THEOREM 1.6 (Cartan). *Let (M,ω) be a symplectic manifold, and let $H \in C^\infty(\mathbb{R} \times M)$ be a function. Set*

$$\omega_H = \pi_2^*(\omega) + dH \wedge dt.$$

Then:

1. $(\mathbb{R} \times M, \omega_H)$ *is an almost contact manifold.*
2. *The field $\widetilde{X}_H$ generates the characteristic distribution of the form ω_H.*
3. *If $\omega = d\theta$ and $\theta_H = \pi_2^*\theta + H\,dt$, then $\omega_H = d\theta_H$; in this case, if $H + \pi_2^*(\theta)(X_H)$ is nowhere zero, $(\mathbb{R} \times M, \theta_H)$ is a contact manifold.*

EXERCISE 1.12. Prove Theorem 1.6.

2. Infinitesimal contact transformations and characteristic fields

In Chapter 1 we saw that passing to the infinitesimal point of view is useful in the theory of ordinary differential equations. In this section, we consider infinitesimal contact transformations.

2.1. Infinitesimal contact transformations.
Let $A_t\colon J^1(M) \to J^1(M)$ be a one-parameter group of finite contact transformations. Consider the vector field $X = d/dt|_{t=0} A_t^*$. By the definition of contact transformations, $A_t^*(U_1) = \lambda_t U_1$, where $\lambda_t \in C^\infty(J^1(M))$, so that

$$X(U_1) = \lambda U_1, \qquad \text{where } \lambda = \left.\frac{d\lambda_t}{dt}\right|_{t=0}. \tag{2.1}$$

DEFINITION 2.1. A vector field X on the manifold $J^1(M)$ is called an *infinitesimal contact transformation*, or a *contact vector field*, if

$$X(U_1) = \lambda U_1, \qquad \text{where } \lambda \in C^\infty(J^1(M)).$$

EXERCISE 2.1. Prove that contact vector fields are closed under the commutator operation.

It can easily be checked that a one-parameter group corresponding to a contact vector field consists of finite contact transformations.

EXAMPLE 2.1. In the special local coordinates $x_1, \ldots, x_n$, u, $p_1, \ldots, p_n$ on $J^1(M)$ the vector fields $\partial/\partial x_1, \ldots, \partial/\partial x_n$ and $\partial/\partial u$ are contact, while the fields $\partial/\partial p_i$ are not.

EXAMPLE 2.2. Let $n = 2$. The vector field $X = x_1 \partial/\partial x_2 - x_2 \partial/\partial x_1 + p_1 \partial/\partial p_2 - p_2 \partial/\partial p_1$ is an infinitesimal contact transformation, since $X(du - p_1\, dx_1 - p_2\, dx_2) = p_2\, dx_1 + p_1\, dx_2 - p_1\, dx_2 - p_2\, dx_1 = 0$.

EXAMPLE 2.3. Consider the vector field $X \in \mathrm{D}(J^1(M))$ in special local coordinates given by

$$X = \sum_{i=1}^{n} \alpha_i x_i \frac{\partial}{\partial x_i} + \beta u \frac{\partial}{\partial u} + \sum_{i=1}^{n} (\beta - \alpha_i) p_i \frac{\partial}{\partial p_i}, \tag{2.2}$$

where $\alpha_1, \ldots, \alpha_n$ and β are constants. It is easily shown that $X(U_1) = \beta U_1$, so that X is an infinitesimal contact transformation.

In §1 we saw that any smooth transformation $A\colon J^0(M) \to J^0(M)$ gives rise to the contact transformation $\bar{A}\colon J^1(M) \to J^1(M)$. The same is true for infinitesimal transformations as well. Namely, let $X \in \mathrm{D}(J^0(M))$, and A_t the one-parameter group of shifts along trajectories of X. Clearly, the vector field $\bar{X} = d/dt|_{t=0}\, \bar{A}_t^* \in \mathrm{D}(J^1(M))$ is contact. This vector field $\bar{X}$ is called the *lifting* of the field X to the manifold of 1-jets $J^1(M)$.

All contact fields discussed in Examples 2.1–2.3 are in fact liftings of fields. The fields $\partial/\partial x_i,\, \partial/\partial u \in \mathrm{D}(J^1(M))$ (see Example 2.1) are liftings of $\partial/\partial x_i$ and $\partial/\partial u$ considered as vector fields on $J^0(M)$. The vector field from Example 2.2 is the lifting of the infinitesimal rotation $x_1 \partial/\partial x_2 - x_2 \partial/\partial x_1$, while the field from Example 2.3 is the lifting of the infinitesimal scale transformation

$$\alpha_1 x_1 \frac{\partial}{\partial x_1} + \cdots + \alpha_n x_n \frac{\partial}{\partial x_n} + \beta u \frac{\partial}{\partial u} \in \mathrm{D}(J^0(M)).$$

This is an immediate consequence of the corresponding coordinate formulas for the liftings of A_t.

EXERCISE 2.2. Check that a contact vector field $X \in \mathrm{D}(J^1(M))$ is the lifting of a field $X^{(0)} \in \mathrm{D}(J^0(M))$ if and only if in the special local coordinates $x_1, \ldots, x_n$, u, $p_1, \ldots, p_n$ on $J^1(M)$ one has $[X, \partial/\partial p_i] = \alpha_{i1} \partial/\partial p_1 + \cdots + \alpha_{in} \partial/\partial p_n$, where the α_{ik} are smooth functions on $J^1(M)$, $1 \leq i, k \leq n$.

EXAMPLE 2.4. The vector field $X = \sum_{i=1}^{n}(2p_i \partial/\partial x_i + p_i^2 \partial/\partial u)$ is contact (because $X(U_1) = 0$); however, $[X, \partial/\partial p_l] = -2\partial/\partial x_l - 2p_l \partial/\partial u$. Hence X cannot be the lifting of a vector field $X^{(0)} \in \mathrm{D}(J^0(M))$.

Let us now give a description of contact vector fields in terms of generating functions [**72, 77**].

THEOREM 2.1. *A contact vector field X on $J^1(M)$ is uniquely determined by the function $f = U_1(X)$. To every function $f \in C^\infty(J^1(M))$ there corresponds a contact field X_f, i.e., $f = U_1(X_f)$. The correspondence $f \mapsto X_f$ is linear over $\mathbb{R}$*

and satisfies the following properties:

$$U_1(X_f) = f,$$
$$X_f(U_1) = X_1(f) \cdot U_1,$$
$$X_{f+g} = X_f + X_g, \qquad g \in C^\infty(J^1(M)),$$
$$X_{fg} = fX_g + gX_f - fgX_1,$$
$$X_f(f) = X_1(f) \cdot f.$$

PROOF. Consider the 2-form dU_1. We claim that for any point $\theta \in J^1(M)$ the form $(dU_1)_\theta$ is nondegenerate on the Cartan plane $\mathcal{C}_\theta$. Indeed, from the coordinate expression for dU_1 it follows that $X_1|_\theta \,\lrcorner\, (dU_1)_\theta = 0$, where $X_1 = \partial/\partial u$, and $(dU_1)_\theta$ is nondegenerate on any hyperplane that does not contain $X_1|_\theta$. Since $(U_1)_\theta(X_1) = 1 \neq 0$, one has $\mathcal{C}_\theta \not\ni X_1|_\theta$.

Thus, the map $Y \mapsto Y \,\lrcorner\, dU_1$ establishes an isomorphism between vector fields that lie in the Cartan distribution (i.e., on which the form U_1 vanishes) and 1-forms that vanish on X_1.

Let X be a contact vector field on $J^1(M)$, $X(U_1) = hU_1$, where the function h is in $C^\infty(J^1(M))$. Decomposing X into the sum

$$X = f \cdot X_1 + Y, \qquad \text{where } f \in C^\infty(J^1(M)), \quad U_1(Y) = 0, \qquad (2.3)$$

we get

$$Y \,\lrcorner\, dU_1 = X \,\lrcorner\, dU_1 = hU_1 - d(X \,\lrcorner\, U_1).$$

Thus,

$$Y \,\lrcorner\, dU_1 = hU_1 - df. \qquad (2.4)$$

Applying both sides of (2.4) to the field X_1, we get $h = X_1(f)$.

The equality $U_1(Y) = 0$ means that Y_θ lies in $\mathcal{C}_\theta$ for any $\theta \in J^1(M)$. Since the 2-form dU_1 is nondegenerate on $\mathcal{C}_\theta$, the vector Y_θ can be recovered by means of (2.4). Hence, the field X is uniquely determined by the function $f = U_1(X)$.

Conversely, for an arbitrary function f on $J^1(M)$ define the field Y by (2.4), with $h = X_1(f)$. Then (2.3) yields $X = X_f$.

The rest of the proof is by direct verification. $\square$

DEFINITION 2.2. Given a contact vector field X, the function $f = U_1(X)$ is called the *generating function* of X.

EXERCISE 2.3. Let $f_1, \ldots, f_k \in C^\infty(J^1(M))$, and let $\varphi \in C^\infty(\mathbb{R}^k)$ be an arbitrary smooth function. Using Theorem 2.1, prove that

$$Y_{\varphi(f_1,\ldots,f_k)} = \sum_{i=1}^{k} \frac{\partial \varphi}{\partial f_i} Y_{f_i}, \qquad (2.5)$$

where $Y_f = X_f - fX_1$.

In view of (2.5), to describe X_f in local coordinates it suffices to obtain the coordinate formulas for Y_{x_i}, Y_u, Y_{p_i} ($i = 1, \ldots, n$). It immediately follows from the definitions that

$$Y_{x_i} = \frac{\partial}{\partial p_i}, \qquad i = 1, \ldots, n, \tag{2.6}$$

$$Y_u = \sum_{i=1}^{n} p_i \frac{\partial}{\partial p_i}, \tag{2.7}$$

$$Y_{p_i} = -\frac{\partial}{\partial x_i} - p_i \frac{\partial}{\partial u}, \qquad i = 1, \ldots, n. \tag{2.8}$$

Using (2.5), we get

$$Y_f = -\sum_{i=1}^{n} \frac{\partial f}{\partial p_i} \frac{\partial}{\partial x_i} - \left(\sum_{i=1}^{n} p_i \frac{\partial f}{\partial p_i}\right) \frac{\partial}{\partial u} + \sum_{i=1}^{n} \left(\frac{\partial f}{\partial x_i} + p_i \frac{\partial f}{\partial u}\right) \frac{\partial}{\partial p_i}. \tag{2.9}$$

Since $X_f = Y_f + f X_1$, we obtain

$$X_f = -\sum_{i=1}^{n} \frac{\partial f}{\partial p_i} \frac{\partial}{\partial x_i} + \left(f - \sum_{i=1}^{n} p_i \frac{\partial f}{\partial p_i}\right) \frac{\partial}{\partial u} + \sum_{i=1}^{n} \left(\frac{\partial f}{\partial x_i} + p_i \frac{\partial f}{\partial u}\right) \frac{\partial}{\partial p_i}. \tag{2.10}$$

REMARK 2.1. Consider the pullback $f = \pi^*(H)$ of a function $H \in C^\infty(T^*(M))$ along the projection $\pi \colon J^1(M) \to T^*(M)$. The projection of the field X_f to $T^*(M)$ is the Hamiltonian field X_H. From the equalities $X_f(U_1) = X_1(f) U_1$ and $dU_1 = -\pi^*(\omega)$, where ω is the symplectic structure on $T^*(M)$, it follows that $X_H \lrcorner d\omega = dH$.

EXERCISE 2.4. The $2n+1$ vector fields (2.6), (2.7), and (2.8) lie in the $2n$-dimensional plane $\mathcal{C}_\theta$ at each point θ, so that they are linearly dependent. Find the linear dependence (over $C^\infty(J^1(M))$) between these fields.

EXERCISE 2.5. Consider the trivial extension of a vector field $Y \in D(M)$ to $J^0(M) = M \times \mathbb{R}$. Find the generating function for the lifting $\bar{Y} \in D(J^1(M))$ of Y.

EXERCISE 2.6. Let $f \in C^\infty(J^0(M))$ be a function, and $\pi^*(f) \in C^\infty(J^1(M))$ its pullback along the projection π defined by (1.1). Prove that the contact field $X_{\pi^*(f)}$ is the lifting of some field $Y \in D(J^0(M))$, and describe Y in coordinates.

The isomorphism between contact vector fields and functions on $J^1(M)$ makes it possible to define a number of brackets on the set of functions.

First, let X_f and X_g be the contact vector fields with the generating functions $f, g \in C^\infty(J^1(M))$. The commutator of these fields is a contact field as well.

DEFINITION 2.3. The generating function $\{f, g\} = U_1([X_f, X_g])$ of the contact field $[X_f, X_g]$ is called the *Jacobi bracket* of f and g.

PROPOSITION 2.2. For $f, g \in C^\infty(J^1(M))$, one has

$$\{f, g\} = X_f(g) - X_1(f) g. \tag{2.11}$$

PROOF. Since $X_f(U_1) = X_f \lrcorner dU_1 + d(X_f \lrcorner U_1) = X_f \lrcorner dU_1 + df$ and $X_f(U_1) = X_1(f) \cdot U_1$, it follows that

$$dU_1(X_f, X_g) = g X_1(f) - X_g(f).$$

Hence,

$$\{f, g\} = U_1([X_f, X_g]) = X_f(U_1(X_g)) - X_g(U_1(X_f)) - dU_1(X_f, X_g)$$
$$= X_f(g) - X_g(f) - g X_1(f) + X_g(f) = X_f(g) - X_1(f) g. \qquad \square$$

THEOREM 2.3. *The set of function on $J^1(M)$ is a Lie algebra over $\mathbb{R}$ with respect to the Jacobi bracket.*

PROOF. It is obvious that the Jacobi bracket is skew-symmetric and bilinear. The Jacobi identity is a direct consequence of (2.11). □

EXERCISE 2.7. Using (2.11), prove that in the special coordinates the Jacobi bracket takes the form

$$\{f,g\} = \sum_{i=1}^n \left(\frac{\partial f}{\partial x_i}\frac{\partial g}{\partial p_i} - \frac{\partial f}{\partial p_i}\frac{\partial g}{\partial x_i}\right) + \sum_{i=1}^n p_i \left(\frac{\partial f}{\partial u}\frac{\partial g}{\partial p_i} - \frac{\partial g}{\partial u}\frac{\partial f}{\partial p_i}\right) + f\frac{\partial g}{\partial u} - g\frac{\partial f}{\partial u}.$$

Another bracket is defined as follows:

DEFINITION 2.4. The *Mayer bracket* $[f,g]$ of two functions $f,g \in C^\infty(J^1(M))$ is defined by $[f,g] = dU_1(X_f, X_g) \in C^\infty(J^1(M))$.

The Mayer bracket is bilinear over $\mathbb{R}$ and skew-symmetric, but instead of the Jacobi identity, one has

$$[[f,g],h] + [[h,f],g] + [[g,h],f] = X_1(f)[g,h] + X_1(g)[h,f] + X_1(h)[f,g].$$

Using

$$[f,g] = Y_f(g) = X_f(g) - fX_1(g),$$

we obtain the following coordinate formula for the Mayer bracket:

$$[f,g] = \sum_{i=1}^n \left(\frac{\partial f}{\partial x_i}\frac{\partial g}{\partial p_i} - \frac{\partial f}{\partial p_i}\frac{\partial g}{\partial x_i}\right) + \sum_{i=1}^n p_i \left(\frac{\partial f}{\partial u}\frac{\partial g}{\partial p_i} - \frac{\partial g}{\partial u}\frac{\partial f}{\partial p_i}\right).$$

Finally, let us define the Poisson bracket.

DEFINITION 2.5. The *Poisson bracket* (f,g) of two functions $f,g \in C^\infty(J^1(M))$ is defined by $(f,g) = X_f(g) \in C^\infty(J^1(M))$.

EXERCISE 2.8. Prove the following properties of the Poisson bracket:
1. Bilinearity over $\mathbb{R}$.
2. $(f,g) + (g,f) = X_1(f)g + X_1(g)f$.
3. $(f,g) = [f,g] + X_1(g)f$.
4. $(f,g) = \{f,g\} + X_1(f)g$.

In coordinates, the Poisson bracket looks like

$$(f,g) = \sum_{i=1}^n \left(\frac{\partial f}{\partial x_i}\frac{\partial g}{\partial p_i} - \frac{\partial f}{\partial p_i}\frac{\partial g}{\partial x_i}\right) + \sum_{i=1}^n p_i \left(\frac{\partial f}{\partial u}\frac{\partial g}{\partial p_i} - \frac{\partial g}{\partial u}\frac{\partial f}{\partial p_i}\right) + f\frac{\partial g}{\partial u}.$$

REMARK 2.2. The restrictions of all the three brackets to functions that are pullbacks of functions on $T^*(M)$ coincide with the Poisson bracket on $T^*(M)$, because $X_1(\pi^*(H)) = 0$ for all $H \in C^\infty(T^*(M))$.

2.2. Infinitesimal symmetries of equations.

DEFINITION 2.6. Let $\mathcal{E} \subset J^1(M)$ be a first-order differential equation. A contact vector field $X \in D(J^1(M))$ is called an *infinitesimal contact symmetry* of the equation $\mathcal{E}$, if at each point $\theta \in \mathcal{E}$ the vector X_θ is tangent to $\mathcal{E}$.

In the sequel, such a symmetry will be called simply a "symmetry". The generating function of a symmetry will also be referred to as a "symmetry" for short.

It is clear that the set of symmetries of $\mathcal{E}$ is a Lie algebra over $\mathbb{R}$ with respect to the commutation operation. In terms of generating functions, the Lie algebra structure is given by the Jacobi bracket.

If the equation $\mathcal{E}$ is (locally) of the form $\mathcal{E} = \{F = 0\}$, $F \in C^\infty(J^1(M))$, then X is a symmetry of $\mathcal{E}$, if $X(F)|_\mathcal{E} \equiv 0$, or, which is equivalent, $X(F) = \mu F$ for some function $\mu \in C^\infty(J^1(M))$.

Note that the equality $X_F(F) = X_1(F) \cdot F$ implies that the equation $\mathcal{E} = \{F = 0\}$ always possesses at least one symmetry, namely, X_F.

As an example, let us discuss equations that are symmetric with respect to the contact fields from Examples 2.1–2.3.

EXAMPLE 2.5. The vector field X given by (2.2) (see Example 2.3) is a symmetry of $\mathcal{E} = \{F = 0\}$, if the function F is quasihomogeneous in the following sense: for any real numbers $\alpha_1, \ldots, \alpha_n, \beta$ there exists $\gamma \in \mathbb{R}$ such that for each $\tau \in \mathbb{R}$ one has

$$F(\tau^{\alpha_1} x_1, \ldots, \tau^{\alpha_n} x_n, \tau^\beta u, \tau^{\beta-\alpha_1} p_1, \ldots, \tau^{\beta-\alpha_n} p_n) \equiv \tau^\gamma F(x_1, \ldots, x_n, u, p_1, \ldots, p_n). \tag{2.12}$$

Indeed, suppose that F satisfies (2.12) and consider the trajectory of X passing through a point $\theta = (x_1^{(0)}, \ldots, x_n^{(0)}, u^{(0)}, p_1^{(0)}, \ldots, p_n^{(0)}) \in \mathcal{E}$:

$$\begin{cases} x_i = x_i^{(0)} e^{\alpha_i t}, \\ u = u^{(0)} e^{\beta t}, \\ p_i = p_i^{(0)} e^{(\beta-\alpha_i)t}, & i = 1, \ldots, n. \end{cases}$$

Set $\tau = e^t$ and compute F at a point $(x_1, \ldots, x_n, u, p_1, \ldots, p_n)$ on the considered trajectory:

$$F(x_1, \ldots, x_n, u, p_1, \ldots, p_n) = \tau^\gamma F(x_1^{(0)}, \ldots, x_n^{(0)}, u^{(0)}, p_1^{(0)}, \ldots, p_n^{(0)}) = 0.$$

Therefore, the trajectory of X passing through θ lies entirely on $\mathcal{E}$. Hence, $X(F) = 0$ on $\mathcal{E}$.

PROPOSITION 2.4. *The contact field $\partial/\partial x_i$ (resp., $\partial/\partial u$) is a symmetry of $\mathcal{E} = \{F = 0\}$, where $F \in C^\infty(J^1(M))$, if and only if F does not depend on x_i (resp., u).*

PROOF. If $\mathcal{E} = \{G = 0\}$ and $\partial G/\partial x_i = \mu G$, $\mu \in C^\infty(J^1(M))$, then $G = e^\nu F$, where ν is a function satisfying $\partial \nu/\partial x_i = \mu$ and F does not depend on x_i. Since e^ν is nowhere zero, we have $\mathcal{E} = \{G = 0\} = \{F = 0\}$. The field $\partial/\partial u$ can be handled in the same way. □

EXERCISE 2.9. Prove that the vector field from Example 2.2 for $n = 2$ is a symmetry of an equation $\mathcal{E}$ if and only if $\mathcal{E}$ is of the form

$$F(u, x_1^2 + x_2^2, p_1^2 + p_2^2, x_1 p_1 + x_2 p_2) = 0.$$

2.3. Characteristic vector fields and integration of first-order equations.

Recall (see Ch. 1, §3) that a vector field X is a characteristic symmetry of a distribution P if X belongs to P and the Lie derivative with respect to X preserves P.

PROPOSITION 2.5. *The Cartan distribution on $J^1(M)$ has no nontrivial characteristic symmetries.*

PROOF. If X is a characteristic symmetry, then X is contact and $U_1(X) = 0$. Hence, by Theorem 2.1, $X = 0$. □

Recall also that from the geometric viewpoint the finding of (generalized) solutions to a differential equation amounts to integration of the Cartan distribution restricted to the locus of this equation. So, the problem of finding all (generalized) solutions to a first-order differential equation $\mathcal{E}$ can be stated as follows:

Find all maximal integral manifolds of the induced Cartan distribution $\mathcal{C}(\mathcal{E})$ on the manifold $\mathcal{E} \subset J^1(M)$.

Now, we claim that although the Cartan distribution on $J^1(M)$ has no characteristic symmetries, any nontrivial first-order equation has a nontrivial characteristic symmetry. Indeed, consider the 1-form $U_1|_{\mathcal{E}}$ that defines the Cartan distribution $\mathcal{C}(\mathcal{E})$. For any point $\theta \in J^1(M)$, the 2-form $(dU_1)_\theta$ is nondegenerate on the Cartan plane $\mathcal{C}_\theta$. If θ is a nonsingular point of the equation $\mathcal{E} = \{F = 0\}$, then

$$\mathcal{C}(\mathcal{E})_\theta = \mathcal{C}_\theta \cap T_\theta(\mathcal{E})$$

is a hyperplane in $\mathcal{C}_\theta$. On this hyperplane, the 2-form $(dU_1)_\theta$ is of rank $2n-2$ and, therefore, is degenerate. Thus, on the everywhere dense set of generic points of $\mathcal{E}$ there exists the direction field $l_\theta \subset T_\theta(\mathcal{E})$, where $l_\theta \subset \mathcal{C}(\mathcal{E})_\theta$ is the degeneration subspace of $(dU_1)_\theta$.

PROPOSITION 2.6. *The direction field l_θ is tangent to every generalized solution of $\mathcal{E}$.*

PROOF. Recall that an n-dimensional plane $R \subset \mathcal{C}_\theta$ is called *Lagrangian* with respect to the 2-form $(dU_1)_\theta$, if the restriction $(dU_1)_\theta|_R$ vanishes. It is not hard to prove that if R is contained in a hyperplane $S \subset \mathcal{C}_\theta$, then R must contain the 1-dimensional degeneration subspace of $(dU_1)_\theta|_S$.

Now, let N be a generalized solution to $\mathcal{E}$, i.e., a maximal n-dimensional integral manifold of the Cartan distribution contained in $\mathcal{E}$. Since $U_1|_N = 0$, we have $(dU_1)_N = 0$, and, therefore, for any $\theta \in N$ the plane $T_\theta(N) \subset \mathcal{C}(\mathcal{E})_\theta \subset \mathcal{C}_\theta$ is Lagrangian with respect to $(dU_1)_\theta$. Hence, $l_\theta \subset T_\theta(N)$. □

The field $\{l_\theta\}$ is called the *characteristic direction field*. Its integral curves are called the *characteristics* of $\mathcal{E}$.

PROPOSITION 2.7. *Let $\mathcal{E} = \{F = 0\}$ be an equation. The characteristics of $\mathcal{E}$ are integral curves of the characteristic vector field Y_F.*

PROOF. Since $U_1(Y_F) = 0$, we have $(Y_F)_\theta \in \mathcal{C}_\theta$ for any point $\theta \in J^1(M)$. From the formula

$$Y_F \,\lrcorner\, dU_1 = X_1(F)U_1 - dF \qquad (2.13)$$

(see (2.4)) it follows that

$$Y_F(F) = dF(Y_F) = X_1(F)U_1(Y_F) - dU_1(Y_F, Y_F) = 0.$$

Therefore Y_F is tangent to every level surface of F. Thus, $(Y_F)_\theta \in \mathcal{C}(\mathcal{E})_\theta$ for all $\theta \in \mathcal{E}$.

By virtue of (2.13), the form $Y_F \lrcorner\, dU_1$ vanishes on the $(2n-1)$-dimensional plane $\mathcal{C}(\mathcal{E})_\theta$. Hence, $(Y_F)_\theta$ is a degeneration vector of $(dU_1)_\theta$ on this plane, i.e., for all $\theta \in \mathcal{E}$ we have $(Y_F)_\theta \in l_\theta$, which proves the theorem. $\square$

PROPOSITION 2.8. *The vector field Y_F is an infinitesimal symmetry of the distribution $\mathcal{C}(\mathcal{E})$.*

PROOF. Since $Y_F = X_F - FX_1$, the vector fields Y_F and X_F coincide on $\mathcal{E} = \{F = 0\}$. To conclude the proof, it remains to recall that X_F is an infinitesimal symmetry of the Cartan distribution. $\square$

Now let us turn to the discussion of the Cauchy problem. By Propositions 2.6 and 2.7, a noncharacteristic Cauchy problem for the equation $\mathcal{E} = \{F = 0\}$ reduces to the solution of the following system of characteristic equations:

$$\begin{cases} \dot{x}_i = -\dfrac{\partial F}{\partial p_i}, \\ \dot{u} = -\displaystyle\sum_{i=1}^n p_i \dfrac{\partial F}{\partial p_i}, \\ \dot{p}_i = \dfrac{\partial F}{\partial x_i} + p_i \dfrac{\partial F}{\partial u}, \qquad i = 1, \ldots, n, \end{cases} \qquad (2.14)$$

describing the trajectories of the field Y_F.

From now on an $(n-1)$-dimensional integral manifold of the Cartan distribution contained in $\mathcal{E}$ will be called *Cauchy data* for $\mathcal{E}$.

EXAMPLE 2.6. Let $\Gamma \subset M$ be a smooth hypersurface, u_0 a smooth function on it, and $\widetilde{u}_0$ an extension of u_0 such that the graph $j_1(\widetilde{u}_0)(\Gamma)$ is contained in $\mathcal{E}$. This graph is an $(n-1)$-dimensional manifold of Cauchy data.

"Almost always" the graph $j_1(\widetilde{u}_0)(\Gamma)$ is determined by the initial function u_0 and does not depend on the choice of $\widetilde{u}_0$. To prove this, observe that for any point $a \in \Gamma$ the n-dimensional plane tangent to the graph of $\widetilde{u}_0$ at the point $(a, u_0(a))$ must contain the $(n-1)$-dimensional plane tangent to the surface $\{(x, u_0(x)) \mid x \in \Gamma\}$. Therefore, all contact elements $j_1(\widetilde{u}_0)(a)$ lie on a 1-dimensional curve, which in a general case intersects $\mathcal{E}$ at one point.

In particular, suppose that in local coordinates the hypersurface $\Gamma \subset M$ is given by $\Gamma = \{x_n = 0\}$ and the equation $\mathcal{E}$ is solved with respect to p_n:

$$p_n = f(x_1, \ldots, x_n, u, p_1, \ldots, p_{n-1}).$$

Then, for any function $u_0 \in C^\infty(\Gamma)$ and any point $a \in \Gamma$, the contact element $j_1(\widetilde{u}_0)(a)$ is uniquely determined and in coordinates has the form

$$\left(a_1, \ldots, a_n, u_0(a), \left.\frac{\partial u_0}{\partial x_1}\right|_a, \ldots, \left.\frac{\partial u_0}{\partial x_{n-1}}\right|_a, f\left(a, u_0(a), \left.\frac{\partial u_0}{\partial x_1}\right|_a, \ldots, \left.\frac{\partial u_0}{\partial x_{n-1}}\right|_a\right)\right).$$

PROPOSITION 2.9. *Let $\mathcal{E} \subset J^1(M)$ be a first-order differential equation, and let $R \subset \mathcal{E}$ be a manifold of Cauchy data such that none of the characteristic directions l_θ, $\theta \in R$, is tangent to R. Then, in a neighborhood of R, there exists a generalized solution of $\mathcal{E}$ containing R.*

PROOF. For each point $\theta \in R$ there exists a segment of a characteristic passing through θ that does not intersect R in other point. Take the union $N = \bigcup_{\theta \in R} \chi_\theta \subset \mathcal{E}$ of these segments. In view of the theorem about smooth dependence of solutions on initial conditions, without loss of generality we can assume that N is a smooth manifold.

Since for $\theta \in R$ the tangent plane $T_\theta(N)$ is generated by the plane $T_\theta(R) \subset \mathcal{C}_\theta(\mathcal{E})$ and the straight line $l_\theta \subset \mathcal{C}_\theta(\mathcal{E})$, we get $T_\theta(N) \subset \mathcal{C}_\theta$.

Consider a point $\theta \in N$ that does not belong to R. There is a shift along trajectories of the characteristic field which takes this point to a point $\theta_0 \in R$. The manifold N can be considered as invariant under this shift, so that the plane $T_\theta(N)$ maps to $T_{\theta_0}(N)$. Since the characteristic field belongs to the Cartan distribution, the shift along its trajectories takes the space $\mathcal{C}_\theta(\mathcal{E})$ to $\mathcal{C}_{\theta_0}(\mathcal{E})$. Hence, $T_\theta(N) \subset \mathcal{C}_\theta$.

Thus, N is the solution we seek for. □

EXAMPLE 2.7. Consider the equation with two independent variables
$$u - \frac{\partial u}{\partial x_1} \frac{\partial u}{\partial x_2} = 0,$$
i.e., $\mathcal{E} = \{u - p_1 p_2 = 0\}$. Let us find the solution whose restriction to the hypersurface $\Gamma = \{x_2 = 0\}$ equals $u_0(x_1) = x_1^2$. The graph of u_0 uniquely determines the following 1-dimensional Cauchy data:
$$x_1 = \tau, \quad x_2 = 0, \quad u = \tau^2, \quad p_1 = 2\tau, \quad p_2 = u/p_1 = \tau/2,$$
where τ is a parameter.

Using (2.9), we get
$$Y_F = p_2 \frac{\partial}{\partial x_1} + p_1 \frac{\partial}{\partial x_2} + 2 p_1 p_2 \frac{\partial}{\partial u} + p_1 \frac{\partial}{\partial p_1} + p_2 \frac{\partial}{\partial p_2}.$$
Parametric equations of trajectories of this field have the form
$$\begin{cases} x_1 = \tau(e^t + 1)/2, \\ x_2 = 2\tau(e^t - 1), \\ u = \tau^2 e^{2t}, \\ p_1 = 2\tau e^t, \\ p_2 = \tau e^t / 2, \end{cases} \quad (2.15)$$
where t is a parameter on trajectories.

Equations (2.15) give the solution in parametric form. Solving the first two equations for τ and t and substituting the expressions obtained into the third equation yields the solution to the Cauchy problem in explicit form: $u = (4x_1 + x_2)^2/16$.

EXAMPLE 2.8. Consider the *Hamilton–Jacobi equation*[4]
$$H\left(q, \frac{\partial u}{\partial q}\right) = C,$$
where $q = (q_1, \ldots, q_n)$, $\partial u/\partial q = (\partial u/\partial q_1, \ldots, \partial u/\partial q_n)$, and C is a constant, i.e., H does not depend on u.

---
[4]Sometimes by the Hamilton–Jacobi equation is meant the one of the form $u_t = \varphi(t, x, u_x)$, where $x = (x_1, \ldots, x_n)$, $u_x = (u_{x_1}, \ldots, u_{x_n})$. Such an equation can be obtained by setting $x_i = q_i$, $t = q_{n+1}$, and $H(q, u_q) = u_t - \varphi(t, x, u_x)$.

Note that the surface $H(q,p) = c$, as well as the Cartan distribution on $J^1(M)$, is invariant under shifts along trajectories of the field $\partial/\partial u$. The corresponding quotient equation can be considered as a submanifold of T^*M. The quotient Cartan distribution is given by the universal form $p\,dq$, since this form coincides with the projection of the form $-U_1$. Therefore, the symplectic structure on T^*M is defined by the projection of $-dU_1$.

The system (2.14) corresponding to the characteristic field takes the form
$$\begin{cases} \dot{q}_i = \dfrac{\partial H}{\partial p_i}, \\ \dot{u} = -\displaystyle\sum_{i=1}^{n} p_i \dfrac{\partial H}{\partial p_i}, \\ \dot{p}_i = -\dfrac{\partial H}{\partial q_i}, & i = 1, \ldots, n. \end{cases}$$

Note that the first and third equations correspond to the Hamiltonian vector field on T^*M, i.e., the characteristics are described by the Hamilton equations.

The invariant theory of the Hamilton–Jacobi equations (including the solution of the Cauchy problem) is fully considered in [143].

2.4. Symmetries and first integrals. As discussed above, solution of a noncharacteristic Cauchy problem for a first-order differential equation in the vicinity of nonsingular point amounts to integration of the characteristic direction field l_θ, i.e., to solution of the characteristic system (2.14).

Recall that a *first integral* for a system of differential equations (resp., a vector field) is a function constant on solutions to this system (resp., trajectories of this field). We shall say that $\varphi \in C^\infty(J^1(M))$ is a *first integral of the equation* $\mathcal{E} = \{F = 0\} \subset J^1(M)$, if φ is a first integral of the characteristic system (2.14) of $\mathcal{E}$. Symmetries and first integrals are intimately related.

PROPOSITION 2.10. *If X_{f_1} and X_{f_2} are two symmetries of $\mathcal{E} = \{F = 0\}$ linearly independent over $C^\infty(J^1(M))$, then the function $f = f_2/f_1$ is a first integral of $\mathcal{E}$.*

PROOF. We have
$$\lambda_i F = X_{f_i}(F) = Y_{f_i}(F) + f_i X_1(F) = f_i X_1(F) - Y_F(f_i)$$
for some functions $\lambda_i \in C^\infty(J^1(M))$, $i = 1, 2$. It follows that $Y_F(f_i) = f_i X_1(F) - \lambda_i F$. Hence
$$Y_F(f_2/f_1) = (Y_F(f_2)f_1 - Y_F(f_1)f_2)/f_1^2$$
$$= (f_1 f_2 X_1(F) - \lambda_2 f_1 F - f_1 f_2 X_1(F) + \lambda_1 f_2 F)/f_1^2 = \left(\frac{\lambda_1 f_2 - \lambda_2 f_1}{f_1^2}\right) F. \quad \square$$

EXERCISE 2.10. Prove that contact vector fields X_{f_1} and X_{f_2} are linearly independent over $C^\infty(J^1(M))$ if and only if they (or, which is the same, their generating functions) are linearly independent over $\mathbb{R}$.

EXAMPLE 2.9. The equation
$$\frac{\partial u}{\partial x_1}\frac{\partial u}{\partial x_2} - x_1^a x_2^b u^c = 0$$

for $c \neq 2$ has the contact symmetries $X_{\alpha u - x_1 p_1}$, where $\alpha = (a+1)/(2-c)$, and $X_{\beta u - x_2 p_2}$, with $\beta = (b+1)/(2-c)$. Therefore, the function

$$f = \frac{\beta u - x_2 p_2}{\alpha u - x_1 p_1}$$

is a first integral of this equation.

PROPOSITION 2.11. *If an equation $\mathcal{E} \subset J^1(M)$ possesses an $(l+1)$-dimensional Abelian algebra of symmetries, then it has l independent first integrals which are mutually in involution with respect to the Mayer bracket.*

PROOF. It can easily be checked that for all $f, g \in C^\infty(J^1(M))$ one has

$$[f, g] = \{f, g\} + f'g - g'f, \tag{2.16}$$

where $f' = X_1(f)$, $g' = X_1(g)$.

Let $X_{f_1}, \ldots, X_{f_{l+1}}$ be mutually commuting symmetries of $\mathcal{E}$. Put

$$g_1 = \frac{f_1}{f_{l+1}}, \ldots, g_l = \frac{f_l}{f_{l+1}}.$$

We claim that the first integrals g_i are mutually in involution on $\mathcal{E}$.

Indeed, $\{f_i, f_k\} = 0$ on $\mathcal{E}$, because the symmetries f_i and f_k commute. Using (2.16), we get

$$[f_i, f_k] = f'_i f_k - f'_k f_i.$$

From (2.5) it easily follows that

$$[g_i, g_k] = \frac{1}{f_{l+1}^3}(f_{l+1}[f_i, f_k] - f_i[f_{l+1}, f_k] - f_k[f_i, f_{l+1}])$$
$$= \frac{1}{f_{l+1}^3}(f_{l+1} f'_i f_k - f_{l+1} f_i f'_k - f_i f'_{l+1} f_k + f_i f_{l+1} f'_k - f_{l+1} f'_i f_k + f'_{l+1} f_i f_k) = 0$$

on $\mathcal{E}$, which proves the proposition. □

Thus, if $X_{f_0}, X_{f_1}, \ldots, X_{f_l}$ is a basis of an Abelian algebra of symmetries of $\mathcal{E}$, then the first integrals $g_i = f_i/f_0$, $i = 1, \ldots, l$, of the characteristic distribution are independent and are mutually in involution.

EXAMPLE 2.10. Any equation of the form $F(u, p_1, \ldots, p_n) = 0$ has the n-dimensional Abelian algebra of symmetries generated by

$$\frac{\partial}{\partial x_1} = -X_{p_1}, \ldots, \frac{\partial}{\partial x_n} = -X_{p_n}.$$

It immediately follows that $p_2/p_1, \ldots, p_n/p_1$ is an involutive system of first integrals.

3. Complete integrals of first-order differential equations

In §2, we saw that the Cauchy problem for a first-order equation amounts to integration of the characteristic system of the equation. This observation gives rise to another method of solving first-order equations, namely, the complete integral method. Complete integrals are dealt with in [20, 113]. Note that the role of symmetries in the construction of complete integrals is quite important.

3. COMPLETE INTEGRALS

3.1. Complete integrals: a coordinate approach. It is a common knowledge that a system of n first-order ordinary differential equations for n unknown functions has, in the vicinity of a nonsingular point, an n-parameter family of solutions. By analogy, let us look for an n-parameter family
$$u = V(x_1, \ldots, x_n, a_1, \ldots, a_n) \tag{3.1}$$
of solutions to a first-order partial differential equation $\mathcal{E}$. Here $a_1, \ldots, a_n$ are parameters.

Differentiating this relation with respect to x_i yields
$$p_1 = \frac{\partial V}{\partial x_1}, \ldots, p_n = \frac{\partial V}{\partial x_n}. \tag{3.2}$$
If one can obtain the initial equation $\mathcal{E}$ through eliminating the constants $a_1, \ldots, a_n$ from (3.1) and (3.2), then $V(x_1, \ldots, x_n, a_1, \ldots, a_n)$ is said to be a complete integral. Writing this condition analytically, we arrive at the following definition.

DEFINITION 3.1 (coordinate). A *complete integral* of a first-order partial differential equation with n independent variables is an n-parameter family
$$u = V(x_1, \ldots, x_n, a_1, \ldots, a_n) \tag{3.3}$$
of solutions to this equation. A complete integral is called *nondegenerate* if the matrix
$$\begin{pmatrix} \partial V/\partial a_1 & \ldots & \partial V/\partial a_n \\ \partial^2 V/\partial a_1 \partial x_1 & \ldots & \partial^2 V/\partial a_n \partial x_1 \\ \cdots\cdots\cdots\cdots\cdots\cdots\cdots\cdots\cdots\cdots\cdots\cdots \\ \partial^2 V/\partial a_1 \partial x_n & \ldots & \partial^2 V/\partial a_n \partial x_n \end{pmatrix} \tag{3.4}$$
is of maximal rank n almost everywhere.

EXAMPLE 3.1. The family
$$u = 2a_1 x_1 x_2^3 + a_1^2 x_2^6 + a_2 x_2^2$$
is a complete integral for the equation
$$\left(\frac{\partial u}{\partial x_1}\right)^2 + x_1 \frac{\partial u}{\partial x_1} + 2u = x_2 \frac{\partial u}{\partial x_2}.$$
In this case, the matrix (3.4) vanishes on the surface $\{x_2 = 0\}$. Outside this surface the complete integral is nondegenerate.

EXAMPLE 3.2. For each equation of the form $F(p_1, \ldots, p_n) = 0$, the family $u = a_1 x_1 + \cdots + a_{n-1} x_{n-1} + c x_n + a_n$, where c is defined by the equality $F(a_1, \ldots, a_{n-1}, c) = 0$, is a complete integral.

EXAMPLE 3.3. For a Clairaut equation of the form $u = x_1 p_1 + \cdots + x_n p_n + f(p_1, \ldots, p_n)$, the family $u = a_1 x_1 + \cdots + a_n x_n + f(a_1, \ldots, a_n)$ is a complete integral.

3.2. The construction of complete integrals using symmetry algebras. The integration method discussed below is based on "integration" of an algebra of symmetries, i.e., on construction of the Lie group of finite contact transformations for which the given algebra is the algebra of infinitesimal generators.

Let M be an n-dimensional smooth manifold, $\mathcal{E} \subset J^1(M)$ an equation of codimension one, and $X_{f_1}, \ldots, X_{f_n}$ a basis of a Lie algebra $\mathfrak{g}$ of infinitesimal symmetries of this equation. Suppose that the Lie group G corresponding to $\mathfrak{g}$ is realized as a group of finite contact transformations, by which we mean an exponential correspondence defined for sufficiently small values of $a_1, \ldots, a_n$ that takes every vector field $a_1 X_{f_1} + \cdots + a_n X_{f_n}$ to a finite transformation $A_{(a_1,\ldots,a_n)}$, with

$$\frac{d}{dt}(A_{(ta_1,\ldots,ta_n)})^* f = (A_{(ta_1,\ldots,ta_n)})^*(a_1 X_{f_1} + \cdots + a_n X_{f_n})f,$$

for any $f \in C^\infty(J^1(M))$.

Now let N be a "sufficiently good" generalized solution of $\mathcal{E}$. In more exact terms, assume that the span of $T_\theta(N)$ and the vectors $(X_{f_1})_\theta, \ldots, (X_{f_n})_\theta$ is of dimension $2n$ almost everywhere in N. Since the finite contact symmetries $A_{(a_1,\ldots,a_n)}$ take generalized solutions of $\mathcal{E}$ to generalized solutions,

$$N_{(a_1,\ldots,a_n)} = A_{(a_1,\ldots,a_n)}(N)$$

is an n-parameter family of generalized solutions, i.e., a complete integral of $\mathcal{E}$.

Let us find out what particular solutions can be used to construct complete integrals.

Let $u = u(x_1, \ldots, x_n)$ be a solution of $\mathcal{E}$. The plane tangent to the graph $N = j_1(u)(M)$ spans the vectors

$$T_i = \frac{\partial}{\partial x_i} + \frac{\partial u}{\partial x_i}\frac{\partial}{\partial u} + \sum_{k=1}^n \frac{\partial^2 u}{\partial x_i \partial x_k}\frac{\partial}{\partial p_k}.$$

Thus, constructing a complete integral requires that the matrix composed of the coefficients of the fields

$$T_1, \ldots, T_n, X_{f_1}, \ldots, X_{f_n} \tag{3.5}$$

be of rank $2n$ almost everywhere in N.

EXAMPLE 3.4. As we mentioned in Example 2.10, any equation of the form $F(u, p_1, \ldots, p_n) = 0$ possesses the n-dimensional algebra of contact symmetries generated by

$$\left\{\frac{\partial}{\partial x_1}, \ldots, \frac{\partial}{\partial x_n}\right\}.$$

The exponential correspondence for this algebra has the form

$$A_{(a_1,\ldots,a_n)}: (x_1, \ldots, x_n) \longmapsto (x_1 + a_1, \ldots, x_n + a_n).$$

Therefore, if $u = \widetilde{u}(x_1, \ldots, x_n)$ is a solution satisfying the above conditions, then

$$V(x_1, \ldots, x_n, a_1, \ldots, a_n) = \widetilde{u}(x_1 - a_1, \ldots, x_n - a_n)$$

is a complete integral for this equation.

Note that the matrix (3.5) is nonsingular if the matrix

$$\begin{pmatrix} \partial\widetilde{u}/\partial x_1 & \partial^2\widetilde{u}/\partial x_1\partial x_1 & \ldots & \partial^2\widetilde{u}/\partial x_1\partial x_n \\ \cdots & \cdots & \cdots & \cdots \\ \partial\widetilde{u}/\partial x_n & \partial^2\widetilde{u}/\partial x_n\partial x_1 & \ldots & \partial^2\widetilde{u}/\partial x_n\partial x_n \end{pmatrix}$$

is of rank n almost everywhere. Consider several particular cases.

a. The equation

$$\frac{\partial u}{\partial x_1}\frac{\partial u}{\partial x_2} = u^a$$

has the solution $\tilde{u} = ((2-a)^2 x_1 x_2)^{1/(2-a)}$ for $a \neq 2$ and the solution $\tilde{u} = e^{2\sqrt{x_1 x_2}}$ for $a = 2$. Using this, we obtain the complete integral

$$u = [(2-a)^2 (x_1 - a_1)(x_2 - a_2)]^{1/(2-a)}$$

in the first case and

$$u = e^{2\sqrt{(x_1 - a_1)(x_2 - a_2)}}$$

in the second case.

b. The equation

$$u = \sum_{i=1}^{n} b_i \left(\frac{\partial u}{\partial x_i}\right)^2$$

has the solution $\tilde{u} = \sum_{i=1}^{n} x_i^2 / 4b_i$, which yields the complete integral

$$u = \sum_{i=1}^{n} \frac{(x_i - a_i)^2}{4b_i}.$$

EXAMPLE 3.5. It is not hard to construct the exponential correspondence for the Abelian algebra consisting of infinitesimal translations and scale transformations (see Examples 2.1 and 2.3). In this case, each infinitesimal transformation can be integrated separately.

The equation

$$2x_1 \frac{\partial u}{\partial x_1} \frac{\partial u}{\partial x_2} - u \frac{\partial u}{\partial x_2} = a$$

possesses the Abelian symmetry algebra with the basis

$$x_1 \frac{\partial}{\partial x_1} - p_1 \frac{\partial}{\partial p_1}, \quad \frac{\partial}{\partial x_2}.$$

The particular solution $\tilde{u}^2 = 2x_2(x_1 - a)$ gives the complete integral

$$V^2(x_1, x_2, a_1, a_2) = 2(x_2 - a_2)(a_1 x_1 - a).$$

Now let us discuss an example of how to construct a complete integral using a non-Abelian algebra of contact symmetries.

EXAMPLE 3.6. Suppose that an equation $F(x_3, u, p_1, p_2, p_3) = 0$, in addition to obvious symmetries $\partial/\partial x_1$ and $\partial/\partial x_2$, has a scale symmetry of the form

$$X = \sum_{i=1}^{3} \alpha_i x_i \frac{\partial}{\partial x_i} + \beta u \frac{\partial}{\partial u} + \sum_{i=1}^{3} (\beta - \alpha_i) p_i \frac{\partial}{\partial p_i}. \tag{3.6}$$

Computing shifts along trajectories of this field, we easily obtain the exponential correspondence

$$A_{(a_1, a_2, a_3)} \colon (x_1, x_2, x_3, u, p_1, p_2, p_3) \longmapsto \left[\left(x_1 + \frac{a_1}{\alpha_1 a_3}\right) e^{\alpha_1 a_3} - \frac{a_1}{\alpha_1 a_3},\right.$$

$$\left.\left(x_2 + \frac{a_2}{\alpha_2 a_3}\right) e^{\alpha_2 a_3} - \frac{a_2}{\alpha_2 a_3}, x_3 e^{\alpha_3 a_3}, u e^{\beta a_3}, p_1 e^{(\beta - \alpha_1) a_3}, p_2 e^{(\beta - \alpha_2) a_3}, p_3 e^{(\beta - \alpha_3) a_3}\right].$$

Take as an example the equation

$$4u = \left(\frac{p_1}{p_3^{\alpha_1}}\right)^2 + \left(\frac{p_2}{p_3^{\alpha_2}}\right)^2 + 4x_3 p_3,$$

which has the symmetry of the form (3.6) with $\alpha_3 = 1$ and $\beta = 0$. Using the particular solution $\tilde{u} = x_1^2 + x_2^2 + x_3$ and the above exponential correspondence, we get the complete integral

$$V = \left[\left(x_1 + \frac{a_1}{\alpha_1 a_3}\right)e^{-\alpha_1 a_3} - \frac{a_1}{\alpha_1 a_3}\right]^2 + \left[\left(x_2 + \frac{a_2}{\alpha_2 a_3}\right)e^{-\alpha_2 a_3} - \frac{a_2}{\alpha_2 a_3}\right]^2 + x_3 e^{-a_3}.$$

3.3. Complete integrals: an invariant approach.

DEFINITION 3.2 (geometric). Let $\mathcal{E} \subset J^1(M)$ be a differential equation with one dependent variable, and let $n = \dim M$. Suppose that a domain $\mathcal{O} \subset \mathcal{E}$ is foliated by an n-parameter family of generalized solutions to the equation (n-dimensional integral manifolds of the distribution $\mathcal{C}(\mathcal{E})$), i.e., there are an n-dimensional smooth manifold N and a regular smooth map $\pi_N \colon \mathcal{O} \to N$ such that inverse images $N_a = \pi_N^{-1}(a)$ of points $a \in N$ are n-dimensional integral manifolds of $\mathcal{C}(\mathcal{E})$; such a foliation is called a *complete integral* of $\mathcal{E}$.

We claim that Definitions 3.1 and 3.2 are equivalent. Indeed, let us prove that the coordinate definition implies the existence of a domain $\mathcal{O} \subset \mathcal{E}$ foliated by solutions of $\mathcal{E}$.

To this end, define a map $\bar{\alpha} \colon M \times \bar{N} \to \mathcal{E}$, where $\bar{N} = \{a = (a_1, \ldots, a_n)\}$ is a parameter domain, by the formula $(b, a) \mapsto j_1(V(x, a))(b)$. Since the matrix (3.4) is nonsingular, this map has rank $2n$ at each point and, therefore, is a local diffeomorphism. Hence, we can choose open domains $U \subset M$, $N \subset \bar{N}$, $\mathcal{O} \subset \mathcal{E}$ such that the restriction

$$\alpha \colon U \times N \to \mathcal{O} \tag{3.7}$$

of $\bar{\alpha}$ is a diffeomorphism and for any $a \in N$ the surface $W_a = \alpha(U \times \{a\})$ is a solution of $\mathcal{E}$ over the domain U. We construct the desired map π_N by putting $\pi_N(\theta) = \beta_2(\theta)$ if $\alpha^{-1}(\theta) = (\beta_1(\theta), \beta_2(\theta))$. Clearly, $W_a = \pi_N^{-1}(a)$.

Using the above procedure yields a system of local coordinates

$$(x_1, \ldots, x_n, a_1, \ldots, a_n), \tag{3.8}$$

in $\mathcal{O} \subset \mathcal{E}$ such that the integral surfaces of the complete integral are given by

$$\begin{cases} a_1 = C_1, \\ \ldots\ldots\ldots \\ a_n = C_n, \end{cases}$$

where $C_1, \ldots, C_n$ are constants.

EXERCISE 3.1. Show that the geometric definition of a complete integral implies the coordinate one.

Given a complete integral, we need not explicitly solve system (2.14) to find the characteristics. Indeed, observe that the map α (see (3.7)) is given by

$$\begin{cases} u = V(x, a), \\ p_1 = V_{x_1}(x, a), \\ \ldots\ldots\ldots\ldots \\ p_n = V_{x_n}(x, a), \end{cases} \tag{3.9}$$

which describe the relation between the special local coordinates x, u, p and the coordinates (3.8). It follows that in the coordinates (3.8) the form U_1 is written as

$$U_1 = V_{a_1}\, da_1 + \cdots + V_{a_n}\, da_n = V_{a_1}\left(da_1 + \frac{V_{a_2}}{V_{a_1}}\, da_2 + \cdots + \frac{V_{a_n}}{V_{a_1}}\, da_n\right).$$

We can assume without loss of generality that locally $V_{a_1} \neq 0$, so that $\mathcal{C}(\mathcal{E})$ is given by the 1-form

$$\omega = da_1 + \frac{V_{a_2}}{V_{a_1}}\, da_2 + \cdots + \frac{V_{a_n}}{V_{a_1}}\, da_n.$$

Taking the functions

$$a_1,\ldots,a_n,\, y_1,\, y_2 = \frac{V_{a_2}}{V_{a_1}},\ldots,y_n = \frac{V_{a_n}}{V_{a_1}},$$

where y_1 is an arbitrary independent function, for coordinates on $\mathcal{E}$, we get

$$\omega = da_1 + y_2\, da_2 + \cdots + y_n\, da_n.$$

Therefore, $\partial/\partial y_1$ defines the degeneration direction of $d\omega$ on $\mathcal{C}(\mathcal{E})$ (i.e., $\partial/\partial y_1 \lrcorner \omega = 0$ and $\partial/\partial y_1 \lrcorner d\omega = 0$), while the functions $a_1,\ldots,a_n, y_2,\ldots,y_n$ are first integrals of the characteristic distribution.

EXAMPLE 3.7. Consider the equation

$$\frac{\partial u}{\partial x_1}\frac{\partial u}{\partial x_2} - x_1 \frac{\partial u}{\partial x_1} - x_2 \frac{\partial u}{\partial x_2} = 0.$$

It can easily be checked that the function $V(x_1, x_2, a_1, a_2) = a_1 + (x_1 + x_2 a_2)^2/2a_2$ is a complete integral for this equation.

We have $V_{a_1} = 1$, $p_1 = (x_1 + x_2 a_2)/a_2$, $p_2 = x_1 + x_2 a_2$, whence $a_2 = p_2/p_1$ and $a_1 = u - p_1 p_2/2$. Substituting a_1 and a_2 into the expression for V_{a_2} yields $y_2 = p_1(p_2 x_2 - p_1 x_1)/2p_2$. The functions a_1, a_2, and y_2 are first integrals of the characteristic distribution.

EXAMPLE 3.8. Consider the equation

$$2x_1 \frac{\partial u}{\partial x_1}\frac{\partial u}{\partial x_2} - u\frac{\partial u}{\partial x_2} - a = 0.$$

A complete integral for this equation can be obtained by means of the relation

$$u^2 = V^2 = 2(a_1 x_1 - a)(x_2 - a_2).$$

It is easily shown that

$$y_2 = \frac{V_{a_2}}{V_{a_1}} = -\frac{a_1 p_2}{x_1 p_1},\quad a_2 = x_2 - \frac{p_1(a_1 x_1 - a)}{a_1 p_2},$$

where a_1 is the integral defined by $2p_1(a_1 x_2 - a)^2 = u^2 p_2 a_1$. The functions a_1, a_2, y_2 constitute a complete set of first integrals of the characteristic distribution on the equation at hand.

3.4. The Lagrange–Charpit method. To construct a complete integral in the case of two independent variables, it suffices to know one nontrivial first integral of the characteristic distribution. The method allowing one to do this is called the *Lagrange–Charpit method* [36, 113].

Let $\mathcal{E} = \{F = 0\}$ be a first-order equation with two independent variables, $F \in C^\infty(J^1(\mathbb{R}^2))$, and let $f \in C^\infty(J^1(\mathbb{R}^2))$ be an integral of the characteristic vector field Y_F, i.e.,
$$Y_F(f) = 0. \tag{3.10}$$
Let us look for generalized solutions of $\mathcal{E}$ satisfying
$$\begin{cases} F(x_1, x_2, u, p_1, p_2) = 0, \\ f(x_1, x_2, u, p_1, p_2) = a_1, \end{cases} \tag{3.11}$$
where a_1 is a constant.

Solving (3.11) for p_1 and p_2 yields
$$\begin{cases} p_1 = g_1(x_1, x_2, u, a_1), \\ p_2 = g_2(x_1, x_2, u, a_1). \end{cases}$$

It is readily seen that under conditions (3.10) and (3.11) the form $U_1 = du - p_1\, dx_1 - p_2\, dx_2$ is closed, so that the system
$$\begin{cases} \dfrac{\partial u}{\partial x_1} = g_1(x_1, x_2, u, a_1), \\ \dfrac{\partial u}{\partial x_2} = g_2(x_1, x_2, u, a_1), \end{cases}$$
is compatible and has a solution $u = V(x_1, x_2, a_1, a_2)$, which is the desired complete integral of $\mathcal{E}$.

EXAMPLE 3.9. Consider the equation
$$x_2 \frac{\partial u}{\partial x_1}\frac{\partial u}{\partial x_2} - u\frac{\partial u}{\partial x_1} + c\frac{\partial u}{\partial x_2} = 0,$$
where a is a nonzero constant. It is easy to prove that p_2 is a first integral of the characteristic distribution of the equation under consideration. Setting $p_2 = a_2$ and solving the equation for p_1, we get $p_1 = ca_2/(u - x_2 a_2)$. Thus, the system of equations $\partial u/\partial x_2 = a_2$ and $\partial u/\partial x_1 = ca_2/(u - x_2 a_2)$ is compatible. Integrating the first equation with respect to x_1, we get $u = a_2 x_2 + v(x_1)$. The second equation yields $v_{x_1} = ca_2/v$, whence $v = \pm\sqrt{2ca_2 x_1 + a_1}$. Therefore, $u = V_\pm(x_1, x_2, a_1, a_2) = a_2 x_2 \pm \sqrt{2ca_2 x_1 + a_1}$ is a complete integral of the equation in hand for either of the two signs.

The Lagrange–Charpit method can be generalized to the case of an arbitrary number of independent variables. To describe this generalization, let us first discuss the problem of the compatibility for an overdetermined system
$$\begin{cases} f_1 = c_1, \\ \dotfill \\ f_r = c_r, \end{cases} \tag{3.12}$$
with $f_i \in C^\infty(J^1(M))$, $i = 1, \ldots, r$, and $c_1, \ldots, c_r$ constants.

PROPOSITION 3.1. *A sufficient condition for system* (3.12) *to be compatible is that for all* $1 \le i, k \le r$ *the Mayer brackets* $[f_i, f_k]$ *vanish on the surface* (3.12).

3. COMPLETE INTEGRALS

PROOF. First of all, note that the characteristic vector fields $Y_{f_1}, \ldots, Y_{f_r}$ are tangent to the $(2n - r + 1)$-dimensional surface in $J^1(M)$ given by (3.12), since $Y_{f_i}(f_k) = [f_i, f_k] = 0$ on this surface.

Further, we have
$$[Y_f, Y_g] = Y_{[f,g]} + g'Y_f - f'Y_g - [f,g]X_1, \qquad (3.13)$$
where $f' = U_1([X_1, X_f])$, $g' = U_1([X_1, X_g])$.

EXERCISE 3.2. Prove (3.13).

Using (3.13), we obtain $[Y_{f_i}, Y_{f_k}] = f'_k Y_{f_i} - f'_i Y_{f_k}$ on the surface (3.12), so that, by the Frobenius theorem, the vector fields $Y_{f_1}, \ldots, Y_{f_r}$ generate a completely integrable distribution Y on this surface.

Next, observe that the manifold of intersection of an n-dimensional integral manifold of the Cartan distribution with the surface (3.12) is at least of dimension $n - r$. Take an $(n-r)$-dimensional integral manifold of the Cartan distribution belonging to the surface (3.12) such that for any $\theta \in R$ the tangent plane $T_\theta(R)$ and the span of $(Y_{f_1})_\theta, \ldots, (Y_{f_r})_\theta$ intersect at zero. Let N be the union of all integral manifolds of the distribution Y passing through points of R. We claim that the n-dimensional manifold N is a generalized solution of (3.12) near R.

Indeed, for each point $\theta \in R$ the space $T_\theta(N)$ is the sum of $T_\theta(R) \subset \mathcal{C}_\theta$ and $Y_\theta \subset \mathcal{C}_\theta$. Thus, $T_\theta(N) \subset \mathcal{C}_\theta$ for all $\theta \in N$ near R. □

Recall that we have been working under the assumption that the form $dF|_{\mathcal{E}}$, where F is the function that defines our equation $\mathcal{E} = \{F = 0\}$, is almost everywhere nonzero. In this case, any function $f \in C^\infty(J^1(M))$ vanishing on $\mathcal{E}$ has the form $f = \nu F$ for some $\nu \in C^\infty(J^1(M))$.

Under this condition, $f \in C^\infty(J^1(M))$ is a first integral of the characteristic distribution of $\mathcal{E}$ if and only if
$$[F, f] = \lambda F \qquad \text{for some } \lambda \in C^\infty(J^1(M)). \qquad (3.14)$$

EXAMPLE 3.10. Consider the differential equation
$$h(x_1^2 + x_2^2)\left[\left(\frac{\partial u}{\partial x_1}\right)^2 + \left(\frac{\partial u}{\partial x_2}\right)^2\right] - \left(x_1\frac{\partial u}{\partial x_1} + x_2\frac{\partial u}{\partial x_2} - u\right)^2 = 0.$$
In this case $F = h(x_1^2 + x_2^2)(p_1^2 + p_2^2) - (x_1p_1 + x_2p_2 - u)^2$. Taking $f = (x_2p_1 - x_1p_2)/u$, we get
$$[F, f] = Y_F(f) = \frac{2(x_2p_1 - x_1p_2)}{u^2}F.$$

Therefore condition (3.14) is satisfied, so that the function f can be used in constructing a complete integral by means of the Lagrange–Charpit method.

System (3.12) is said to be *in involution* if $[f_i, f_k] = 0$, $1 \leq i, k \leq r$, by virtue of this system.

Let us now state the Lagrange–Charpit method in the general case.

THEOREM 3.2. *Let $\mathcal{E} = \{F = 0\} \subset J^1(M)$ be a first order differential equation with n independent variables, and let $f_2, \ldots, f_n$ be smooth functions on $J^1(M)$ such that the following conditions hold:*
1. $[F, f_k] = \lambda_k F$, $k = 2, \ldots, n$, $\lambda_k \in C^\infty(J^1(M))$.
2. $[f_i, f_k] = \lambda_{i,k} F$, $2 \leq i, k \leq n$, $\lambda_{i,k} \in C^\infty(J^1(M))$.

3. The functions $F, f_2, \ldots, f_n$ are functionally independent.

Then a complete integral of $\mathcal{E}$ can be computed by the following system

$$\begin{cases} F = 0, \\ f_2 = a_2, \\ \cdots\cdots\cdots \\ f_n = a_n. \end{cases} \qquad (3.15)$$

PROOF. The system (3.15) defines an $(n+1)$-dimensional surface P_a, $a = (a_2, \ldots, a_n)$, in $\mathcal{E}$. Let $\mathcal{C}(P_a)$ denote the induced Cartan distribution on this surface, i.e., the distribution given by the 1-form $\omega_a = U_1|_{P_a}$ (equivalently, $\mathcal{C}(P_a)_\theta = \mathcal{C}_\theta \cap T_\theta(P_a)$ for all $\theta \in P_a$).

The distribution $\mathcal{C}(P_a)$ is completely integrable. Indeed, since $\mathcal{C}(P_a)$ is generated by $Y_F, Y_{f_2}, \ldots, Y_{f_n}$, this fact is a part of the proof of Proposition 3.1.

Let f_1 be a first integral of this distribution. The system

$$\begin{cases} F = 0, \\ f_1 = b_1, \\ \cdots\cdots\cdots \\ f_n = b_n \end{cases}$$

defines a generalized solution of $\mathcal{E}$ for arbitrary constants $b_1, \ldots, b_n$. Thus, we have constructed a complete integral for $\mathcal{E}$ (see Definition 3.2). Regularity of the parameter map can be proved using the Frobenius theorem. $\square$

EXAMPLE 3.11. Consider the following equation with n independent variables:

$$u \frac{\partial u}{\partial x_1} \cdot \frac{\partial u}{\partial x_2} \cdot \ldots \cdot \frac{\partial u}{\partial x_n} - x_1 x_2 \cdots x_n = 0.$$

The functions

$$f_1 = \frac{p_1 u^{1/n}}{x_1}, \quad f_2 = \frac{p_2 u^{1/n}}{x_2}, \ldots, f_n = \frac{p_n u^{1/n}}{x_n}$$

are first integrals of the characteristic distribution for the considered equation $\mathcal{E} = \{up_1 p_2 \cdots p_n - x_1 \cdots x_n = 0\}$.

It can easily be checked that the integrals $f_1, \ldots, f_n$ are in involution, so that the system

$$\begin{cases} p_1 u^{1/n} = A_1 x_1, \\ \cdots\cdots\cdots\cdots\cdots \\ p_n u^{1/n} = A_n x_n \end{cases}$$

is compatible for all $A_1, \ldots, A_n = $ const, being compatible with the initial equation $\mathcal{E}$ for $A_1 \cdot \ldots \cdot A_n = 1$. Solutions of this system are written as

$$u = \left(\frac{n+1}{2n} \sum_{i=1}^n A_i x_i^2 + A_n \right)^{n/(n+1)}, \quad A_n = \text{const}.$$

In particular, putting $A_n = (A_1 \cdots A_{n-1})^{-1}$, we get a complete integral of $\mathcal{E}$.

CHAPTER 3

The Theory of Classical Symmetries

The methods of Chapter 2 were based on representation of a first order differential equation as a submanifold in the space $J^1(M)$ endowed with the Cartan distribution. In this chapter we extend the approach to arbitrary systems of nonlinear partial differential equations. We consider here the space of k-jets and the most important geometric structures related to it. We define classical symmetries of differential equations and, using some examples of equations in mathematical physics, illustrate methods for computing these symmetries. Also, we discuss applications of classical symmetries to construction of exact solutions.

1. Equations and the Cartan distribution

Consider a system of nonlinear differential equations of order k:

$$\begin{cases} F_1(\boldsymbol{x}, \boldsymbol{u}, \boldsymbol{p}) = 0, \\ \dots\dots\dots\dots\dots \\ F_r(\boldsymbol{x}, \boldsymbol{u}, \boldsymbol{p}) = 0, \end{cases} \tag{1.1}$$

where the F_l are smooth functions, $\boldsymbol{x} = (x_1, \dots, x_n)$ are independent variables, $\boldsymbol{u} = (u^1, \dots, u^m)$ are unknown functions, while $\boldsymbol{p}$ denotes the set of partial derivatives $p_\sigma^j = \partial^{|\sigma|} u^j / \partial x_1^{i_1} \cdots \partial x_n^{i_n}$, $\sigma = (i_1, \dots, i_n)$ being a multi-index, $|\sigma| = i_1 + \cdots + i_n \leq k$.

Geometric study of system (1.1) consists in treating the equations $F_i(\boldsymbol{x}, \boldsymbol{u}, \boldsymbol{p}) = 0$, $1 \leq i \leq r$, not as conditions on the functions $\boldsymbol{u}$ themselves, but on the k-th order Taylor series of these functions. This approach allows us to introduce a finite-dimensional space whose coordinates correspond to the values of the functions $\boldsymbol{u}$ and of their derivatives.

Thus we shall consider the variables

$$x_1, \dots, x_n, \; u^1, \dots, u^m, \; p_\sigma^i, \qquad |\sigma| \leq k,$$

as coordinates in some space $J^k(n, m)$ whose dimension is

$$\dim J^k(n, m) = n + m \sum_{i=0}^{k} \binom{n+i-1}{n-1} = n + m \binom{n+k}{k}.$$

Relations (1.1) determine in $J^k(n, m)$ a surface $\mathcal{E}$ of codimension r which is the geometric image corresponding to the given system of nonlinear differential equations. This surface in $J^k(n, m)$ will be called a *differential equation of order k with n independent and m dependent variables*. (We shall make this definition more precise below.) The surface $\mathcal{E}$ is a coordinate-free object, unlike its representation as a system of the form (1.1), since the same equation can possess different analytical representations.

The fact that the variable p^j_σ, $\sigma = (i_1, \ldots, i_n)$, corresponds to the partial derivative of u^j with respect to $x_1, \ldots, x_n$ is expressed geometrically in the following way. Consider on $J^k(n,m)$ the *Cartan distribution* $\mathcal{C} = \mathcal{C}^k(n,m)$ determined by the basic 1-forms

$$\omega^j_\sigma = dp^j_\sigma - \sum_{i=1}^n p^j_{\sigma+1_i} dx_i, \qquad 1 \leq j \leq m, \quad |\sigma| \leq k-1, \tag{1.2}$$

where $1_i = (0, \ldots, 0, 1, 0, \ldots, 0)$, with 1 in the i-th place[1]. A straightforward computation shows that the number of basic 1-forms is equal to

$$N = m \sum_{i=0}^{k-1} \binom{n+i-1}{n-1} = m \binom{n+k-1}{k-1}.$$

It is easily checked that the forms ω^j_σ are linearly independent at every point and consequently the number N coincides with the codimension of the Cartan distribution $\mathcal{C}$. Hence the dimension of the Cartan distribution is

$$\dim \mathcal{C}^k(n,m) = \dim J^k(n,m) - \operatorname{codim} \mathcal{C}^k(n,m) = n + m \binom{n+k-1}{n-1}.$$

Consider the projection

$$\pi_k \colon J^k(n,m) \to \mathbb{R}^n, \qquad \pi_k(\boldsymbol{x}, \boldsymbol{u}, \boldsymbol{p}) = \boldsymbol{x}. \tag{1.3}$$

In the preceding chapters (see §2 of Ch. 1 and §1 of Ch. 2) we showed that the Cartan distributions on the manifolds $J^k(1,1)$ and $J^1(n,1)$ allow one to distinguish between all sections of the mapping (1.3) and those that correspond to jets of smooth functions.

In general, the following theorem is valid:

THEOREM 1.1. *Let $Q \subset J^k(n,m)$ be an n-dimensional surface nondegenerately projecting to the space $\mathbb{R}^n$ of independent variables in a neighborhood of a point $y \in Q$. The surface Q is an integral manifold of the Cartan distribution $\mathcal{C}$ in the neighborhood under consideration if and only if it can be given by relations of the form*

$$\begin{cases} u^1 = f^1(x_1, \ldots, x_n), \\ \cdots \cdots \cdots \cdots \cdots \\ u^m = f^m(x_1, \ldots, x_n), \\ p^j_\sigma = \dfrac{\partial^{|\sigma|} f^j}{\partial x_1^{i_1} \ldots \partial x_n^{i_n}}(x_1, \ldots, x_n). \end{cases} \tag{1.4}$$

for some smooth functions $f^1, \ldots, f^m$, where $\sigma = (i_1, \ldots, i_n)$ and $|\sigma| \leq k$.

PROOF. In fact, nondegeneracy of the projection $\pi_k|_Q$ in a neighborhood of the point y (see Figure 3.1) means that locally the equation of Q can be represented as

$$\begin{cases} u^j = f^j(x_1, \ldots, x_n), \\ p^j_\sigma = f^j_\sigma(x_1, \ldots, x_n), \end{cases}$$

---

[1] Here and below we formally set $p^j_{(0,\ldots,0)} = u^j$.

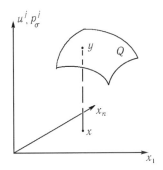

FIGURE 3.1. Nondegenerate projection of Q to $\mathbb{R}^n$

$j = 1, \ldots, m$, $|\sigma| \leq k$, where f^j and f^j_σ are smooth functions. The surface Q is an integral manifold for the distribution $\mathcal{C}$ if the basis 1-forms vanish on this surface, i.e.,

$$\omega^j_\sigma\big|_Q = df^j_\sigma - \sum f^j_{\sigma+1_i} dx_i = \sum \left(\frac{\partial f^j_\sigma}{\partial x_i} - f^j_{\sigma+1_i} \right) dx_i = 0,$$

which is equivalent to the equalities

$$f^j_\sigma = \frac{\partial^{|\sigma|} f^j}{\partial x_1^{i_1} \ldots \partial x_n^{i_n}}$$

for all $j = 1, \ldots, m$ and $|\sigma| \leq k$. □

DEFINITION 1.1. The surface Q given by the relations (1.4) will be denoted by $\Gamma^k_{\boldsymbol{f}}$ and called the *graph of the k-jet* of the vector function $\boldsymbol{f} = (f^1, \ldots, f^m)$.

From Theorem 1.1 it follows that a solution of a given equation $\mathcal{E} \subset J^k(n, m)$ is exactly an n-dimensional integral manifold of the Cartan distribution $\mathcal{C}^k$ nondegenerately projecting to the space $\mathbb{R}^n$ and lying on the surface $\mathcal{E}$:

$$\omega^j_\sigma\big|_Q = 0, \quad \ker d\pi_k\big|_y = 0, \quad y \in Q, \quad Q \subset \mathcal{E}.$$

Solutions of the equation $\mathcal{E}$ can be defined without using the fact that $\mathcal{E}$ is a surface in $J^k(n, m)$. Namely, consider the manifold $\mathcal{E}$ together with the distribution $\mathcal{C}(\mathcal{E})$ induced on $\mathcal{E}$ by the distribution $\mathcal{C}$. The plane of the distribution $\mathcal{C}(\mathcal{E})$ at every point $y \in \mathcal{E}$ is the intersection of the plane $\mathcal{C}_y$ with the tangent plane $T_y\mathcal{E}$ to the surface $\mathcal{E}$. Obviously, integral manifolds of the distribution $\mathcal{C}(\mathcal{E})$ nondegenerately projecting to the space $\mathbb{R}^n$ of independent variables coincide with integral manifolds of the distribution $\mathcal{C}$ lying in $\mathcal{E}$ and possessing the same property. Therefore, one can say that solutions are n-dimensional manifolds of the distribution $\mathcal{C}(\mathcal{E})$ diffeomorphically projecting to the space $\mathbb{R}^n(x_1, \ldots, x_n)$. Recall that this is the approach to solutions of equations with which we started Chapter 1. Considering arbitrary n-dimensional maximal integral manifolds, we arrive at the concept of generalized solutions.

EXAMPLE 1.1 (the Burgers equation). Consider the Burgers equation

$$u_t - uu_x - u_{xx} = 0.$$

In the 8-dimensional space of jets $J^2(2,1)$, the surface $\mathcal{E}$ corresponds to this equation and is described in standard coordinates
$$x = x_1, \; t = x_2, \; u, \; p_{(1,0)}, \; p_{(0,1)}, \; p_{(1,1)}, \; p_{(0,2)}, \; p_{(2,0)}$$
by the relation
$$p_{(0,1)} - u p_{(1,0)} - p_{(2,0)} = 0.$$
The Cartan distribution on $J^2(2,1)$ is determined by the basis 1-forms
$$\omega_{(0,0)} = du - p_{(1,0)} dx_1 - p_{(0,1)} dx_2,$$
$$\omega_{(1,0)} = dp_{(1,0)} - p_{(2,0)} dx_1 - p_{(1,1)} dx_2,$$
$$\omega_{(0,1)} = dp_{(0,1)} - p_{(1,1)} dx_1 - p_{(0,2)} dx_2.$$
Solutions of this equation are two-dimensional integral manifolds of this distribution lying in $\mathcal{E}$.

Choosing the functions $x = x_1$, $t = x_2$, u, $p_{(1,0)}$, $p_{(0,1)}$, $p_{(1,1)}$, $p_{(0,2)}$ for coordinates on the surface $\mathcal{E}$ and replacing $p_{(2,0)}$ by $p_{(0,1)} - u p_{(1,0)}$ everywhere, we shall obtain the distribution on the 7-dimensional space given by the basis forms
$$\widetilde{\omega}_{(0,0)} = du - p_{(1,0)} dx_1 - p_{(0,1)} dx_2,$$
$$\widetilde{\omega}_{(1,0)} = dp_{(1,0)} - (p_{(0,1)} - u p_{(1,0)}) dx_1 - p_{(1,1)} dx_2, \qquad (1.5)$$
$$\widetilde{\omega}_{(0,1)} = dp_{(0,1)} - p_{(1,1)} dx_1 - p_{(0,2)} dx_2.$$
Two-dimensional integral manifolds of this distribution, where the functions $u, p_{i,j}$ are expressed via x_1, x_2 (which is equivalent to nondegeneracy of the differential of the projection to $\mathbb{R}^2(x,t)$), correspond to solutions of the equation. For example, the two-dimensional surface
$$u = -\frac{x}{t}, \; p_{(1,0)} = -\frac{1}{t}, \; p_{(0,1)} = \frac{x}{t^2}, \; p_{(1,1)} = \frac{1}{t^2}, \; p_{(0,2)} = -\frac{2x}{t^3}$$
corresponds to the solution $u = -x/t$.

Note that, contrary to the case of systems of ordinary differential equations ($n = 1$), the Cartan distribution $\mathcal{C}$ on the manifolds $J^k(n,m)$, as well as the distributions $\mathcal{C}(\mathcal{E})$ obtained by restricting $\mathcal{C}$ to a partial differential equation $\mathcal{E}$, is not, in general, completely integrable. Nevertheless, for some overdetermined systems the distribution $\mathcal{C}(\mathcal{E})$ may be completely integrable.

EXERCISE 1.1. Let $m = k = 1$, $n = r = 2$. Consider the system of equations
$$\begin{cases} u_x = f(x,y,u), \\ u_y = g(x,y,u). \end{cases}$$
Prove that if this system is compatible, then the Cartan distribution restricted to the corresponding surface is completely integrable.

2. Jet manifolds and the Cartan distribution

In this section, we consider the basic objects important for the geometric theory of differential equations. These are jet manifolds and the Cartan distributions on them. We already met these objects when studying ordinary differential equations and first-order partial differential equations. Here we deal with the most general definitions and study basic properties.

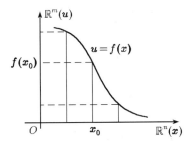

FIGURE 3.2

2.1. Geometric definition of the jet spaces. Up to now we have considered equations on a vector function $u = (u^1, \ldots, u^m)$ depending on the variables $x = (x_1, \ldots, x_n)$. From the geometric point of view, the function $u = f(x)$ is an analytical representation for a section of the projection $\mathbb{R}^m \times \mathbb{R}^n \to \mathbb{R}^n$. In fact, to choose a function $f(x)$ means to put into correspondence to every point $x_0 \in \mathbb{R}^n$ another point $f(x_0) \in \mathbb{R}^m$, which can be considered as a point of the fiber $\mathbb{R}^m$ over the point x_0 (see Figure 3.2).

To deal with equations on an arbitrary manifold, this construction needs to be generalized. Consider an m-dimensional locally trivial bundle $\pi \colon E \to M$ over an n-dimensional manifold M. Recall that a section of the bundle π is a mapping $s \colon M \to E$ such that $\pi \circ s$ is the identity mapping of the base M. In other words, s takes a point $x \in M$ to some point of the fiber E_x. In the particular case of the trivial bundle $M \times N \to M$, sections are the mappings $M \to N$. In what follows, for the sake of simplicity we shall always assume that all bundles under consideration are vector bundles, i.e., their fibers are vector spaces and the gluing functions are linear transformations. Nevertheless, almost all constructions considered below are valid for arbitrary locally trivial bundles [**60, 101**].

Let $\mathcal{U} \subset M$ be a neighborhood over which the bundle π becomes trivial, i.e., such that $\pi^{-1}(\mathcal{U}) \cong \mathcal{U} \times \mathbb{R}^m$. If $e_1, \ldots, e_m$ is a basis in the fiber of the bundle π, i.e., in the space $\mathbb{R}^m$, then any section can be represented over $\mathcal{U}$ in the form $s = f^1 e_1 + \cdots + f^m e_m$, where f^i are smooth functions on $\mathcal{U}$. If $\mathcal{U}$ is a coordinate neighborhood on the manifold M with local coordinates $x_1, \ldots, x_n$, then any point of the fiber is determined by its projection to $\mathcal{U}$ and by its coordinates $u^1, \ldots, u^m$ with respect to the chosen basis. The functions $x_1, \ldots, x_n, u^1, \ldots, u^m$ are coordinates in $\pi^{-1}(\mathcal{U})$, and are called *adapted* coordinates for the bundle under consideration. Thus, any section is represented in adapted coordinates by a vector function $f = (f^1, \ldots, f^m)$ in the variables $x_1, \ldots, x_n$.

DEFINITION 2.1. Two sections φ_1 and φ_2 of the bundle π will be called *tangent* with order k over the point $x_0 \in M$, if the vector functions $u = f_1(x)$, $u = f_2(x)$ corresponding to these sections have the same partial derivatives up to order k at the point x_0.

Obviously, this condition is equivalent to the fact that the k-th order Taylor series of the sections coincide. Since the functions themselves may be considered as their partial derivatives of zero order, the tangency condition for $k = 0$ reduces to coincidence of $f_1(x_0)$ and $f_2(x_0)$, i.e., the graphs of the sections s_1 and s_2 must intersect the fiber E_{x_0} at the same point (Figure 3.3a). On Figures 3.3b and 3.3c

74   3. THE THEORY OF CLASSICAL SYMMETRIES

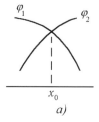 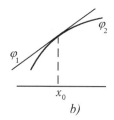 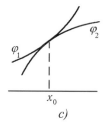

FIGURE 3.3

the cases of tangency with orders 1 and 2 are shown: a straight line tangent to a curve and an arc of the osculating circle.

EXERCISE 2.1. 1. Prove that Definition 2.1 is invariant, i.e., it does not depend on the choice of adapted coordinates in the bundle π.

2. Prove that two sections have tangency of order k if their graphs (see Definition 1.1) are tangent of the same order k.

Tangency of sections with order k at a point x is an equivalence relation which will be denoted by $s_1 \overset{k,x}{\sim} s_2$. The set of equivalence classes of sections, i.e., the set of k-th order Taylor series, will be denoted by J_x^k and called the *space of k-jets* of the bundle π at the point x. A point of this space (the equivalence class of a section s) will be denoted by $[s]_x^k$. Thus, if $s_1 \overset{k,x}{\sim} s_2$, then $[s_1]_x^k = [s_2]_x^k$. The *space of k-jets of the bundle π* is the union of J_x^k over all points $x \in M$:

$$J^k(\pi) = \bigcup_{x \in M} J_x^k.$$

For any point $\theta = [s]_x^k \in J^k(\pi)$ let us set $\pi_k(\theta) = x$. Thus we have the projection $\pi_k \colon J^k(\pi) \to M$, and $\pi_k^{-1}(x) = J_x^k$.

For the case $k = 0$ we have $J^0(\pi) = \bigcup_{x \in M} E_x = E$, i.e., the space $J^0(\pi)$ coincides with the total space of the bundle π.

Let us define local coordinates on the space of k-jets of the bundle π. To this end, take the functions x_i, u^j, and p_σ^j, corresponding to dependent and independent variables and to partial derivatives of the latter with respect to the former. In fact, let $x_1, \ldots, x_n, u^1, \ldots, u^m$ be an adapted coordinate system in the bundle π over a neighborhood $\mathcal{U}$ of the point $x \in M$. Consider the set $\pi_k^{-1}(\mathcal{U}) \subset J^k(\pi)$. Let us complete local coordinates $x_1, \ldots, x_n, u^1, \ldots, u^m$ by the functions p_σ^j defined by the formula

$$p_\sigma^j([s]_x^k) = \frac{\partial^{|\sigma|} s^j}{\partial x_1^{i_1} \cdots \partial x_n^{i_n}}, \qquad j = 1, \ldots, m, \quad |\sigma| \leq k.$$

In what follows, we shall call the coordinates p_σ^j *canonical* (or *special*) *coordinates* associated to the adapted coordinate system (x_i, u^j).

For a given bundle π one can consider all jet manifolds $J^k(\pi)$, $k = 0, 1, \ldots$, arranging them one over another as the tower

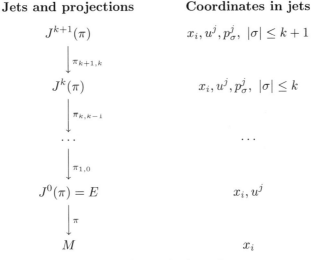

The projections $\pi_{t+1,t}$ are defined by the formula
$$\pi_{t+1,t}\colon J^{t+1}(\pi) \longrightarrow J^t(\pi), \qquad \pi_{t+1,t}([s]_x^{t+1}) = [s]_x^t,$$
where $t = 0, \ldots, k$. Since the equivalence class $[s]_x^{t+1} \in J^{t+1}(\pi)$ determines the class $[s]_x^t \in J^t(\pi)$ uniquely, the projections $\pi_{t+1,t}$ are well defined.

EXERCISE 2.2. 1. Prove that the family of neighborhoods $\pi_k^{-1}(\mathcal{U})$ together with the coordinate functions x_i, u^j, p_σ^j determines a smooth manifold structure in $J^k(\pi)$.

2. Prove that the projection $\pi_k \colon J^k(\pi) \to M$ is a smooth locally trivial vector bundle.

EXERCISE 2.3. Prove that the projections $\pi_{t+1,t}$ are smooth locally trivial bundles. Prove also that $\pi_t \circ \pi_{t+1,t} = \pi_{t+1}$.

2.2. The Cartan distribution. Now let us introduce the basic geometric structure on the manifold $J^k(\pi)$. This structure is the *Cartan distribution*. First of all, let us note that if s is a section of the bundle π, then for any point $x \in M$ one can define the element $j_k(s)(x) = [s]_x^k \in J_x^k(\pi)$. Obviously, the mapping $j_k(s)\colon M \to J^k(\pi)$ is a smooth section of the bundle $\pi_k \colon J^k(\pi) \to M$. It is called the *k-jet* of the section s. The graph of the k-jet in the space $J^k(\pi)$ is denoted by Γ_s^k.

We say that an n-dimensional plane in the tangent space $T_\theta(J^k(\pi))$, $\theta \in J^k(\pi)$, is an *R-plane* if it is tangent to the graph of the k-jet of some section of the bundle π. Obviously, any R-plane is horizontal with respect to the projection $\pi_k \colon J^k(M) \to M$.

Note that a point $\theta' \in J^{k+1}(\pi)$ can be considered as the pair consisting of the point $\theta = \pi_{k+1,k}(\theta') \in J^k(\pi)$ and an R-plane $L_{\theta'} \subset T_\theta(J^k(\pi))$ defined as the plane tangent to the graph of the k-jet of some section s such that $[s]_x^{k+1} = \theta'$. (It is easily seen that this plane is uniquely determined by the jet $[s]_x^{k+1}$.) In other words, θ' is the set of values of derivatives up to order $k+1$, while the plane $L_{\theta'} \subset T_\theta J^k(\pi)$ is determined by the values of first derivatives of k-th derivatives.

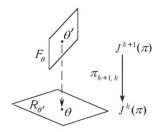

FIGURE 3.4. The fiber of the projection $\pi_{k+1,k}$

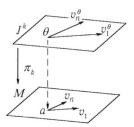

FIGURE 3.5. A basis of the plane $L_{\theta'}$

EXERCISE 2.4. Write down the conditions for an n-dimensional plane in the space $T_\theta(J^k(\pi))$ to be an R-plane.

Let us fix a point $\theta \in J^k(\pi)$ and consider various points $\theta' \in J^{k+1}(\pi)$ projecting to θ under the mapping $\pi_{k+1,k}$. These points form the fiber of the bundle $\pi_{k+1,k} \colon J^{k+1}(\pi) \to J^k(\pi)$ over the point θ. We shall denote this fiber by $F_{k+1,k}(\theta)$ or by F_θ (see Figure 3.4). In other words, we fix the value of some vector function and of its derivatives up to order k and let the other derivatives vary in an arbitrary way. When a point θ' moves along the fiber F_θ, the corresponding n-dimensional plane $L_{\theta'} \subset T_\theta(J^k(\pi))$ rotates somehow around the point θ, but always remains horizontal with respect to the projection $\pi_k \colon J^k(\pi) \to M$.

DEFINITION 2.2. The *Cartan plane* $\mathcal{C}_\theta = \mathcal{C}_\theta^k$ at the point $\theta \in J^k(\pi)$ is the span of all planes $L_{\theta'}$ for $\theta' \in F_\theta$, i.e., the span of all planes tangent to the graphs Γ_s^k of all k-jets of sections for which $[s]_x^k = \theta$. The correspondence

$$\mathcal{C} \colon \theta \longmapsto \mathcal{C}_\theta^k$$

is called the *Cartan distribution* on $J^k(\pi)$.

Let us give a coordinate description of the Cartan distribution. To do this, we shall explicitly write down a basis of the plane $L_{\theta'}$ corresponding to the point $\theta' \in J^{k+1}(\pi)$ in an adapted coordinate system $x_1, \ldots, x_n, u^1, \ldots, u^m$. The coordinates of the point θ' will be denoted by $x_i(\theta')$, $u^j(\theta')$, $p_\sigma^j(\theta')$ (see Figure 3.5).

Let s be a section such that $[s]_a^{k+1} = \theta'$, let $\Gamma_s^k \subset J^k(\pi)$ be its graph, and let the R-plane $L_{\theta'}$ be tangent to Γ_s^k. Let us choose a basis in L_θ consisting of the vectors $v_1^\theta, \ldots, v_n^\theta$ whose projections to M coincide with $v_1 = \partial/\partial x_1|_a, \ldots, v_n = \partial/\partial x_n|_a$. From equations (1.4) determining the surface Γ_s^k it is easy to deduce formulas for

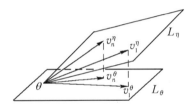

FIGURE 3.6. R-planes at the point θ

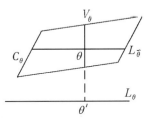

FIGURE 3.7. Structure of the Cartan plane

the basis vectors v_i^θ:

$$v_i^\theta = \frac{\partial}{\partial x_i} + \sum_{|\sigma|\le k}\sum_{j=1}^m p_{\sigma+1_i}^j(\theta')\frac{\partial}{\partial p_\sigma^j}. \tag{2.1}$$

It is easily seen that the above summation over σ, $|\sigma| < k$, is determined by the point θ and is independent of the choice of the point θ' projecting to θ. Thus any plane L_η corresponding to $\eta \in F_\theta$ can be obtained from the given plane L_θ by rotating in the "vertical direction". More exactly, the position of L_η with respect to L_θ can be determined by the set of n shift vectors $\delta_i = v_i^\eta - v_i^\theta$ (see Figure 3.6) vertical with respect to the projection $\pi' = \pi_{k,k-1}$. Denote the space of vertical vectors at the point θ by $V_\theta = T_\theta(F_{\theta''})$, where $\theta'' = \pi_{k,k-1}(\theta)$; the vectors $\partial/\partial p_\sigma^j$, $|\sigma| = k$, form a basis in this space. Note that any vertical tangent vector $v \in V_\theta$ can be considered as a shift vector: for a basis vector $\partial/\partial p_\sigma^j$, $|\sigma| = k$, it suffices to take the point $\eta \in F_\theta$ so that all its coordinates, except for one, coincide with the corresponding coordinates of the point θ' while $p_{\sigma+1_i}^j(\eta) = p_{\sigma+1_i}^j(\theta') + 1$. Then $\delta_i = v_i^\eta - v_i^\theta = \partial/\partial p_\sigma^j$.

From (2.1) it follows that the Cartan distribution on $J^k(\pi)$ is determined by the set of 1-forms $\omega_\sigma^j = dp_\sigma^j - \sum_{i=1}^n p_{\sigma+1_i}^j dx_i$, $|\sigma| \le k-1$, which are called the *Cartan forms*.

The geometric structure of Cartan planes (Figure 3.7) is described by the following theorem:

THEOREM 2.1. *Let $\mathcal{C}$ be the Cartan distribution on the manifold $J^k(\pi)$. Then the following statements are valid:*

1. *The Cartan plane $\mathcal{C}_\theta \subset T_\theta(J^k(\pi))$ is the direct sum of the vertical and horizontal subspaces*

$$\mathcal{C}_\theta = V_\theta \oplus L_{\widetilde{\theta}}, \tag{2.2}$$

where V_θ is the tangent space to the fiber of the projection $\pi_{k,k-1}$ and $L_{\tilde\theta}$ is the R-plane corresponding to some point $\tilde\theta$ in the fiber F_θ over the point θ.
2. The Cartan plane $\mathcal{C}_\theta$ consists of tangent vectors at the point θ that project to L_θ under the mapping $\pi_{k,k-1}$, i.e.,

$$\mathcal{C}_\theta = (\pi_{k,k-1})_*^{-1}(L_\theta). \tag{2.3}$$

PROOF. The first statement follows from the above considerations. Let us note only that in the decomposition (2.2) the first summand, unlike the second one, is determined by the point θ uniquely.

To prove the second statement, note that the projection $(\pi_{k,k-1})_*$ takes the subspace V_θ to zero while the plane $L_{\tilde\theta}$ is taken to L_θ bijectively. Therefore, $\mathcal{C}_\theta \subset (\pi_{k,k-1})_*^{-1}(L_\theta)$. The inverse embedding follows from the fact that the inverse image of the point $(\pi_{k,k-1})_*^{-1}(\theta'')$ coincides with $V_\theta \subset \mathcal{C}_\theta$. □

COROLLARY 2.2. *Any horizontal (with respect to the projection $\pi_{k,k-1}$) subspace of the Cartan plane $\mathcal{C}_\theta$ cannot be of dimension greater than $n = \dim M$.*

COROLLARY 2.3. *A plane $P \subset \mathcal{C}_\theta$ is horizontal with respect to the projection $\pi_{k,k-1}$ if and only if it is horizontal with respect to the projection $\pi_k\colon J^k(\pi) \to M$ (i.e., degeneracy of the Cartan plane under the mappings*

$$J^k(\pi) \longrightarrow J^{k-1}(\pi) \longrightarrow \cdots \longrightarrow J^0(\pi) \longrightarrow M$$

may occur at the first step only).

Let us find a coordinate representation for a vector field X lying in the Cartan distribution $\mathcal{C}^k$ on the manifold $J^k(\pi)$. To do this, we shall use the splitting (2.2). The set of vertical vector fields $\partial/\partial p^j_\sigma$, $j = 1, \ldots, m$, $|\sigma| = k$, forms a basis of the vertical subspace V_θ at the point θ. A basis of the horizontal subspace can be chosen as the set of truncated operators of total derivatives

$$D_i^{(k)} = \frac{\partial}{\partial x_i} + \sum_{j=1}^m \sum_{|\sigma|<k} p^j_{\sigma+1_i} \frac{\partial}{\partial p^j_\sigma}. \tag{2.4}$$

Thus any vector field X lying in the Cartan distribution can be decomposed in the chosen basis:

$$X = \sum_{i=1}^m a^i D_i^{(k)} + \sum_{|\sigma|=k} b^j_\sigma \frac{\partial}{\partial p^j_\sigma}. \tag{2.5}$$

The distribution $\mathcal{C}^k$ is not completely integrable, since, for example, for $|\sigma| = k-1$ the commutator

$$\left[\frac{\partial}{\partial p^j_{\sigma+1_i}}, D_i^{(k)}\right] = \frac{\partial}{\partial p^j_\sigma}$$

is not of the form (2.5). Consequently, maximal integral manifolds of the distribution $\mathcal{C}^k$ are of dimension less than $\dim \mathcal{C}^k$.

We can finally give a definition of a differential equation of order k similar to that given for differential equations of first order.

DEFINITION 2.3. A *differential equation of order k in the bundle* $\pi\colon E \to M$ is a submanifold $\mathcal{E} \subset J^k(\pi)$ endowed with the Cartan distribution $\mathcal{C}(\mathcal{E})\colon \theta \mapsto \mathcal{C}_\theta(\mathcal{E}) = \mathcal{C}^k_\theta \cap T_\theta(\mathcal{E})$, $\theta \in \mathcal{E}$. A maximal integral manifold (of dimension $\dim M$) of the Cartan distribution is called a *(generalized) solution* of the equation $\mathcal{E}$.

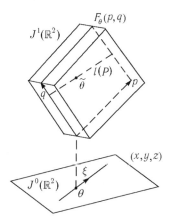

FIGURE 3.8. The ray manifold

EXERCISE 2.5. Let $\pi\colon E \to M$ be a fiber bundle and $\pi_{1,0}\colon J^1(\pi) \to E$. Show that sections of the bundle $\pi_{1,0}$ are connections in the bundle π, while the condition of zero curvature determines a first order equation in the bundle $\pi_{1,0}$.

2.3. Integral manifolds of the Cartan distribution. In Theorem 1.1, we showed that the graphs Γ_f^k of k-jets of sections are integral manifolds of the Cartan distribution. Here we shall describe the structure of arbitrary maximal integral manifolds.

DEFINITION 2.4. Let $P \subset \mathcal{C}_\theta^{k-1}$ be a plane of dimension s, $s \leq \dim M$. The subset
$$l(P) = \{\widetilde{\theta} \in F_\theta \mid L_{\widetilde{\theta}} \supset P\}$$
of the fiber over the point θ is called the *ray submanifold* (or simply *ray*) corresponding to the plane P (see Figure 3.8). If $N \subset J^{k-1}(\pi)$ is a submanifold, then the set
$$L(N) = \bigcup_{q \in N} l(T_q N)$$
is called the *prolongation* (or *lifting*) of the submanifold N.

EXERCISE 2.6. Show that if $\Gamma_f^{k-1} \subset N$, then $\Gamma_f^k \subset L(N)$.

EXAMPLE 2.1. Consider the manifold of 1-jets $J^1(2,1)$ of scalar functions in two variables with the coordinates x, y, z, p, q, where z is a function of x and y, while p and q are its derivatives with respect to x and y.

Let us fix a point $\theta \in J^0(\mathbb{R}^2)$ with the coordinates x, y, z. The fiber F_θ of the bundle $J^1(\mathbb{R}^2) \to J^0(\mathbb{R}^2)$ over this point is the plane with coordinates p, q. Let $\xi = X\partial/\partial x + Y\partial/\partial y + Z\partial/\partial z$ be a nonzero vector at the point θ, and P the straight line determined by this vector. Let us deduce the equations describing the corresponding ray manifold $l(P)$.

Let p_0, q_0 be the coordinates of the point $\widetilde{\theta} \in F_\theta$. Then the corresponding plane $L_{\widetilde{\theta}}$ is described by the equation
$$dz - p_0\, dx - q_0\, dy = 0$$

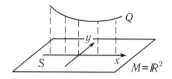

FIGURE 3.9

while the condition $L_{\widetilde{\theta}} \supset P$, i.e., $\xi \in L_{\widetilde{\theta}}$, is of the form

$$Xp_0 + Yq_0 = Z.$$

Consequently, if the values of X, Y, and Z are fixed while p and q vary, the submanifold $l(P) \subset F_\theta$ is described by the equation

$$Xp + Yq = Z.$$

In generic position, $l(P)$ is a straight line in the plane F_θ. Note that the direction of this line is determined by the projection (X, Y) of the vector ξ to the plane $\mathbb{R}^2_{x,y}$. The position of its initial point also depends on Z. In the exceptional case, when the vector ξ is vertical (i.e., when $X = Y = 0$ and $Z \neq 0$), the set $l(P)$ is empty, since an R-plane cannot contain vertical vectors.

Consider now a submanifold $N \subset J^0(\mathbb{R}^2)$ nondegenerately projecting to $\mathbb{R}^2_{x,y}$. Depending on the dimension of N, its prolongation $L(N)$ is described in the following way:

1. N consists of one point θ only. Then $L(N) = F_\theta$, the fiber over θ.
2. N is a curve parametrized as $(\alpha(t), \beta(t), \gamma(t))$. Then $L(N)$ is a two-dimensional surface in the five-dimensional space $J^1(\mathbb{R}^2)$ given by the equations

$$\begin{cases} x = \alpha(t), \\ y = \beta(t), \\ z = \gamma(t), \\ \alpha'(t)p + \beta'(t)q = \gamma'(t). \end{cases} \quad (2.6)$$

The curve $N \subset J^0(\mathbb{R}^2)$ can be considered as the graph of some function $z(x, y)$ defined on a curve $x = \alpha(t)$, $y = \beta(t)$ lying in the plane $\mathbb{R}^2_{x,y}$. The last equation of (2.6) shows how the partial derivatives $p = z_x$, $q = z_y$ of this function are related to each other on the curve.

3. N is a surface of the form $z = f(x, y)$. Then every point $\theta \in N$ uniquely determines a point $\widetilde{\theta} \in F_\theta$ such that $L_{\widetilde{\theta}} = T_\theta(N)$. In fact, the coordinates p, q of the point $\widetilde{\theta}$ must coincide with the slopes of the tangent plane $T_\theta(N)$. Hence, $\widetilde{\theta} = [f]^1_a$, while $L(N) = \Gamma^1_f$ is the graph of the 1-jet of the function f.

EXAMPLE 2.2. Consider an integral curve Q of the Cartan distribution on $J^1(2, 1)$ and its lifting to $J^2(2, 1)$. From Corollaries 2.2 and 2.3 it follows that the lifting of Q is not empty if and only if the curve Q is horizontal with respect to the projection $\pi_1 \colon J^1(\mathbb{R}^2) \to \mathbb{R}^2$. In this case, one can choose coordinates x, y on $\mathbb{R}^2$ such that the projection S of the curve Q in a neighborhood of some point is of the form $y = 0$ (see Figure 3.9). The curve itself in this case can be described by the relations of the form

$$\begin{cases} y = 0, \\ z = \alpha(x), \\ p = \beta(x), \\ q = \gamma(x). \end{cases} \tag{2.7}$$

Since this curve is an integral manifold of the Cartan distribution given by the equation $(dz - p\,dx - q\,dy)|_Q = 0$, one has $\beta = \alpha'$. Conditions (2.7) mean that we know the values of the function $z(x,y)$ and of its derivative $z_y(x,y)$ for $y = 0$. Thus we obtain the standard formulation for the Cauchy problem for second order equations. Hence, in this situation the problem of lifting is equivalent to describing the graphs of the jets for functions satisfying Cauchy conditions.

Let $\theta(x,y,z,p,q)$ be a point of the curve Q and $\xi = (X,Y,Z,P,R)$ a tangent vector. Let us describe the set $l(\xi)$ in the fiber F_θ. Denote by r,s,t coordinates in the fiber of the projection $J^2(\mathbb{R}^2) \to J^1(\mathbb{R}^2)$ (they correspond to the derivatives z_{xx}, z_{xy}, z_{yy} respectively). Then the equations

$$\begin{cases} Xr + Ys = P, \\ Xs + Yt = R \end{cases} \tag{2.8}$$

describe the line $l(\xi)$.

Note that the classical concept of a characteristic for second order equations is based on consideration of system (2.8) together with the initial equation $\mathcal{E} \subset J^2(\mathbb{R}^2)$.

Now note that for an integral manifold $N \subset J^{k-1}(\pi)$ nondegenerate with respect to the projection to the base (in what follows, such manifolds will be called *horizontal*) the corresponding manifold $L(N) \subset J^k(\pi)$ is also integral one. To see this, it suffices, applying Theorem 2.1, to check that at any point $\widetilde{\theta} \in L(N)$ the image of the tangent space $T_{\widetilde{\theta}}(L(N))$ under the projection $\pi_{k,k-1}$ lies in $L_{\widetilde{\theta}}$. By the definitions and by the construction of $L(N)$, we have $(\pi_{k,k-1})_{*,\widetilde{\theta}}(L(N)) = T_{\widetilde{\theta}}(L(N)) = T_\theta(N) \subset L_{\widetilde{\theta}}$. Thus $L(N)$ is an integral manifold. Therefore, we obtain a method to construct integral manifolds in $J^k(\pi)$ starting with the ones in $J^{k-1}(\pi)$.

As will be proved in Theorem 2.7, the above construction is of a universal nature. We shall now describe a local structure of horizontal integral manifolds N and deduce a formula to compute the dimension of $L(N)$ from the dimension of N.

PROPOSITION 2.4. *Any horizontal integral manifold N of the Cartan distribution on $J^k(\pi)$ locally lies on the graph Γ_f^k of the k-jet of some section f.*

PROOF. Let $r = \dim N \leq n$. Let us introduce, in a neighborhood of the point $a = \pi_k(\theta)$, $\theta \in N$, coordinates $x_1, \ldots, x_n$ such that $\widetilde{N} = \pi_k(N)$ is given by the equations $x_{r+1} = \cdots = x_n = 0$. Since the manifold N is horizontal, it is nondegenerate with respect to the projection $\pi_k \colon J^k(\pi) \to M$. Consequently, N is described by the equations

$$x_i = 0, \quad i > r, \qquad p_\sigma^j = \widetilde{f}_\sigma^j(x_1, \ldots, x_r), \quad j = 1, \ldots, m, \ |\sigma| \leq k.$$

Restricting the Cartan forms to N (cf. the proof of Theorem 1.1), we obtain

$$\widetilde{f}_{\sigma+1_i}^j = \frac{\partial \widetilde{f}_\sigma^j(x_1, \ldots, x_r)}{\partial x_i}, \qquad |\sigma| \leq k-1, \ i = 1, \ldots, r.$$

Thus, the functions $\tilde{f}^1 \ldots, \tilde{f}^m$ are defined on $\tilde{N}$. They satisfy the above conditions, and the values of all their normal derivatives are given along $\tilde{N}$. Hence, there exist functions $f^1, \ldots, f^m$ such that $f^j|_{\tilde{N}} = \tilde{f}^j$ and with the given values of normal derivatives on $\tilde{N}$. Consequently, $\boldsymbol{f}(x) = (f^1, \ldots, f^m)$ is the section we need. □

PROPOSITION 2.5. *Let $\pi \colon E \to M$ be an m-dimensional bundle over an n-dimensional manifold M, and $N \subset J^k(\pi)$ a horizontal integral manifold of the Cartan distribution of dimension $r \leq n$. Then*[2]

$$\dim L(N) = r + m\binom{n-r+k}{n-r-1}.$$

PROOF. Let us fix a point $\theta \in N$ and denote by P_θ the tangent space $T_\theta(N)$. We introduce a coordinate system $x_1, \ldots, x_n$ in a neighborhood of the point $\pi_k(\theta)$ such that $\pi_k(N)$ is described by the equations $x_{r+1} = \cdots = x_n = 0$. Then the vectors

$$\xi_i = \left(\frac{\partial}{\partial x_i} + \sum_{j=1}^{m} \sum_{|\sigma| \leq k} a^j_{\sigma,i} \frac{\partial}{\partial p^j_\sigma}\right)\bigg|_\theta, \qquad 1 \leq i \leq r, \tag{2.9}$$

form a basis of the plane P_θ, where $a^j_{\sigma,i} = \partial^{|\sigma|+1} f^j / \partial x_\sigma \partial x_i$, for the section $\boldsymbol{f} = (f^1, \ldots, f^m)$ constructed in Proposition 2.4.

The condition that a point $\widetilde{\theta}$ with the coordinates $\widetilde{p}^j_\tau$ and lying in the fiber over θ belongs to $l(P_\theta)$ can be written in the form

$$\widetilde{p}^j_{\sigma+1_i} = a^j_{\sigma,i}, \qquad 1 \leq i \leq r, \quad 1 \leq j \leq m.$$

This system of linear inhomogeneous equations on the unknowns $\widetilde{p}^j_\tau$ is compatible, i.e., for all i, l, j, μ the equalities $a^j_{\mu+1_i,l} = a^j_{\mu+1_l,i}$ are valid. Therefore, relations (2.9) make it possible to find unique coordinates $\widetilde{p}^j_\tau$ such that the multi-index τ, $|\tau| = k+1$, is not equal to zero in the first r components.

There are no relations for other coordinates; the number of these coordinates for any fixed $j = 1, \ldots, m$ equals the number of all indices τ, $|\tau| = k+1$, whose components $r+1, \ldots, n$, do not vanish. Hence, this number is $\binom{n-r+k}{n-r-1}$.

Thus, $\dim l(P_\theta) = m\binom{n-r+k}{n-r-1}$, and since the manifold $L(N)$ is fibered over N with the fiber $l(P_\theta)$ over the point $\theta \in N$, we have $\dim L(N) = r + m\binom{n-r+k}{n-r-1}$. □

COROLLARY 2.6. *If $r_1 = \dim N_1 < r_2 = \dim N_2$, then $\dim L(N_1) \geq \dim L(N_2)$, and equality is achieved in the following cases only:*
(a) $m = n = 1$.
(b) $k = 0$, $m = 1$.
(c) $m = 1$, $r_1 = n - 1$, $r_2 = n$.

All maximal integral manifolds of the Cartan distribution are described in the following theorem:

THEOREM 2.7. *An integral manifold Q of the Cartan distribution on $J^k(\pi)$ is maximal if and only if everywhere, except possibly for a manifold of lesser dimension, it is locally of the form $L(N)$ for some horizontal integral manifold $N \subset J^{k-1}(\pi)$.*

---

[2] For $\beta < 0$ we set $\binom{\alpha}{\beta} = 0$.

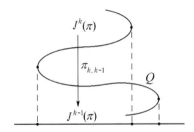

FIGURE 3.10. Regular and singular points of the projection $\pi_{k,k-1}$

PROOF. Let Q be a maximal integral manifold in $J^k(\pi)$. We shall prove the statement of the theorem for any point $\theta \in Q$ in whose neighborhood the rank of the mapping $\pi' = \pi_{k,k-1} \colon J^k(\pi) \to J^{k-1}(\pi)$ is constant (clearly, such points form an open everywhere dense set in Q; see Figure 3.10).

In what follows, we shall need a lemma which is a consequence of the implicit function theorem and is given here without proof.

LEMMA 2.8. *Let A, B be smooth manifolds, and let $f \colon A \to B$ be a smooth mapping of constant rank p in a neighborhood of a point $a \in A$. Then there exists a neighborhood $\mathcal{U} \ni a$ such that the set $f(\mathcal{U})$ is a submanifold in a neighborhood of $b = f(a)$. Moreover, its tangent space at b is the image of the tangent space of A at a: $T_b(f(\mathcal{U})) = f_*(T_a(A))$.*

Let us come back to the proof of the theorem. Let $\theta \in Q$ and $N = \pi'(U)$ be the projection of the corresponding neighborhood $U \subset Q$ to $J^{k-1}(\pi)$. By Lemma 2.8, N is a submanifold in $J^{k-1}(\pi)$. Let $\theta' = \pi'(\theta)$. By the same lemma and by Theorem 2.1, one has $T_{\theta'}(N) = T_{\theta'}(\pi'(Q)) = \pi'_*(T_\theta(Q)) \subset \pi'_*(C_\theta) = L_\theta$. It follows that N is a horizontal manifold and $\dim N \leq n$, i.e., we can apply Proposition 2.4. The embedding $L_\theta \supset T_{\theta'}(N)$, by definition, means that $\theta \in L(N)$. Hence, $U \subset L(N)$, i.e., $Q \subset L(N)$ locally. Therefore, by the maximality condition, $Q = L(N)$.

Conversely, consider a manifold of the form $L(N)$. As we have just proved, it is contained in the maximal integral manifold $L(N_1)$, and $\dim L(N) \leq \dim L(N_1)$. But $N = \pi'(L(N))$, $N_1 = \pi'(L(N_1))$. Therefore, $N \subset N_1$ and $\dim N \leq \dim N_1$. By Corollary 2.6, one has $\dim L(N_1) \leq \dim L(N)$, i.e., $L(N) = L(N_1)$. □

Note that if N is the graph of the $(k-1)$-jet of some section of the bundle π, then the manifold $L(N)$ is the graph of the k-jet of the same section. Hence, the graphs of jets are maximal integral manifolds of the Cartan distribution.

From Theorem 2.7 and Corollary 2.6 we obtain the following statement:

COROLLARY 2.9. *Except for the cases $m = n = 1$ and $k = m = 1$, the manifolds $L(N)$, where N is zero-dimensional, are of the maximal dimension among all integral manifolds of the Cartan distribution on $J^k(\pi)$. In other words, the fibers of the projection $\pi_{k,k-1}$ are integral manifolds of maximal dimension.*

This fact plays a fundamental role in the description of transformations preserving the Cartan distribution. The next section deals with these transformations.

3. Lie transformations

Informally speaking, Lie transformations are transformations of independent and dependent variables, together with derivatives of the latter with respect to the former, which preserve differential relations between them. From a geometric viewpoint, Lie transformations are diffeomorphisms of the jet manifolds $J^k(\pi)$ preserving the Cartan distribution, i.e., preserving the structure which contains information on differential relations between variables. Thus, Lie transformations are symmetries of the Cartan distribution. The symmetries on $J^1(M)$, i.e., contact transformations, were studied in Chapter 2. By analogy, Lie transformations of the manifold $J^k(\pi)$ are naturally called contact transformations of order k. We shall show below that in the case of one dependent variable such transformations reduce to the usual contact transformations, while in the case of several dependent variables they reduce to point transformations, i.e., to changes of independent and dependent variables.

3.1. Finite Lie transformations. Assume that we have n independent variables $x_1, \ldots, x_n$ and m dependent variables $u^1, \ldots, u^m$. Suppose that a change of variables is given by the formulas

$$\begin{cases} \overline{x}_i = f^i(\boldsymbol{x}, \boldsymbol{u}), \\ \overline{u}^j = g^j(\boldsymbol{x}, \boldsymbol{u}). \end{cases} \tag{3.1}$$

Then we can express partial derivatives $\overline{p}^j_\sigma = \partial^{|\sigma|}\overline{u}^j/\partial \overline{x}^\sigma$ via x_i, u^j, and $p^j_\tau = \partial^{|\tau|}u^j/\partial x^\tau$.

EXAMPLE 3.1. Consider the scale transformation

$$\overline{x}_i = \alpha_i x_i, \qquad \overline{u}^j = \beta_j u^j.$$

The derivatives $\overline{p}^j_i = \partial \overline{u}^j/\partial \overline{x}_i$ are transformed as

$$\overline{p}^j_i = \frac{\beta_j}{\alpha_i} p^j_i$$

in this case. One can also write down prolongations of this transformation to derivatives of arbitrary order:

$$\overline{p}^j_\sigma = \frac{\beta_j}{\alpha^\sigma} p^j_\sigma,$$

where $\alpha^{(i_1, \ldots, i_n)} = \alpha_1^{i_1} \ldots \alpha_n^{i_n}$

EXAMPLE 3.2. The translation along a constant vector in the space of dependent and independent variables is described by the formulas

$$\overline{x}_i = x_i + \xi_i, \qquad \overline{u}^j = u^j + \eta^j.$$

It acts as the identity for the variables p^j_σ,

$$\overline{p}^j_\sigma = p^j_\sigma.$$

EXAMPLE 3.3. The Galilean transformation (pass to a new frame of reference moving with a constant velocity) is given by the formulas

$$\overline{t} = t, \qquad \overline{x} = x - vt, \qquad \overline{u} = u.$$

Prolongation of this transformation to first-order derivatives is of the form

$$\overline{u}_t = u_t + v u_x, \qquad \overline{u}_x = u_x.$$

EXAMPLE 3.4. An arbitrary diffeomorphism of the plane (x, u) (i.e., change of one dependent and one independent variable)

$$\begin{cases} \bar{x} = \lambda(x, u), \\ \bar{u} = \mu(x, u) \end{cases} \quad (3.2)$$

induces a linear-fractional (in p) transformation of the three-dimensional space with coordinates x, u, p. It is easily checked that $\bar{p} = d\bar{u}/d\bar{x}$, and thus

$$\bar{p} = \frac{d\bar{u}}{d\bar{x}} = \frac{\mu_x dx + \mu_u du}{\lambda_x dx + \lambda_u du} = \frac{\mu_x + \mu_u p}{\lambda_x + \lambda_u p}. \quad (3.3)$$

EXERCISE 3.1. Let $m = n$, i.e., the number of independent variables coincides with the number of dependent ones. Consider the change of variables

$$\bar{x}_i = u^i, \quad \bar{u}^j = x_j, \quad i, j = 1, \ldots, n = m.$$

This transformation is called the *hodograph transformation*. Derive formulas for the action of this transformation on first derivatives.

Consider now a general geometric construction corresponding to changes of variables and to prolongation of these changes to derivatives over independent variables.

Let $\pi \colon E \to M$ be a fiber bundle. Then formulas (3.1) can be interpreted as coordinate representation of a diffeomorphism in the space E. Recall that such transformations are called *point transformations*.

For any $k \geq 1$, a diffeomorphism $A \colon E \to E$ generates the diffeomorphism $A^{(k)} \colon J^k(\pi) \to J^k(\pi)$ defined in the following way. Let θ be a point of the manifold $J^k(\pi)$, and let b and a be its projections to E and M respectively. Choose a section φ such that $\theta = [\varphi]_a^k$, and consider its graph $\Gamma_\varphi \subset E$. Under action of the transformation A, the point b is taken to some point $b' \in E$, while the graph Γ_φ is taken to the submanifold $A(\Gamma_\varphi) \subset E$. If, in a neighborhood of the point b', this submanifold is the graph of some section φ', we shall set $\theta' = [\varphi']_{a'}^k$, where $a' = \pi(b')$ and $A^{(k)}(\theta) = \theta'$.

EXERCISE 3.2. Show that the point θ' is independent of the choice of the section φ: it is uniquely determined by the k-jet of φ at a, i.e., by the point θ.

The transformation $A^{(k)}$ is called the *k-th lifting* of the point transformation A. In general, this lifting is not defined on the whole manifold $J^k(\pi)$, but (as will be seen below) on an open everywhere dense domain in $J^k(\pi)$. This geometric fact corresponds to the following analytic observation: when one transforms derivatives, the denominator may vanish, and these are exactly the points where the transformation is undefined (cf. equation (3.3)).

EXAMPLE 3.5. Consider the hodograph transformation of one dependent and one independent variable

$$\bar{x} = u, \quad \bar{u} = x.$$

Let $p = du/dx$, $\bar{p} = d\bar{u}/d\bar{x}$. Then, as follows from Exercise 3.1, $\bar{p} = 1/p$. Note that though the initial transformation A was an everywhere defined diffeomorphism of the space $\mathbb{R}^3 = J^1(\mathbb{R})$, the transformation $A^{(1)}$ is defined in the space $\mathbb{R}^3 = J^1(\mathbb{R})$ everywhere except for the plane $p = 0$.

From the above construction of the transformation $A^{(k)}$ it follows that it is a symmetry of the Cartan $\mathcal{C}^k$ distribution in its domain. In fact, the subspaces $\mathcal{C}^k_\theta$ are spans of R-planes $L_{\widetilde{\theta}}$, $\widetilde{\theta} \in J^{k+1}$, while the differential $A^{(k)}_*$ takes such planes to each other. In fact, if $A(\Gamma_\varphi) = \Gamma_{\varphi'}$, $\theta = [\varphi]^k_a$, $\widetilde{\theta} = [\varphi]^{k+1}_a$, $\widetilde{\theta}' = [\varphi']^{k+1}_{a'}$, then $A^{(k)}_{*,\theta}(L_{\widetilde{\theta}}) = L_{\widetilde{\theta}'}$.

DEFINITION 3.1. A diffeomorphism $F \colon J^k(\pi) \to J^k(\pi)$ is called a *Lie transformation* if $F_{*,\theta}(\mathcal{C}^k_\theta) = \mathcal{C}^k_{F(\theta)}$ for any point $\theta \in J^k(\pi)$.

Since locally the Cartan distribution is described by formulas (1.2), a diffeomorphism F is a Lie transformation if and only if the equalities

$$F^*(\omega^j_\sigma) = \sum_{l=1}^m \sum_{|\tau| \leq k-1} \lambda^{j,\tau}_{\sigma,l} \omega^l_\tau,$$

are locally valid for all $j = 1, \ldots, m$ and $|\sigma| \leq k-1$, where $\lambda^{j,\tau}_{\sigma,l}$ are smooth functions on $J^k(\pi)$.

EXAMPLE 3.6. Let us once again obtain the formulas for lifting the point trasnsformation to $J^1(\pi)$, using the above formulated property of Lie transformations. We shall look for the function $\bar{p}(x, u, p)$ using the condition that the image of the distribution $du - p\,dx = 0$ under the mapping (3.2) coincides with the distribution $d\bar{u} - \bar{p}\,d\bar{x} = 0$. This condition means that when one replaces all variables in the expression $d\bar{u} - \bar{p}\,d\bar{x}$ by the corresponding functions in x, u, p, one must obtain a form proportional to $du - p\,dx$. After computations, we obtain

$$d\bar{u} - \bar{p}\,d\bar{x} = \bar{u}_x\,dx + \bar{u}_u\,du - \bar{p}(\bar{x}_x\,dx + \bar{x}_u\,du)$$
$$= (\bar{u}_u - \bar{p}\bar{x}_u)\,du + (\bar{u}_x - \bar{p}\bar{x}_x)\,dx = \lambda(du - p\,dx),$$

which yields $\bar{u}_x - \bar{p}\bar{x}_x = -p(\bar{u}_u - \bar{p}\bar{x}_u)$ and consequently

$$\bar{p} = \frac{\bar{u}_x + \bar{u}_u p}{\bar{x}_x + \bar{x}_u p},$$

as it is to be by (3.3).

In a similar manner, we shall obtain the formulas for lifting an arbitrary point transformation (3.1) to $J^1(\pi)$. Namely, the functions[3] $\bar{p}^j_i$, $i = 1, \ldots, n$, $j = 1, \ldots, m$, are determined by m systems of linear equations (for $j = 1, \ldots, m$)

$$\begin{pmatrix} D_1(\bar{x}_1) & \ldots & D_1(\bar{x}_n) \\ \ldots\ldots\ldots\ldots\ldots\ldots\ldots \\ D_n(\bar{x}_1) & \ldots & D_n(\bar{x}_n) \end{pmatrix} \begin{pmatrix} \bar{p}^j_1 \\ \ldots \\ \bar{p}^j_n \end{pmatrix} = \begin{pmatrix} D_1(\bar{u}^j) \\ \ldots\ldots \\ D_n(\bar{u}^j) \end{pmatrix}, \qquad (3.4)$$

where D_i denotes the *total derivative operator* along x_i, which acts on the functions $f(x_1, \ldots, x_n, u^1, \ldots, u^m)$ by the formula

$$D_i(f) = \frac{\partial f}{\partial x_i} + \sum_{j=1}^m p^j_i \frac{\partial f}{\partial u^j}.$$

The determinant of the matrix on the left-hand side of (3.4) can reasonably be called the "total Jacobian" of the system of functions $\bar{x}_1, \ldots, \bar{x}_n$. Note that vanishing of this Jacobian on some open set implies functional dependence of the functions

---
[3] In the next formula, and in all similar cases, we shall write p_i instead of p_{1_i}.

$\overline{x}_i$. Since the transformation (3.1) is a local diffeomorphism, the functions $\overline{x}_i$ are independent and thus equations (3.4) uniquely determine the values of $\overline{p}_i^j$ almost everywhere.

EXERCISE 3.3. Let $A\colon J^0(\pi) \to J^0(\pi)$ be a point transformation. Prove that if the lifting $A^{(1)}$ is defined in a neighborhood of a point $\theta \in J^1(\pi)$, then the lifting $A^{(k)}$ is defined in a neighborhood of any point θ' projecting to θ under the mapping $\pi_{k,1}\colon J^k(\pi) \to J^1(\pi)$.

The lifting procedure can be applied not only to diffeomorphisms of the manifold of dependent and independent variables $E = J^0(\pi)$, but to Lie transformations of arbitrary jet manifolds $J^k(\pi)$. In fact, let a Lie transformation $A\colon J^k(\pi) \to J^k(\pi)$ be given. As we know, the manifold $J^{k+1}(\pi)$ of contact elements of order $k+1$ can be understood as the set of all horizontal n-dimensional integral planes (R-planes) in $J^k(\pi)$. The planes which are not overturned by the transformation A (i.e., the planes L such that $A_*(L)$ is horizontal again) form an open everywhere dense set in this manifold. The lifted transformation $A^{(1)}$ acts on the set of these planes, taking a plane L to $A_*(L)$.

EXAMPLE 3.7. Consider the Legendre transformation (see also §1 of Ch. 2) in the five-dimensional space $J^1(2,1)$

$$\begin{cases} \overline{x} = -p, \\ \overline{y} = -q, \\ \overline{u} = u - xp - yq, \\ \overline{p} = x, \\ \overline{q} = y. \end{cases} \tag{3.5}$$

As we showed in Chapter 2, this is a contact transformation and, consequently, a Lie transformation. To describe the lifting $A^{(1)}$ of this transformation to the 8-dimensional space $J^2(2,1)$, it is necessary to express the second derivatives

$$\overline{r} = \frac{\partial^2 \overline{u}}{\partial \overline{x}^2}, \quad \overline{s} = \frac{\partial^2 \overline{u}}{\partial \overline{x} \partial \overline{y}}, \quad \overline{t} = \frac{\partial^2 \overline{u}}{\partial \overline{y}^2}$$

via x, y, u, p, q, r, s, t. Since the Cartan distribution is invariant with respect to Lie transformations, the 1-forms

$$\begin{aligned} d\overline{p} - \overline{r}\, d\overline{x} - \overline{s}\, d\overline{y} &= dx + \overline{r}\, dp + \overline{s}\, dq, \\ d\overline{q} - \overline{s}\, d\overline{x} - \overline{t}\, d\overline{y} &= dy + \overline{s}\, dp + \overline{t}\, dq \end{aligned} \tag{3.6}$$

are to be represented as linear combinations of the 1-forms

$$du - p\, dx - q\, dy, \quad dp - r\, dx - s\, dy, \quad dq - s\, dx - t\, dy.$$

Therefore, (3.6) identically vanishes when one replaces the form dp by $dx + s\, dy$ and dq by $s\, dx + t\, dy$. Thus, we obtain the system of equations

$$\begin{pmatrix} r & s \\ s & t \end{pmatrix} \begin{pmatrix} \overline{r} & \overline{s} \\ \overline{s} & \overline{t} \end{pmatrix} = \begin{pmatrix} -1 & 0 \\ 0 & -1 \end{pmatrix},$$

which implies

$$\overline{r} = -\frac{t}{rt - s^2}, \quad \overline{s} = \frac{s}{rt - s^2}, \quad \overline{t} = -\frac{r}{rt - s^2}. \tag{3.7}$$

EXERCISE 3.4. Let $A\colon J^k(\pi) \to J^k(\pi)$ be a Lie transformation. Prove that at the points where the lifting $A^{(l+s)}\colon J^{k+l+s}(\pi) \to J^{k+l+s}(\pi)$, $l, s \in \mathbb{N}$, is defined one has $(A^{(l)})^{(s)} = A^{(l+s)}$.

We finish this subsection by a complete description of Lie transformations.

THEOREM 3.1. *Any Lie transformation X of the jet manifold $J^k(\pi)$, $k \geq 1$, is described as follows*:
1. *For $\dim \pi = 1$, the transformation X is the $(k-1)$-st lifting of some (arbitrary) contact transformation of the space $J^1(\pi)$.*
2. *For $m = \dim \pi > 1$, the transformation X is the k-th lifting of some (arbitrary) diffeomorphism of the space $J^0(\pi)$ of dependent and independent variables (i.e., all Lie transformations are point transformations in this case).*

REMARK 3.1. From the theorem it follows that in the case $m = 1$ contact geometry of order k is richer, since contact transformations of $J^1(\pi)$ include both arbitrary diffeomorphisms of $J^0(\pi)$ and transformations of a more general nature. The Legendre transformation (3.5) is an example of a transformation which is contact but not point.

PROOF. The proof of this theorem is based on the above established fact: the fibers of the projection $\pi_{k,k-1}$ are integral manifolds of the Cartan distribution on $J^k(\pi)$ of maximal dimension (Corollary 2.9). The case $m = n = 1$ needs a special treatment.

Assume that $mn > 1$. Let $A\colon J^k(\pi) \to J^k(\pi)$ be a Lie transformation and $k \geq 2$ if $m = 1$; $k \geq 1$ if $m > 1$. Since the transformation A preserves the Cartan distribution, it must take its maximal integral manifolds to maximal integral manifolds. Since A is a diffeomorphism, the dimension of these manifolds is also preserved. Consequently, A takes a fiber of the projection $\pi_{k,k-1}$ to another fiber. Hence, A induces a transformation A_1 of the manifold $J^{k-1}(\pi)$: if $A(\pi_{k,k-1}^{-1}(\theta)) = \pi_{k,k-1}^{-1}(\eta)$, then $A_1(\theta) = \eta$.

Since A_1 is the projection of the Lie transformation A, the transformation A_1 itself is also a Lie transformation. Let us consider A and the lifting $(A_1)^{(1)}$ of A_1. They both are symmetries of the Cartan distribution $\mathcal{C}^k$, preserve the fibers of the projection $\pi_{k,k-1}\colon J^k(\pi) \to J^{k-1}(\pi)$, and induce the same transformation of $J^{k-1}(\pi)$. Therefore, the diffeomorphism $A' = A^{(1)} \circ A^{-1}$ is a Lie transformation of $J^k(\pi)$ inducing the identity mapping of $J^{k-1}(\pi)$.

EXERCISE 3.5. Prove that Lie transformation of the space $J^k(\pi)$ projecting to the identity mapping of $J^{k-1}(\pi)$ is itself the identity.

Consequently $A = A_1^{(1)}$. One can apply the same reasoning to the transformation A_1 and continue in the same way till we obtain some transformation of the space $J^1(\pi)$ (in the case $m = 1$) or of $J^0(\pi)$ (for $m > 1$). To conclude the proof, it suffices to use Exercise 3.4.

In the case $m = n = 1$, the proof of the theorem is implied by the following considerations. Let x, p_0, p_1, ..., p_k be canonical coordinates on $J^k(1,1)$. The Cartan distribution $\mathcal{C}$ on $J^k(1,1)$ is two-dimensional: it is generated by the vector fields $Z = \partial/\partial p_k$ and $D = \partial/\partial x + \sum_{i=0}^{k-1} p_{i+1}\partial/\partial p_i$. It suffices to prove that any Lie transformation A takes the field Z to itself, i.e., is a symmetry of the one-dimensional distribution V spanned by the field Z; all further reasoning will be similar to the above.

LEMMA 3.2. *Let* $\mathcal{C}' = [\mathcal{C}, \mathcal{C}]$ *be the distribution on the manifold* $J^k(1,1)$, $k \geq 2$, *spanned by the commutators of the vector fields lying in the Cartan distribution* $\mathcal{C}$. *Then, if* $\mathrm{D}(\mathcal{P})$ *denotes the set of all vector fields lying in some distribution* $\mathcal{P}$, *and* $\mathcal{P}_\mathrm{D}$ *denotes the set of fields* X *such that* $[X, \mathrm{D}(\mathcal{P})] \subset \mathrm{D}(\mathcal{P})$, *one has*

$$\mathrm{D}(V) = \mathrm{D}(\mathcal{C}) \cap \mathcal{C}'_\mathrm{D}.$$

PROOF OF THE LEMMA. Obviously, the distribution $\mathcal{C}'$ is three-dimensional. It is spanned by the fields Z, D, and $[Z,D] = \partial/\partial p_{k-1} = Y$. It is easily seen that $[Z,Y] = 0$, $[Y,D] = X = \partial/\partial p_{k-2}$. Let the field $S = \alpha Z + \beta D$ be a symmetry of the distribution $\mathcal{C}'$. Then $[Y,S] \in \mathrm{D}(\mathcal{C}')$, i.e. $[Y,S]$ is represented as a linear combination of the fields Z, D, Y. But since

$$[Y,S] = [Y, \alpha Z + \beta D] = Y(\alpha) Z + Y(\beta) D + \beta X$$

and the fields X, D, Z are linear independent at all points of the manifold J^k, we obtain $\beta = 0$. Consequently, $S \in \mathrm{D}(V)$, and this completes the proof of the lemma. □

Now let $A \colon J^k(1,1) \to J^k(1,1)$. Then, obviously, $A_* \mathrm{D}(\mathcal{C}) = \mathrm{D}(\mathcal{C})$, $A_* \mathcal{C}' = \mathcal{C}'$, and $A_* \mathcal{C}'_\mathrm{D} = \mathcal{C}'_\mathrm{D}$. Then, by the lemma, one has $A_* \mathrm{D}(V) = \mathrm{D}(V)$. The theorem is proved. □

3.2. Lie fields.

DEFINITION 3.2. A vector field X on the manifold $J^k(\pi)$ is called a *Lie field*, if shifts along its trajectories are Lie transformations.

Let X be a Lie field on $J^k(\pi)$ and $\{A_t\}$ the corresponding one-parameter group of transformations of $J^k(\pi)$. By definition, A_t is a Lie transformation, and consequently its liftings $A_t^{(l)}$ to the manifolds $J^{k+l}(\pi)$ are defined and are also Lie transformations. The field $X^{(l)}$ corresponding to the one-parameter group $\{A_t^{(l)}\}$ is called the *lifting* of the field X. Similarly to the case of liftings of Lie transformations, for any natural numbers l and s the equality $(X^{(l)})^{(s)} = X^{(l+s)}$ holds.

THEOREM 3.3. *Any Lie field* $J^k(\pi)$, $k \geq 1$, *is of the form*:
  (a) $X^{(k)}$ *for* $\dim \pi > 1$, *where* X *is a vector field on the space* $J^0(\pi)$.
  (b) $X^{(k-1)}$ *for* $\dim \pi = 1$, *where* X *is a contact vector field on the space* $J^1(\pi)$.

Note that by the already discussed properties of Lie transformations, any vector field on $J^0(\pi)$ and any contact vector field on $J^1(\pi)$ can be lifted to a Lie field on $J^k(\pi)$.

The infinitesimal point of view has two important advantages. The first is that the lifting of a Lie field, unlike that of a finite Lie transformation, is always defined on the whole manifold. In fact, to compute a vector of the field $X^{(1)}$ at some point $\theta \in J^{k+1}(\pi)$, one needs to know an arbitrarily small part of the trajectory passing through this point. The R-plane L_θ in $J^k(\pi)$ corresponds to the point θ, and this plane projects to the base M nondegenerately. It remains to note that for sufficiently small transformations the image of this plane will still project to the base nondegenerately and thus define a point in $J^{k+1}(\pi)$.

The second advantage is that there exist explicit computational formulas expressing the components of the vector field $X^{(k)}$ via the components of X. In addition, it becomes possible to use the techniques of generating functions (see below).

Let us consider a bundle $\pi\colon E \to M$ and canonical coordinates in a neighborhood of a point $\theta \in J^k(\pi)$. Note that a field X on $J^k(\pi)$ is a Lie field if and only if for any Cartan form $\omega_\sigma^j = dp_\sigma^j - \sum_{i=1}^n p_{\sigma+1_i}^j\, dx_i$ on $J^k(\pi)$ one locally has

$$X(\omega_\sigma^j) = \sum_{l=1}^m \sum_{|\tau| \leq k-1} \lambda_{\sigma,l}^{j,\tau} \omega_\tau^l.$$

THEOREM 3.4. *If*

$$X = \sum_{i=1}^n a_i \frac{\partial}{\partial x_i} + \sum_{j=1}^m \sum_{0 \leq |\sigma| \leq k} b_\sigma^j \frac{\partial}{\partial p_\sigma^j}, \tag{3.8}$$

is a Lie field, then the coefficients b_σ^j are computed by the recurrent formulas

$$b_{\sigma+1_i}^j = D_i(b_\sigma^j) - \sum_{s=1}^n p_{\sigma+1_s}^j D_i(a_s), \tag{3.9}$$

where $0 \leq |\sigma| \leq k-1$ and

$$D_i = \frac{\partial}{\partial x_i} + \sum_\sigma p_{\sigma+1_i}^j \frac{\partial}{\partial p_\sigma^j}$$

is the total derivative operator with respect to x_i.

PROOF. Let X be a Lie field of the form (3.8). Then, by invariance of the Cartan distribution, the forms

$$X(\omega_\sigma^j) = db_\sigma^j - \sum_{i=1}^n (b_{\sigma+1_i}^j\, dx_i + p_{\sigma+1_i}^j\, da_i) \tag{3.10}$$

are linear combinations of the Cartan forms.

EXERCISE 3.6. Show that any 1-form on $J^k(\pi)$ can be uniquely represented as

$$\omega = \sum_{i=1}^n \varphi_i\, dx_i + \omega_\mathcal{C}, \tag{3.11}$$

where φ_i are functions on $J^{k+1}(\pi)$ while $\omega_\mathcal{C}$ is a linear combination of Cartan forms. In particular, for any smooth function φ on $J^k(\pi)$ the equality

$$d\varphi = \sum_i D_i(\varphi)\, dx_i + \sum_{j,\sigma} \frac{\partial \varphi}{\partial p_\sigma^j} \omega_\sigma^j \tag{3.12}$$

holds.

Note that the form ω represented by (3.11) vanishes on the Cartan distribution if and only if the coefficients φ_i are trivial. Applying this observation to (3.10), we obtain

$$X(\omega_\sigma^j) = \sum_i \left(D_i(b_\sigma^j) - b_{\sigma+1_i}^j - \sum_s p_{\sigma+1_s}^j D_i(a_s) \right) dx_i + \omega_\mathcal{C},$$

which implies (3.9). □

EXERCISE 3.7. Check that equalities (3.9) comply with the representation of contact fields on $J^1(M)$ in the form

$$X_f = -\sum_i \frac{\partial f}{\partial p_i}\frac{\partial}{\partial x_i} + \left(f - \sum_i p_i \frac{\partial f}{\partial p_i}\right)\frac{\partial}{\partial u} + \sum_i \left(\frac{\partial f}{\partial x_i} + p_i \frac{\partial f}{\partial u}\right)\frac{\partial}{\partial p_i}.$$

Let us now introduce the concept of the *generating section* of a Lie field. To do this, consider evolution of the graphs of sections of the bundle π under the action of the corresponding one-parameter group of transformations.

Let $X = \sum_i a_i \partial/\partial x_i + \sum_j b^j \partial/\partial u^j + \cdots$ be a Lie field on $J^k(\pi)$, and $\{A_t\}$ the corresponding one-parameter group of local transformations of the manifold $J^k(\pi)$. Consider a section $\boldsymbol{f}$ of the bundle π. The graph of its k-jet $\Gamma^k_{\boldsymbol{f}} \subset J^k(\pi)$ is an n-dimensional integral manifold of the Cartan distribution on $J^k(\pi)$ nondegenerately projecting to the base M. As already pointed out, for sufficiently small t the manifold $A_t(\Gamma^k_{\boldsymbol{f}})$ is also the graph of the k-jet for some other section $\boldsymbol{f}_t$: $A_t(\Gamma^k_{\boldsymbol{f}}) = \Gamma^k_{\boldsymbol{f}_t}$.

Thus the transformations A_t generate an evolution $\boldsymbol{f}_t$ of the section $\boldsymbol{f}$. Let us find the velocity $d/dt|_{t=0}\boldsymbol{f}_t$ of this evolution at the starting moment. To compute this derivative, let us represent the field X as the sum of two fields, one of them being vertical and the other one tangent to the manifold $\Gamma^k_{\boldsymbol{f}}$.

If the section $\boldsymbol{f}$ is locally presented by the functions $u^j = f^j(x_1,\ldots,x_n)$, $j = 1,\ldots,m$, then the vector fields tangent to the manifold $\Gamma^k_{\boldsymbol{f}}$ can be written in the following form (note that on the surface $\Gamma^k_{\boldsymbol{f}}$ the equalities $p^j_\sigma = \partial^{|\sigma|} f^j/\partial x^\sigma$, $|\sigma| \le k$, hold):

$$D_i^{(\boldsymbol{f})} = \frac{\partial}{\partial x_i} + \sum_{j=1}^m \left(\sum_{|\sigma|<k} p^j_{\sigma+1_i}\frac{\partial}{\partial p^j_\sigma} + \sum_{|\sigma|=k} \frac{\partial^{|\sigma|+1} f^j}{\partial x^\sigma \partial x_i}\frac{\partial}{\partial p^j_\sigma}\right).$$

Therefore, a field tangent to the graph $\Gamma^k_{\boldsymbol{f}}$ is of the form

$$X_2 = \sum_{i=1}^n a_i D_i^{(\boldsymbol{f})},$$

while the vertical component of the Lie field X equals

$$X_1 = X - \sum_{i=1}^n a_i D_i^{(\boldsymbol{f})} = \sum_j \left(b^j - \sum_i a_i p^j_i\right)\frac{\partial}{\partial u^j} + \cdots$$

Since the field X_2 shifts the manifold $\Gamma^k_{\boldsymbol{f}}$ along itself, it does not influence the evolution. On the other hand, the coefficients of the field X_1 at $\partial/\partial u^j$ exactly coincide with the velocities of the components $\boldsymbol{f}_t = (f^1_t,\ldots,f^m_t)$:

$$\left.\frac{d}{dt}\right|_{t=0} f^j_t = \left(b^j - \sum_{i=1}^n a_i p^j_i\right)\bigg|_{\Gamma^k_{\boldsymbol{f}}}. \tag{3.13}$$

The vector function $\varphi = (\varphi^1,\ldots,\varphi^m)$, where

$$\varphi^j = b^j - \sum_{i=1}^n a_i p^j_i, \tag{3.14}$$

is called the *generating section* (or, in the case $m = 1$, the *generating function*) of the Lie field

$$X = \sum_{i=1}^{n} a_i \frac{\partial}{\partial x_i} + \sum_{j=1}^{m} b^j \frac{\partial}{\partial u^j} + \dots . \tag{3.15}$$

EXERCISE 3.8. Show that in the case of the trivial bundle $\pi\colon \mathbb{R} \times M \to M$, the above definition of generating function coincides with that for a contact field given in Chapter 2.

From this exercise and from Theorem 3.3 we obtain the following

PROPOSITION 3.5. *Any Lie field X on $J^k(\pi)$ is uniquely determined by its generating section $\varphi = (\varphi^1, \dots, \varphi^m)$. The components of the generating section can be computed using the formulas*

$$\varphi^j = X \lrcorner \omega^j_{(0,\dots,0)},$$

where $\omega^j_{(0,\dots,0)} = du^j - \sum_{i=1}^{n} p_i^j \, dx_i$.

Note that in the case $m > 1$ the generating section $\varphi = (\varphi^1, \dots, \varphi^m)$ linearly depends on the variables p_i^j, and for any i, j, l the equalities

$$\frac{\partial \varphi^j}{\partial p_i^j} = \frac{\partial \varphi^l}{\partial p_i^l}, \qquad \frac{\partial \varphi^j}{\partial p_i^l} = 0, \qquad l \neq j,$$

hold.

Arbitrary generating sections lead to the theory of higher symmetries, which will be considered in Chapter 4.

4. Classical symmetries of equations

4.1. Defining equations. Let $\mathcal{E} \subset J^k(\pi)$ be an equation of order k.

DEFINITION 4.1. A Lie transformation $A\colon J^k(\pi) \to J^k(\pi)$ such that $A(\mathcal{E}) \subset \mathcal{E}$ is called a *classical (finite) symmetry* of the equation $\mathcal{E} \subset J^k(\pi)$.

DEFINITION 4.2. A Lie field is called a *classical infinitesimal symmetry* of the equation $\mathcal{E} \subset J^k(\pi)$, if it is tangent to $\mathcal{E}$.

Obviously, these definitions are natural generalizations of the basic constructions considered in Chapters 1 and 2.

A direct consequence of the definitions and considerations of §§1–3 is the following[4]

PROPOSITION 4.1. 1. *Let $A\colon J^k(\pi) \to J^k(\pi)$ be a symmetry of the equation $\mathcal{E} \subset J^k(\pi)$, and let f be a solution of this equation, i.e., a section of the bundle π such that its graph Γ_f^k lies in $\mathcal{E}$. Then $A(\Gamma_f^k)$ is a generalized solution of the equation $\mathcal{E}$. In particular, if the manifold $A(\Gamma_f^k)$ is of the form $\Gamma_{f'}^k$ for some section $f' = A^* f$, then f' is a solution of $\mathcal{E}$ as well.*

2. If X is an infinitesimal symmetry of the equation $\mathcal{E}$ and f is its solution, then for any point $\theta \in \Gamma_f^k$ there exist a neighborhood $\mathcal{U} \ni \theta$ and $\varepsilon > 0$ such that for any $t \in [-\varepsilon, \varepsilon]$ the manifold $A_t(\mathcal{U})$, where A_t is the one-parameter group of

---

[4] Everywhere below in this chapter we use the term "symmetry" in the sense of a classical symmetry.

transformations corresponding to X, is locally of the form $\Gamma^k_{A^*_t(f)}$, i.e., X determines a flow on the set of solutions of the equation $\mathcal{E}$.

From a practical viewpoint, finite symmetries are by all means preferable, but it is not simple to find them. There are no general algorithms to search for them, and they can be found only accidentally or by using a sort of "physical reasons". On the contrary, a search for infinitesimal symmetries is governed by a certain algorithm leading, from a technical point of view, to solving (usually, overdetermined) systems of linear differential equations. The reason is that Lie fields are efficiently described by the corresponding generating functions. That is why we work below with infinitesimal symmetries only and often use the word "symmetry" to mean the generating function of an infinitesimal symmetry. The latter identification does not cause ambiguities, since the correspondence between Lie fields and their generating functions is one-to-one.

Let φ and ψ be symmetries of the equation $\mathcal{E} \subset J^k(\pi)$, i.e., the Lie fields $X^{(s)}_\varphi$ and $X^{(s)}_\psi$ are tangent to the manifold $\mathcal{E}$ ($s = k$ for $m > 1$ and $s = k-1$ for $m = 1$). Then obviously the commutator $[X^{(s)}_\varphi, X^{(s)}_\psi]$ is also a symmetry of the equation $\mathcal{E}$ and consequently is of the form $X^{(s)}_\rho$ for some generating section ρ. This section is called the *Jacobi bracket* of the generating sections φ and ψ, and is denoted by $\{\varphi, \psi\}$.

Evidently, the set of symmetries forms a Lie $\mathbb{R}$-algebra with respect to the Jacobi bracket.

If $x_1, \ldots, x_n, u^1, \ldots, u^m, \ldots, p^j_\sigma, \ldots$ are canonical coordinates in $J^k(\pi)$, then from the definition of the commutator of vector fields and from equations (3.9) it follows that the j-component of the Jacobi bracket is of the form

$$\{\varphi, \psi\}^j = \sum_{\alpha=1}^m \left(\varphi^\alpha \frac{\partial \psi^j}{\partial u^\alpha} - \psi^\alpha \frac{\partial \varphi^j}{\partial u^\alpha} \right)$$
$$+ \sum_{i=1}^n \sum_{\alpha=1}^m \left(\left(\frac{\partial \varphi^\alpha}{\partial x_i} + \sum_{\beta=1}^m p^\beta_i \frac{\partial \varphi^\alpha}{\partial u^\beta} \right) \frac{\partial \psi^j}{\partial p^\alpha_i} - \left(\frac{\partial \psi^\alpha}{\partial x_i} + \sum_{\beta=1}^m p^\beta_i \frac{\partial \psi^\alpha}{\partial u^\beta} \right) \frac{\partial \varphi^j}{\partial p^\alpha_i} \right). \quad (4.1)$$

Now let us deduce local conditions for a Lie field X to be a symmetry of the equation $\mathcal{E} \subset J^k(\pi)$. To do this, let us choose canonical coordinates in $J^k(\pi)$ and assume that the submanifold $\mathcal{E}$ in these coordinates is described by the relations

$$F^\alpha = 0, \quad \alpha = 1, \ldots, r, \quad (4.2)$$

where F^α are smooth functions on $J^k(\pi)$. Suppose that the system (4.2) is locally of maximal rank. Then the condition that a field X is tangent to the manifold $\mathcal{E} = \{\boldsymbol{F} = 0\}$, where $\boldsymbol{F} = (F^1, \ldots, F^r)$, is of the form

$$X(F^\alpha) = \sum_{\beta=1}^r \lambda^\alpha_\beta F^\beta, \quad \alpha = 1, \ldots, r, \quad (4.3)$$

for some smooth functions λ^α_β, or equivalently

$$X(F^\alpha)|_\mathcal{E} = 0, \quad \alpha = 1, \ldots, r. \quad (4.4)$$

Let us represent the field X in the form $X^{(s)}_\varphi$, where $s = k$ or $s = k-1$, depending on the dimension of the bundle π, and express the coefficients of the lifting of the field X_φ via generating functions using equations (3.9). Then from (4.3) or (4.4) we shall obtain a system of equations on φ called *defining equations*.

EXAMPLE 4.1. Consider in $J^2(2,1)$ a general second order equation in two independent and one dependent variables:

$$F(x_1, x_2, u, p_1, p_2, p_{(2,0)}, p_{(1,1)}, p_{(0,2)}) = 0.$$

Let

$$X_\varphi = -\sum_{i=1}^n \varphi_{p_i} \frac{\partial}{\partial x_i} + \left(\varphi - \sum_{i=1}^n p_i \varphi_{p_i}\right) \frac{\partial}{\partial u} + \sum_{i=1}^n (\varphi_{x_i} + p_i \varphi_u) \frac{\partial}{\partial p_i} \quad (4.5)$$

be a contact vector field in $J^1(2,1)$. Then the coefficients $b_{(i,j)}$ at $\partial/\partial p_{(i,j)}$, $i+j=2$, of the lifting of the field to $J^2(2,1)$ are of the following form:

$$b_{(2,0)} = p_{(2,0)} \varphi_u + \varphi_{x_1 x_1} + 2 p_1 \varphi_{x_1 u} + 2 p_{(2,0)} \varphi_{x_1 p_1} + 2 p_{(1,1)} \varphi_{x_1 p_2}$$
$$+ p_1^2 \varphi_{uu} + 2 p_1 p_{(2,0)} \varphi_{u p_1} + 2 p_1 p_{(1,1)} \varphi_{u p_2} + p_{(2,0)}^2 \varphi_{p_1 p_1}$$
$$+ 2 p_{(2,0)} p_{(1,1)} \varphi_{p_1 p_2} + p_{(1,1)}^2 \varphi_{p_2 p_2},$$

$$b_{(1,1)} = p_{(1,1)} \varphi_u + \varphi_{x_1 x_2} + p_2 \varphi_{x_1 u} + p_1 \varphi_{x_2 u} + p_{(1,1)} \varphi_{x_1 p_1} + p_{(0,2)} \varphi_{x_1 p_2}$$
$$+ p_{(2,0)} \varphi_{x_2 p_1} + p_{(1,1)} \varphi_{x_2 p_2} + p_1 p_2 \varphi_{uu} + (p_1 p_{(1,1)} + p_2 p_{(2,0)}) \varphi_{u p_1}$$
$$+ (p_1 p_{(0,2)} + p_2 p_{(1,1)}) \varphi_{u p_2} + p_{(2,0)} p_{(1,1)} \varphi_{p_1 p_1}$$
$$+ (p_{(2,0)} p_{(0,2)} + p_{(1,1)}^2) \varphi_{p_1 p_2} + p_{(1,1)} p_{(0,2)} \varphi_{p_2 p_2},$$

$$b_{(0,2)} = p_{(0,2)} \varphi_u + \varphi_{x_2 x_2} + 2 p_2 \varphi_{x_2 u} + 2 p_{(1,1)} \varphi_{x_2 p_1} + 2 p_{(0,2)} \varphi_{x_2 p_2}$$
$$+ p_2^2 \varphi_{uu} + 2 p_2 p_{(1,1)} \varphi_{u p_1} + 2 p_2 p_{(0,2)} \varphi_{u p_2}$$
$$+ p_{(1,1)}^2 \varphi_{p_1 p_1} + 2 p_{(1,1)} p_{(0,2)} \varphi_{p_1 p_2} + p_{(0,2)}^2 \varphi_{p_2 p_2}.$$

Then the equation $X_\varphi^{(1)} F = \lambda F$, where the coefficients of the field $X_\varphi^{(1)}$ are computed by the above formulas, is the defining equation in the situation under consideration.

EXERCISE 4.1. Deduce similar formulas for a system of two second order equations in two dependent variables.

4.2. Invariant solutions and reproduction of solutions. Let X be an infinitesimal symmetry of the equation $\mathcal{E} \subset J^k(\pi)$, $\boldsymbol{f}_0$ its solution, and $\Gamma^k_{\boldsymbol{f}_0} = j_k(\boldsymbol{f}_0)(M) \subset \mathcal{E}$. Let $\{A_t\}$ be a one-parameter group of diffeomorphisms corresponding to the field X. Then the submanifold $A_t(\Gamma^k_{\boldsymbol{f}_0}) \subset J^k(\pi)$ is an integral manifold of the Cartan distribution. In the case when $A_t(\Gamma^k_{\boldsymbol{f}_0})$ and $\Gamma^k_{\boldsymbol{f}_0}$ are horizontal with respect to the projection to the base M, then $A_t(\Gamma^k_{\boldsymbol{f}_0}) = \Gamma^k_{\boldsymbol{f}_t}$ for some section $\boldsymbol{f}_t$. In addition, by the definition of a symmetry, $A_t(\Gamma^k_{\boldsymbol{f}_0}) \subset \mathcal{E}$. Thus, at least locally, $\boldsymbol{f}_t$ is a one-parameter family of solutions of $\mathcal{E}$: $j_k(\boldsymbol{f}_t)(M) = \Gamma^k_{\boldsymbol{f}_t} \subset \mathcal{E}$. A passage from the initial solution $\boldsymbol{f}$ to the family $\boldsymbol{f}_t$ is called *reproduction* of the solution $\boldsymbol{f}$ by means of the symmetry X.

If φ is the generating section of the field $X = X_\varphi$, then searching for the family $\boldsymbol{f}_t$ reduces to solving the system of equations

$$\frac{\partial \boldsymbol{f}}{\partial t}(\boldsymbol{x}, t) = \varphi|_{\Gamma^k_{\boldsymbol{f}}}(\boldsymbol{x}, t), \qquad \boldsymbol{f}(\boldsymbol{x}, 0) = \boldsymbol{f}_0(\boldsymbol{x}), \quad (4.6)$$

or, component-wise,

$$\frac{\partial f^j}{\partial t}(\boldsymbol{x},t) = \varphi^j|_{\Gamma^k_{\boldsymbol{f}}}(\boldsymbol{x},t), \quad f^j(\boldsymbol{x},0) = f_0^j(\boldsymbol{x}), \qquad j = 1, \ldots, m,$$

as follows from (3.13).

If $\varphi|_{\Gamma^k_{\boldsymbol{f}}}(\boldsymbol{x},t) = 0$, then $\boldsymbol{f}(\boldsymbol{x},t) = \boldsymbol{f}(\boldsymbol{x},0)$ for all admissible values of t, i.e., $\boldsymbol{f}_0$ is a fixed point for the one-parameter group of transformations $\{A_t\}$ when it acts on sections. In other words, the manifold $\Gamma^k_{\boldsymbol{f}_0}$ is invariant under the field X_φ.

DEFINITION 4.3. If $\Gamma^k_{\boldsymbol{f}_0}$ is invariant with respect to the field X_φ, then $\boldsymbol{f}_0$ is called a φ-*invariant solution* of the equation $\mathcal{E}$.

The concepts introduced above can be generalized as follows. Let G be a Lie group whose Lie algebra $\mathfrak{g}$ is realized as a subalgebra in the Lie algebra of infinitesimal symmetries (or, equivalently, in the Lie algebra of generating functions of symmetries) of the equation $\mathcal{E}$. Then, starting from an arbitrary solution $\boldsymbol{f}_0$, we shall obtain a $(\dim \mathfrak{g})$-parameter family of solutions $\{\,\boldsymbol{f}_g \mid g(\Gamma^k_{\boldsymbol{f}_0}) = \Gamma^k_{\boldsymbol{f}_g}, g \in G\,\}$ of the equation $\mathcal{E}$ (provided the manifolds $g(\Gamma^k_{\boldsymbol{f}_0})$ are horizontal with respect to the projection to the base M).

DEFINITION 4.4. If $\boldsymbol{f}_g = \boldsymbol{f}_0$ for all $g \in G$, then $\boldsymbol{f}_0$ is called a G-*invariant solution* of the equation $\mathcal{E}$ (or $\mathfrak{g}$-*invariant*, if we consider the corresponding Lie algebra).

Let $\varphi_1, \ldots, \varphi_s$ be generating sections and assume that they are generators of the Lie algebra $\mathfrak{g}$. Then to find $\mathfrak{g}$-invariant solutions means to solve the overdetermined system of differential equations

$$\begin{cases} F_1(\theta) = 0, \\ \ldots\ldots\ldots \\ F_r(\theta) = 0, \\ \varphi_1(\theta) = 0, \\ \ldots\ldots\ldots \\ \varphi_s(\theta) = 0, \end{cases} \qquad (4.7)$$

$\theta \in J^k(\pi)$, where $F_1, \ldots, F_r$ are the functions determining the equation $\mathcal{E}$. In particular, X_φ-invariant solutions can be found from the system (4.7), where the only equation in addition to $\mathcal{E}$ is $\varphi = 0$.

Let us stress one important point. If an equation possesses a classical symmetry X, one can diminish the number of independent variables by 1. In fact, consider, for simplicity, the case $\dim \pi = 1$. Then the equations $\varphi_1 = 0, \ldots, \varphi_s = 0$ can be considered as a system of equations in the unknowns $p_1, \ldots, p_n$. It is natural to consider the case when this system is of maximal rank. Then, without loss of generality, one may assume that it is of the form

$$\begin{cases} p_{n-s+1} = \widetilde{\varphi}_1(\boldsymbol{x}, u, p_1, \ldots, p_{n-s}), \\ \ldots\ldots\ldots\ldots\ldots\ldots\ldots\ldots\ldots\ldots\ldots \\ p_n = \widetilde{\varphi}_s(\boldsymbol{x}, u, p_1, \ldots, p_{n-s}). \end{cases}$$

Substituting these equalities into the initial equation, we obtain an equation that does not contain the derivatives with respect to the last s variables. Obviously, in the case of several dependent variables the number of dependent ones is diminished in the same manner.

When an equation possesses two symmetries, the number of independent variables diminishes by 2, etc. In particular, if $\dim \mathfrak{g} = n - 1$, where n is the number of independent variables, invariant solutions can be found by solving ordinary differential equations, and for $\dim \mathfrak{g} = n$ the equations become "algebraic", i.e. do not contain derivatives. Some examples will be discussed below.

We finish this section with some remarks.

1. In general, even for small t, one cannot expect the manifolds $A_t(\Gamma^k_{f_0})$ to be horizontal. Therefore, when reproducing a regular solution, we shall obtain general integral $\dim M$-dimensional manifolds of the Cartan distribution, i.e., "generalized" solutions, or solutions with "singularities". At some points, a tangent plane to such a solution degenerately projects to the base (and this corresponds to the fact that some derivatives of $\boldsymbol{f}_t(\boldsymbol{x})$ become infinite); therefore a solution may become multivalued. Such generalized solutions are usual in the analysis of discontinuity propagation, shock waves, in catastrophe theory, etc. The physical meaning of these solutions is determined by the appropriate context.

2. For many nonlinear equations it is often convenient to find an invariant solution with respect to some symmetry and then to reproduce it by means of other symmetries. Sometimes this is the only way to find explicit formulas.

3. The concept of invariant solutions includes, as a particular case, the notion of a *self-similar solution*: a solution is called self-similar if it is invariant with respect to the scale symmetries, i.e., symmetries of the form

$$X = \sum_i \alpha_i x_i \frac{\partial}{\partial x_i} + \sum_j \beta_j u^j \frac{\partial}{\partial u^j}.$$

5. Examples of computations

Many examples of symmetry computations, of constructing invariant solutions, and of other applications can be found in [**44, 45, 46, 91, 114, 137**].

5.1. The Burgers equation. The Burgers equation is of the form

$$u_t = u u_x + u_{xx}. \tag{5.1}$$

This equation describes the motion of weakly nonlinear waves in gases, when dissipative effects are sufficiently small to be considered in the first approximation only. When dissipation tends to zero, this equation gives an adequate description of waves in a nonviscous medium. Initially proposed by Burgers to describe one-dimensional turbulence, this equation was later used to study other wave phenomena. The Burgers equation is linearized by the substitution $u = y_x/y$. This fact indicates existence of a large symmetry algebra.

Rewriting equation (5.1) in canonical coordinates

$$x_1 = x, \ x_2 = t, \ p_0 = u, \ p_1 = u_x, \ p_2 = u_t, \ p_{(2,0)} = u_{xx}$$

on $J^2(2,1)$, we obtain

$$p_2 = p_0 p_1 + p_{(2,0)}.$$

Then the defining equations for symmetries obtained in Example 4.1 acquire the form

$$(\varphi - p_1\varphi_{p_1} - p_2\varphi_{p_2})p_1 + p_0(\varphi_{x_1} + p_1\varphi_{p_0}) + p_{(2,0)}\varphi_{p_0} + \varphi_{x_1x_1} + 2p_1\varphi_{x_1p_0} + 2p_{(2,0)}\varphi_{x_1p_1}$$
$$+ 2p_{(1,1)}\varphi_{x_1p_2} + p_1^2\varphi_{p_0p_0} + 2p_1p_{(2,0)}\varphi_{p_0p_1} + 2p_1p_{(1,1)}\varphi_{p_0p_2} + p_{(2,0)}^2\varphi_{p_1p_1}$$
$$+ 2p_{(2,0)}p_{(1,1)}\varphi_{p_1p_2} + p_{(1,1)}^2\varphi_{p_2p_2} - \varphi_{x_2} - p_2\varphi_{p_0} = \lambda(p_0p_1 + p_{(2,0)} - p_2), \quad (5.2)$$

where λ is a smooth function on $J^2(2,1)$. Analysis of this equation shows that the function φ is of the form

$$\varphi = A(x_1, x_2, p_0)p_1 + B(x_1, x_2, p_0)p_2 + C(x_1, x_2, p_0).$$

From here, using (5.2), we see that the functions A, B, and C satisfy the equation

$$p_1p_2A_{p_0} + p_1A_{x_2} + B_{x_2}p_2 + C_{x_2} + C_{p_0}p_2 - p_0p_1(A_{x_1} + p_1A_{p_0}) - p_0(C_{x_1} + p_1C_{p_0})$$
$$- p_1C - 2D_x(A)p_{(2,0)} - D_x(A_{x_1} + A_{p_0})p_1 - D_x(C_{x_1} + C_{p_0}) = 0, \quad (5.3)$$

where D_x is the total derivative over $x = x_1$.

The last equation is polynomial in $p_{(2,0)}$. Therefore, the coefficient at $p_{(2,0)}$, i.e., $4A_{p_0}p_1 + B_{x_2} - 2A_{x_1}$, has to vanish. In turn, this coefficient is polynomial in p_1; hence, the coefficients of this polynomial are also trivial[5].

Thus, $A_{p_0} = 0$ and $B_{x_2} - 2A_{x_1} = 0$. Substituting these equalities into (5.3), we obtain

$$p_1A_{x_2} + p_0p_1A_{x_1} + C_{x_2} - p_0C_{x_1} - p_1C - p_1A_{x_1x_1}$$
$$- C_{x_1x_1} - 2p_1C_{x_1p_0} - p_1^2C_{p_0p_0} = 0 \quad (5.4)$$

The last equality is quadratic in p_1, and so

$$C_{p_0p_0} = 0,$$
$$A_{x_2} + p_0A_{x_1} - C - A_{x_1x_1} - 2C_{x_1p_0} = 0, \quad (5.5)$$
$$C_{x_2} - p_0C_{x_1} - C_{x_1x_1} = 0.$$

Differentiating the second equation of (5.5) with respect to p_0, we obtain[6]

$$A_{x_1} = C_{p_0}. \quad (5.6)$$

From the first equation in (5.5) it follows that C is linear in p_0:

$$C = r(x_1, x_2)p_0 + s(x_1, x_2).$$

Substituting this into the last equation of (5.5), we obtain

$$r_{x_2}p_0 + s_{x_2} - p_0^2 r_{x_1} - p_0 s_{x_1} - p_0 r_{x_1x_1} - s_{x_1x_1} = 0.$$

Equating the coefficients at the powers of p_0 to zero, we obtain the relations

$$r_{x_1} = 0, \quad r_{x_2} - s_{x_1} = 0, \quad s_{x_2} - s_{x_1x_1} = 0. \quad (5.7)$$

The first of these means that $C_{p_0x_1} = 0$. Taking (5.6) into account, we get

$$A_{x_1x_1} = 0.$$

---

[5]This recursive way of reasoning, reducing the order of the higher derivatives, is typical for the first stage of solving defining equations.

[6]In general, at the second step, it is typical to obtain differential consequences and to use compatibility conditions.

Differentiating the second equation in (5.7) with respect to x_1 and using the first equation, we have $s_{x_1x_1} = 0$, i.e.,
$$s = w(x_2)x_1 + v(x_2).$$
Taking the third equation of (5.7) into account, we get $s_{x_2} = 0$, i.e.,
$$s = wx_1 + v, \quad w, v \in \mathbb{R}. \tag{5.8}$$
Taking the derivative of the second equation (5.7) with respect to x_2 and using (5.8), we obtain $r_{x_2x_2} = 0$, and consequently
$$r = mx_2 + n, \quad m, n \in \mathbb{R}$$
(recall that r is independent of x_1 by the first equation of (5.7)). Hence, from the second equation of (5.7) we obtain $m = w$.

As a result, we get
$$A = (wx_2 + v)x_1 + nx_2 + k, \quad B = wx_2^2 + 2wx_2 + l, \quad C = (wx_2 + v)p_0 + vx_1 + n,$$
where $w, v, n, k, l \in \mathbb{R}$. It follows that the functions
$$x_1x_2p_1 + x_2^2p_2 + x_2p_0 + x_1, \quad x_1p_1 + 2x_2p_2 + p_0, \quad x_2p_1 + 1, \quad p_1, \quad p_2 \tag{5.9}$$
form an $\mathbb{R}$-basis in the space of symmetries of the Burgers equation.

Coming back to "physical" notation, these functions can be represented in the form
$$xtu_x + t^2u_t + tu + x, \quad xu_x + 2tu_t + u, \quad tu_x + 1, \quad u_x, \quad u_t. \tag{5.10}$$

Let us finally write down the Lie fields on $J^0(2,1)$ corresponding to the above listed generating functions φ:
$$(tu + x)\frac{\partial}{\partial u} - tx\frac{\partial}{\partial x} - t^2\frac{\partial}{\partial t},$$
$$u\frac{\partial}{\partial u} - x\frac{\partial}{\partial x} - 2t\frac{\partial}{\partial t} \quad \text{(scale symmetry)},$$
$$\frac{\partial}{\partial u} - t\frac{\partial}{\partial x}, \quad \text{(Galilean symmetry)}, \tag{5.11}$$
$$\frac{\partial}{\partial x} \quad \text{(translation along } x\text{)},$$
$$\frac{\partial}{\partial t} \quad \text{(translation along } t\text{)}.$$

5.2. The Korteweg–de Vries equation. Another well-known equation widely used in studying nonlinear phenomena is the Korteweg–de Vries equation,
$$u_t - 6uu_x + u_{xxx} = 0.$$

The summands uu_x and u_{xxx} correspond to dissipative and dispersive phenomena in nonlinear wave processes. This equation was suggested by Korteweg and de Vries in 1895 to describe waves of small (but finite) amplitude in long time periods. Other applications of this famous equation include description of rotating fluid flow in a pipe, the theory of ion-sound or magnetohydrodynamic waves in low-temperature plasma, longitudinal waves in bars, etc.

From the mathematical point of view, the interest in the Korteweg–de Vries equation was caused by special properties both of the equation itself, and of its solutions. In particular, it possesses solitary wave solutions (solitons), as well as an infinite number of commuting conservation laws (see Chapter 5) and an infinite-dimensional algebra of (higher) symmetries (Chapter 4).

Skipping technical computations, which differ from those for the Burgers equation in details only, we shall give the final answer here. An $\mathbb{R}$-basis of the symmetry Lie algebra is formed by the following generating functions:

$$xu_x + 3tu_t + 2u, \quad 6tu_x + 1, \quad u_x, \quad u_t. \tag{5.12}$$

The Lie fields corresponding to these functions are

$$\begin{aligned}
&2u\frac{\partial}{\partial u} - x\frac{\partial}{\partial x} - 3t\frac{\partial}{\partial t} &&\text{(scale symmetry)}, \\
&\frac{\partial}{\partial u} - 6t\frac{\partial}{\partial x} &&\text{(Galilean symmetry)}, \\
&\frac{\partial}{\partial x} &&\text{(translation along } x), \\
&\frac{\partial}{\partial t} &&\text{(translation along } t).
\end{aligned} \tag{5.13}$$

5.3. The Khokhlov–Zabolotskaya equation. Propagation of a bounded three-dimensional acoustic beam in a nonlinear medium is described by the equation with the following dimensionless form:

$$-\frac{\partial^2 u}{\partial q_1 \partial q_2} + \frac{1}{2}\frac{\partial^2 (u^2)}{\partial q_1^2} + \frac{\partial^2 u}{\partial q_3^2} + \frac{\partial^2 u}{\partial q_4^2} = 0, \tag{5.14}$$

where $q_1, q_2 > 0$ and $-\infty < q_3, q_4 < +\infty$. In this equation, the value of u is proportional to the deviation of the medium's density from the balanced density, while the dimensionless variables q_1, q_2, q_3, q_4 are expressed via the time t and space variables x, y, z in the following way:

$$q_1 = \frac{t - x/c_0}{\sqrt{\gamma + 1}}\sqrt{\rho_0 c_0}, \quad q_2 = \mu x, \quad q_3 = \sqrt{\frac{2\mu}{c_0}} y, \quad q_4 = \sqrt{\frac{2\mu}{c_0}} z,$$

where c_0 is the sound velocity in the media, γ is the isentropic exponent, x is the coordinate in the direction of beam propagation, μ is a small parameter, and ρ_0 is the balanced density. In the coordinates q_i, p_σ, the equation (5.14) acquires the form

$$-p_{(1,1,0,0)} + up_{(2,0,0,0)} + p_{(1,0,0,0)}^2 + p_{(0,0,2,0)} + p_{(0,0,0,2)} = 0. \tag{5.15}$$

This equation is called the Khokhlov–Zabolotskaya equation.

THEOREM 5.1. *The algebra of all classical symmetries for the Khokhlov–Zabolotskaya equation* (5.15) *is generated by the following symmetries*[7]:

$$\begin{cases} f(A) = A'q_3 p_1 + 2A p_3 + A''q_3, \\ g(B) = B'q_4 p_1 + 2B p_4 + B''q_4, \\ h(C) = C p_1 + C', \\ T_2 = p_2, \\ L = q_1 p_1 + q_2 p_2 + q_3 p_3 + q_4 p_4, \\ M = 2q_1 p_1 + 4q_2 p_2 + 3q_3 p_3 + 3q_4 p_4 + 2u, \\ M_{34} = q_4 p_3 - q_3 p_4, \end{cases} \quad (5.16)$$

where A, B, C *are arbitrary smooth functions in* q_2.

5.3.1. "Physically meaningful" symmetries. Let us single out of all classical symmetries of equation (5.15) the "physically meaningful" ones, i.e., those preserving the condition of solution decay at infinity.

Consider the symmetry $f(A)$. Then

$$X_{f(A)} = A'q_3 \frac{\partial}{\partial q_1} - 2A \frac{\partial}{\partial q_3} + A''q_3 \frac{\partial}{\partial u}.$$

It is easily seen that the flow of this field is described by the system of equations

$$\begin{cases} q_1 = AA'\tau^2 + A'q_3^0 \tau + q_1^0, \\ q_2 = q_2^0, \\ q_3 = 2A\tau + q_3^0, \\ q_4 = q_4^0, \\ u = -AA''\tau^2 - A''q_3^0 \tau + u^0, \end{cases} \quad (5.17)$$

where u^0, q_1^0, q_2^0, q_3^0, q_4^0 are the initial values.

Consider an arbitrary solution $u^0 = u^0(q_1^0, q_2^0, q_3^0, q_4^0)$ of equation (5.15) decaying at infinity, i.e., satisfying the condition $u^0 \to 0$ as $\|(q_1^0, q_2^0, q_3^0, q_4^0)\| \to \infty$. Then the flow (5.17) takes this function to the solution

$$u = -AA''\tau^2 - A''q_3^0 \tau + u^0(q_1^0, q_2^0, q_3^0, q_4^0)$$
$$= u(AA'\tau^2 + A'q_3^0 \tau + q_1^0, q_2^0, 2A\tau + q_3^0, q_4^0).$$

Let $\|(AA'\tau^2 + A'q_3^0\tau + q_1^0, q_2^0, 2A\tau + q_3^0, q_4^0)\| \to \infty$ for a fixed value of the parameter τ. If $A'' \neq 0$, then the condition of decay of u at infinity will not hold, since $u \to \infty$ as $q_3^0 \to \infty$ in this case. Hence, $A'' = 0$. Obviously, this is also sufficient for the decay of a solution at infinity.

From the condition $A'' = 0$ it follows that $A = C_1 q_2 + C_2$, where C_1, C_2 are constants. Thus, among the infinite number of symmetries $f(A)$, only the following two are physically meaningful:

$$T_3 = p_3, \qquad R_3 = q_3 p_1 + 2q_2 p_3.$$

---

[7] Note that the computations needed to prove the theorem are rather laborious. This is why we use symbolic computation programs in this field. A detailed analysis of computations can be found in [**73, 75, 109**]

5. EXAMPLES OF COMPUTATIONS

In the same way, the symmetries of the form $g(B)$ produce two physically meaningful symmetries
$$T_4 = p_4, \qquad R_4 = q_4 p_1 + 2q_2 p_4,$$
while the symmetries $h(C)$ produce the sole symmetry
$$T_1 = p_1.$$
It is easily checked that the symmetries T_1, T_2, M, and M_{34} preserve the condition of decay at infinity, while the symmetry L does not.

On the other hand, it is easily checked that the following linear combinations of L and M are physically meaningful:
$$M_1 = 2(M + 2L) = 2q_1 p_1 + q_3 p_3 + q_4 p_4 - 2u,$$
$$M_2 = (M - L) = 2q_2 p_2 + q_3 p_3 + q_4 p_4 + 2u.$$

Thus, the subalgebra of physically meaningful symmetries is generated by the following generating functions:
$$T_i = p_i, \; i = 1,2,3,4 \quad \text{(translations)},$$
$$M_{34} = q_4 p_3 - q_3 p_4 \quad \text{(rotation)},$$
$$\left.\begin{array}{l} M_1 = 2q_1 p_1 + q_3 p_3 + q_4 p_4 - 2u \\ M_2 = 2q_2 p_2 + q_3 p_3 + q_4 p_4 + 2u \end{array}\right\} \quad \text{(scale symmetries)},$$
$$R_3 = q_3 p_1 + 2q_2 p_3,$$
$$R_4 = q_4 p_1 + 2q_2 p_4.$$

Translations $T_i = p_i$, $i = 1, 2, 3, 4$, are responsible for homogeneity of the four-dimensional space-time, the rotation M_{34} corresponds to isotropy of the space in the plane perpendicular to the beam propagation direction, and M_1 and M_2 correspond to invariance of the Khokhlov–Zabolotskaya equation with respect to the scaling transformations
$$A^{(1)}_\tau : u(q) \longmapsto e^{-2\tau} u(e^{2\tau} q_1, q_2, e^\tau q_3, e^\tau q_4),$$
$$A^{(2)}_\tau : u(q) \longmapsto e^{2\tau} u(q_1, e^{2\tau} q_2, e^\tau q_3, e^\tau q_4).$$
Finally, R_3 and R_4 correspond to invariance of the equation with respect to the transformations
$$A^{(3)}_\tau : u(q) \longmapsto u(q_1 + \tau q_3 + \tau^2 q_2, q_2, q_3 + 2\tau q_2, q_4),$$
$$A^{(4)}_\tau : u(q) \longmapsto u(q_1 + \tau q_4 + \tau^2 q_2, q_2, q_3, q_4 + 2\tau q_2).$$

5.3.2. Invariant solutions. Let us find solutions invariant with respect to some subalgebras of the Lie algebra of all classical symmetries of the Khokhlov–Zabolotskaya equation (5.15).

Consider the subalgebra of symmetries $\mathcal{G}$ generated by the symmetries M_1, M_2, R_3, R_4. The Lie algebra structure in $\mathcal{G}$ with respect to the Jacobi bracket in the basis $e_1 = (M_1 + M_2)/2$, $e_2 = (-M_2 + M_1)/2$, $e_3 = R_3$, $e_4 = R_4$ is given by the relations
$$[e_2, e_3] = e_3, \quad [e_2, e_4] = e_4, \quad [e_3, e_4] = 0, \quad [e_1, e_i] = 0, \quad i = 2, 3, 4.$$
Thus $\mathcal{G}$ is a solvable Lie algebra with the commutator subalgebra
$$\mathcal{G}^{(1)} = [\mathcal{G}, \mathcal{G}] = \mathbb{R} e_3 \oplus \mathbb{R} e_4.$$

Consider an arbitrary three-dimensional subalgebra $\mathcal{H} \subset \mathcal{G}$ containing $\mathcal{G}^{(1)}$. Any such subalgebra is generated by the elements e_3, e_4, $\lambda_1 e_1 + \lambda_2 e_2$.

Let us describe $\mathcal{H}$-invariant solutions of equation (5.15). If $\lambda_1 + \lambda_2 \neq 0$, then $\mathcal{H}$-invariant solutions $u(q)$ are of the form

$$u(q) = x^\lambda v(x^{-2-\lambda} y),$$

where $x = q_2$, $y = 4q_1 q_2 - q_3^2 - q_4^2$, $\lambda = -2\lambda_2/(\lambda_1 + \lambda_2)$. Substituting this expression into the Khokhlov–Zabolotskaya equation, we obtain the equation for $v(t)$:

$$vv'' + (v')^2 + \alpha t v'' = 0, \tag{5.18}$$

where $t = x^{-2-\lambda} y$, $\alpha = -(\lambda_2 - \lambda_1)/4(\lambda_1 + \lambda_2)$.

Since the left-hand side of (5.18) is of the form $(vv' + \alpha t v' - \alpha v)'$, we obtain the equation

$$vv' + \alpha t v' - \alpha v = \mathrm{const}.$$

In the case when the constant in the left-hand side vanishes, this equation possesses the solution

$$v \ln v + av = \alpha t,$$

where a is a constant.

5.4. The Kadomtsev–Pogutse equations. This system is a simplification of the general magnetohydrodynamic (MHD) system, where some inessential (from the point of view of maintaining high-temperature plasma in TOKAMAK-like systems) details are omitted. The authors of the paper [**48**] started from the ideal MHD equations, because in this context characteristic times of the most important physical processes are essentially less than the distinctive dissipation time. Dissipation here is caused by viscosity and electric resistance of plasma. Besides, it was taken into account that plasma stability needs: (a) plasma pressure and the cross-component of magnetic field pressure were much less than the pressure created by the longitudinal component of the magnetic field, and (b) the smaller TOKAMAK radius was much less then the external radius.

As a result, a system of two scalar equations was obtained which in an adequate norming is of the form

$$\begin{cases} \dfrac{\partial \psi}{\partial t} + [\nabla_\perp \varphi, \nabla_\perp \psi]_z = \dfrac{\partial \varphi}{\partial z}, \\ \dfrac{\partial}{\partial t} \Delta_\perp \varphi + [\nabla_\perp \varphi, \nabla_\perp \Delta_\perp \varphi]_z = \dfrac{\partial}{\partial t} \Delta_\perp \psi + [\nabla_\perp \psi, \nabla_\perp \Delta_\perp \psi]_z, \end{cases} \tag{5.19}$$

where $\nabla_\perp = \{\partial/\partial x, \partial/\partial y\}$, $\Delta_\perp = \partial^2/\partial x^2 + \partial^2/\partial y^2$, and $[u, v]_z = u_x v_y - u_y v_x$ is the z-component of the vector product $[u, v]$ of the vectors u and v.

Here the functions φ and ψ are the potentials of the velocity and of the cross-component of the magnetic field respectively (they also may be understood as the potential of the electric field and the z-component of the vector potential of the magnetic field). The coordinate system is chosen in such a way that the z-axis is directed along the TOKAMAK axis.

We shall call equations (5.19) the Kadomtsev–Pogutse equations. They are also called reduced MHD equations.

The relative simplicity of the Kadomtsev–Pogutse equations made it possible to construct an experimentally confirmed theory of kink instability; they were also used in the quantitative analysis of instability [**147**].

5.4.1. *Computation of symmetries.* Here we replace general notation for coordinates in jet spaces by specific notation, to make formulas more understandable for reading. The coordinates in the base will be denoted by x, y, z, t, while for coordinates in the fiber of $J^0(4,2)$ the notation φ and ψ will be used. We shall write $\varphi_{x^i y^j z^k t^l}$ and $\psi_{x^i y^j z^k t^l}$ instead of p_σ^1, p_σ^2 respectively, where $\sigma = (i,j,k,l)$, $|\sigma| = i+j+k+l$. In this notation, (5.19) acquires the form

$$F_1 = \psi_t + \varphi_x \psi_y - \varphi_y \psi_x - \varphi_z = 0, \tag{5.20}$$

$$\begin{aligned}F_2 &= \varphi_{x^2 t} + \varphi_{y^2 t} + \varphi_x \varphi_{x^2 y} + \varphi_x \varphi_{y^3} - \varphi_y \varphi_{x^3} - \varphi_y \varphi_{xy^2} \\ &\quad - \psi_{x^2 z} - \psi_{y^2 z} - \psi_x \psi_{xy^2} - \psi_x \psi_{y^3} + \psi_y \psi_{x^3} + \psi_y \psi_{xy^2} = 0.\end{aligned} \tag{5.21}$$

The system of Kadomtsev–Pogutse equations contains two dependent variables φ and ψ, and consequently the generating function of a symmetry is two-component in this case:

$$\begin{aligned}\mathcal{S} &= S + A\varphi_x + B\varphi_y + C\varphi_z + E\varphi_t, \\ \mathcal{T} &= T + A\psi_x + B\psi_y + C\psi_z + E\psi_t.\end{aligned} \tag{5.22}$$

Here A, B, C, E, S, T are functions on $J^0(4,2)$, i.e., functions in x, y, z, t.

THEOREM 5.2 ([98]). *The algebra of classical symmetries of the Kadomtsev–Pogutse equations as a linear space over $\mathbb{R}$ is generated by symmetries with the generating functions of the form*

$$\mathcal{A}_\alpha = \begin{pmatrix} \alpha'(\varphi+\psi) + \alpha(\varphi_z + \varphi_t) \\ \alpha'(\varphi+\psi) + \alpha(\psi_z + \psi_t) \end{pmatrix}, \quad \mathcal{B}_\beta = \begin{pmatrix} \beta'(\varphi-\psi) + \beta(\varphi_z - \varphi_t) \\ -\beta'(\varphi-\psi) + \beta(\psi_z - \psi_t) \end{pmatrix},$$

$$\mathcal{C}_\gamma = \begin{pmatrix} \gamma'(x^2+y^2) + 2\gamma(y\varphi_x - x\varphi_y) \\ \gamma'(x^2+y^2) + 2\gamma(y\psi_x - x\psi_y) \end{pmatrix}, \quad \mathcal{D}_\delta = \begin{pmatrix} -\delta'(x^2+y^2) + 2\delta(y\varphi_x - x\varphi_y) \\ \delta'(x^2+y^2) + 2\delta(y\psi_x - x\psi_y) \end{pmatrix},$$

$$\mathcal{G}_G = \begin{pmatrix} yG_t + G\varphi_x \\ yG_z + G\psi_x \end{pmatrix}, \quad \mathcal{H}_H = \begin{pmatrix} -xH_t + H\varphi_y \\ -xH_z + H\psi_y \end{pmatrix}, \quad \mathcal{K}_K = \begin{pmatrix} K_t \\ K_z \end{pmatrix},$$

$$\mathcal{L} = \begin{pmatrix} \varphi + z\varphi_z + t\varphi_t \\ \psi + z\psi_z + t\psi_t \end{pmatrix}, \quad \mathcal{M} = \begin{pmatrix} x\varphi_x + y\varphi_y + 2z\varphi_z + 2t\varphi_t \\ x\psi_x + y\psi_y + 2z\psi_z + 2t\psi_t \end{pmatrix}.$$

Here $\alpha = \alpha(z+t)$, $\beta = \beta(z-t)$, $\gamma = \gamma(z+t)$, $\delta = \delta(z-t)$, $G = G(z,t)$, $H = H(z,t)$, $K = K(z,t)$ are arbitrary functions and

$$\alpha'(\zeta) = \frac{d\alpha(\zeta)}{d\zeta}, \quad \beta'(\eta) = \frac{d\beta(\eta)}{d\eta}, \quad \gamma'(\xi) = \frac{d\gamma(\xi)}{d\xi}, \quad \delta'(\theta) = \frac{d\delta(\theta)}{d\theta}.$$

The vector fields on $J^0(4,2)$ corresponding to the above listed generating functions are of the form

$$X_{\mathcal{A}_\alpha} = \alpha'(\varphi+\psi)\left(\frac{\partial}{\partial\varphi}+\frac{\partial}{\partial\psi}\right) - \alpha\left(\frac{\partial}{\partial z}+\frac{\partial}{\partial t}\right),$$

$$X_{\mathcal{B}_\beta} = \beta'(\varphi-\psi)\left(\frac{\partial}{\partial\varphi}-\frac{\partial}{\partial\psi}\right) - \beta\left(\frac{\partial}{\partial z}-\frac{\partial}{\partial t}\right),$$

$$X_{\mathcal{C}_\gamma} = \gamma'(x^2+y^2)\left(\frac{\partial}{\partial\varphi}+\frac{\partial}{\partial\psi}\right) - 2\gamma\left(y\frac{\partial}{\partial x}-x\frac{\partial}{\partial y}\right),$$

$$X_{\mathcal{D}_\delta} = \delta'(x^2+y^2)\left(\frac{\partial}{\partial\varphi}-\frac{\partial}{\partial\psi}\right) - 2\delta\left(y\frac{\partial}{\partial x}-x\frac{\partial}{\partial y}\right),$$

$$X_{\mathcal{G}_G} = yG_t\frac{\partial}{\partial\varphi}+yG_z\frac{\partial}{\partial\psi}-G\frac{\partial}{\partial x},$$

$$X_{\mathcal{H}_H} = -xH_t\frac{\partial}{\partial\varphi}-xH_z\frac{\partial}{\partial\psi}-H\frac{\partial}{\partial y},$$

$$X_{\mathcal{K}_K} = K_t\frac{\partial}{\partial\varphi}+K_z\frac{\partial}{\partial\psi}$$

$$X_\mathcal{L} = \varphi\frac{\partial}{\partial\varphi}+\psi\frac{\partial}{\partial\psi}-z\frac{\partial}{\partial z}-t\frac{\partial}{\partial t},$$

$$X_\mathcal{M} = -x\frac{\partial}{\partial x}-y\frac{\partial}{\partial y}-2z\frac{\partial}{\partial z}-2t\frac{\partial}{\partial t}$$

The above listed symmetries have the following physical meanings: $\mathcal{L}$ and $\mathcal{M}$ are scale symmetries, $\mathcal{C}_\gamma$ and $\mathcal{D}_\delta$ are generalized rotations in the (x,y)-plane (they become real rotations when $\gamma = \delta = $ const), $\mathcal{A}_\alpha$ and $\mathcal{B}_\beta$ are generalized translations along z and t respectively (in the case $\alpha = \beta = 1$ real translations are obtained as their linear combinations), the symmetries $\mathcal{G}_G$ and $\mathcal{H}_H$ are generalized translations along x and y respectively (real translations correspond to the case $G = H = 1$), and, finally, $\mathcal{K}_K$ is the gauge by the trivial solution.

The Lie algebra structure of symmetries is described in Table 1.

5.4.2. *Invariant solutions.*

$\mathcal{A}_\alpha$**-invariant solutions.** The invariance conditions are of the form

$$\alpha'(\varphi+\psi)+\alpha(\varphi_z+\varphi_t) = 0, \quad \alpha'(\varphi+\psi)+\alpha(\psi_z+\psi_t) = 0.$$

Solving this system, we obtain

$$\varphi = (\alpha(z+t))^{-1}A(x,y,z-t)+B(x,y,z-t),$$

$$\psi = (\alpha(z+t))^{-1}A(x,y,z-t)+B(x,y,z-t),$$

where the functions A and B satisfy the following conditions implied by (5.20) and (5.21):

$$\begin{aligned} A_\eta &= B_xA_y - A_xB_y, \\ \Delta_\perp A_\eta &= B_x\Delta_\perp A_y + A_x\Delta_\perp B_y - A_y\Delta_\perp B_x - B_y\Delta_\perp A_x, \end{aligned} \qquad (5.23)$$

where $\eta = z - t$. Note that equations (5.23) are the same for all α.

$\mathcal{G}_G$**-invariant solutions.** Invariance conditions in this case lead to the following equations:

$$\varphi = -\frac{G_t}{G}xy + A(y,z,t), \quad \psi = -\frac{G_z}{G}xy + B(y,z,t),$$

5. EXAMPLES OF COMPUTATIONS

TABLE 1. Commutation relations in the algebra of classical symmetries for the Kadomtsev–Pogutse equations

[,]	$\mathcal{A}_{\widehat{\alpha}}$	$\mathcal{B}_{\widehat{\beta}}$	$\mathcal{C}_{\widehat{\gamma}}$	$\mathcal{D}_{\widehat{\delta}}$	$\mathcal{G}_{\widehat{G}}$
$\mathcal{A}_\alpha$	$2\mathcal{A}_{\widehat{\alpha}\alpha'-\widehat{\alpha}'\alpha}$	0	$-2\mathcal{G}_{\widehat{\gamma}'\alpha}$	0	$-\mathcal{G}_{\alpha(\widehat{G}_t+\widehat{G}_z)}$
$\mathcal{B}_\beta$	0	$2\mathcal{B}_{\beta'\widehat{\beta}-\widehat{\beta}'\beta}$	0	$-2\mathcal{D}_{\beta\widehat{\delta}'}$	$\mathcal{G}_{\beta(\widehat{G}_t-\widehat{G}_z)}$
$\mathcal{C}_\gamma$	$2\mathcal{C}_{\gamma'\widehat{\alpha}}$	0	0	0	$-2\mathcal{H}_{\delta\widehat{G}}$
$\mathcal{D}_\delta$	0	$2\mathcal{D}_{\delta'\widehat{\beta}}$	0	0	$-2\mathcal{G}_{\widehat{\delta}H}$
$\mathcal{G}_G$	$\mathcal{G}_{\widehat{\alpha}(G_t+G_z)}$	$-\mathcal{G}_{\widehat{\beta}(G_t-G_z)}$	$2\mathcal{H}_{\widehat{\gamma}G}$	$2\mathcal{H}_{\widehat{\delta}G}$	0
$\mathcal{H}_H$	$\mathcal{H}_{\widehat{\alpha}(H_t+H_z)}$	$\mathcal{H}_{\widehat{\beta}(H_t-H_z)}$	$-2\mathcal{G}_{\widehat{\gamma}H}$	$-2\mathcal{G}_{\widehat{\delta}H}$	$-\mathcal{K}_{\widehat{G}H}$
$\mathcal{K}_K$	$-c\mathcal{K}_{\widehat{\alpha}(K_t+K_z)}$	$-c\mathcal{K}_{\widehat{\beta}(K_t-K_z)}$	0	0	0
$l\mathcal{L}$	$-l\mathcal{A}_{(t+z)\widehat{\alpha}'-\widehat{\alpha}}$	$-l\mathcal{B}_{(z-t)\widehat{\beta}'-\widehat{\beta}}$	$-l\mathcal{C}_{(z+t)\widehat{\gamma}'}$	$-l\mathcal{D}_{(z-t)\widehat{\delta}'}$	$-l\mathcal{G}_{t\widehat{G}_t+z\widehat{G}_z}$
$m\mathcal{M}$	0	0	0	0	$m\mathcal{H}_{\widehat{H}}$

[,]	$\mathcal{H}_{\widehat{H}}$	$-\mathcal{K}_{\widehat{K}}$	$\widehat{l}\mathcal{L}$	$\widehat{m}\mathcal{M}$
$\mathcal{A}_\alpha$	$-\mathcal{H}_{\alpha(\widehat{H}_t+\widehat{H}_z)}$	$\mathcal{H}_{\alpha(\widehat{K}_t+\widehat{K}_z)}$	$\widehat{l}\mathcal{A}_{(t+x)\alpha'-\alpha}$	0
$\mathcal{B}_\beta$	$\mathcal{H}_{\beta(\widehat{H}_t-\widehat{H}_z)}$	$\mathcal{K}_{\beta(\widehat{K}_t-\widehat{K}_z)}$	$\widehat{l}\mathcal{B}_{(z-t)\beta'-\beta}$	0
$\mathcal{C}_\gamma$	$2\mathcal{G}_{\gamma\widehat{H}}$	0	$\widehat{l}\mathcal{C}_{(t+z)\gamma'}$	0
$\mathcal{D}_\delta$	$2\mathcal{G}_{\gamma\widehat{H}}$	0	$\widehat{l}\mathcal{D}_{(z-t)\delta'}$	0
$\mathcal{G}_G$	$\mathcal{K}_{G\widehat{H}}$	0	$\widehat{l}\mathcal{G}_{zG_z+tG_t}$	$-\widehat{m}\mathcal{G}_G$
$\mathcal{H}_H$	0	0	$\widehat{l}\mathcal{H}_{tH_t+zH_z}$	$-\widehat{m}\mathcal{H}_H$
$\mathcal{K}_K$	0	0	$\widehat{l}\mathcal{K}_{tK_t+zK_z}$	$-2\widehat{m}\mathcal{K}_K$
$l\mathcal{L}$	$-l\mathcal{H}_{t\widehat{H}_t+z\widehat{H}_z}$	$-l\mathcal{K}_{t\widehat{K}_t+z\widehat{K}_z}$	0	0
$m\mathcal{M}$	$m\mathcal{H}_{\widehat{H}}$	$2m\mathcal{K}_{\widehat{K}}$	0	0

where A and B satisfy the *linear* system

$$-B_t + A_z = \frac{y}{G}(G_z A - G_t B)_y,$$
$$(A_t - B_z)_{y^2} = \frac{y}{G}(G_t A - G_z B)_{y^3}.$$

$\mathcal{C}_\gamma$**-invariant solutions.** From the invariance conditions we obtain

$$\varphi = r^2\theta\gamma'/2\gamma + A(r,z,t), \quad \psi = r^2\theta\gamma'/2\gamma + B(r,z,t),$$

where r are θ are the polar coordinates on the plane x,y. The functions A and B satisfy the linear system

$$-B_t + A_z = (r\gamma'/2\gamma)(A-B)_r,$$
$$\Delta_\perp(A_t - B_z) = [(r\gamma'/2\gamma)\Delta_\perp + (2\gamma'/r\gamma)](A-B)_r.$$

It is interesting to note that $\mathcal{C}_\gamma$-invariant solutions are uniquely defined only in the sector $|\theta - \theta_0| < \pi$ for some θ_0. Discontinuous solutions can be constructed out of such pieces.

Let us consider examples of solutions invariant with respect to some two-dimensional subalgebras.

$\mathcal{A}_\alpha, \mathcal{B}_\beta$-**invariant solutions.** These solutions are of the form

$$\varphi = (\alpha(z+t))^{-1} A(x,y) + (\beta(z-t))^{-1} B(x,y),$$
$$\psi = (\alpha(z+t))^{-1} A(x,y) - (\beta(z-t))^{-1} B(x,y). \tag{5.24}$$

Thus these solutions represent an arbitrary (the so-called Alfvén) wave propagating along the z-axis with the velocity $c_A = 1$ and with the amplitude depending on x and y. The functions A and B must satisfy the system of equations

$$\begin{cases} A_x B_y - A_y B_x = 0, \\ A_x \Delta B_y - A_y \Delta B_x + B_x \Delta A_y - B_y \Delta A_x = 0. \end{cases}$$

From the first equation of the system it follows that $B = f(A)$. From the second we obtain

$$f'(A)(A_x \Delta A_y - A_y \Delta A_x) + f''(A)[A_{xy}(A_x^2 - A_y^2) - A xy(A_{x^2} - A_{y^2})] = 0.$$

In particular, for $B = A$ we have

$$\Delta A = F(A). \tag{5.25}$$

Other examples of invariant solutions for the Kadomtsev–Pogutse equations, including the case of three-dimensional subalgebras, can be found in [**37**].

5.4.3. *Reproduction of solutions.* Let (φ_0, ψ_0) be a solution of system (5.20), (5.21), and let $Q = (Q^1, Q^2)$ be its symmetry. Recall that solving the system

$$\frac{\partial}{\partial \tau} \begin{pmatrix} \varphi \\ \psi \end{pmatrix} = \begin{pmatrix} Q^1 \\ Q^2 \end{pmatrix}$$

with the initial values $(\varphi, \psi)|_{\tau=0} = (\varphi_0, \psi_0)$ gives a family of solutions for equations (5.20), (5.21) depending on τ. Thus, the solution (φ_0, ψ_0) is "reproduced" or "deformed" by the symmetry Q. Consider some examples.

$\mathcal{C}_\gamma$-**reproduction.** In this case we need to solve the equations

$$\varphi_\tau = \gamma' r^2/2 + \gamma \varphi_\theta, \qquad \varphi|_{\tau=0} = \varphi_0(r, \theta, z, t),$$
$$\psi_\tau = \gamma' r^2/2 + \gamma \psi_\theta, \qquad \psi|_{\tau=0} = \varphi_0(r, \theta, z, t). \tag{5.26}$$

The solutions are

$$\varphi_\tau = \gamma' r^2 \tau/2 + \varphi_0(r, \theta + \tau\gamma, z, t), \qquad \psi_\tau = \gamma' r^2 \tau/2 + \psi_0(r, \theta + \tau\gamma, z, t).$$

It can be noticed that these solutions are single-valued, while invariant solutions for this symmetry are multivalued (see above).

From the formulas for the family φ, ψ one can see that the phase θ can be deviated from the axis z by means of the Alfvén wave propagation $\gamma(z+t)$, but this causes the correction $\gamma r^2 \tau$ to the solution. Note that the $\mathcal{D}_\delta$-reproduction leads to similar formulas (and to the deviation $\tau \delta(z-t)$ of the phase). If the solution is periodic in z (for example, the TOKAMAK z-axis is closed and is of length l), then the "quantization conditions" for the parameter τ arise. For example, taking the phase deviation $\tau z = \tau[(z+t) + (z-t)]/2$, we shall obtain that $\tau l = 2\pi n$, $\tau = 2\pi n/l$. But this effect may not appear if deviation is periodic (for example, if one has the "phase trembling" of the form $\gamma = \sin[\pi N(z+t)/l]$).

$\mathcal{G}_G$-**reproduction.** In this case
$$\varphi = y\tau G_t + \varphi_0(x+\tau G, y, z, t), \quad \psi = y\tau G_z + \psi_0(x+\tau G, y, z, t).$$
Here the solution deviates from the x-axis, and the correction $(y\tau G_t, y\tau G_z)$ to the initial solution arises. Quite similarly, the $\mathcal{H}_H$-reproduction is a shift along the y-axis by τH. Combining different shifts, one can bend the solution in a helical way.

Nevertheless, it should be noted that the indicated freedom in deformation choice is in a sense illusive: from the physical point of view, strong deformation seems to lead to necessity of taking the plasma pressure into account, and consequently of adding new terms to the initial equation.

$\mathcal{A}_\alpha$-**reproduction.** To solve the system
$$\varphi_\tau = \alpha(\varphi_z + \varphi_t) + \alpha'(\varphi + \psi),$$
$$\psi_\tau = \alpha(\psi_z + \psi_t) + \alpha'(\varphi + \psi),$$
let us add and subtract its equations:
$$(\varphi+\psi)_\tau = \alpha[(\varphi+\psi)_z + (\varphi+\psi)_t] + 2\alpha'(\varphi+\psi),$$
$$(\varphi-\psi)_\tau = \alpha[(\varphi-\psi)_z + (\varphi-\psi)_t].$$

The transformed system is easy to solve with respect to the new unknown functions $(\varphi+\psi), (\varphi-\psi)$. The answer is
$$\varphi = \frac{\alpha(\Sigma)+\alpha(z+t)}{2\alpha(z+t)}\varphi_0(x,y,Z,T) + \frac{\alpha(\Sigma)-\alpha(z+t)}{2\alpha(z+t)}\psi_0(x,y,Z,T),$$
$$\psi = \frac{\alpha(\Sigma)-\alpha(z+t)}{2\alpha(z+t)}\varphi_0(x,y,Z,T) + \frac{\alpha(\Sigma)+\alpha(z+t)}{2\alpha(z+t)}\psi_0(x,y,Z,T).$$

Here
$$\Sigma = \Gamma^{-1}(\tau + \Gamma(z+t)), \quad \Gamma(z+t) = \frac{1}{2}\int^{z+t}\frac{d\xi}{\alpha(\xi)},$$
$$Z = \tfrac{1}{2}(z-t+\Sigma), \quad T = \tfrac{1}{2}(t-z+\Sigma),$$
and Γ^{-1} is the inverse function. One can obtain explicit formulas for particular values of α. For example, for $\psi_0 \equiv 0$ and $\alpha = z+t$ we have the "rumpled" solution
$$\varphi = (1+e^\tau)\frac{\varphi_0(x,y,e^\tau(z+t),z-t)}{2}.$$

It is interesting to note that if one takes a discontinuous symmetry, then the initially smooth solution includes in a family of discontinuous ones. For example, if $\psi_0 \equiv 0, \alpha = \xi^{-1}, \xi = z+t$, then
$$\varphi = 5(1+|\xi|(\tau+\xi^2)^{-1/2})\frac{\varphi_0(x,y,\text{sgn}(\xi)\sqrt{\tau+\xi^2},z-t)}{2}.$$
For $\xi = 0, \tau \neq 0$ a traveling break arises proportional to
$$\varphi_0(x,y,\sqrt{\tau},z-t) - \varphi_0(x,y,-\sqrt{\tau},z-t).$$
Though the break of the potential φ does not admit a straightforward physical interpretation, when the function φ_0 is even a gradient break arises, and it makes sense physically.

Consider another example of using invariant solutions for reproduction [**37**]. Choose an $\mathcal{A}_\alpha, \mathcal{B}_\beta$-invariant solution (5.24), taking $\alpha = 2, \beta = -2$ and functions A

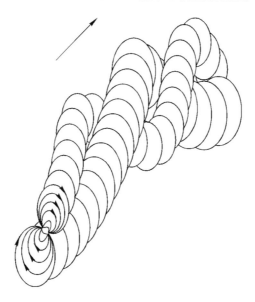

FIGURE 3.11

and B such that $A = B$ and $\Delta A = \exp A$ (cf. (5.25)). For the solution, we shall take the function

$$A(z, \theta) = -\ln\left[\frac{r^4}{2}\sinh^2\left(\frac{\cos\theta}{r} + 1\right)\right].$$

Then the corresponding invariant solution is of the form

$$\begin{cases} \varphi = 0, \\ \psi = A. \end{cases}$$

This solution is static (independent of time) and constant along the z-axis. Let us apply the $\mathcal{C}_\alpha$-deformation to it, with $\alpha(z + t) = \exp(-(z + t))/2$. For $\tau = 1$, from (5.26) we obtain

$$\varphi = -\frac{r^2}{2}e^{-(z+t)},$$

$$\psi = \frac{r^2}{2}e^{-(z+t)} - \ln\left\{\frac{r^4}{2}\sinh^2\left[\frac{\cos\theta - e^{-(z+t)}}{r} + 1\right]\right\}.$$

In Figure 3.11 one can see the magnetic level surface $\psi(x, y, z, 0) = 5$. The arrow shows the direction of Alfvén velocity along the z-axis. Thus, after deformation the solution lost its static property and became inhomogeneous along the z-axis.

In Figure 3.12 the same solution is shown, deformed first by $\mathcal{G}_G$-deformation and then by $\mathcal{H}_H$-deformation, where $G = (\sin 20z)/20$ and $H = (\cos 20z)/20$.

6. Factorization of equations by symmetries

The symmetry group of a differential equation consists of diffeomorphisms of the jet space preserving the equation itself and taking solutions to solutions. What will be the quotient object of this action, i.e., the orbit space? More generally, what will be the orbit space when we consider the action of a subgroup of the symmetry group?

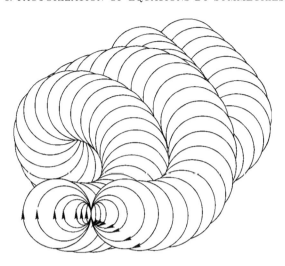

FIGURE 3.12

Consider an equation $\mathcal{E} \subset J^k(\pi)$, whose symmetry group is a Lie group G, and assume that $H \subset G$ is a Lie subgroup. The Cartan distribution $\mathcal{C} = \mathcal{C}^k$ is invariant with respect to symmetries, and the solutions are n-dimensional maximal integral manifolds of the distribution $\mathcal{C}^k(\mathcal{E})$. Consider the orbit space of the group H and denote by μ the factorization mapping[8]:

$$\begin{array}{ccc} \mathcal{E} & \longrightarrow & J^k(\mathbb{R}^n) \\ \mu|_{\mathcal{E}} \downarrow & & \downarrow \mu \\ \mathcal{E}' & \longrightarrow & J^k(\mathbb{R}^n)/H \end{array}$$

The quotient distribution $\mathcal{C}'$ is defined on $\mathcal{E}'$, and almost everywhere it is of the same dimension as $\mathcal{C}$. In fact, let $\mathcal{C}'_a = \mu_*(\mathcal{C}_{\mu^{-1}(a)})$ for $a \in J^k(\pi)/H$. It is a well-defined construction, since the distribution $\mathcal{C}$ is invariant with respect to H. Besides, vectors tangent to the orbits of H almost everywhere do not belong to the distribution $\mathcal{C}$. In fact, let $b \in J^k(\pi)$, and let H_b be the orbit of the point b under the action of the group H. Then the mapping $H \to H_b$, $h \in H \mapsto h(b) \in H_b$, generates an epimorphism of the corresponding Lie algebra $\mathcal{H}$ to the tangent space $T_b(H_b)$. Therefore, any vector $X_b \in T_b(H_b)$ is of the form $X_\varphi|_b$, where φ is the generating section of some symmetry from $\mathcal{H}$. Let $X_b \in \mathcal{C}_b$. This means, in particular, that

$$(X_\varphi \,\lrcorner\, (du^j - \sum_{i=1}^n p_i^j \, dx_i))_b = \varphi^j(b) = 0.$$

Therefore, for almost all $b \in J^k(\pi)$ one has $\mathcal{C}_b \cap \ker \mu_*|_b = 0$, and consequently $\dim \mathcal{C} = \dim \mathcal{C}'$.

Solutions of the equation $\mathcal{E}$ are maximal integral n-dimensional manifolds of the distribution $\mathcal{C}$, and in general they project to n-dimensional manifolds of the distribution $\mathcal{C}'$ lying in $\mathcal{E}'$. It should be stressed that we are not interested in all integral manifolds of $\mathcal{C}'$, but in those which are images of solutions of the initial equation (since this equation is the ultimate object of our study). Therefore, the

---

[8] We assume that the orbit space is a smooth manifold and μ is a smooth bundle.

desired condition defining in $\mathcal{E}'$ the integral manifolds of $\mathcal{C}'$ will be the quotient equation[9].

Let $L' \subset \mathcal{E}'$ be the image of some solution L of the equation $\mathcal{E}$. Then

$$\mu^{-1}(L') = H(L) = \bigcup_{h \in H} h(L). \tag{6.1}$$

Denote by $i_{L'}$ the immersion $\mu^{-1}(L') \to J^k(\pi)$. Then the restriction of the Cartan distribution to $\mu^{-1}(L')$ is determined by the differential system $\mathcal{D}_{L'} = i_{L'}^* \mathcal{D}$, where $\mathcal{D} \subset \Lambda^1(J^k(\pi))$ is the differential system given by the Cartan forms and determining $\mathcal{C}$. It is easily seen that the distribution $\mathcal{D}_{L'}$ is completely integrable: any $h(L)$ is its integral manifold and the union of these manifolds is $\mu^{-1}(L')$. The Frobenius complete integrability condition $d\mathcal{D}_{L'} \subset \mathcal{D}_{L'}$ for the distribution $\mathcal{D}_{L'}$ is exactly the condition defining those L', which are the images of solutions of the initial equation under factorization with respect to H. Thus, precisely these conditions constitute the quotient equation.

In the examples considered below such quotient equations are represented explicitly in coordinate form due to convenient relations between the dimensions of the objects under consideration; in more general situations, it is often difficult to obtain a coordinate description.

6.1. Second-order equations in two independent variables. Consider a second order equation in two independent and one dependent variables, where a two-dimensional Lie group H acts without fixed points. More exactly, let $\mathcal{E} \subset J^2(2,1)$, and let q_1, q_2, u, p_1, p_2, $p_{(2,0)}$, $p_{(1,1)}$, $p_{(0,2)}$ be canonical coordinates in $J^2(2,1)$. Then $\dim \mathcal{E} = 7$, $\dim \mathcal{E}/H = 5$. In the same coordinates, a basis of the differential system $\mathcal{D}$ determining the Cartan distribution is

$$\omega_0 = du - p_1 \, dq_1 - p_2 \, dq_2,$$
$$\omega_1 = dp_1 - p_{(2,0)} \, dq_1 - p_{(1,1)} \, dq_2,$$
$$\omega_2 = dp_2 - p_{(1,1)} \, dq_1 - p_{(0,2)} \, dq_2,$$

and consequently the dimension of Cartan planes at generic points is 4.

As already explained, the vector fields forming the Lie algebra $\mathcal{H}$ of the Lie group H do not intersect the Cartan distribution almost everywhere, and therefore the quotient Cartan distribution on the five-dimensional manifold $\mathcal{E}/H$ is also four-dimensional. Hence, the differential system $\mathcal{D}'$ determining this distribution is one-dimensional (at least locally) and the distribution $\mathcal{C}'$ can be described by one differential form w. In the generic situation, a one-form on an even-dimensional manifold determines a contact structure and by the Darboux lemma it can be written as

$$w = dv - y_1 dx_1 - y_2 dx_2$$

in appropriate coordinates x_1, x_2, v, y_1, y_2 on $\mathcal{E}/H$. Thus, the quotient manifold $\mathcal{E}/H$ itself can be (locally) identified with $J^1(\mathbb{R}^2)$.

In these coordinates (also locally), "almost all" maximal integral manifolds of the form w can be represented as 1-jets of the graphs of functions in two variables:

$$w = g(x_1, x_2), \quad y_1 = \frac{\partial g}{\partial x_1}, \quad y_2 = \frac{\partial g}{\partial x_2}.$$

---

[9]Of course, if it may be realized, at least locally, as a submanifold in the jet space $J^{k'}(\pi')$ for some π' and k'.

Let us denote such a manifold by V_g. The Frobenius conditions $d\mathcal{D}_{V_g} \subset \mathcal{D}_{V_g}$ of complete integrability are first-order partial differential equations for the functions determining the graph of the 1-jet $j_1(g)$. Hence, they are second order equations for the function g. Actually, in the examples below a scalar second order equation arises, and this is the desired quotient equation. Its solutions, as follows by construction, correspond to two-parameter families of solutions of the initial equation.

EXAMPLE 6.1. Consider the Laplace equation $\mathcal{E} = \{p_{(2,0)} + p_{(0,2)} = 0\}$. Let the group H be generated by translations along q_1 and q_2; respectively, its Lie algebra $\mathcal{H}$ possesses the basis $\partial/\partial q_1$, $\partial/\partial q_2$.

The form ω determining the contact structure on $\mathcal{E}/H$ is uniquely, up to a factor, defined by the conditions

$$\frac{\partial}{\partial q_1} \lrcorner \mu^*\omega = 0, \quad \frac{\partial}{\partial q_2} \lrcorner \mu^*\omega = 0,$$
$$\frac{\partial}{\partial q_1}(\mu^*\omega) = 0, \quad \frac{\partial}{\partial q_2}(\mu^*\omega) = 0,$$
(6.2)

where $\mu^*\omega = \alpha_0 \omega_0 + \alpha_1 \omega_1 + \alpha_2 \omega_2$.

By the last condition, equations (6.2) imply

$$(\alpha_0 p_1 + \alpha_1 p_{(2,0)} + \alpha_2 p_{(1,1)})\big|_\mathcal{E} = 0,$$
$$(\alpha_0 p_2 + \alpha_1 p_{(1,1)} + \alpha_2 p_{(0,2)})\big|_\mathcal{E} = 0,$$

from which, up to a factor, one obtains

$$\alpha_0 = (p_{(1,1)}^2 - p_{(2,0)} p_{(0,2)})\big|_\mathcal{E}, \quad \alpha_1 = (p_1 p_{(0,2)} - p_2 p_{(1,1)})\big|_\mathcal{E},$$
$$\alpha_2 = (p_2 p_{(2,0)} - p_1 p_{(1,1)})\big|_\mathcal{E}.$$

The second pair of conditions (6.2) means that $\mu^*\omega$ is independent of q_1, q_2. Therefore, the factorization mapping μ can be identified with the projection to the plane $\{q_1 = q_2 = 0\} \subset J^2(\mathbb{R}^2)$. Thus,

$$\mu^*\omega = f \cdot [(p_{(1,1)}^2 - p_{(2,0)} p_{(0,2)}) \, du + (p_1 p_{(0,2)} - p_2 p_{(1,1)}) \, dp_1$$
$$+ (p_2 p_{(2,0)} - p_1 p_{(1,1)}) \, dp_2]\big|_\mathcal{E},$$

where f is a function (coefficient of proportionality).

Set $f = (p_{(1,1)}^2 - p_{(2,0)} p_{(0,2)})^{-1}$. Since the equality $p_{(0,2)} = -p_{(2,0)}$ holds on the equation, we have

$$\omega = d\bar{u} - \frac{\bar{p}_2 \bar{p}_{(1,1)} + \bar{p}_1 \bar{p}_{(2,0)}}{\bar{p}_{(2,0)}^2 + \bar{p}_{(1,1)}^2} d\bar{p}_1 - \frac{\bar{p}_1 \bar{p}_{(1,1)} - \bar{p}_2 \bar{p}_{(2,0)}}{\bar{p}_{(2,0)}^2 + \bar{p}_{(1,1)}^2} d\bar{p}_2,$$

where $\bar{\xi}$ denotes the restriction of ξ to the equation. Taking into account that $\omega = dv - y_1 \, dx_1 - y_2 \, dx_2$, we obtain

$$v = \bar{u}, \quad x_i = \bar{p}_i, \quad y_1 = \frac{\bar{p}_2 \bar{p}_{(1,1)} + \bar{p}_1 \bar{p}_{(2,0)}}{\bar{p}_{(2,0)}^2 + \bar{p}_{(1,1)}^2}, \quad y_2 = \frac{\bar{p}_1 \bar{p}_{(1,1)} - \bar{p}_2 \bar{p}_{(2,0)}}{\bar{p}_{(2,0)}^2 + \bar{p}_{(1,1)}^2}.$$

From this, in turn, we can see that

$$\bar{u} = v, \quad \bar{p}_i = x_i, \quad \bar{p}_{(2,0)} = \frac{x_1 y_1 - x_2 y_2}{y_1^2 + y_2^2}, \quad \bar{p}_{(1,1)} = \frac{x_1 y_2 + x_2 y_1}{y_1^2 + y_2^2}.$$

In these new coordinates, restrictions of the Cartan forms to the equation are of the form

$$\begin{aligned}
\overline{\omega}_0 &= dv - x_1\, dq_1 - x_2\, dq_2,\\
\overline{\omega}_1 &= dx_1 - \frac{x_1 y_1 - x_2 y_2}{y_1^2 + y_2^2}\, dq_1 - \frac{x_1 y_2 + x_2 y_1}{y_1^2 + y_2^2}\, dq_2,\\
\overline{\omega}_2 &= dx_2 - \frac{x_1 y_2 + x_2 y_1}{y_1^2 + y_2^2}\, dq_1 + \frac{x_1 y_1 - x_2 y_2}{y_1^2 + y_2^2}\, dq_2.
\end{aligned} \tag{6.3}$$

Consider now in $\mathcal{E}/H$ a submanifold of the form V_g, which is the graph of the 1-jet of the function $g(x_1, x_2)$:

$$v = g(x_1, x_2), \quad y_i = \frac{\partial g}{\partial x_i}.$$

The same equations describe the manifold $\widetilde{V}_g = \mu^{-1} V_g \subset J^2(\mathbb{R}^2)$. Introduce the notation $\omega_i^g = \omega_i|_{\widetilde{V}_g}$. Since $\omega|_{V_g} = 0$, we have

$$\alpha_0 \omega_0^g + \alpha_1 \omega_1^g + \alpha_2 \omega_2^g = 0,$$

i.e., the forms $\{\omega_i^g\}$ are linearly independent[10]. Therefore, one can assume, for example, that in the domain $\alpha_2 = -y_2 \neq 0$ the restriction of the Cartan distribution to $\widetilde{V}_g$ is determined by the forms ω_0^g, ω_1^g. Consequently, integrability conditions for this distribution are

$$d\omega_0^g \wedge \omega_0^g \wedge \omega_1^g = 0, \qquad d\omega_1^g \wedge \omega_0^g \wedge \omega_1^g = 0. \tag{6.4}$$

On the other hand, using (6.3), one can obtain the following expressions for the basis forms ω_0^g, ω_1^g:

$$\begin{aligned}
\omega_0^g &= g_{x_1}\, dx_1 + g_{x_2}\, dx_2 - x_1\, dq_1 - x_2\, dq_2,\\
\omega_1^g &= dx_1 - \frac{x_1 y_1 - x_2 y_2}{y_1^2 + y_2^2}\, dq_1 - \frac{x_1 y_2 + x_2 y_1}{y_1^2 + y_2^2}\, dq_2.
\end{aligned}$$

Rather cumbersome computations show that when one substitutes these expressions into (6.4), the first condition in (6.4) holds identically while the second is equivalent to the equation $y_2(g_{x_1 x_1} + g_{x_2 x_2}) = 0$. But in the domain we chose we have $y_2 \neq 0$, and therefore the equality

$$g_{x_1 x_1} + g_{x_2 x_2} = 0$$

coincides with the condition describing our functions. In other words, this is exactly the quotient equation for the Laplace equation[11].

Thus, the factorization procedure leads us to the Laplace equation again (note that the Laplace equation is linear and consequently its total symmetry algebra is infinite dimensional). From the above theoretical reasons and from coincidence of the initial equation with the quotient equation it follows that to any solution of the Laplace equation there corresponds a two-parameter family of solutions of the same equation.

---

[10]It is easily seen that $\alpha_0 = 1$, $\alpha_1 = -y_1$, $\alpha_2 = -y_2$, since $\omega = dv - y_1\, dx_1 - y_2\, dx_2$.

[11]Note that the result is independent of the choice of the domain $y_2 \neq 0$, which was convenient for computations; for another domain the quotient equation will be the same.

In fact, if $g(x_1, x_2)$ is a solution of the Laplace equation, then the manifold $\widetilde{V}_g$ is an integral manifold of the Cartan distribution, and the restrictions $D_{q_i}^g$ of the total derivatives D_{q_i} to $\widetilde{V}_g$ are of the form

$$\begin{aligned}
D_{q_1}^g &= \left(\frac{\partial}{\partial q_1} + p_1\frac{\partial}{\partial u} + p_{(2,0)}\frac{\partial}{\partial p_1} + p_{(1,1)}\frac{\partial}{\partial p_2}\right)\bigg|_{\widetilde{V}_g} \\
&= \frac{\partial}{\partial q_1} + \frac{x_1 y_1 - x_2 y_2}{y_1^2 + y_2^2}\frac{\partial}{\partial x_1} + \frac{x_1 y_2 + x_2 y_1}{y_1^2 + y_2^2}\frac{\partial}{\partial x_2}, \\
D_{q_2}^g &= \left(\frac{\partial}{\partial q_2} + p_1\frac{\partial}{\partial u} + p_{(1,1)}\frac{\partial}{\partial p_1} + p_{(0,2)}\frac{\partial}{\partial p_2}\right)\bigg|_{\widetilde{V}_g} \\
&= \frac{\partial}{\partial q_2} + \frac{x_1 y_2 + x_2 y_1}{y_1^2 + y_2^2}\frac{\partial}{\partial x_1} - \frac{x_1 y_1 - x_2 y_2}{y_1^2 + y_2^2}\frac{\partial}{\partial x_2},
\end{aligned} \qquad (6.5)$$

where $y_i = \partial g/\partial x_i$. We shall look for integral manifolds of the form

$$q_1 = \varphi(x_1, x_2), \qquad q_2 = \psi(x_1, x_2)$$

in $\widetilde{V}_g$. The functions φ are ψ obtained from the conditions

$$D_{q_i}^g(q_1 - \varphi) = 0, \quad D_{q_i}^g(q_2 - \psi) = 0, \qquad i = 1, 2.$$

The last system can be rewritten explicitly using formulas (6.5). Solving it with respect to $\partial\varphi/\partial x_i$, $\partial\psi/\partial x_i$, we obtain the following formulas:

$$\frac{\partial \varphi}{\partial x_1} = \left(x_1\frac{\partial g}{\partial x_1} - x_2\frac{\partial g}{\partial x_2}\right)r, \qquad \frac{\partial \varphi}{\partial x_2} = \left(x_1\frac{\partial g}{\partial x_2} + x_2\frac{\partial g}{\partial x_1}\right)r,$$

$$\frac{\partial \psi}{\partial x_1} = \left(x_1\frac{\partial g}{\partial x_2} + x_2\frac{\partial g}{\partial x_1}\right)r, \qquad \frac{\partial \psi}{\partial x_2} = \left(x_2\frac{\partial g}{\partial x_2} - x_1\frac{\partial g}{\partial x_1}\right)r,$$

where $r = (x_1^2 + x_2^2)^{-1}$. Thus, when g is given, the functions φ, ψ are obtained by straightforward integration.

For example, if $g = x_2$, then $\varphi = \arctan(x_2/x_1) + C_1$, $\psi = \ln(x_1^2 + x_2^2)/2 + C_2$. Coming back to the standard coordinates on $J^2(\mathbb{R}^2)$, we shall have the following relations valid on $\widetilde{V}_g$ (where $g = x_2$):

$$u = p_2, \quad p_{(2,0)} = -p_{(0,2)}, \quad p_{(2,0)} = -p_2, \quad p_{(1,1)} = p_1,$$

$$C_1 + q_1 = \arctan\frac{p_2}{p_1}, \quad C_2 + q_2 = \frac{\ln(p_1^2 + p_2^2)}{2}, \quad C_1, C_2 \in \mathbb{R}.$$

Expressing $p_2 = u$ via q_i, we shall obtain the desired two-parameter family of solutions of the Laplace equation:

$$u = e^{q_2 + C_2}\sin(q_1 + C_1).$$

EXAMPLE 6.2. Consider the Laplace equation $\mathcal{E} = \{p_{(2,0)} + p_{(0,2)} = 0\}$ again. Let the group H correspond to the Lie algebra $\mathcal{H}$ generated by the fields $X_{q_1} = \partial/\partial p_1 + q_1\partial/\partial u$ and $X_{q_2} = \partial/\partial p_2 + q_2\partial/\partial u$.

Acting in the same way as we did in Example 6.1, we shall see that the quotient equation again coincides with the Laplace equation. To any solution g of the quotient equation there corresponds a two-parameter family of the initial equation.

These families can be found by solving the following system of equations:

$$\begin{cases} u = g(q_1, q_2) + q_1 p_1 + q_2 p_2, \\ p_{(0,2)} = -p_{(2,0)}, \\ p_{(2,0)} = -\dfrac{q_1 g_{q_1} - q_2 g_{q_2}}{q_1^2 + q_2^2}, \\ p_{(1,1)} = -\dfrac{q_2 g_{q_1} + q_1 g_{q_2}}{q_1^2 + q_2^2}, \\ p_1 = \varphi(q_1, q_2) + C_1, \\ p_2 = \psi(q_1, q_2) + C_2, \end{cases}$$

where C_1 and C_2 are arbitrary constants. In turn, the functions φ and ψ are determined by the following relations:

$$g_{q_1} \, dq_1 + g_{q_2} \, dq_2 + q_1 \, dp_1 + q_2 \, dp_2 = q_1 \, d(p_1 - \varphi) + q_2 \, d(p_2 - \psi),$$

$$(q_1^2 + q_2^2) \, dp_1 + (g_{q_1} q_1 - g_{q_2} q_2) \, dq_1 + (g_{q_2} q_1 + g_{q_1} q_2) \, dq_2 = (q_1^2 + q_2^2) \, d(p_1 - \varphi).$$

In particular, for $g = q_1$ we obtain the following two-parameter family of solutions of the Laplace equation:

$$u = q_1 \left(C_1 - \frac{\ln(q_1^2 + q_2^2)}{2} \right) + q_2 \left(C_2 + \arctan \frac{q_2}{q_1} \right),$$

EXAMPLE 6.3. Consider the wave equation $p_{(2,0)} - p_{(0,2)} = 0$ and the group H of translations along the independent variables. The corresponding Lie algebra $\mathcal{H}$ is generated by the fields $\partial/\partial q_1$ and $\partial/\partial q_2$.

In this case, the quotient equation also coincides with the initial wave equation. Any solution g of the quotient equation corresponds to a two-parameter family of solutions of the initial equation. This family can be found from the relations

$$\begin{cases} u = g(p_1, p_2), \\ p_{(2,0)} = p_{(0,2)}, \\ p_{(2,0)} = \dfrac{p_1 g_{p_1} - p_2 g_{p_2}}{g_{p_1}^2 - g_{p_2}^2}, \\ p_{(1,1)} = \dfrac{p_2 g_{p_1} - p_1 g_{p_2}}{g_{p_1}^2 - g_{p_2}^2}, \\ q_1 = C_1 + \varphi(p_1, p_2), \\ q_2 = C_2 + \psi(p_1, p_2), \end{cases}$$

where C_1 and C_2 are arbitrary real constants. In turn, the functions φ and ψ are determined from the relations

$$d\varphi = \frac{p_2 g_{p_2} - p_1 g_{p_1}}{p_2^2 - p_1^2} \, dp_1 + \frac{p_2 g_{p_1} - p_1 g_{p_2}}{p_2^2 - p_1^2} \, dp_2,$$

$$d\psi = \frac{p_2 g_{p_1} - p_1 g_{p_2}}{p_2^2 - p_1^2} \, dp_1 + \frac{p_2 g_{p_2} - p_1 g_{p_1}}{p_2^2 - p_1^2} \, dp_2.$$

In particular, for $g = q_1$ we obtain the following two-parameter family of solutions of the wave equation:

$$u = e^{q_1 + C_1} \cosh(q_2 + C_2).$$

EXAMPLE 6.4. Let us consider the heat equation $p_1 = p_{(0,2)}$. Choose the group H of translations along the independent variables with the Lie algebra $\mathcal{H}$ generated by the fields $\partial/\partial q_1$ and $\partial/\partial q_2$.

In suitable coordinates x, y, v, the quotient equation is a quasilinear parabolic equation of the form
$$(xv_x)^2 v_{xx} - 2(xv_x)(y - xv_y)v_{xy} + (y - xv_y)^2 v_{yy} = 0.$$

EXERCISE 6.1. Show that the quotient equation of the heat equation over the group of scale symmetries $u \mapsto \tau u$ coincides with the Burgers equation.

EXAMPLE 6.5. Consider the Burgers equation $p_2 + up_1 + p_{(2,0)} = 0$. Let H be the group whose Lie algebra is generated by the fields $\partial/\partial q_1$ and $-q_2 \partial/\partial q_1 + \partial/\partial u$, i.e., by translation along q_1 and by Galilean symmetry.

In suitable coordinates x, y, v the quotient equation is of the form
$$v_{xx} - \frac{v_x^2}{2v} + v_x \frac{x^2}{2v} + \frac{v_y}{2v} - 3x = 0.$$

EXAMPLE 6.6 (evolution equations in one space variable). We expound here the results of [**116**], where evolution equations with one space variable were studied. Most of the examples considered there are particular cases of the above general construction. But specifics of the equation $u_t = F(t, x, u, u_x, u_{xx}, \dots)$ allow one to obtain complete answers related to the factorization procedure.

The functions constant on the orbits of the group H or of its action on higher order jets are called *scalar differential invariants* [**3, 121, 138**]. In [**116**], it was proved that if $\dim \mathcal{H} = m$, then in the generic situation the number of independent differential invariants on $J^m(\mathbb{R}^2)$ equals 3.

These three invariants are chosen as one dependent and two independent variables (i.e., as coordinates in $J^0(\mathbb{R}^2)$) for the quotient equation. For total derivatives along new coordinates "invariant derivations" are chosen, i.e., linear combinations $pD_t + qD_x$ such that $[pD_t + qD_x, h] = 0$ for an arbitrary generating function $h \in \mathcal{H}$. Naturally, the choice both of invariants and of invariant derivations is not unique, but reasonable rules are given to obtain quotient equations of the simplest form. In particular, using this approach, the following results were obtained.

1. Let a Lie algebra $\mathcal{G}_0$ consisting of symmetries of the equation $\mathcal{E}$ contain an ideal $\mathcal{G}_1$. If μ_i denotes the factorization determined by the algebra $\mathcal{G}_i$ and $\mathcal{E}_i$ is the corresponding quotient equation, then the algebra of all symmetries of the equation $\mathcal{E}_0$ contains a subalgebra $\mathcal{H} \simeq \mathcal{G}_0/\mathcal{G}_1$. For the factorization μ over the subalgebra $\mathcal{H}$ one has $\mu_0 = \mu \circ \mu_1$.

2. The result of a second order evolution equation factorization is either an evolution equation, or an equation which can be transformed to the equation
$$v_{tt} + 2vv_{xt} + v^2 v_{xx} + \Phi(v_t, v_x, v, x, t) = 0$$
by point transformations. In particular, it is always parabolic.

3. Equations $u_t = u_{xx} + h(t, x)u$ always possess symmetries of the form $\alpha(t,x)\partial/\partial u$, where $\alpha(t,x)$ is a solution (gauge by a solution). Factorization over any finite-dimensional algebra consisting of such symmetries leads to a linear equation of the same type, but, in general, with a different potential h. In particular, when the algebra is one-dimensional, the quotient equation is of the form
$$v_t = v_{xx} + (h + 2(\ln \alpha)_{xx})v.$$

It turns out that factorization in this case coincides with the "dressing procedure" used in the inverse scattering problem. The passage from h to $h + 2(\ln \alpha)_{xx}$ is exactly the Darboux transformation.

7. Extrinsic and intrinsic symmetries

Up to now, we have understood symmetries as Lie transformations in $J^k(\pi)$. Such an approach corresponds to studying equations as if from "outside". Conceptually, the other approach is more reasonable. It is based on study of the restriction $\mathcal{C}(\mathcal{E})$ of the Cartan distribution to the equation $\mathcal{E} \subset J^k(\pi)$ and of diffeomorphisms of $\mathcal{E}$ preserving $\mathcal{C}(\mathcal{E})$. This point of view corresponds to the intrinsic geometry of the equation and seems to be more adequate to study individual equations, though far less convenient with regard to computations. Luckily, for a very broad class of equations these two approaches happen to be equivalent. In this section, we describe sufficient conditions for this equivalence. We say that equations for which the equivalence holds are *rigid*. The exposition below is based on results from [60, 130].

Let us introduce some notions needed below. Recall that the distribution $\mathcal{C}(\mathcal{E})$ is defined at each point as follows: $\mathcal{C}(\mathcal{E})_\theta = \mathcal{C}^k \cap T_\theta(\mathcal{E})$.

DEFINITION 7.1. A diffeomorphism of the manifold $\mathcal{E}$ is called an *intrinsic symmetry* if it preserves the distribution $\mathcal{C}(\mathcal{E})$. A vector field on $\mathcal{E}$ is called an *infinitesimal intrinsic symmetry* if the corresponding one-parameter group preserves $\mathcal{C}(\mathcal{E})$.

Denote by $\operatorname{Sym}_e(\mathcal{E})$ the group of *extrinsic* symmetries defined in §4 and by $\operatorname{Sym}_i(\mathcal{E})$ the group of intrinsic symmetries of the equation $\mathcal{E}$. The restriction operation leads to the natural mapping

$$\operatorname{Sym}_e \mathcal{E} \longrightarrow \operatorname{Sym}_i \mathcal{E}.$$

Equations for which this mapping is an epimorphism are called *common*. An equation is rigid if the mapping is bijective. Of course, rigid equations are common, but the converse is not true (for example, first order equations in one dependent variable; see below).

DEFINITION 7.2. An equation $\mathcal{E} \subset J^k(\pi)$ is called $\mathcal{C}$-*general*, if:
1. The set $\pi_{k,k-1}(\mathcal{E})$ is everywhere dense in $J^{k-1}(\pi)$.
2. The fibers of the bundle $\pi_\mathcal{E} = \pi_{k,k-1}|_\mathcal{E} : \mathcal{E} \to \pi_{k,k-1}(\mathcal{E})$ are connected, and they and only they are integral manifolds of maximal dimension for the distribution $\mathcal{C}(\mathcal{E})$.

In fact, any intrinsic symmetry of a $\mathcal{C}$-general equation $\mathcal{E} \subset J^k(\pi)$ determines a transformation in the space $J^{k-1}(\pi)$. Namely, the following fact is valid:

PROPOSITION 7.1. *If $\mathcal{E} \subset J^k(\pi)$ is a $\mathcal{C}$-general equation and A is an intrinsic symmetry of $\mathcal{E}$, then there exists a diffeomorphism A' of the space $J^{k-1}(\pi)$ such that the diagram*

$$\begin{array}{ccc} \mathcal{E} & \xrightarrow{A} & \mathcal{E} \\ \pi_\mathcal{E} \downarrow & & \downarrow \pi_\mathcal{E} \\ J^{k-1}(\pi) & \xrightarrow{A'} & J^{k-1}(\pi) \end{array} \qquad (7.1)$$

is commutative.

PROOF. Let $\theta_{k-1} \in \pi_{k,k-1}(\mathcal{E})$. Then, by definition, the set $\pi_{k,k-1}^{-1}(\theta_{k-1})$ is an integral manifold of maximal dimension for the distribution $\mathcal{C}(\mathcal{E})$. Since A is an automorphism of the distribution $\mathcal{C}(\mathcal{E})$, the set $A(\pi_{k,k-1}^{-1}(\theta_{k-1}))$ is also an integral manifold of maximal dimension and consequently is of the form $\pi_{k,k-1}^{-1}(\theta'_{k-1})$ for some point $\theta'_{k-1} \in \pi_{k,k-1}(\mathcal{E})$. Let us set $A'(\theta_{k-1}) = \theta'_{k-1}$. Commutativity of the diagram is obvious. □

Let us now discover when the mapping A' is a Lie transformation of the space $J^{k-1}(\pi)$. To this end, we shall need two more notions.

DEFINITION 7.3. An equation $\mathcal{E} \subset J^k(\pi)$ is called $\mathcal{C}$-*complete* if for any point $\theta_{k-1} \in \pi_{k,k-1}(\mathcal{E})$ the span of the set[12]

$$\bigcup_{\theta_k \in \pi_\mathcal{E}^{-1}(\theta_{k-1})} L_{\theta_k} \subset T_{\theta_{k-1}} J^{k-1}(\pi)$$

coincides with the Cartan plane $\mathcal{C}_{\theta_{k-1}}$.

Thus, an equation is $\mathcal{C}$-complete if it completely determines Cartan planes at the points of $\pi_{k,k-1}(\mathcal{E})$.

Denote by $\mathcal{E}^{(l)} \subset J^{k+l}(\pi)$ the l-*prolongation*[13] of the equation $\mathcal{E}$, i.e., the equation consisting of all *differential consequences* of order $\leq l$ of the equation $\mathcal{E}$.

DEFINITION 7.4. An equation $\mathcal{E} \subset J^k(\pi)$ is called l-*solvable* (at a point $\theta_k \in \mathcal{E}$) if the mapping $\pi_{l+k,k} \colon \mathcal{E}^{(l)} \to \mathcal{E}$ is a surjection (resp., if $\mathcal{E}^{(l)} \cap \pi_{l+k,k}^{-1}(\theta_k) \neq \varnothing$).

PROPOSITION 7.2. *If a $\mathcal{C}$-general equation is 1-solvable and $\mathcal{C}$-complete, then the above mapping A' is a Lie transformation. Moreover, the 1-lifting of A' restricted to $\mathcal{E}$ coincides with A.*

PROOF. Since the equation $\mathcal{E}$ is $\mathcal{C}$-complete, to prove the first statement it suffices to show that the conditions $v \in L_{\theta_k}$, $\theta_k \in \pi_\mathcal{E}^{-1}(\theta_{k-1})$, $\theta_{k-1} \in \pi_{k,k-1}(\mathcal{E})$ imply that $A'_*(v) \in \mathcal{C}_{A'(\theta_{k-1})}$.

Let a point $\theta_{k+1} \in \mathcal{E}^{(1)}$ be such that $\pi_{k+1,k}(\theta_{k+1}) = \theta_k$. Then $L_{\theta_{k+1}} \subset T_{\theta_k}(\mathcal{E})$ and $(\pi_{k,k-1})_*(L_{\theta_{k+1}}) = L_{\theta_k}$. Therefore, there exists a vector $v_1 \in L_{\theta_{k+1}}$ such that $(\pi_\mathcal{E})_*(v_1) = v$. Then $A'_*(v) = A'_*((\pi_\mathcal{E})_*(v_1)) = (\pi_\mathcal{E})_*(A_*(v_1))$. But A is an automorphism of the distribution $\mathcal{C}(\mathcal{E})$, and consequently $A_*(v_1) \in \mathcal{C}_{A(\theta_k)}(\mathcal{E}) \subset \mathcal{C}_{A(\theta_k)}$.

On the other hand, $(\pi_{k,k-1})_*(\mathcal{C}_{A(\theta_k)}) = L_{A(\theta_k)}$, and so $A'_*(v) \in L_{A(\theta_k)} \subset \mathcal{C}_{A'(\theta_{k-1})}$.

The diffeomorphism A coincides with the restriction of the lifting of the Lie transformation A' to $\mathcal{E}$. This means that $A'_*(L_{\theta_k}) = L_{A(\theta_k)}$, where $\theta_k \in \mathcal{E}$. But the above arguments show that $A'_*(L_{\theta_k}) \subset L_{A(\theta_k)}$. Since A' is a diffeomorphism and $\dim L_{\theta_k} = \dim L_{A(\theta_k)} = n$, the second statement is also proved. □

Equations satisfying the assumptions of Proposition 7.2 will be called *normal*.

Let us explain why normal equations are rigid. If an equation is 1-solvable, the condition of $\mathcal{C}$-completeness can be replaced by the following one: the span of the spaces $(\pi_{k,k-1})_*(\mathcal{C}_{\theta_k}(\mathcal{E})) \subset \mathcal{C}_{\theta_{k-1}}$, where $\theta_k \in \pi_\mathcal{E}^{-1}(\theta_{k-1})$, coincides with $\mathcal{C}_{\theta_{k-1}}$. In fact, in this case we have $L_{\theta_{k+1}} \subset \mathcal{C}_{\theta_k}(\mathcal{E})$ for $\theta_{k+1} \in \mathcal{E}^{(1)}$, and $(\pi_{k,k-1})_*(L_{\theta_{k+1}}) = L_{\theta_k} = (\pi_{k,k-1})_*(\mathcal{C}_{\theta_k})$. Therefore, $(\pi_{k,k-1})_*(\mathcal{C}_{\theta_k}(\mathcal{E})) = L_{\theta_k}$.

---

[12] Recall that by L_{θ_k} we denote the R-plane in $J^{k-1}(\pi)$ determined by the point $\theta_k \in J^k(\pi)$ (see §2).

[13] For the geometric definition, see §3 of Chapter 4.

The last remark shows that normality is an "intrinsic" property, i.e., it can be formulated only in terms of the manifold $\mathcal{E}$ and of the distribution $\mathcal{C}(\mathcal{E})$ on this manifold. In fact, the integral manifolds of maximal dimension determine in this case a bundle ν whose base B can be endowed with the distribution

$$B \ni b \longmapsto \mathcal{C}_b = \text{span of the planes } \{ \nu_*(\mathcal{C}_y(\mathcal{E})) \mid y \in \nu^{-1}(b) \}.$$

In the situation under consideration, the base B and the distribution described are $J^{k-1}(\pi)$ and the Cartan distribution, respectively. If now one considers the manifold B^1 of all n-dimensional maximal integral planes of the distribution constructed on B, then $\mathcal{E} \subset B^1$ and $B^1 = J^k(\pi)$. This procedure shows that normality allows one to reconstruct the "environment" of the equation, i.e., to reconstruct the embedding $\mathcal{E} \subset J^k(\pi)$ using the distribution $\mathcal{C}(\mathcal{E})$ on $\mathcal{E}$ only. Thus we have proved the following theorem.

THEOREM 7.3. *Normal equations are rigid.*

Let us now write down explicitly the conditions of 1-solvability and $\mathcal{C}$-generality constituting the normality property.

Let the equation $\mathcal{E}$ be given as the zero set of the function $\boldsymbol{y} = \boldsymbol{F}(\boldsymbol{\theta}, \boldsymbol{u}, \boldsymbol{p}_\sigma)$, or $\mathcal{E} = \{ F_i(\boldsymbol{\theta}, \boldsymbol{u}, \boldsymbol{p}_\sigma) = 0 \mid 1 \leq i \leq r \}$. Then locally $\boldsymbol{F}$ can be considered as the mapping $\boldsymbol{F} \colon J^k(\pi) \to J^0(\pi')$, where π' is an r-dimensional bundle over the same base. Let a point $\theta_{k+1} \in J^{k+1}(\pi)$ be represented as a pair (θ, L), where L is an R-plane at the point $\theta_k = \pi_{k+1,k}(\theta_{k+1}) \in J^k(\pi)$. Then for almost all θ_{k+1} the plane $\boldsymbol{F}_*(L)$ will be an R-plane at the point $G(\theta_k)$. Consequently, setting $\boldsymbol{F}^{(1)}(\theta_{k+1}) = (F(\theta_k), F_*(L))$, we obtain a mapping $J^{k+1}(\pi) \to J^1(\pi')$. By definitions, the following diagram is commutative:

$$\begin{array}{ccccc} J^{k+1}(\pi) & \xrightarrow{\pi_{k+1,k}} & J^k(\pi) & \xrightarrow{\pi_k} & M \\ {\scriptstyle F^{(1)}} \downarrow & & {\scriptstyle F} \downarrow & & \downarrow {\scriptstyle \text{id}} \\ J^1(\pi') & \xrightarrow{\pi'_{1,0}} & J^0(\pi') & \xrightarrow{\pi'_0} & M \end{array} \qquad (7.2)$$

From commutativity of this diagram it follows that the intersection $\mathcal{E}^{(1)} \cap \pi_{k+1,k}^{-1}(\theta_k)$ is not empty, if the restriction of $F^{(1)}$ to the fiber,

$$\boldsymbol{F}^{(1)} \colon \pi_{k+1,k}^{-1}(\theta_k) \longrightarrow (\pi'_{1,0})^{-1}(\boldsymbol{F}(\theta_k)),$$

is surjective. But this restriction is exactly the leading symbol of the mapping $\boldsymbol{F}$. In coordinates, it is represented by the matrix $\partial G_i / \partial p_\sigma^j$, $|\sigma| = k$. Hence, the equation is 1-solvable if the leading symbol of the determining vector function is surjective.

If the equation is determined, i.e., the number of equations coincides with the number of dependent variables, then the dimension of fibers of the projection $J^{k+1}(\pi) \to J^k(\pi)$ is greater than the dimension of fibers of the projection $J^1(\pi') \to J^0(\pi')$. Therefore, in a generic situation we always have surjectivity. If, for example, $\dim \pi = \dim \pi' = 1$ and $\mathcal{E} = \{g = 0\}$, then the symbol is surjective provided the inequality $\partial g / \partial p_\sigma \neq 0$ holds for at least one multi-index σ, $|\sigma| = k$. Thus, 1-solvability is a rather weak condition.

The $\mathcal{C}$-generality condition is related to simple dimensional reasons.

PROPOSITION 7.4. *Let* $\mathcal{E} \subset J^k(\pi)$, $\dim \pi = m$, $\dim M = n$, *and let the fibers of the projection* $\pi_{\mathcal{E}}$ *be connected. Then* $\mathcal{E}$ *is a* $\mathcal{C}$-*general equation if the inequality*

$$\operatorname{codim} \mathcal{C} \leq m \frac{(n+k-2)!}{(k-1)!(n-1)!} - 2$$

holds.

PROOF. Indeed, the number

$$\lambda = m \frac{(n+k-2)!}{(k-1)!(n-1)!} - 1$$

equals the difference between dimensions of integral manifolds projecting to $J^{k-1}(\pi)$ with dimensions $r = 0$ and 1 (see Proposition 2.5). Therefore, when restricting the Cartan distribution to $\mathcal{E}$, dimensions of maximal integral manifolds with $r = 0$ will be diminished at least by λ. □

If $\mathcal{E}$ is a determined equation, then the assumptions of the previous proposition are satisfied in all cases except for: (a) $k = 1$, (b) $n = 1$, (c) $m = 1$, $k = n = 2$.

Let us consider examples of equations where extrinsic and intrinsic symmetries do not coincide.

EXAMPLE 7.1. Let $\theta_k \in J^k(\pi)$, where $k \geq 0$ for $\dim \pi > 1$ or $k \geq 1$ for $\dim \pi = 1$. Let $\mathcal{E} = \pi_{k+1,k}^{-1}(\theta_k)$. Then $\mathcal{E}$ is not a rigid and even not a common equation.

In fact, for any point $\theta_{k+1} \in \mathcal{E}$ the tangent plane $T_{\theta_{k+1}}(\mathcal{E})$ lies in the Cartan distribution, and consequently $\mathcal{C}_{\theta_{k+1}}(\mathcal{E}) = T_{\theta_{k+1}}(\mathcal{E})$. Therefore, the set of intrinsic symmetries of this equation coincides with the group of its diffeomorphisms. But any extrinsic symmetry of the equation $\mathcal{E}$ is determined by some Lie transformation of the space $J^k(\pi)$ leaving the point θ_k fixed.

EXAMPLE 7.2. First order equations in one dependent variable are common, but not rigid.

Let $k = \dim \pi = 1$. We confine ourselves to the case of determined equations, i.e., when $\operatorname{codim} \mathcal{E} = 1$. Let us first recall some facts from Chapter 2.

The contact structure on $J^1(\pi)$ is determined by any 1-form $\omega_\lambda = \lambda U_1(\pi)$, where λ is a function from $C^\infty(J^1(\pi))$ vanishing nowhere. To escape topological problems, we shall assume that there exists a function λ such that the form $d\,\omega_\lambda|_\mathcal{E}$ is nondegenerate. This is always true locally. In what follows, we fix λ and set $\omega = \omega_\lambda$, $\widetilde{\omega} = \omega_\lambda|_\mathcal{E}$.

It is now not difficult to describe the structure of intrinsic symmetries of the equation $\mathcal{E}$. Obviously, $X \in \operatorname{Sym}_i \mathcal{E}$ if and only if $X(\widetilde{\omega}) = \mu\widetilde{\omega}$, where $\mu \in C^\infty(\mathcal{E})$. Since $X(\widetilde{\omega}) = X \,\lrcorner\, d\widetilde{\omega} + d(X \,\lrcorner\, \widetilde{\omega})$, it can be easily seen that $X \in \operatorname{Sym}_i \mathcal{E}$ if and only if

$$\mu\widetilde{\omega} = X \,\lrcorner\, d\widetilde{\omega} + d\widetilde{f}, \qquad (7.3)$$

where $\widetilde{f}$ denotes $X \,\lrcorner\, \widetilde{\omega}$.

The form $d\widetilde{\omega}$ is nondegenerate, and therefore the mapping

$$G \colon \mathrm{D}(\mathcal{E}) \longrightarrow \Lambda^1(\mathcal{E}), \qquad G(Y) = -Y \,\lrcorner\, d\widetilde{\omega},$$

is an isomorphism of $C^\infty(\mathcal{E})$-modules of vector fields and 1-forms on $\mathcal{E}$. Applying the mapping G^{-1} to equation (7.3), we find that $X \in \operatorname{Sym}_i \mathcal{E}$ if and only if

$$\mu Y_{\widetilde{\omega}} = Y_{d\widetilde{f}} - X,$$

where $Y_\theta = G^{-1}(\theta)$, $\theta \in \Lambda^1(\mathcal{E})$.

Note now that $Y_\theta \lrcorner \theta = -Y_\theta \lrcorner (Y_\theta \lrcorner d\widetilde{\omega}) = 0$. Therefore,

$$(\mu Y_{\widetilde{\omega}})(\widetilde{\omega}) = (\mu Y_{\widetilde{\omega}}) \lrcorner d\widetilde{\omega} + d(\mu Y_{\widetilde{\omega}} \lrcorner \widetilde{\omega}) = -\mu\widetilde{\omega},$$

i.e., $\mu Y_{\widetilde{\omega}} \in \mathrm{Sym}_i \mathcal{E}$ (as is easily seen, the fields of the form $\mu Y_{\widetilde{\omega}}$ are directed along the characteristics of the equation $\mathcal{E}$). Consequently, $X \in \mathrm{Sym}_i \mathcal{E}$ if and only if $Y_{d\widetilde{f}} \in \mathrm{Sym}_i \mathcal{E}$. But

$$Y_{d\widetilde{f}}(\widetilde{\omega}) = Y_{d\widetilde{f}} \lrcorner d\widetilde{\omega} + d(Y_{d\widetilde{f}} \lrcorner \widetilde{\omega}) = d((X + \mu Y_{\widetilde{\omega}}) \lrcorner \widetilde{\omega}) - d\widetilde{f} = d(X \lrcorner \widetilde{\omega}) - d\widetilde{f} = 0.$$

Hence, $Y_{d\widetilde{f}} \in \mathrm{Sym}_i \mathcal{E}$ if and only if $Y_{d\widetilde{f}} \lrcorner \widetilde{\omega} = \widetilde{f}$.

Let us rewrite the last condition more clearly:

$$Y_{d\widetilde{f}} \lrcorner \widetilde{\omega} = -Y_{d\widetilde{f}} \lrcorner Y_{\widetilde{\omega}} \lrcorner d\widetilde{\omega} = Y_{\widetilde{\omega}} \lrcorner Y_{d\widetilde{f}} \lrcorner d\widetilde{\omega} = Y_{\widetilde{\omega}} \lrcorner (-d\widetilde{f}) = -Y_{\widetilde{\omega}}(\widetilde{f}).$$

Thus, we have proved the following result:

PROPOSITION 7.5. *A field $X \in \mathrm{D}(\mathcal{E})$ is an intrinsic symmetry of the equation $\mathcal{E}$ if and only if it is of the form*

$$X = \mu Y_{\widetilde{\omega}} + Y_{d\widetilde{f}},$$

where μ is a function and $Y_{\widetilde{\omega}}(\widetilde{f}) = -\widetilde{f}$. We also have $X(\widetilde{\omega}) = \mu\widetilde{\omega}$.

Recall that the 2-form $d\widetilde{\omega}$ determines a symplectic structure on $\mathcal{E}$ and the field Y_{dh}, $h \in C^\infty(\mathcal{E})$, is a Hamiltonian vector field with the Hamiltonian h. Therefore, a Hamiltonian field on $\mathcal{E}$ is an intrinsic symmetry if the Hamiltonian h satisfies the condition $Y_{\widetilde{\omega}}(h) = -h$.

Let us now describe the relations between the algebras $\mathrm{Sym}_i \mathcal{E}$ and $\mathrm{Sym}_e \mathcal{E}$. Recall that a contact field X on $J^1(\pi)$ is uniquely determined by its generating function $f = X \lrcorner \omega$ and is a unique solution of the system consisting of the following two equations:

$$X \lrcorner d\omega + df = X_1(f)\omega, \qquad f = X \lrcorner \omega, \tag{7.4}$$

where the field $X_1 = \partial/\partial x_1$ is uniquely determined by the following conditions:

$$X_1 \lrcorner \omega = 1, \qquad X_1 \lrcorner d\omega = 0.$$

THEOREM 7.6. *A contact field X_f is an extrinsic symmetry of a first order equation $\mathcal{E}$ in one dependent variable if and only if $Y_{\widetilde{\omega}}(\widetilde{f}) = -\widetilde{f}$.*

PROOF. From (7.4) one obtains

$$X_g(f) + X_g \lrcorner X_f \lrcorner d\omega = X_g(df + X_f \lrcorner d\omega) = X_g \lrcorner X_1(f)\omega = gX_1(f).$$

In a similar way, the equality $X_f(g) + X_f \lrcorner X_g \lrcorner d\omega = fX_1(g)$ holds. Since $X_f \lrcorner X_g \lrcorner d\omega + X_g \lrcorner X_f \lrcorner d\omega = 0$, we have

$$X_f(g) + X_g(f) = fX_1(g) + gX_1(f) = X_1(fg).$$

Let locally $\mathcal{E} = \{g = 0\}$, $dg_\theta \neq 0$, $\theta \in \mathcal{E}$. Then, if X is tangent to $\mathcal{E}$, from the last equality it follows that

$$X_f(g)|_{\mathcal{E}} = -\widetilde{X}_g(\widetilde{f}) + \widetilde{f}\, X_1(g)|_{\mathcal{E}}, \tag{7.5}$$

where $\widetilde{X}$ denotes the restriction of the field X to $\mathcal{E}$ (recall that the field X_g is always tangent to the hypersurface $\{g = 0\}$). Restricting the equality $-X_g \, \lrcorner \, d\omega = dg - X_1(g)\omega$ to $\mathcal{E}$, we obtain

$$-\widetilde{X}_g \, \lrcorner \, d\widetilde{\omega} = \alpha\widetilde{\omega}, \qquad \alpha = -\, X_1(g)|_{\mathcal{E}}\,.$$

Therefore, in the above notation,

$$\widetilde{X}_g = \alpha Y_{\widetilde{\omega}}$$

and consequently (7.5) can be rewritten in the form

$$X_f(g)|_{\mathcal{E}} = -\alpha Y_{\widetilde{\omega}}(\widetilde{f}) - \alpha \widetilde{f}.$$

Thus, the equality $X_f(g)|_{\mathcal{E}} = 0$ (i.e., the condition that X_f is tangent to the manifold $\mathcal{E}$) will hold if and only if $Y_{\widetilde{\omega}}(\widetilde{f}) = -\widetilde{f}$, since $\alpha = -X_1(g)|_{\mathcal{E}} \ne 0$. The last fact is implied by $X_1|_\theta \notin T_\theta(\mathcal{E})$, since $X_1 \in \ker d\omega|_\theta$, while the form $d\omega|_{\mathcal{E}}$ is nondegenerate. $\square$

COROLLARY 7.7. *For first order equations in one dependent variable, the mapping* $\mathrm{Sym}_e\,\mathcal{E} \to \mathrm{Sym}_i\,\mathcal{E}$ *is always epimorphic.*

PROOF. Consider the function $h = f + \lambda g$, $\lambda \in C^\infty(J^1(\pi))$, $\lambda \ne 0$, and let us try to find the function λ to satisfy the equality $\widetilde{X}_h = \mu Y_{\widetilde{\omega}} + Y_{d\widetilde{f}}$. By Theorem 7.6, the field $\widetilde{X}_f$ is an intrinsic symmetry of the equation $\mathcal{E}$, and consequently, by Proposition 7.5, the equality

$$\widetilde{X}_f = \nu Y_{\widetilde{\omega}} + Y_{d\widetilde{f}}, \qquad \nu \in C^\infty(\mathcal{E}),$$

holds. Further, as was shown in Chapter 2, $X_{\varphi\psi} = \varphi X_\psi + \psi X_\varphi - \varphi\psi X_1$. From this equality and from (7.5) we obtain

$$\widetilde{X}_{\lambda g} = \widetilde{\lambda} \widetilde{X}_g = \alpha \widetilde{\lambda} Y_{\widetilde{\omega}}, \qquad \widetilde{\lambda} = \lambda|_{\mathcal{E}}\,.$$

Therefore, choosing λ to satisfy

$$\widetilde{\lambda} = \frac{\mu - \nu}{\alpha}, \qquad \alpha = -\, X_1(g)|_{\mathcal{E}},$$

we can assume that $\widetilde{X}_h = \mu Y_{\widetilde{\omega}} + Y_{d\widetilde{f}}$. $\square$

Let $\widetilde{X}_f = \mu Y_{\widetilde{\omega}} + Y_{d\widetilde{f}}$. Then, restricting the equality

$$X_f(\omega) = X_f \, \lrcorner \, d\omega + df = X_1(f)\omega$$

to the equation $\mathcal{E}$ and taking into account that X_f is tangent to $\mathcal{E}$, we obtain

$$\widetilde{X}_f(\widetilde{\omega}) = \widetilde{X}_f \, \lrcorner \, d\widetilde{\omega} + d\widetilde{f} = \beta\widetilde{\omega}, \qquad \beta = X_1(f)|_{\mathcal{E}}\,.$$

On the other hand, as Proposition 7.5 shows, if $Y = \mu Y_{\widetilde{\omega}} + Y_{d\widetilde{f}}$, then $Y(\widetilde{\omega}) = \mu\widetilde{\omega}$. Therefore, $\beta = \mu$. Thus, the following result is valid:

COROLLARY 7.8. *Let* $Y = \mu Y_{\widetilde{\omega}} + Y_{d\widetilde{f}} \in \mathrm{Sym}_i\,\mathcal{E}$. *Then* $\widetilde{X}_f = Y$ *if and only if* $f|_{\mathcal{E}} = \widetilde{f}$ *and* $X_1(f)|_{\mathcal{E}} = -\mu$.

The last two corollaries imply the following result:

COROLLARY 7.9. *First order equations in one dependent variable are common but not rigid.*

CHAPTER 4

Higher Symmetries

In Chapter 3 we considered nonlinear partial differential equations as submanifolds of spaces $J^k(\pi)$ endowed with the Cartan distribution. Based on this approach, we constructed the symmetry theory. In fact, if one considers a differential equation not "as is" but together with all its differential consequences[1], the symmetry theory can be naturally generalized. These differential consequences form the so-called *infinite prolongation* of an equation, while the spaces of *infinite jets* are a natural environment for these prolongations. The Cartan distribution exists on infinite jets as well and, contrary to the finite case, is completely integrable. Studying automorphisms of this distribution, one comes to the concept of a *higher symmetry* of a differential equation. A widely known example of such symmetries is the series of the higher Korteweg–de Vries equations and similar series for other integrable nonlinear systems (see for example [**22**]).

Spaces of infinite jets and infinite prolongations of most equations are infinite-dimensional manifolds. For this reason, we start our exposition with a description of basic differential geometric constructions (such as smooth functions, vector fields, differential forms, etc.) on these manifolds (§1). In §2, the Cartan distribution on the space of infinite jets is introduced and infinitesimal automorphisms of this distribution are studied. A complete description of these automorphisms is given in terms of *evolutionary derivations* (a generalization of Lie fields) and the corresponding generating functions. The next section deals with the geometric definition of infinite prolongations. Defining equations for higher symmetries are also obtained here. Finally, in §4, some computational examples are considered.

1. Spaces of infinite jets
and basic differential geometric structures on them

The space of infinite jets of sections of a fiber bundle $\pi\colon P \to M$ is the inverse limit of the tower of finite jets with respect to the projections $\pi_{k+1,k}\colon J^{k+1}(\pi) \to J^k(\pi)$. The aim of this section is to introduce the basic elements of calculus and differential geometry on $J^\infty(\pi)$ (smooth functions, vector fields, differential forms, etc.) needed to construct the theory of higher symmetries. Solving this problem meets with methodological difficulties, because $J^\infty(\pi)$ is infinite dimensional. To overcome these difficulties, we pass to a "dual", algebraic language of smooth function rings on these manifolds. The algebraic-geometrical dualism considerably simplifies the definition of necessary concepts and makes all considerations sufficiently transparent. A good way to check efficiency of the algebraic language is to try to rewrite the material below using a purely geometrical approach.

---

[1] Note that in §7 of Ch. 3, when discussing extrinsic and intrinsic symmetries, we already found it necessary to consider differential consequences of an equation.

1.1. The manifolds $J^\infty(\pi)$. Let M be a smooth manifold of dimension n, and let $\pi\colon P \to M$ be a smooth locally trivial vector bundle over M whose fiber is of dimension m.

Let us consider the chain of projections (see Chapter 3)

$$M \xleftarrow{\pi} P \xleftarrow{\pi_{1,0}} J^1(\pi) \longleftarrow \cdots \longleftarrow J^k(\pi) \xleftarrow{\pi_{k+1,k}} J^{k+1}(\pi) \longleftarrow \cdots \quad (1.1)$$

and for any point $x \in M$ let us choose a sequence of points $\theta_l \in J^l(\pi)$, $l = 0, 1, \ldots, k, \ldots$, such that $\pi_{l+1,l}(\theta_{l+1}) = \theta_l$ and $\pi(\theta_0) = x$. Due to these equalities and the Borel lemma [80], using the definition of the spaces $J^l(\pi)$ one can choose a local section s of the bundle π such that $\theta_k = [s]_x^l$ for any l. Thus any point θ_l is determined by the partial derivatives up to order l of the section s at the point x, while the whole sequence of points $\{\theta_l\}$ contains information on all partial derivatives of the section s at x. Denote by $J^\infty(\pi)$ the set of all such sequences. Points of the space $J^\infty(\pi)$ may be obviously understood as classes of sections of the bundle π tangent to each other with infinite order or, which is the same, as infinite Taylor series of these sections.

For any point $\theta_\infty = \{x, \theta_k\}_{k\in\mathbb{N}} \in J^\infty(\pi)$, let us set $\pi_{\infty,k}(\theta_\infty) = \theta_k$ and $\pi_\infty(\theta_\infty) = x$. Then for all $k \geq l \geq 0$ one has the following commutative diagrams:

$$\begin{array}{ccc} J^\infty(\pi) \xrightarrow{\pi_{\infty,k}} J^k(\pi) & \quad & J^\infty(\pi) \xrightarrow{\pi_{\infty,k}} J^k(\pi) \\ \searrow_{\pi_\infty} \swarrow_{\pi_k} & & \searrow_{\pi_{\infty,l}} \swarrow_{\pi_{k,l}} \\ M & & J^l(\pi) \end{array} \quad (1.2)$$

i.e., the equalities $\pi_k \circ \pi_{\infty,k} = \pi_\infty$ and $\pi_{k,l} \circ \pi_{\infty,k} = \pi_{\infty,l}$ are valid. In addition, if s is a section of the bundle π, then the mapping $j_\infty(s)\colon M \to J^\infty(\pi)$ is defined by the equality $j_\infty(s)(x) = \{x, [s]_x^k\}_{k\in\mathbb{N}}$. One has the following identities: $\pi_{\infty,k} \circ j_\infty(s) = j_k(s)$ and $\pi_\infty \circ j_\infty(s) = \mathrm{id}_M$, where id_M is the identity diffeomorphism of the manifold M.

DEFINITION 1.1. The section $j_\infty(s)$ of the bundle $\pi_\infty\colon J^\infty(\pi) \to M$ is called the *infinite jet of the section* $s \in \Gamma(\pi)$.

Similarly to the spaces $J^k(\pi)$, $k \leq \infty$, the set $J^\infty(\pi)$ is endowed with a natural structure of a smooth manifold, but, in contrast to the former, it is infinite-dimensional. Local coordinates arising in $J^\infty(\pi)$ over a neighborhood $\mathcal{U} \subset M$ are $x_1, \ldots, x_n$ together with all functions p_σ^j, where $|\sigma|$ is of an arbitrary (but finite) value.

DEFINITION 1.2. The bundle $\pi_\infty\colon J^\infty(\pi) \to M$ is called the *bundle of infinite jets*, while the space $J^\infty(\pi)$ is called the *manifold of infinite jets* of the bundle π.

Our first aim is to construct on $J^\infty(\pi)$ the analogs of the basic differential-geometric concepts one meets in calculus over finite-dimensional manifolds. When doing this, we shall use the following informal but highly important principle: any natural construction on $J^\infty(\pi)$ is to "remember" the fact that the manifold of infinite jets is the inverse limit of the tower of projections of finite-dimensional manifolds (1.1). Let us begin with the notion of a smooth function on $J^\infty(\pi)$.

1.2. Smooth functions on $J^\infty(\pi)$. Let M be a smooth manifold and $C^\infty(M)$ the set of smooth functions on it. If M' is another smooth manifold and $G\colon M' \to M$ is a smooth mapping, then the latter generates the mapping $G^*\colon C^\infty(M) \to$

$C^\infty(M')$ defined by $G^*(f)(x') = f(G(x'))$, $f \in C^\infty(M)$, $x' \in M'$, and satisfying the identities $G^*(f_1 + f_2) = G^*(f_1) + G^*(f_2)$, $G^*(f_1 f_2) = G^*(f_1) G^*(f_2)$, and $G^*(\alpha f) = \alpha G^*(f)$ for any $f_1, f_2, f \in C^\infty(M)$ and $\alpha \in \mathbb{R}$. In other words, G^* is a homomorphism of $\mathbb{R}$-algebras. If G is a submersion, then it is easily seen that G^* is a monomorphism, i.e., $\ker G^* = 0$. Let us consider now the sequence of submersions (1.1) and denote by $\mathcal{F}_k(\pi)$ the ring of smooth functions on the manifold $J^k(\pi)$, $k \geq 0$, and by $\mathcal{F}_{-\infty}(\pi)$ the ring $C^\infty(M)$. Then (1.1) determines the sequence of $\mathbb{R}$-algebra embeddings

$$\mathcal{F}_{-\infty}(\pi) \xrightarrow{\nu} \mathcal{F}_0(\pi) \xrightarrow{\nu_{1,0}} \cdots \longrightarrow \mathcal{F}_k(\pi) \xrightarrow{\nu_{k+1,k}} \mathcal{F}_{k+1}(\pi) \longrightarrow \cdots,$$

where $\nu = \pi^*$ and $\nu_{k+1,k} = \pi_{k+1,k}^*$.

Now let us define the algebra $\mathcal{F} = \mathcal{F}(\pi)$ of smooth functions on $J^\infty(\pi)$. From the existence of the projections $\pi_{\infty,k} \colon J^\infty(\pi) \to J^k(\pi)$ it follows that any algebra $\mathcal{F}_k = \mathcal{F}_k(\pi)$ is embedded in the algebra $\mathcal{F}$ by some homomorphism ν_k. From the equalities $\pi_l \circ \pi_{k,l} = \pi_k$, $k \geq l$ (see Chapter 3), and from the commutative diagram (1.2) it follows that the diagram

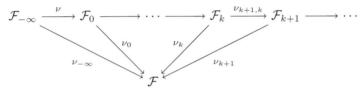

is also commutative. Consequently, the algebra $\mathcal{F}$ must contain the union of all algebras $\mathcal{F}_k$, $k = -\infty, 0, 1, \ldots$. On the other hand, since $J^\infty(\pi)$ is completely determined by its projections to the manifolds $J^k(\pi)$, it is natural to assume that $\mathcal{F}$ is exhausted by this union. Thus, let us set $\mathcal{F}(\pi) = \bigcup_k \mathcal{F}_k(\pi)$. From this definition it follows that any function φ on $J^\infty(\pi)$ is uniquely determined by its restriction to some manifold $J^k(\pi)$ with a sufficiently big k. This number k is called the *filtration degree* of the function φ and is denoted by $\deg(\varphi)$. Obviously, the set $\mathcal{F}$ is a commutative $\mathbb{R}$-algebra, the operations in which are related to filtrations in the following way:

$$\begin{aligned} \deg(\varphi_1 + \varphi_2) &\leq \max(\deg(\varphi_1), \deg(\varphi_2)), \\ \deg(\varphi_1 \varphi_2) &= \max(\deg(\varphi_1), \deg(\varphi_2)), \\ \deg(\alpha \varphi) &= \deg(\varphi), \end{aligned} \qquad (1.3)$$

where $\varphi_1, \varphi_2, \varphi \in \mathcal{F}$, $\alpha \in \mathbb{R}$, $\alpha \neq 0$.

The algebras $\mathcal{F}_k$ are subalgebras of $\mathcal{F}$ determined by the conditions $\mathcal{F}_k = \{\, \varphi \in \mathcal{F} \mid \deg(\varphi) \leq k \,\}$. The algebra $\mathcal{F}$, together with the function $\deg \colon \mathcal{F} \to \mathbb{Z}$ taking integer values (and, if necessary, the values $\pm\infty$) and possessing the properties (1.3), is called *filtered*, or an *algebra with filtration*. Filtration of the algebra $\mathcal{F}$ is the algebraic counterpart of the tower of projections (1.1). We can now revise the above-mentioned informal principle:

All natural differential-geometric constructions on $J^\infty(\pi)$ are to be consistent with the structure of a filtered algebra in $\mathcal{F}(\pi)$.

EXAMPLE 1.1. Let $G \colon J^k(\pi) \to J^k(\pi)$ be a Lie transformation of the space of k-jets. Then, as was shown in the previous chapter, G is the k-th lifting of some diffeomorphism G_0 of the space $J^0(\pi)$ if $\dim(\pi) > 1$, and is the $(k-1)$-st lifting of some contact transformation G_1 of the manifold $J^1(\pi)$ if $\dim(\pi) = 1$. The family

$\{G_\varepsilon^{(l)}\}$ of all liftings of the mapping G_ε (ε equals 0 or 1, depending on the dimension of π) is compatible with the projections $\pi_{l,l-1}$ and determines an automorphism G_ε^* of the algebra $\mathcal{F}$. Moreover, if $\varphi \in \mathcal{F}$ and $\deg(\varphi) \geq \varepsilon$, then

$$\deg G(\varphi) = \deg(\varphi).$$

Therefore one can assume that any Lie transformation (together with all its liftings) determines a diffeomorphism of the space $J^\infty(\pi)$.

DEFINITION 1.3. A mapping $G \colon J^\infty(\pi) \to J^\infty(\xi)$, where $\pi \colon E \to V$ and $\xi \colon Q \to M$ are vector bundles, is called *smooth* if for any smooth function $\varphi \in \mathcal{F}(\xi)$ the element $G^*(\varphi) = \varphi \circ G \in \mathcal{F}(\pi)$ is a smooth function on the space $J^\infty(\pi)$ and there exist integers l and l_0 such that

$$\deg G^*(\varphi) = \deg(\varphi) + l$$

for all $\varphi \in \mathcal{F}(\xi)$ with $\deg(\varphi) \geq l_0$.

Below we shall describe a broad class of smooth mappings which, unlike the mappings described in Example 1.1, raise filtration degree. To do this, we shall need the following interpretation of elements $\varphi \in \mathcal{F}(\pi)$.

Let $\varphi \in \mathcal{F} = \mathcal{F}(\pi)$ and $\deg(\varphi) = k$, i.e., $\varphi \in \mathcal{F}_k$. Consider an arbitrary section s of the bundle π. Then, as we already know, the section $j_k(s)$ of the bundle π_k corresponds to s, while the composition $\varphi \circ j_k(s) = \varphi(s)$ is a smooth function on the manifold M. In other words, the function φ puts into correspondence to smooth sections of the bundle π elements of $C^\infty(M)$ and, since the points lying on the graph of the section $j_k(s)$ are the Taylor expansions of order k for this section, the values of the function $\varphi(s)$ are determined by the partial derivatives of the section s up to order k. Consequently, the function φ determines in a canonical way a scalar nonlinear differential operator on the set of sections of the bundle π, and the order of this operator coincides with the filtration degree of the element φ. This correspondence becomes even clearer when one passes to coordinate representation: if $\mathcal{U}$ is a neighborhood in M and $x_1, \ldots, x_n, \ldots, p_\sigma^j, \ldots$ are the corresponding local coordinates in $J^\infty(\pi)$, then φ is a function in the variables x_i, p_σ^j, $i = 1, \ldots, n$, $j = 1, \ldots, m$, $|\sigma| \leq k$, and if $s = (s^1(x), \ldots, s^m(x))$, then $\varphi(s) = \varphi(x_1, \ldots, x_n, \ldots, \partial^{|\sigma|} s^j / \partial x_1^{i_1} \cdots \partial x_n^{i_n}, \ldots)$. Sometimes it is convenient to distinguish between functions $\varphi \in \mathcal{F}$ and the corresponding differential operators. In this case, the operator determined by the function φ will be denoted by Δ_φ, while the function corresponding to the operator Δ will be denoted by φ_Δ.

Let us construct now a similar correspondence for matrix differential operators. Locally, over a neighborhood $\mathcal{U} \subset M$, any matrix operator defined on sections of the bundle $\pi|_\mathcal{U}$ is to take its values in vector functions, i.e., in sections of the direct product $\pi'|_\mathcal{U} \colon \mathcal{U} \times \mathbb{R}^{m'} \to \mathcal{U}$, and these values should be compatible on the intersection $\mathcal{U} \cap \mathcal{U}'$ of two such neighborhoods. Passing to the global point of view, i.e., taking into consideration operators defined on the entire manifold M, one needs to consider a bundle π' whose restriction to $\mathcal{U}$ coincides with the bundles $\pi'|_\mathcal{U} \colon \mathcal{U} \times \mathbb{R}^{m'} \to \mathcal{U}$. In other words, global counterparts of matrix differential operators are operators acting from sections of a locally trivial vector bundle π to sections of another locally trivial vector bundle π' over the same manifold M.

Let Δ be a matrix differential operator of order k, s a section of the bundle π, and $x \in M$. Then the value of the section $\Delta(s)$ at x is determined by the values of partial derivatives of the section s, or by the k-th jet of s at x. In other

1. SPACES OF INFINITE JETS

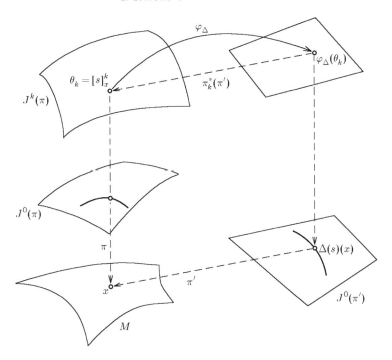

FIGURE 4.1. Construction of the section φ_Δ

words, the operator Δ makes it possible to put into correspondence to every point $\theta_k \in J^k(\pi)$ a point of the fiber of the bundle π' growing over $x = \pi_k(\theta_k)$. This can be understood in two ways.

First, this means that the operator Δ determines a section φ_Δ of the bundle $\pi_k^*(\pi')$ over $J^k(\pi)$. This bundle is the pullback of the bundle π' with respect to the projection π_k. In fact, the total space of the pullback $\pi_k^*(\pi')$ consists of points $(\theta_k, \theta'_0) \in J^k(\pi) \times J^0(\pi')$ such that $\pi_k(\theta_k) = \pi'(\theta'_0)$, and we can set $\varphi_\Delta(\theta_k) = (\theta_k, (\Delta(s))(x))$, where $\theta_k = [s]_x^k$, $s \in \Gamma(\pi)$ (see Figure 4.1).

Second, from the above it follows that a mapping $\Phi_\Delta \colon J^k(\pi) \to J^0(\pi')$ corresponds to the operator Δ and this mapping is compatible with the projections to the manifold M, i.e., $\pi' \circ \Phi_\Delta = \pi_k$. If $\theta_k = [s]_x^k \in J^k(\pi)$, $s \in \Gamma(\pi)$, then we set $\Phi(\theta_k) = [\Delta(s)]_x^0 = (\Delta(s))(x)$. Thus we have the following commutative diagram:

$$
\begin{array}{ccc}
J^k(\pi) & \xrightarrow{\Phi_\Delta} & J^0(\pi') \\
& \searrow{\pi_k} \quad \swarrow{\pi'} & \\
& M &
\end{array}
\qquad (1.4)
$$

i.e., Φ_Δ is a morphism of the bundle π_k to the bundle π'. Note that for any operator $\Delta \colon \Gamma(\pi) \to \Gamma(\pi')$ of order k the identity

$$\Delta = \Phi_\Delta^* \circ j_k$$

holds, and it can be taken for the definition. Here $\Phi_\Delta^* \colon \Gamma(\pi_k) \to \Gamma(\pi')$ is the mapping induced by the morphism Φ_Δ.

EXAMPLE 1.2. The operator j_k constructed above is an operator of order k acting from the bundle π to the bundle $\pi_k \colon J^k(\pi) \to M$. The corresponding section

φ_{j_k} puts into correspondence to any point $\theta_k \in J^k(\pi)$ the same point considered as an element of the fiber of the pullback $\pi_k^*(\pi_k)$ growing over θ_k. In other words, φ_{j_k} is the diagonal. It is also easily seen that the mapping $\Phi_{j_k} \colon J^k(\pi) \to J^k(\pi) = J^0(\pi_k)$ is the identity.

1.3. Prolongations of differential operators. Let $\Delta \colon \Gamma(\pi) \to \Gamma(\pi')$ and $\Delta' \colon \Gamma(\pi') \to \Gamma(\pi'')$ be two differential operators, of orders k and k' respectively. Then their composition $\Delta' \circ \Delta$ is a differential operator of order $\leq k + k'$. To establish this fact (which is obvious locally), let us note the following.

First, for any morphism Φ of the bundles $\pi \colon P \to M$ and $\xi \colon Q \to M$ its liftings

$$\Phi^{(k)} \colon J^k(\pi) \to J^k(\xi), \quad k \geq 0, \qquad \Phi^{(k)}([s]_x^k) = [\Phi \circ s]_x^k$$

are defined and the following equalities are valid:

$$\Phi^{(0)} = \Phi, \qquad \Phi^{(l)} \circ \pi_{k,l} = \xi_{k,l} \circ \Phi^{(k)}, \qquad \xi_k \circ \Phi^{(k)} = \pi_k, \quad k \geq l.$$

Second, for any $k, k' \geq 0$ one can construct the map

$$\Phi_{k,k'} \colon J^{k+k'}(\pi) \to J^{k'}(\pi_k), \qquad \Phi_{k,k'}([s]_x^{k+k'}) = [j_k(s)]_{\theta_k}^{k'},$$

where $\theta_k = [s]_x^k$. Then one has $\pi_k = \Phi_{k,k'} \circ (\pi_k)_{k'}$, and, as we can easily see, $\Phi_{k,k'}(j_{k+k'}(s)) = j_{k'}(j_k(s))$. Thus, $\Phi_{k,k'} \circ j_{k+k'} = j_{k'} \circ j_k$. In other words, $\Phi_{k,k'} = \Phi_{j_{k'} \circ j_k}$, which proves that the composition $j_{k'} \circ j_k$ is a differential operator of order $k + k'$.

Let us return to the case of general operators Δ, Δ' and consider the diagram

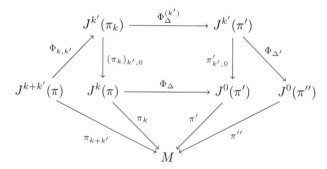

From commutativity of this diagram it follows that for any section $s \in \Gamma(\pi)$ the equality

$$\Delta'(\Delta(s)) = \Phi_{\Delta'}(\Phi_\Delta^{(k')}(\Phi_{k,k'}(j_{k+k'}(s))))$$

holds, i.e.,

$$\Phi_{\Delta' \circ \Delta} = \Phi_{\Delta'} \circ \Phi_\Delta^{(k')} \circ \Phi_{k,k'},$$

which proves the desired result.

In particular, for any differential operator Δ its composition Δ_l with the operator $j_l \colon \Gamma(\pi') \to \Gamma(\pi'_l)$, $l \geq 0$, is defined. Let $\Phi_\Delta^l = \Phi_{\Delta_l} \colon J^{k+l}(\pi) \to J^l(\pi')$ be the corresponding mapping of jet manifolds. Then, as is easily seen, the diagram (1.4) can be extended to the following commutative diagram:

$$
\begin{array}{ccccccc}
 & J^k(\pi) & \longleftarrow \cdots \longleftarrow & J^{k+l}(\pi) & \xleftarrow{\pi_{k+l+1,k+l}} & J^{k+l+1}(\pi) & \longleftarrow \cdots \\
\pi_k \nearrow & & & & & & \\
M & \Big\downarrow \Phi_\Delta = \Phi_\Delta^0 & & \Big\downarrow \Phi_\Delta^l & & \Big\downarrow \Phi_\Delta^{l+1} & \\
\pi' \searrow & & & & & & \\
 & J^0(\pi') & \longleftarrow \cdots \longleftarrow & J^l(\pi') & \xleftarrow{\pi'_{l+1,l}} & J^{l+1}(\pi') & \longleftarrow \cdots
\end{array}
$$

The set of mappings $\Phi_\Delta = \{\Phi_\Delta^l\}_{l>0}$ determines a homomorphism $\Phi_\Delta^* \colon \mathcal{F}(\pi') \to \mathcal{F}(\pi)$ possessing the property

$$\deg \Phi_\Delta^*(\varphi) = \deg(\varphi) + k, \qquad \varphi \in \mathcal{F}(\pi').$$

Thus, Φ_Δ is a smooth mapping of $J^\infty(\pi)$ to $J^\infty(\pi')$.

DEFINITION 1.4. The operator Δ_l is called the *l-th prolongation* of the nonlinear differential operator Δ.

As we shall see below, this notion plays quite an important role in studying differential equations.

Let $\mathcal{U} \subset M$ be a coordinate neighborhood in M, $x_1, \ldots, x_n, u^1, \ldots, u^m, \ldots,$ $p_\sigma^j, \ldots$ the corresponding coordinates in $\pi_\infty|_\mathcal{U}$, and $x_1, \ldots, x_n, v^1, \ldots, v^{m'}, \ldots,$ $q_\sigma^{j'}, \ldots$ the coordinates in $\pi'_\infty|_\mathcal{U}$. If

$$(\Delta(s))^{j'} = \varphi^{j'}\left(x_1, \ldots, x_n, s^1(x), \ldots, s^m(x), \ldots, \frac{\partial^{|\sigma|} s^j}{\partial x_1^{i_1} \cdots \partial x_n^{i_n}}, \ldots\right), \tag{1.5}$$

where $s \in \Gamma(\pi|_\mathcal{U})$, is the coordinate representation of the operator Δ and

$$j_k(s)(x_1, \ldots, x_n) = \left(x_1, \ldots, x_n, \ldots, \frac{\partial^{|\sigma|} s^j}{\partial x_1^{i_1} \cdots \partial x_n^{i_n}}, \ldots\right) \tag{1.6}$$

is that of the operator j_k, then, as is easily checked, the first prolongation of Δ is described by (1.5) and the relations

$$(\Delta(s))_i^{j'} = \frac{\partial}{\partial x_i} \varphi^{j'}\left(x_1, \ldots, x_n, \ldots, \frac{\partial^{|\sigma|} s^j}{\partial x_1^{i_1} \cdots \partial x_n^{i_n}}, \ldots\right)$$

$$= \frac{\partial \varphi^{j'}}{\partial x_i} + \sum_{j,\sigma} \frac{\partial \varphi^{j'}}{\partial p_\sigma^j} \frac{\partial}{\partial x_i} \frac{\partial^{|\sigma|} s^j}{\partial x_1^{i_1} \cdots \partial x_n^{i_n}},$$

$i = 1, \ldots, n$, or, which is equivalent, by the relations

$$\begin{cases} v^{j'} = \varphi^{j'}(x, \ldots, p_\sigma^j, \ldots), \\ q_i^{j'} = D_i \varphi^{j'}, \quad i = 1, \ldots, n, \end{cases}$$

where $D_i = \partial/\partial x_i + \sum_{|\sigma|=0}^k \sum_{j=1}^m p_{\sigma+1_i}^j \partial/\partial p_\sigma^j$ are the operators of total derivatives.

In a similar manner, the l-th prolongation of the operator Δ is described by the system of relations

$$q_\tau^{j'} = D_\tau \varphi^{j'}, \qquad j' = 1, \ldots, m', \quad |\tau| = 0, \ldots, l, \tag{1.7}$$

where $D_\tau = D_1^{l_1} \circ \cdots \circ D_n^{l_n}$, if $\tau = (l_1, \ldots, l_n)$.

EXAMPLE 1.3. Let Δ be the operator determining the Burgers equation (see Chapter 3) and represented in coordinate form by

$$v = p_{(0,1)} - u p_{(1,0)} - p_{(2,0)}. \tag{1.8}$$

Then its first prolongation is of the form

$$v = p_{(0,1)} - u p_{(1,0)} - p_{(2,0)},$$
$$q_{(1,0)} = p_{(1,1)} - p_{(1,0)}^2 - u p_{(2,0)} - p_{(3,0)},$$
$$q_{(0,1)} = p_{(0,2)} - p_{(1,0)} p_{(0,1)} - u p_{(1,1)} - p_{(2,1)},$$

while the l-th prolongation is represented as

$$q_{(i,j)} = D_1^i D_2^j (p_{(0,1)} - p_{(0,0)} p_{(1,0)} - p_{(2,0)})$$
$$= p_{(i,j+1)} - p_{(i+2,j)} - \sum_{\alpha=0}^{i} \binom{i}{\alpha} \sum_{\beta=0}^{j} \binom{j}{\beta} p_{(\alpha,\beta)} p_{(i+1-\alpha, j-\beta)},$$

where $i + j \leq l$.

EXERCISE 1.1. Describe the l-th prolongation of the operator j_k.

When one considers matrix differential operators, mappings of objects of a more general nature than the algebras $\mathcal{F}(\pi)$ arise. Let us recall that any differential operator $\nabla \colon \Gamma(\pi) \to \Gamma(\xi)$ of order k is identified with the section φ_∇ of the bundle $\pi_k^*(\xi)$. Denote the set of such sections by $\mathcal{F}_k(\pi, \xi)$. Similarly to the case of the algebras $\mathcal{F}(\pi)$, the following chain of embeddings takes place:

$$\mathcal{F}_{-\infty}(\pi, \xi) = \Gamma(\xi) \xrightarrow{\nu} \mathcal{F}_0(\pi, \xi) \longrightarrow \cdots$$
$$\longrightarrow \mathcal{F}_k(\pi, \xi) \xrightarrow{\nu_{k+1,k}} \mathcal{F}_{k+1}(\pi, \xi) \longrightarrow \cdots,$$

and consequently the set $\mathcal{F}(\pi, \xi) = \bigcup_{k=-\infty}^{\infty} \mathcal{F}_k(\pi, \xi)$ filtered by the subsets $\mathcal{F}_k(\pi, \xi)$ is defined. The elements of $\mathcal{F}(\pi, \xi)$ can be added to each other and multiplied by the elements of $\mathcal{F}(\pi)$, and these operations are compatible with the filtration, i.e., $\mathcal{F}(\pi, \xi)$ is a *filtered module* over the filtered algebra $\mathcal{F}(\pi)$. If now one considers an operator $\Delta \colon \Gamma(\pi) \to \Gamma(\pi')$ of order l and its composition with another operator $\nabla' \colon \Gamma(\pi') \to \Gamma(\xi)$ of order k, then one will obtain the operator of order $k+l$ acting from $\Gamma(\pi)$ to $\Gamma(\xi)$. Thus, we have the system of mappings $\Phi_{\Delta,\xi}^k \colon \mathcal{F}_k(\pi', \xi) \to \mathcal{F}_{k+l}(\pi, \xi)$ or, which is the same, the mapping $\Phi_{\Delta,\xi}^* \colon \mathcal{F}(\pi', \xi) \to \mathcal{F}(\pi, \xi)$, raising filtration degree by l. The following equalities are valid:

$$\Phi_{\Delta,\xi}^*(\varphi_1 + \varphi_2) = \Phi_{\Delta,\xi}^*(\varphi_1) + \Phi_{\Delta,\xi}^*(\varphi_2),$$
$$\Phi_{\Delta,\xi}^*(\varphi \varphi_1) = \Phi_{\Delta}^*(\varphi) \Phi_{\Delta,\xi}^*(\varphi_1),$$

where $\varphi_1, \varphi_2 \in \mathcal{F}(\pi', \xi)$, $\varphi \in \mathcal{F}(\pi)$. In other words, for any fiber bundle ξ the mapping $\Phi_{\Delta,\xi}^*$ is a homomorphism of filtered modules acting over the homomorphism Φ_Δ^* of filtered algebras.

EXERCISE 1.2. Construct the homomorphism $\Phi_{\Delta,\xi}^*$ starting from the smooth mapping $\Phi_\Delta \colon J^\infty(\pi) \to J^\infty(\pi')$.

1. SPACES OF INFINITE JETS

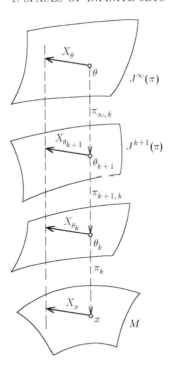

FIGURE 4.2. A tangent vector to the manifold $J^\infty(\pi)$

1.4. Vector fields on $J^\infty(\pi)$. Let us now analyze the concept of a vector field on the manifold $J^\infty(\pi)$. Let X_θ be a tangent vector at a point $\theta \in J^\infty(\pi)$. Then, for natural reasons, the projections $\pi_{\infty,k}$ and π_∞ are to determine the sequence of tangent vectors $(\pi_{\infty,k})_*(X_\theta)$ at the points $\theta_k = \pi_{\infty,k}(\theta)$ together with the tangent vector $\pi_{\infty,*}(X_\theta) \in T_x(M)$, $x = \pi_\infty(\theta)$. Now, using already familiar reasoning, we shall define a tangent vector X_θ to the manifold $J^\infty(\pi)$ at the point θ as the set $\{X_x, X_{\theta_k}\}$ of the tangent vectors to the manifolds M and $J^k(\pi)$ at the points $x = \pi_\infty(\theta)$ and $\theta_k = \pi_{\infty,k}(\theta)$ respectively, such that $(\pi_{k+1,k})_*(X_{\theta_{k+1}}) = X_{\theta_k}$ and $(\pi_k)_*(X_{\theta_k}) = X_x$ (see Figure 4.2).

If $\mathcal{U} \subset M$ is a coordinate neighborhood of the point $x \in M$ and $x_1, \ldots, x_n, \ldots, p_\sigma^j, \ldots$ are canonical coordinates in $\pi_\infty^{-1}(\mathcal{U})$, then any vector X_θ tangent to the manifold $J^\infty(\pi)$ at the point θ is represented in the form of an infinite sum

$$X_\theta = \sum_{i=1}^n a_i \frac{\partial}{\partial x_i} + \sum_{|\sigma| \geq 0} \sum_{j=1}^m b_\sigma^j \frac{\partial}{\partial p_\sigma^j}, \tag{1.9}$$

in which the coefficients a_i and b_σ^j are real numbers. One also has

$$(\pi_{\infty,k})_*(X_\theta) = \sum_{i=1}^n a_i \frac{\partial}{\partial x_i} + \sum_{|\sigma|=0}^k \sum_{j=1}^m b_\sigma^j \frac{\partial}{\partial p_\sigma^j}, \quad (\pi_\infty)_*(X_\theta) = \sum_{i=1}^n a_i \frac{\partial}{\partial x_i}.$$

Let X_θ be a tangent vector to $J^\infty(\pi)$ at θ, and X_{θ_k} its projection to $T_{\theta_k}(J^k(\pi))$. Then X_{θ_k} can be understood as a derivation of the algebra of smooth functions on $J^k(\pi)$ with values in the field of constants $\mathbb{R}$. In other words, $X_{\theta_k} \colon \mathcal{F}_k \to \mathbb{R}$ is a mapping possessing the property

$$X_{\theta_k}(\varphi_1\varphi_2) = \varphi_1(\theta_k)X_{\theta_k}(\varphi_2) + \varphi_2(\theta_k)X_{\theta_k}(\varphi_1),$$

where $\varphi_1, \varphi_2 \in \mathcal{F}_k$. The condition of compatibility of the vectors X_{θ_k}, X_x with the projections $\pi_{\infty,k}$ and π_k means that $X_{\theta_{k+1}} \circ \nu_{k+1,k} = X_{\theta_k}$ and $X_{\theta_0} \circ \nu = X_x$, i.e., that the diagram

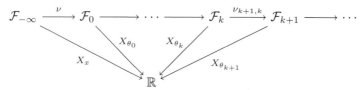

is commutative. In other words, similarly to the case of finite-dimensional manifolds, a tangent vector X_θ is interpreted as a derivation of the algebra $\mathcal{F}$ with values in the field of real numbers:

$$X_\theta(\varphi_1\varphi_2) = \varphi_1(\theta)X_\theta(\varphi_2) + \varphi_2(\theta)X_\theta(\varphi_1), \tag{1.10}$$

where $\varphi_1, \varphi_2 \in \mathcal{F}$.

If we now take a family $X = \{X_\theta\}$ of tangent vectors on $J^\infty(\pi)$ parametrized by points of $\theta \in J^\infty(\pi)$, i.e., a vector field on $J^\infty(\pi)$, then (1.10) will become the relation

$$X(\varphi_1\varphi_2) = \varphi_1 X(\varphi_2) + \varphi_2 X(\varphi_1), \tag{1.11}$$

valid for all smooth functions φ_1, φ_2 on $J^\infty(\pi)$. To complete our arguments, it remains to recall that $\mathcal{F}(\pi)$ is filtered and to define a vector field on the manifold $J^\infty(\pi)$ as a derivation of $\mathcal{F}(\pi)$ (i.e., an $\mathbb{R}$-linear mapping $X \colon \mathcal{F} \to \mathcal{F}$ satisfying the Leibniz rule (1.11)), such that

$$\deg X(\varphi) = \deg(\varphi) + k, \quad \varphi \in \mathcal{F},$$

where k is an integer independent of φ and denoted by $\deg(X)$. Denote by $\mathrm{D}(\pi)$ the set of all vector fields on $J^\infty(\pi)$.

DEFINITION 1.5. A field $X \in \mathrm{D}(\pi)$ is called *vertical* (or π-*vertical*), if the equality $X(\pi_\infty^*(\varphi)) = 0$ holds for any function $\varphi \in C^\infty(M) \subset \mathcal{F}$.

The set of all vertical fields is denoted by $\mathrm{D}^v(\pi)$.

EXERCISE 1.3. Show that if $X, Y \in \mathrm{D}(\pi)$ and $\varphi \in \mathcal{F}$, then $X + Y$, φX and $[X, Y] = X \circ Y - Y \circ X$ are also vector fields on $J^\infty(\pi)$, and

$$[X, Y] + [Y, X] = 0,$$
$$[X, \varphi Y] = X(\varphi)Y + \varphi[X, Y],$$
$$[X, Y + Z] = [X, Y] + [X, Z],$$
$$[X, [Y, Z]] + [Y, [Z, X]] + [Z, [X, Y]] = 0,$$

where $Z \in \mathrm{D}(\pi)$, i.e., the set of vector fields on $J^\infty(\pi)$ possesses the same algebraic properties as vector fields on finite-dimensional manifolds. In particular, they form a Lie algebra over $\mathbb{R}$. Note that $\mathrm{D}^v(\pi)$ is a subalgebra of this Lie algebra.

EXAMPLE 1.4 (liftings of vector fields). Let X be a vector field on the manifold M, φ a smooth function on $J^\infty(\pi)$ with $\deg(\varphi) = k$, and $\Delta = \Delta_\varphi \colon \Gamma(\pi) \to C^\infty(M)$ the differential operator associated to this function. Then, since X is a first order differential operator acting from $C^\infty(M)$ to $C^\infty(M)$, the composition $X \circ \Delta \colon \Gamma(\pi) \to C^\infty(M)$ is defined and is a differential operator of order $k+1$. Denote by $\widehat{X}(\varphi)$ the function on $J^\infty(\pi)$ corresponding to the operator $X \circ \Delta$.

EXERCISE 1.4. Show that the correspondence $\widehat{X} \colon \varphi \mapsto \widehat{X}(\varphi)$ is a vector field on $J^\infty(\pi)$ with $\deg(\widehat{X}) = 1$.

Let us give a geometric version for the definition of the field $\widehat{X}$. Let $x \in M$ be a point in M and θ_k a point of $J^k(\pi)$ lying in the fiber of the bundle π_k over x. Then one has $\theta_k = [s]_x^k$ for a section $s \in \Gamma(\pi)$. If φ is a smooth function on $J^k(\pi)$ in a neighborhood of the point θ_k, then we set $\widehat{X}_{\theta_k}(\varphi) = X_x(s^*(\varphi))$. Clearly, the right-hand side of the last expression is independent of the representative s of the point θ_k. Consequently, X_{θ_k} is a tangent vector to $J^k(\pi)$ at the point θ_k. If now $\{x, \theta_k\}$ is the sequence of points representing the point $\theta \in J^\infty(\pi)$, then it is easily seen that $(\pi_{k+1,k})_*(\widehat{X}_{\theta_{k+1}}) = \widehat{X}_{\theta_k}$ and $(\pi_k)_*(\widehat{X}_{\theta_k}) = X_x$, i.e., the sequence of vectors $\{X_x, \widehat{X}_{\theta_k}\}$ determines a tangent vector to the manifold $J^\infty(\pi)$ at the point θ.

DEFINITION 1.6. The field $\widehat{X}$ on $J^\infty(\pi)$ is called the *lifting of the vector field* $X \in \mathrm{D}(M)$ to the space of infinite jets.

The following properties of the lifting operation can be deduced directly from the definition:

$$\widehat{fX + gY} = f\widehat{X} + g\widehat{Y}, \tag{1.12a}$$

$$\widehat{[X,Y]} = [\widehat{X}, \widehat{Y}], \tag{1.12b}$$

$$\widehat{X}(f\varphi) = X(f)\varphi + f\widehat{X}(\varphi), \tag{1.12c}$$

where $f, g \in C^\infty(M)$, $\varphi \in \mathcal{F}$, $X, Y \in \mathrm{D}(M)$. Equalities (1.12a)–(1.12b) mean that the lifting operation is a homomorphism of the Lie algebra of vector fields over M to $\mathrm{D}(\pi)$, while from (1.12c) it follows that the projection to M of the lifting to $J^\infty(\pi)$ of any field $X \in \mathrm{D}(M)$ coincides with the initial field. In other words, the correspondence $X \mapsto \widehat{X}$ is a flat (integrable) connection in the bundle π_∞. This fact plays a fundamental role in the geometry of the manifolds $J^\infty(\pi)$, and we shall frequently refer to it in the sequel. For reasons which will become clear in §2, this connection is called the *Cartan connection*.

If $x_1, \ldots, x_n, u^1, \ldots, u^m, \ldots, p_\sigma^j, \ldots$ are coordinates in $J^\infty(\pi)$ over a neighborhood $\mathcal{U} \subset M$, then any field $X \in \mathrm{D}(M)$ is representable as the infinite sum

$$X = X_1 \frac{\partial}{\partial x_1} + \cdots + X_n \frac{\partial}{\partial x_n} + \cdots + X_\sigma^j \frac{\partial}{\partial p_\sigma^j} + \cdots, \tag{1.13}$$

where $X_i, X_\sigma^j \in \mathcal{F}(\pi|_\mathcal{U})$, and the equalities $\deg(X_\sigma^j) = |\sigma| + k$, $k = \deg(X)$, hold for all j and σ such that $|\sigma| \geq k_0 \geq 0$. Vertical fields are characterized by the fact that all their coefficients at $\partial/\partial x_i$ vanish. Note that the infinite number of summands in the right-hand side of (1.13) does not cause computational difficulties (such as check of convergence, etc.), since, by definition of the algebra $\mathcal{F}(\pi)$, any function $\varphi \in \mathcal{F}(\pi)$ may depend on a finite number of variables only, and consequently the number of summands in the expression for $X(\varphi)$ for a particular function φ is always finite as well.

Let us now derive a coordinate representation for the liftings of vector fields lying in $D(M)$. As (1.12a) shows, it is sufficient to do this for the fields of the form $\partial/\partial x_i$. Let $\varphi = \varphi(x_1, \ldots, x_n, \ldots, p_\sigma^j, \ldots) \in \mathcal{F}$ and $s = (s^1, \ldots, s^m) \in \Gamma(\pi)$. Then

$$\left(\frac{\partial}{\partial x_i} \circ \Delta_\varphi\right)(s) = \frac{\partial}{\partial x_i}\varphi\left(x_1, \ldots, x_n, \ldots, \frac{\partial^{|\sigma|}s^j}{\partial x^\sigma}, \ldots\right)$$

$$= \frac{\partial \varphi}{\partial x_i} + \sum_{j,\sigma} \frac{\partial^{|\sigma|+1} s^j}{\partial x_i \partial x^\sigma} \frac{\partial \varphi}{\partial p_\sigma^j}.$$

Hence

$$\widehat{\frac{\partial}{\partial x_i}} = D_i = \frac{\partial}{\partial x_i} + \sum_{|\sigma|=0}^{\infty} \sum_{j=1}^{m} p_{\sigma+1_i}^j \frac{\partial}{\partial p_\sigma^j}. \tag{1.14}$$

Thus liftings of the basic vector fields $\partial/\partial x_i$ coincide with the operators D_i of the total derivatives along x_i. Equality (1.12b) shows that $[D_{i_1}, D_{i_2}] = 0$ for all $i_1, i_2 = 1, \ldots, n$. Note that the fields D_i take functions on k-th jets to those on $(k+1)$-st jets and thus determine no vector field on manifolds of finite jets. It means that treating "truncated" total derivatives as vector fields (see the preceding chapter) is incorrect (though a deceptive appearance of their coordinate representation may cause opposite conclusions). This is one illustration of why it is useful to pass to infinite jet manifolds.

REMARK 1.1. The lifting construction takes place in a more general setting. Let $\nabla \colon \Gamma(\xi) \to \Gamma(\xi')$ be a linear differential operator and $\varphi \in \mathcal{F}(\pi, \xi)$. Consider the operator $\Delta = \Delta_\varphi \colon \Gamma(\pi) \to \Gamma(\xi)$ and the composition $\nabla \circ \Delta \colon \Gamma(\pi) \to \Gamma(\xi')$. Then, setting $\widehat{\nabla}(\varphi) = \varphi_{\nabla \circ \Delta}$, one obtains the map $\widehat{\nabla} = \widehat{\nabla}_\pi \colon \mathcal{F}(\pi, \xi) \to \mathcal{F}(\pi, \xi')$, which is an $\mathbb{R}$-linear operator.

EXERCISE 1.5. Give a pointwise definition of the operator $\widehat{\nabla}$ similar to the one given above for liftings of vector fields. Deduce a coordinate representation of the operator $\widehat{\nabla}$.

1.5. Differential forms on $J^\infty(\pi)$. Let us now discuss the concept of a differential form on $J^\infty(\pi)$. Let $\Lambda^i(\pi_k) \overset{\text{def}}{=} \Lambda^i(J^k(\pi))$ be the module of i-forms on $J^k(\pi)$. The projections π and $\pi_{k+1,k}$ generate the infinite sequence of embeddings

$$\Lambda^i(M) \xrightarrow{\nu} \Lambda^i(\pi_0) \xrightarrow{\nu_{1,0}} \Lambda^i(\pi_1) \longrightarrow \cdots \longrightarrow \Lambda^i(\pi_k) \xrightarrow{\nu_{k+1,k}} \Lambda^i(\pi_{k+1}) \longrightarrow \cdots.$$

Skipping the already familiar motivations, we shall define the module $\Lambda^i(\pi)$ of i-forms on $J^\infty(\pi)$ by setting $\Lambda^i(\pi) = \bigcup_{k=0}^{\infty} \Lambda^i(\pi_k)$. In particular, $\Lambda^0(\pi) = \mathcal{F}(\pi)$. Let us also set $\Lambda^*(\pi) = \bigoplus_{i=0}^{\infty} \Lambda^i(\pi)$. As above, the modules $\Lambda^i(\pi)$ and $\Lambda^*(\pi)$ are filtered by their submodules $\Lambda^i(\pi_k)$ and $\Lambda^*(\pi_k)$ respectively[2]. By the above definition, any element of $\Lambda^*(\pi)$ is in fact a form on some manifold of finite jets. Hence, the operation $\wedge$ of the wedge product and the differential d are defined for these elements and possess their usual properties.

---

[2]To be more exact, the embeddings $\nu_{k+1,k}$ are module homomorphisms over the corresponding algebra embeddings $\mathcal{F}_k \to \mathcal{F}_{k+1}$. To obtain a filtration of the module $\Lambda^*(\pi)$ by a system of $\mathcal{F}$-submodules, we need to consider the submodules in $\Lambda^*(\pi)$ generated by the elements of $\Lambda^*(\pi_k)$, i.e., consisting of differential forms on $J^k(\pi)$ whose coefficients may be arbitrary functions belonging to $\mathcal{F}(\pi)$.

Let $X \in \mathrm{D}(\pi)$ be a vector field and $\omega \in \Lambda^i(\pi)$ be a differential form on $J^\infty(\pi)$. Let us define the *inner product* operation $\lrcorner$ which puts into correspondence to these two objects the form $X \lrcorner \omega \in \Lambda^{i-1}(\pi)$. Let $\theta = \{x, \theta_k\} \in J^\infty(\pi)$ be a point of $J^\infty(\pi)$, and $X_\theta = \{X_x, X_{\theta_k}\}$ the vector of the field X at this point. Consider a number k such that $\omega \in \Lambda^i(\pi_k)$. This number always exists, by the definition of the module $\Lambda^i(\pi)$. Let us set $(X \lrcorner \omega)_\theta = X_{\theta_k} \lrcorner \omega_{\theta_k}$. For any $k' \geq k$ the equality $(\pi_{k',k})_*(X_{\theta_{k'}}) = X_{\theta_k}$ is valid; hence, one has $X_{\theta_{k'}} \lrcorner (\pi_{k'k}^* \omega)_{\theta_{k'}} = X_{\theta_k} \lrcorner \omega_{\theta_k}$, which means that the operation $\lrcorner$ is well defined. If in addition $\deg(X) = l$, then the vector X_{θ_k} is determined by the point $\theta_{k+l} \in J^{k+l}(\pi)$, i.e., $X \lrcorner \omega \in \Lambda^{i-1}(\pi_{k+l}) \subset \Lambda^{i-1}(\pi)$. In particular, the inner product of a vector field with a 1-form determines the isomorphism

$$\mathrm{D}(\pi) \simeq \mathrm{Hom}^0_{\mathcal{F}(\pi)}(\Lambda^1(\pi), \mathcal{F}(\pi)), \tag{1.15}$$

where $\mathrm{Hom}^0_{\mathcal{F}(\pi)}$ denotes the set of homomorphisms preserving filtration. This isomorphism puts into correspondence to any field $X \in \mathrm{D}(\pi)$ the $\mathcal{F}(\pi)$-homomorphism $f_X \colon \Lambda^1(\pi) \to \mathcal{F}(\pi)$ acting as $f_X(d\varphi) = X(\varphi)$, $\varphi \in \mathcal{F}(\pi)$.

Now we can define the *Lie derivative* $L_X\omega$ of a form $\omega \in \Lambda^*(\pi)$ along a vector field $X \in \mathrm{D}(\pi)$. To do this, taking into account the infinitesimal Stokes formula, let us set

$$L_X\omega = X \lrcorner d\omega + d(X \lrcorner \omega). \tag{1.16}$$

Obviously, if $\omega \in \Lambda^i(\pi_k)$ and $\deg(X) = l$, then $L_X\omega \in \Lambda^i(\pi_{k+l})$. We shall also use the notation $X(\omega)$ for the Lie derivative of a form ω along X.

The operations introduced above on the set $\Lambda^*(\pi)$ (wedge product, the de Rham differential, inner product, and Lie derivative) possess all properties they possess in the finite-dimensional case. The same is valid for their coordinate representation. If $\mathcal{U}$ is a coordinate neighborhood in the manifold M and $(x_1, \ldots, x_n, \ldots, p_\sigma^j, \ldots)$ are the corresponding coordinates in $\pi_\infty^{-1}(\mathcal{U})$, then any form $\omega \in \Lambda^i(\pi)$ can be represented as

$$\omega = \sum_{\alpha+\beta=i} \varphi_{i_1,\ldots,i_\alpha,j_1,\ldots,j_\beta}^{\sigma_1,\ldots,\sigma_\beta} dx_{i_1} \wedge \cdots \wedge dx_{i_\alpha} \wedge dp_{\sigma_1}^{j_1} \wedge \cdots \wedge dp_{\sigma_\beta}^{j_\beta}, \tag{1.17}$$

where $|\sigma_1|, \ldots, |\sigma_\beta| \leq k$ while $\varphi_{i_1,\ldots,i_\alpha,j_1,\ldots,j_\beta}^{\sigma_1,\ldots,\sigma_\beta}$ are smooth functions on $\pi_\infty^{-1}(\mathcal{U})$. Then the coordinate representation of the de Rham differential is of the form

$$d\omega = \sum_{\alpha+\beta=i} d\varphi_{i_1,\ldots,i_\alpha,j_1,\ldots,j_\beta}^{\sigma_1,\ldots,\sigma_\beta} \wedge dx_{i_1} \wedge \cdots \wedge dx_{i_\alpha} \wedge dp_{\sigma_1}^{j_1} \wedge \cdots \wedge dp_{\sigma_\beta}^{j_\beta},$$

where

$$d\varphi_{i_1,\ldots,i_\alpha,j_1,\ldots,j_\beta}^{\sigma_1,\ldots,\sigma_\beta} = \sum_\gamma \frac{\partial \varphi_{i_1,\ldots,i_\alpha,j_1,\ldots,j_\beta}^{\sigma_1,\ldots,\sigma_\beta}}{\partial x_\gamma} dx_\gamma + \sum_{\delta,\tau} \frac{\partial \varphi_{i_1,\ldots,i_\alpha,j_1,\ldots,j_\beta}^{\sigma_1,\ldots,\sigma_\beta}}{\partial p_\tau^\delta} dp_\tau^\delta.$$

EXERCISE 1.6. Write down coordinate formulas for the operation of inner product and for the Lie derivative on $J^\infty(\pi)$.

1.6. The horizontal de Rham complex. We shall now describe a special class of forms on $J^\infty(\pi)$ which will be denoted by $\Lambda_0^*(\pi)$. Elements of this class will be called *horizontal forms*.

DEFINITION 1.7. A form $\omega \in \Lambda^*(\pi)$ is called *horizontal* if $X \lrcorner \omega = 0$ for any vertical field $X \in \mathrm{D}^v(\pi)$.

Since any vertical field is locally of the form

$$X = \sum_{j,\sigma} X_\sigma^j \frac{\partial}{\partial p_\sigma^j},$$

from the representation (1.17) it follows that a form ω is horizontal if and only if it is representable in coordinates as

$$\omega = \sum \varphi_{i_1,\ldots,i_\alpha} dx_{i_1} \wedge \cdots \wedge dx_{i_\alpha}, \qquad \varphi_{i_1,\ldots,i_\alpha} \in \mathcal{F}(\pi). \qquad (1.18)$$

Hence locally any horizontal form is a linear combination of forms on the manifold M with coefficients in $\mathcal{F}$: $\omega = \varphi^1 \omega_1 + \cdots + \varphi^l \omega_l$, where $\varphi^1, \ldots, \varphi^l \in \mathcal{F}(\pi)$, $\omega_1, \ldots, \omega_l \in \Lambda^*(M)$. Let $\Delta^1, \ldots, \Delta^l$ be the nonlinear differential operators corresponding to the functions $\varphi^1, \ldots, \varphi^l$ and acting from sections of the bundle π to $C^\infty(M)$. Then one can associate with the form ω the operator Δ_ω acting as

$$\Delta_\omega(s) = \Delta^1(s)\omega_1 + \cdots + \Delta^l(s)\omega_l, \qquad s \in \Gamma(\pi),$$

and taking sections of the bundle π to differential forms on the manifold M. Conversely, to any such operator there corresponds a form (1.18). Consequently, horizontal forms on $J^\infty(\pi)$ coincide with nonlinear differential operators on the bundle π with values in differential forms on the manifold M. The module of horizontal i-forms on $J^\infty(\pi)$ is denoted by $\Lambda_0^i(\pi)$.

If $t^* \colon T^*(M) \to M$ denotes the cotangent bundle of the manifold M and $\Lambda^*(t^*)$ denotes the bundle $\bigoplus_{i=1}^n \Lambda^i(t^*)$, then the following identifications are a corollary of the above:

$$\Lambda_0^i(\pi) = \mathcal{F}(\pi, \Lambda^i(t^*)), \qquad \Lambda_0^*(\pi) = \mathcal{F}(\pi, \Lambda^*(t^*)).$$

Note now that the operator $d = d_M$ of exterior differentiation (the de Rham differential) on the manifold M is a linear differential operator of first order acting from sections of the bundle $\Lambda^i(t^*)$ to those of the bundle $\Lambda^{i+1}(t^*)$, $i = 0, \ldots, n$. Using the definition of the lifting of differential operators (see Remark 1.1), we shall obtain the operator $\widehat{d} \colon \mathcal{F}(\pi, \Lambda^i(t^*)) \to \mathcal{F}(\pi, \Lambda^{i+1}(t^*))$. Since $d \circ d = 0$, we have $\widehat{d}^2(\omega) = 0$ for any horizontal form $\omega \in \Lambda_0^*(\pi)$ as well. In fact, $\widehat{d}^2(\omega) = (\widehat{d} \circ \widehat{d})(\omega) = (\widehat{d \circ d})(\omega) = 0$. Hence we have the sequence

$$\begin{aligned}0 \longrightarrow \mathcal{F}(\pi) \xrightarrow{\widehat{d}} \Lambda_0^1(\pi) \longrightarrow \cdots \\ \longrightarrow \Lambda_0^i(\pi) \xrightarrow{\widehat{d}} \Lambda_0^{i+1}(\pi) \longrightarrow \cdots \longrightarrow \Lambda_0^n(\pi) \longrightarrow 0,\end{aligned} \qquad (1.19)$$

in which the composition of any two adjacent operators is trivial.

DEFINITION 1.8. The sequence (1.19) is called the *horizontal de Rham complex* of the bundle π.

Let us write down the operator $\widehat{d}$ in local coordinates. It is sufficient to do this for horizontal 0-forms, i.e., for functions on $J^\infty(\pi)$. Let us consider a function $\varphi = \varphi(x_1, \ldots, x_n, \ldots, p_\sigma^j, \ldots) \in \mathcal{F}(\pi|_\mathcal{U})$, where $\mathcal{U}$ is a coordinate neighborhood in

M, and a section $s \in \Gamma(\pi)$. Then

$$(\widehat{d\varphi})(s) = d(\varphi(s)) = d\varphi\left(x_1, \ldots, x_n, \ldots, \frac{\partial^{|\sigma|} s^j}{\partial x^\sigma}, \ldots\right)$$

$$= \sum_{\alpha=1}^n \left(\frac{\partial \varphi}{\partial x_\alpha} + \sum_{\sigma,j} \frac{\partial^{|\sigma|+1} s^j}{\partial x_\alpha \partial x^\sigma} \frac{\partial \varphi}{\partial p_\sigma^j}\right) dx_\alpha = \sum_{\alpha=1}^n D_\alpha(\varphi)(s)\, dx_\alpha.$$

Hence,

$$\widehat{d}(\varphi) = \sum_{\alpha=1}^n D_\alpha(\varphi)\, dx_\alpha, \tag{1.20}$$

where D_α, $\alpha = 1, \ldots, n$, are the above operators of total derivatives.

1.7. Distributions on $J^\infty(\pi)$ and their automorphisms. The last problem we intend to discuss in this section is the definition of distributions on the manifolds $J^\infty(\pi)$. Similarly to the finite-dimensional case (see §3 of Ch. 1), a distribution $\mathcal{P}$ on $J^\infty(\pi)$ is to be understood as a correspondence $\mathcal{P}: \theta \mapsto \mathcal{P}_\theta \subset T_\theta(J^\infty(\pi))$ which takes any point $\theta \in J^\infty(\pi)$ to a subspace of the tangent space "smoothly depending" on θ. But because of specifics of the manifold $J^\infty(\pi)$ caused by its infinite dimension, this definition needs to be clarified. Assume that for any $k \geq 0$ a distribution $\mathcal{P}^k \colon \theta_k \mapsto \mathcal{P}^k_{\theta_k} \subset T_{\theta_k}(J^k(\pi))$ is given on the manifold $J^k(\pi)$, and

$$(\pi_{k+1,k})_*(\mathcal{P}^{k+1}_{\theta_{k+1}}) \subset \mathcal{P}^k_{\pi_{k+1,k}(\theta_{k+1})} \tag{1.21}$$

for all $k \geq 0$ and all points $\theta_{k+1} \in J^{k+1}(\pi)$.

DEFINITION 1.9. A system of distributions $\mathcal{P} = \{\mathcal{P}^k\}$ satisfying the condition (1.21) will be called a *predistribution on $J^\infty(\pi)$*. We shall say that two distributions $\mathcal{P}$ and $\bar{\mathcal{P}}$ are *equivalent*, if there exists a $k_0 \geq 0$ such that $\mathcal{P}^k = \bar{\mathcal{P}}^k$ for all $k \geq k_0$. A class of equivalent predistributions will be called a *distribution on $J^\infty(\pi)$*.

We say that a vector $X_\theta \in T_\theta(J^\infty(\pi))$ lies in a distribution, if $(\pi_{\infty,k})_*(X_\theta) \in \mathcal{P}^k_{\pi_{\infty,k}(\theta)}$ for all k, θ and some representative $\mathcal{P}$ of this distribution.

EXERCISE 1.7. Show that if a vector lies in some representative of a distribution, then it lies in any representative of the same distribution.

Consider the subset $\mathcal{P}^k \Lambda^1$ in the module $\Lambda^1(\pi_k)$ consisting of all 1-forms annihilated by vectors of the distribution $\mathcal{P}^k$; let us also set $\mathcal{P}^k \Lambda^* = \Lambda^*(\pi_k) \wedge \mathcal{P}^k \Lambda^1$. The elements of the last set are of the form $\omega = \sum_\alpha \Omega_\alpha \wedge \omega_\alpha$, where $\Omega_\alpha \in \Lambda^*(\pi_k)$ and $\omega_\alpha \in \mathcal{P}^k \Lambda^1$. The set $\mathcal{P}^k \Lambda^*$ is closed with respect to addition and to multiplication by an arbitrary form from $\Lambda^*(\pi_k)$, and is called the *ideal of the distribution $\mathcal{P}^k$*. From the embeddings (1.21) we get the chain of embeddings

$$\mathcal{P}^0 \Lambda^* \subset \mathcal{P}^1 \Lambda^* \subset \cdots \subset \mathcal{P}^k \Lambda^* \subset \mathcal{P}^{k+1} \Lambda^* \subset \cdots,$$

and this chain determines the ideal $\mathcal{P}\Lambda^* = \bigcup_{k \geq 0} \mathcal{P}^k \Lambda^*$ in $\Lambda^*(\pi)$. This ideal is called the ideal of the distribution $\mathcal{P}$. Conversely, any ideal in $\Lambda^*(\pi)$ of the above form determines a distribution on $J^\infty(\pi)$. Using now the isomorphism (1.15), we can give a dual definition for a distribution on $J^\infty(\pi)$. Namely, a distribution $\mathcal{P}\mathrm{D}(\pi)$ consists of vector fields X on $J^\infty(\pi)$ such that $X \,\lrcorner\, \omega = 0$ for any form $\omega \in \mathcal{P}\Lambda^1(\pi)$.

EXERCISE 1.8. Prove that $\mathcal{P}\Lambda^*$ and $\mathcal{P}\mathrm{D}(\pi)$ do not depend on the choice of a representative in the distribution $\mathcal{P}$.

Let us recall that a submanifold N of a finite-dimensional manifold M with a distribution $\mathcal{P}'$ is called *integral* if for any point $x \in N$ one has an embedding $T_x(N) \subset \mathcal{P}'_x$. It is called locally maximal if it is not contained in any other integral manifold of greater dimension. A distribution $\mathcal{P}'$ is called completely integrable if for any point there exists a unique locally maximal integral manifold passing through this point. The classical Frobenius theorem [**115**] says that a distribution $\mathcal{P}'$ is *completely integrable* if for any form $\omega \in \mathcal{P}'\Lambda^1(N)$ the equality $d\omega = \Omega \wedge \omega'$ is valid for some $\omega' \in \mathcal{P}'\Lambda^1(N)$ and a form Ω on the manifold N. In other words, the operator d takes the ideal $\mathcal{P}'\Lambda^*$ to itself, i.e., this ideal is differentially closed. Using these considerations, we shall say that a distribution $\mathcal{P}$ on $J^\infty(\pi)$ is *completely integrable*, if its ideal $\mathcal{P}\Lambda^*$ is differentially closed.

EXERCISE 1.9. Show that a distribution $\mathcal{P}$ on $J^\infty(\pi)$ is completely integrable if and only if the set $\mathcal{P}\,\mathrm{D}(\pi)$ is closed with respect to the commutator of vector fields, i.e. is a Lie subalgebra in $\mathrm{D}(\pi)$.

EXAMPLE 1.5. Let $\mathcal{P}'$ be a distribution on M. Let us consider the submodule $\widehat{\mathcal{P}'}\,\mathrm{D}(\pi)$ in $\mathrm{D}(\pi)$ consisting of linear combinations (with coefficients in $\mathcal{F}(\pi)$) of the fields $\widehat{X}$, $X \in \mathcal{P}'\,\mathrm{D}(M)$. Then the set $\mathcal{P}'\,\mathrm{D}(\pi)$ determines a distribution $\mathcal{P} = \widehat{\mathcal{P}'}$ on $J^\infty(\pi)$. If $\mathcal{P}'$ is a completely integrable distribution, then $\mathcal{P}$ is completely integrable as well. In fact,
$$[\varphi\widehat{X}, \psi\widehat{Y}] = \varphi\widehat{X}(\psi)\widehat{Y} - \psi\widehat{Y}(\varphi)\widehat{X} + \varphi\psi[\widehat{X}, \widehat{Y}].$$
But $[\widehat{X}, \widehat{Y}] = \widehat{[X,Y]}$ by (1.12), and consequently the right-hand side of the last equality lies in $\mathcal{P}\,\mathrm{D}(\pi)$, if $X, Y \in \mathcal{P}'\,\mathrm{D}(M)$ and $\varphi, \psi \in \mathcal{F}$.

Now let $\mathcal{P}$ be a completely integrable distribution on $J^\infty(\pi)$. As in §3 of Ch. 1, a vector field $X \in \mathrm{D}(\pi)$ will be called an (*infinitesimal*) *automorphism* of the distribution $\mathcal{P}$ if
$$L_X \mathcal{P}\Lambda^*(\pi) \subset \mathcal{P}\Lambda^*(\pi).$$
In a dual way, automorphisms of the distribution $\mathcal{P}$ can be defined by the condition
$$[X, \mathcal{P}\,\mathrm{D}(\pi)] \subset \mathcal{P}\,\mathrm{D}(\pi). \tag{1.22}$$
Denote by $\mathrm{D}_\mathcal{P}(\pi)$ the set of automorphisms of the distribution $\mathcal{P}$.

EXERCISE 1.10. Prove that $\mathrm{D}_\mathcal{P}(\pi)$ is a Lie algebra and $\mathcal{P}\,\mathrm{D}(\pi) \subset \mathrm{D}_\mathcal{P}(\pi)$ is its ideal.

The elements of the quotient Lie algebra
$$\mathrm{sym}(\mathcal{P}) = \mathrm{D}_\mathcal{P}(\pi)/\mathcal{P}\,\mathrm{D}(\pi)$$
are called (*infinitesimal*) *symmetries* of the completely integrable distribution $\mathcal{P}$.

2. The Cartan distribution on $J^\infty(\pi)$ and its infinitesimal automorphisms

In this section, we shall define the Cartan distribution on $J^\infty(\pi)$ and study its maximal integral manifolds, infinitesimal automorphisms, and symmetries. This study will lead us to the concept of an evolutionary derivation on $J^\infty(\pi)$, which plays a key role in the theory of symmetries of nonlinear differential equations.

2. THE CARTAN DISTRIBUTION ON $J^\infty(\pi)$

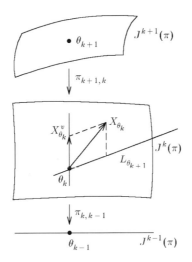

FIGURE 4.3. Construction of the element U_1

2.1. The Cartan distribution. As above, let $\pi\colon P \to M$ be a smooth vector bundle, and $J^\infty(\pi)$ the manifold of its infinite jets. Consider a point $\theta = \{x, \theta_k\}$ in $J^\infty(\pi)$ and the Cartan planes $\mathcal{C}^k_{\theta_k}$ at every point $\theta_k \in J^k(\pi)$. Then, if the point θ is represented as the infinite jet $[s]_x^\infty$ of a section $s \in \Gamma(\pi)$ at the point $x \in M$, the plane $\mathcal{C}^{k+1}_{\theta_{k+1}}$ projects to the R-plane $L_{\theta_{k+1}} \subset \mathcal{C}^k_{\theta_k} \subset T_{\theta_k}(J^k(\pi))$ under the mapping $\pi_{k+1,k}$, where $L_{\theta_{k+1}}$ is the tangent plane to the graph of the jet $j_k(s)$ at the point θ_k. Thus for all k we have the embeddings $(\pi_{k+1,k})_*(\mathcal{C}^{k+1}_{\theta_{k+1}}) \subset \mathcal{C}^k_{\theta_k}$, i.e., the system of distributions $\mathcal{C}^k$ determines the distribution $\mathcal{C} = \mathcal{C}(\pi)$ on $J^\infty(\pi)$.

DEFINITION 2.1. The distribution $\mathcal{C}(\pi)$ is called the *Cartan distribution* on the manifold of infinite jets.

From the above it follows that a vector X_θ tangent to the manifold $J^\infty(\pi)$ lies in the plane $\mathcal{C}_\theta$ if and only if its projection $X_{\theta_k} = (\pi_{\infty,k})_*(X_\theta)$ lies in the space L_{θ_k}. Using this remark, we shall give a more efficient description of the Cartan distribution on $J^\infty(\pi)$. Namely, if $X_\theta = \{X_x, X_{\theta_k}\}$ is an arbitrary tangent vector and $\varphi \in \mathcal{F}(\pi)$ is a smooth function on $J^\infty(\pi)$, we shall consider the projections $X^v_{\theta_k}$ of every vector X_{θ_k} along the plane $L_{\theta_{k+1}}$ to the fiber of the bundle $\pi_{k,k-1}\colon J^k(\pi) \to J^{k-1}(\pi)$ over the point θ_{k-1} (see Figure 4.3). Then $X^v_\theta = \{0, X^v_{\theta_k}\}$ is also a tangent vector at the point θ, and by applying it to the function φ one will obtain a real number. If now $X \in D(\pi)$ is a vector field, one can repeat the described construction at every point $\theta \in J^\infty(\pi)$ and obtain a vertical field $X^v \in D^v(\pi)$, which can be applied to the function $\varphi \in \mathcal{F}$. Thus for any function $\varphi \in \mathcal{F}$ we obtain a 1-form on $J^\infty(\pi)$ denoted by $U_1(\varphi)$ and defined by the equality

$$X \,\lrcorner\, U_1(\varphi) = X^v(\varphi), \qquad X \in D(\pi), \quad \varphi \in \mathcal{F}(\pi). \tag{2.1}$$

From (2.1) we see that the correspondence $U_1\colon \mathcal{F}(\pi) \to \Lambda^1(\pi)$ is a derivation of the algebra $\mathcal{F}$ with values in the $\mathcal{F}$-module $\Lambda^1(\pi)$, i.e., $U_1(\varphi_1 \varphi_2) = \varphi_1 U_1(\varphi_2) + \varphi_2 U_1(\varphi_1)$ for any functions $\varphi_1, \varphi_2 \in \mathcal{F}$.

Coming back to the starting point of our arguments, we see that the following statement is valid.

PROPOSITION 2.1. *A field $X \in \mathrm{D}(\pi)$ lies in the Cartan distribution $\mathcal{C}$ on $J^\infty(\pi)$ if and only if*

$$X \, \lrcorner \, U_1(\varphi) = 0$$

for all functions $\varphi \in \mathcal{F}(\pi)$. In other words, the ideal $\mathcal{C}\Lambda^*(\pi) \subset \Lambda^*(\pi)$ of the Cartan distribution is generated by the forms $U_1(\varphi)$.

The forms $U_1(\varphi)$ will be called the *Cartan forms*[3].

In what follows, we shall need a more convenient (from the computational point of view) description of the operator U_1 and, as a consequence, of Cartan forms. To obtain such a description, note the following. From the above constructions it follows that any vector field $X \in \mathrm{D}(\pi)$ admits the canonical representation

$$X = X^v + X^h, \qquad (2.2)$$

where, as already mentioned, the field X^v is vertical while the field $X^h \stackrel{\text{def}}{=} X - X^v$ is tangent to all manifolds of the form $j_\infty(s)(M)$, i.e., to the graphs of infinite jets of sections s of the bundle π. For this reason, (2.1) can be rewritten in the form $X \, \lrcorner \, U_1(\varphi) = (X - X^h)(\varphi) = X(\varphi) - X^h(\varphi)$. Note that the first summand in the right-hand side of the last equality is $X \, \lrcorner \, d\varphi$. To rewrite the second summand in the form we need, let us compare the expression for $X^h(\varphi)$ with the definition of the form $\widehat{d}\varphi$. By the pointwise definition of the lifting (see §1.4), the restriction of the form $\widehat{d}\varphi$ to the graph of the infinite jet of a section $s \in \Gamma(\pi)$ is represented as $d\Delta_\varphi(s)$. Since the form $\widehat{d}\varphi$ is horizontal, it follows that $X \, \lrcorner \, \widehat{d}\varphi = X^h \, \lrcorner \, \widehat{d}\varphi$. Finally, since the field X^h is tangent to every submanifold $j_\infty(s)(M)$ and at least one such a manifold passes through any point $\theta \in J^\infty(\pi)$[4], the equality $X^h(\widehat{d}\varphi) = X^h(\varphi)$ holds. Thus we arrive at the following important equality:

$$U_1 = d - \widehat{d}. \qquad (2.3)$$

Consequently, the operator U_1 "measures" to what extent the de Rham differential of a function $\varphi \in \mathcal{F}$ differs from the horizontal one.

We shall now describe the set $\mathcal{C}D(\pi)$, i.e., vector fields $X \in \mathrm{D}(\pi)$ annihilating the Cartan forms on $J^\infty(\pi)$ or, in other words, the ones which satisfy the equalities $X(\varphi) = X \, \lrcorner \, \widehat{d}\varphi$ for all smooth functions $\varphi \in \mathcal{F}$. Consider a field $X \in \mathrm{D}(\pi)$ and its restriction $X_M = X|_{\mathcal{F}_{-\infty}(\pi)} : C^\infty(M) \to \mathcal{F}_k(\pi)$, which is a derivation of the ring $C^\infty(M)$ with values in the algebra $\mathcal{F}_k(\pi)$ (such a k exists, since X is a derivation compatible with the filtration of $\mathcal{F}(\pi)$). Informally, the derivation X_M may be understood as a vector field on M whose coefficients are functions from $\mathcal{F}_k$. More exactly, the field X_M, at least locally, can be represented in the form

$$X_M = \varphi_1 X_1 + \cdots + \varphi_l X_l, \qquad (2.4)$$

where $X_1, \ldots, X_l \in \mathrm{D}(M)$ are "real" vector fields on the manifold M and the functions $\varphi_1, \ldots, \varphi_l$ lie in $\mathcal{F}_k(\pi)$. Using the representation (2.4), let us define the field $\widehat{X}_M \in \mathrm{D}(\pi)$ by setting $\widehat{X}_M = \varphi_1 \widehat{X}_1 + \cdots + \varphi_l \widehat{X}_l$. Since, by the pointwise definition of the lifting operation, the vectors $(\widehat{X}_l)_\theta$ are tangent to the graph of the infinite jet of a section $s \in \Gamma(\pi)$ at any point $\theta = [s]_x^\infty \in J^\infty(\pi)$, the same is true

---

[3]Below we shall see that this definition generalizes the coordinate representation of Cartan forms considered in the preceding chapters.

[4]This is a consequence of the Borel lemma [80].

for the vector $(\widehat{X}_M)_\theta = \varphi_1(\theta)(\widehat{X}_1)_\theta + \cdots + \varphi_l(\theta)(\widehat{X}_l)_\theta$. Therefore $\widehat{X}_M^v = 0$, and consequently $\widehat{X}_M \,\lrcorner\, U_1(\varphi) = \widehat{X}_M^v(\varphi) = 0$ for any function $\varphi \in \mathcal{F}$.

Let us consider (also locally) the field $X' = X - \widehat{X}_M$. Since, by definition, $\widehat{X}_M|_{C^\infty(M)} = X_M = X|_{\mathcal{F}_{-\infty}(\pi)}$, the field X' is vertical, and so $X' \,\lrcorner\, U_1(\varphi) = X'(\varphi)$ for any function $\varphi \in \mathcal{F}$. On the other hand, as we established just now, $\widehat{X}_M \,\lrcorner\, U_1(\varphi) = 0$, and if $X \in \mathcal{CD}(\pi)$, then $X \,\lrcorner\, U_1(\varphi) = 0$ as well. Hence, $X'(\varphi) = 0$ for any function $\varphi \in \mathcal{F}$. Therefore $X' = 0$, and consequently $X = \widehat{X}_M$. Combining the above, we obtain the following result.

PROPOSITION 2.2. *The set $\mathcal{CD}(\pi)$ of vector fields lying in the Cartan distribution is generated by the liftings to $J^\infty(\pi)$ of the fields on the manifold M:*

$$\mathcal{CD}(\pi) = \left\{ \varphi_1 \widehat{X}_1 + \cdots + \varphi_l \widehat{X}_l \mid \varphi_i \in \mathcal{F}(\pi), \ X_i \in \mathrm{D}(M), \ l = 1, 2, \ldots \right\}.$$

2.2. Integral manifolds. From Proposition 2.2 and Example 1.5 it follows that the Cartan distribution on $J^\infty(\pi)$ is completely integrable. Let study its maximal integral manifolds. To do this, let us first stress two facts: (a) the Cartan plane $\mathcal{C}_\theta$ at a point $\theta \in J^\infty(\pi)$ does not contain nonzero vertical vectors—this follows from the constructions used at the beginning of this section to define the forms $U_1(\varphi)$; (b) dimension of the space $\mathcal{C}_\theta$ coincides with that of the manifold M—this is a consequence of the proposition just proved. Let $\mathcal{R}^\infty$ be a maximal integral manifold of the Cartan distribution. Then from (a) it follows that the projections $\pi_{\infty,k}|_{\mathcal{R}^\infty}: \mathcal{R}^\infty \to \mathcal{R}^k \stackrel{\mathrm{def}}{=} \pi_{\infty,k}(\mathcal{R}^\infty) \subset J^k(\pi)$, $k = 0, 1, \ldots$, and $\pi_\infty|_{\mathcal{R}^\infty}: \mathcal{R}^\infty \to \mathcal{R} \stackrel{\mathrm{def}}{=} \pi_\infty(\mathcal{R}^\infty) \subset M$ are local diffeomorphisms. On the other hand, it follows from (b) that $\dim \mathcal{R}^k = \dim \mathcal{R}^\infty \leq n$. Consider the projection $\pi|_{\mathcal{R}^0}: \mathcal{R}^0 \to \mathcal{R}$. Since it is a local diffeomorphism, there exists a mapping $s': \mathcal{R} \to \mathcal{R}^0$ such that the composition $\pi|_{\mathcal{R}^0} \circ s'$ is the identity on $\mathcal{R}$. In other words, s' is a partially defined section of the bundle π. Obviously, there exists a section $s \in \Gamma(\pi)$ such that $s|_\mathcal{R} = s'$. Assume now that the embedding $\mathcal{R} \subset M$ is strict, i.e., that $\dim \mathcal{R}^k < n$. Then the conditions that the manifolds $\Gamma_s^k = j_k(s)(M) \subset J^k(\pi)$, $k = 0, 1, \ldots, \infty$, contain the manifolds $\mathcal{R}^k$ for the corresponding values of k are the conditions for the normal (with respect to $\mathcal{R}$) derivatives of the section s. By the Whitney theorem on extension for smooth functions [80], it follows that one can always construct a section satisfying these conditions. This contradicts the maximality of $\mathcal{R}^\infty$. Hence, the following statement is valid:

PROPOSITION 2.3. *Any maximal integral manifold of the Cartan distribution on $J^\infty(\pi)$ is locally the graph of the infinite jet of a section of the bundle π.*

Thus, at least one maximal integral manifold of the Cartan distribution passes through any point $\theta \in J^\infty(\pi)$, and we have a complete description of these manifolds. Of course, such a manifold is not uniquely defined. For example, if $\pi: \mathbb{R} \times \mathbb{R} \to \mathbb{R}$ is the trivial bundle, the graph of infinite jet of the function $u = \exp(-1/x^2)$ passes through the origin in $J^\infty(\pi)$ together with the graph of the infinite jet of the zero function. In general, if π is an arbitrary bundle and $\theta = [s]_x^\infty \in J^\infty(\pi)$, then the graphs of the infinite jets of all sections of the bundle π pass through the point θ differing from s by a section flat at the point $x \in M$. It is useful to compare the results obtained here with what we know about maximal integral manifolds in the spaces of finite jets. As we noted in the previous chapter, the Cartan distribution on $J^k(\pi)$, $k < \infty$, contains vertical vectors. Namely, any vector tangent to the fiber of

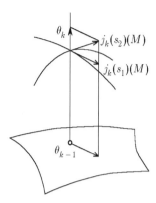

FIGURE 4.4. "Blowing-up" of an R-plane

the bundle $\pi_{k,k-1}$ lies in the Cartan plane $\mathcal{C}^k_{\theta_k}$. This leads to the fact that the integral manifolds passing through the point $\theta_k \in J^k(\pi)$ can be classified by types and the type of a manifold describes the degree of its degeneracy under the projection to $J^{k-1}(\pi)$ (or the dimension of the space of vertical vectors tangent the integral manifold under consideration at the given point). In particular, the graphs of jets of sections of the bundle π are the manifolds of type 0 and are characterized by the fact that they project to $J^{k-1}(\pi)$ locally diffeomorphically. The reasons why the cases of finite and infinite jets are different become clear, if one takes the analytical point of view. In fact, fixing a point $\theta_k \in J^k(\pi)$ for $k < \infty$, we obtain information on all partial derivatives of sections $s \in \Gamma(\pi)$ at the point $x = \pi_k(\theta_k) \in M$ up to order k. Drawing through the point θ_k graphs of different sections, we have freedom of choice which is measured infinitesimally by vectors tangent to the fiber of the projection $\pi_{k,k-1}$ (see Figure 4.4). This leads to the fact that an R-plane at the point θ_k "blows up" by the set of $\pi_{k,k-1}$-vertical vectors. Unlike the situation described above, the choice of a point $\theta \in J^\infty(\pi)$ fixes all partial derivatives of a section $s \in \Gamma(\pi)$ at the point x, and, for obvious reasons, our freedom is limited now by the class of flat functions. Note also that the picture we observe on $J^\infty(\pi)$ is related to existence of the Cartan connection on π_∞ and to integrability of this connection.

We now start to study automorphisms and symmetries of the Cartan distribution on the manifolds $J^\infty(\pi)$. Before starting to discuss the general facts, we shall carry out some coordinate computations that will help us get a kind of intuition on the results to be obtained.

2.3. A computational experiment. We shall need here (as well as everywhere below) a coordinate representation of the Cartan forms $U_1(\varphi)$. Let $x \in M$, and let $\mathcal{U} \ni x$ be a coordinate neighborhood. Consider, as usual, local coordinates $x_1, \ldots, x_n, \ldots, p^j_\sigma, \ldots$ in the neighborhood $\pi_\infty^{-1}(\mathcal{U})$. Then, using formulas (1.20) and (2.3) together with the fact that the operator $U_1 \colon \mathcal{F}(\pi) \to \Lambda^1(\pi)$ is a derivation, it is easily seen that the Cartan module $\mathcal{C}\Lambda^*(\pi)|_{\pi^{-1}(\mathcal{U})}$ is generated by the forms

$$\omega^j_\sigma = U_1(p^j_\sigma) = dp^j_\sigma - \sum_{\alpha=1}^n p^j_{\sigma+1_\alpha}\, dx_\alpha, \tag{2.5}$$

where $j = 1, \ldots, m$, $|\sigma| = 0, 1, \ldots$, i.e., by the Cartan forms defined in Chapter 3. In particular,

$$\omega_\varnothing^j = U_1(u^j) = du^j - \sum_{\alpha=1}^n p_{1_\alpha} \, dx_\alpha. \tag{2.6}$$

Consider the one-dimensional bundle $\pi \colon \mathbb{R} \times \mathbb{R} \to \mathbb{R}$ over the real line. Let $x = x_1$ be the coordinate in the base and $p_{(0)} = u, \ldots, p_{(k)} = p_k, \ldots$ the coordinates in $J^\infty(\pi)$. By (2.5), in this case the Cartan module $\mathcal{C}\Lambda^*(\pi)$ is generated by the forms

$$\omega_k = dp_k - p_{k+1}\, dx, \tag{2.7}$$

where $k = 0, 1, \ldots$. Let the field $X = a\partial/\partial x + \sum_{k=0}^\infty b_k \partial/\partial p_k \in \mathrm{D}(\pi)$, $a, b_k \in \mathcal{F}(\pi)$, be an automorphism of the Cartan distribution. This is equivalent to the fact that $X(\omega_k) = \sum_i \lambda_k^i \omega_i$ for all k or, which is the same, to

$$db_k - b_{k+1}\, dx - p_{k+1}\, da = \sum_i \lambda_k^i (dp_i - p_{i+1}\, dx). \tag{2.8}$$

Writing down the differential at the left-hand side of (2.8) explicitly and collecting similar terms, we obtain the following system of equations:

$$\left(\frac{\partial b_k}{\partial x} - b_{k+1} - p_{k+1}\frac{\partial a}{\partial x}\right) dx + \sum_i \left(\frac{\partial b_k}{\partial p_i} - p_{k+1}\frac{\partial a}{\partial p_i}\right) dp_i$$
$$= \sum_i \lambda_k^i (dp_i - p_{i+1}\, dx),$$

which is equivalent to the system

$$\begin{cases} \dfrac{\partial b_k}{\partial x} - b_{k+1} - p_{k+1}\dfrac{\partial a}{\partial x} = -\sum_i \lambda_k^i p_{i+1}, \\[2mm] \dfrac{\partial b_k}{\partial p_i} - p_{k+1}\dfrac{\partial a}{\partial p_i} = \lambda_k^i, \quad i, k \geq 0. \end{cases} \tag{2.9}$$

Substituting the expressions obtained for λ_k^i to the first equation in (2.9), we see that

$$b_{k+1} = \frac{\partial b_k}{\partial x} - p_{k+1}\frac{\partial a}{\partial x} + \sum_i p_{i+1}\left(\frac{\partial b_k}{\partial p_i} - p_{k+1}\frac{\partial a}{\partial p_i}\right) = D(b_k) - p_{k+1}D(a), \tag{2.10}$$

where $D = D_1$ is the operator of the total derivative along x. As above, let us split the field X into horizontal and vertical components by setting $X^h = \widehat{a\partial/\partial x} = aD$ and $X^v = X - X^h$. Let $X^v = \sum_k b_k^v \partial/\partial p_k$; then $b_k = b_k^v + ap_{k+1}$ for all $k \geq 0$. Substituting this into (2.10) and keeping in mind that $D(p_k) = p_{k+1}$, we obtain the following expressions for the coefficients b_k^v:

$$b_{k+1}^v = D(b_k^v), \qquad k = 0, 1, \ldots,$$

or

$$b_k^v = D^k(b_0^v), \qquad k = 0, 1, \ldots$$

Thus any infinitesimal automorphism of the Cartan distribution on $J^\infty(\pi)$ is representable in the form

$$X = aD + \sum_{k=0}^\infty D^k(b_0^v)\frac{\partial}{\partial p_k}, \tag{2.11}$$

where $b_0^v = b_0 - ap_1$, and consequently is uniquely determined by its restriction to the subalgebra $C^\infty(M) \subset \mathcal{F}(\pi)$ (the first summand in (2.11)) and by some function $b_0^v \in \mathcal{F}(\pi)$ (the second summand). It turns out that the result is of a general nature.

2.4. Evolutionary derivations. Let us again return to an arbitrary vector bundle $\pi\colon P \to M$ and consider an infinitesimal automorphism $X \in \mathrm{D}_\mathcal{C}(\pi)$ of the Cartan distribution on $J^\infty(\pi)$.

PROPOSITION 2.4. *Any vector field $X \in \mathrm{D}_\mathcal{C}(\pi)$ is uniquely determined by its restriction to the subalgebra $\mathcal{F}_0(\pi) = C^\infty(J^0(\pi)) \subset \mathcal{F}(\pi)$.*

PROOF. Since for any two fields $X, X' \in \mathrm{D}_\mathcal{C}(\pi)$ their difference lies in $\mathrm{D}_\mathcal{C}(\pi)$ as well, it is sufficient to show that if $X|_{\mathcal{F}_0} = 0$, then $X|_{\mathcal{F}_k} = 0$ for all $k \geq 0$. Let us prove this by induction on k. The first step of induction ($k = 0$) is the assumptions of the proposition.

Now let $k > 0$ and $X|_{\mathcal{F}_k} = 0$. We must prove that under this condition the derivation $X|_{\mathcal{F}_{k+1}}$ is also trivial. Take $\varphi \in \mathcal{F}_k$. Then

$$X(U_1(\varphi)) = X(d\varphi - \widehat{d\varphi}) = dX(\varphi) - X(\widehat{d\varphi}),$$

or, by the induction hypothesis,

$$X(U_1(\varphi)) = -X(\widehat{d\varphi}).$$

On the other hand, since $X \in \mathrm{D}_\mathcal{C}(\pi)$, there exist functions $f_\alpha, g_\alpha \in \mathcal{F}$, $\alpha = 1, \ldots, l$, such that $X(U_1(\varphi)) = \sum_\alpha f_\alpha U_1(g_\alpha)$, i.e.,

$$X(\widehat{d\varphi}) = -\sum_\alpha f_\alpha U_1(g_\alpha). \tag{2.12}$$

The form on the right-hand side of (2.12) belongs to the Cartan module, and so its restriction to the graph of the jet of any section $s \in \Gamma(\pi)$ vanishes.

Let us use now the fact that for any function $\varphi \in \mathcal{F}(\pi)$ the form $\widehat{d\varphi}$ is horizontal, i.e., can be represented as $\widehat{d\varphi} = \sum_i \varphi_i \, dh_i$, where $\varphi_i \in \mathcal{F}$ and $h_i \in C^\infty(M) \subset \mathcal{F}_0$. Consequently,

$$X(\widehat{d\varphi}) = \sum_i \left(X(\varphi_i) \, dh_i + \varphi_i \, dX(h_i) \right) = \sum_i X(\varphi_i) \, dh_i,$$

since by the induction hypothesis $X(h_i) = 0$. Thus $X(\widehat{d\varphi})$ is also a horizontal form. From (2.12) and from the above it follows that its restrictions to the graphs of jets of sections are trivial. Since horizontal forms are determined by their restrictions to the graphs of jets (see §1), we obtain that

$$X(\widehat{d\varphi}) = 0 \tag{2.13}$$

for any function $\varphi \in \mathcal{F}_k$.

Consider an arbitrary coordinate neighborhood in M and the corresponding local coordinates $(x_1, \ldots, x_n, \ldots, p_\sigma^j, \ldots)$ in $J^\infty(\pi)$. Let us choose one of the coordinate functions p_σ^j, where $|\sigma| \leq k$, for φ. Then, by (2.13),

$$X(\widehat{dp_\sigma^j}) = X\left(\sum_{i=1}^n p_{\sigma+1_i}^j \, dx_i \right) = \sum_{i=1}^n X(p_{\sigma+1_i}^j) \, dx_i = 0.$$

Hence, $X(p_{\sigma+1_i}^j) = 0$ for all σ, $|\sigma| \leq k$, $i = 1, \ldots, n$, and $j = 1, \ldots, m$. In other words, the action of the field X on all coordinate functions on the manifold $J^{k+1}(\pi)$

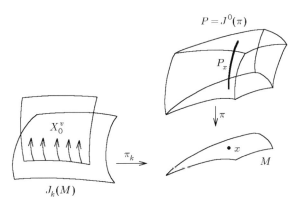

FIGURE 4.5. Vertical field X_0^v

is trivial, i.e., $X|_{\mathcal{F}_{k+1}} = 0$. This proves the induction step and consequently the statement as a whole. □

Let $X \in D_\mathcal{C}(\pi)$ still be an automorphism of the Cartan distribution on $J^\infty(\pi)$. Consider the restriction X_M of the field X to the subalgebra $C^\infty(M) \subset \mathcal{F}$ and the corresponding lifting $\widehat{X}_M \in D(\pi)$. Since, as we already know, $\widehat{X}_M \in \mathcal{C}D(\pi) \subset D_\mathcal{C}(\pi)$, the field $X^v = X - \widehat{X}_M$ is also an automorphism and, moreover, is vertical. Then the restriction $X_0^v = X^v|_{\mathcal{F}_0}$ is a derivation of the algebra $\mathcal{F}_0$ with values in the algebra $\mathcal{F}_k$ for some finite k. Since the field X^v is vertical, the derivation X_0^v can be identified with a family of vectors parametrized by points of the space $J^k(\pi)$ and directed along the fibers of the pullback $\pi_k^*(\pi)$ (see Figure 4.5). On the other hand, since π is a linear bundle, tangent vectors to its fibers can be identified with the points of these fibers. Combining the above constructions, we obtain the mapping

$$D_\mathcal{C}(\pi) \longrightarrow \mathcal{F}(\pi, \pi), \qquad (2.14)$$

and, as Propositions 2.2 and 2.4 show, a field X maps to zero under this correspondence if and only if $X \in \mathcal{C}D(\pi)$. Let us study the image of the mapping (2.14).

Let φ be a section of the bundle $\pi_k^*(\pi)$. Consider a coordinate neighborhood $\mathcal{U} \subset M$ and the neighborhood $\mathcal{U}^k = \pi_k^{-1}(\mathcal{U}) \subset J^k(\pi)$. Similarly to the results obtained in the computational experiment above, let us define in the neighborhood $\mathcal{U}^k$ the field[5]

$$\Im_{\varphi,\mathcal{U}} = \sum_{j,\sigma} D_\sigma(\varphi^j) \frac{\partial}{\partial p_\sigma^j}, \qquad (2.15)$$

where φ^j is the j-th component of the restriction of the section φ to the neighborhood $\mathcal{U}^k$ while D_σ is the composition of the total derivatives corresponding to the multi-index σ. Let us show that the derivation (2.15) in the neighborhood under consideration is an automorphism of the Cartan distribution. In fact, let $\psi \in \mathcal{F}$.

---

[5] In this notation we use the Cyrillic letter Э, which is pronounced like "e" in "ten". It stands for the Russian word meaning "evolutionary".

Then
$$\Im_{\varphi,\mathcal{U}}(U_1(\psi)) = \Im_{\varphi,\mathcal{U}}(d\psi - \widehat{d\psi}) = d(\Im_{\varphi,\mathcal{U}}(\psi)) - \Im_{\varphi,\mathcal{U}}(\widehat{d\psi}). \tag{2.16}$$

By relations (1.20) and (2.15), we have

$$\Im_{\varphi,\mathcal{U}}(\widehat{d\psi}) = \sum_{i,j,\sigma} D_\sigma(\psi^j) \frac{\partial}{\partial p_\sigma^j} D_i(\psi) \, dx_i.$$

But, as is easily seen,

$$\frac{\partial}{\partial p_\sigma^j} D_i(\psi) = \delta(i,\sigma) \frac{\partial \psi}{\partial p_{\sigma-1_i}^j} + D_i\left(\frac{\partial \psi}{\partial p_\sigma^j}\right), \tag{2.17}$$

where

$$\delta(i,\sigma) = \begin{cases} 1 & \text{if the } i\text{-th component of } \sigma \text{ differs from } 0, \\ 0 & \text{otherwise.} \end{cases}$$

Therefore

$$\Im_{\varphi,\mathcal{U}}(\widehat{d\psi}) = \sum_{i,j,\sigma} D_\sigma(\varphi^j) \frac{\partial}{\partial p_\sigma^j} D_i(\psi) \, dx_i$$

$$= \sum_{i,j,\sigma} D_\sigma(\varphi^j) \left(\delta(i,\sigma) \frac{\partial \psi}{\partial p_{\sigma-1_i}^j} + D_i\left(\frac{\partial \psi}{\partial p_\sigma^j}\right) \right) dx_i$$

$$= \sum_{i,j,\sigma} \left(D_\sigma(\varphi^j) \delta(i,\sigma) \frac{\partial \psi}{\partial p_{\sigma-1_i}^j} + D_i\left(D_\sigma(\varphi^j) \frac{\partial \psi}{\partial p_\sigma^j}\right) - D_{\sigma+1_i}(\varphi^j) \frac{\partial \psi}{\partial p_\sigma^j} \right) dx^i.$$

Here the first and last summands annihilate each other, and thus

$$\Im_{\varphi,\mathcal{U}}(\widehat{d\psi}) = \sum_{i,j,\sigma} D_i\left(D_\sigma(\varphi^j) \frac{\partial \psi}{\partial p_\sigma^j}\right) dx_i = \widehat{d}(\Im_{\varphi,\mathcal{U}}(\psi)).$$

Coming back to (2.16), we see that

$$\Im_{\varphi,\mathcal{U}}(U_1(\psi)) = d(\Im_{\varphi,\mathcal{U}}(\psi)) - \widehat{d}(\Im_{\varphi,\mathcal{U}}(\psi)) = U_1 \Im_{\varphi,\mathcal{U}}(\psi). \tag{2.18}$$

Equality (2.18) shows that the derivations of the form $\Im_{\varphi,\mathcal{U}}$ commute with the operator U_1 and consequently are infinitesimal automorphisms of the Cartan distribution. The restriction of $\Im_{\varphi,\mathcal{U}}$ to the subalgebra $\mathcal{F}_0(\pi|_{\mathcal{U}})$ is identified with the section φ. Thus we have a procedure (so far, local) for constructing an automorphism of the Cartan distribution determined by a section φ. Now let $\mathcal{U}, \mathcal{U}' \subset M$ be two coordinate neighborhoods in M. Then the restrictions of the fields $\Im_{\varphi,\mathcal{U}}$ and $\Im_{\varphi,\mathcal{U}'}$ to the algebra $\mathcal{F}_0(\pi|_{\mathcal{U} \cap \mathcal{U}'})$ coincide, and so, by Proposition 2.4, it follows that the fields coincide over the intersection $\mathcal{U} \cap \mathcal{U}'$ as well. Hence our constructions in a well-defined way determine a vertical vector field $\Im_\varphi$ on the entire space $J^\infty(\pi)$, and this field is an infinitesimal automorphism of the Cartan distribution. In other words, the mapping (2.14) is epimorphic, and therefore the following theorem is proved.

THEOREM 2.5. *Any infinitesimal automorphism X of the Cartan distribution on $J^\infty(\pi)$ can be uniquely represented in the form*

$$X = \Im_\varphi + \widehat{X}_M,$$

where φ is a section from $\mathcal{F}(\pi,\pi)$. The algebra $\operatorname{sym}\mathcal{C}(\pi) = \mathrm{D}_\mathcal{C}(\pi)/\mathcal{C}\mathrm{D}(\pi)$ of symmetries of the Cartan distribution is identified with the module $\mathcal{F}(\pi,\pi)$ of the sections of the pullback $\pi_\infty^*(\pi)$,

$$\operatorname{sym}\mathcal{C}(\pi) \simeq \mathcal{F}(\pi,\pi).$$

Let us consider the fields ∂_φ in more details. In the preceding chapter we saw that Lie fields, being automorphisms of the Cartan distribution on the manifolds $J^k(\pi)$ of finite jets, generate flows on the set of sections of the bundle π. The same is valid for the fields of the form ∂_φ, though in general they do not possess one-parameter groups of transformations (see below) on the infinite-dimensional manifold $J^\infty(\pi)$. In fact, let $s \in \Gamma(\pi)$ be a section and $\varphi \in \mathcal{F}_k(\pi,\pi)$. Let us restrict the field ∂_φ to the graph of the k-th jet of the section s. Then, under identification of the tangent spaces to the fibers of π with the fibers themselves, the components of this restriction will be of the form $(j_k(s))^*(\varphi^\alpha)$, $\alpha = 1,\ldots,m$. Therefore, the motion of the section s along the trajectories of the field ∂_φ (provided these trajectories exist) should be governed by the equations

$$\frac{\partial s^j}{\partial t} = \varphi^j\left(x,\ldots,\frac{\partial^{|\sigma|}s^l}{\partial x^\sigma},\ldots\right), \qquad j,l=1,\ldots,m, \quad |\sigma| \leq k,$$

where t is the parameter along the trajectory. In other words, the derivation ∂_φ determines evolution of sections of the bundle π, while the "flow" corresponding to ∂_φ is determined by evolution equations of the form

$$\frac{\partial u^j}{\partial t} = \varphi^j\left(x,\ldots,\frac{\partial^{|\sigma|}u^l}{\partial x^\sigma},\ldots\right), \qquad j,l=1,\ldots,m, \quad |\sigma| \leq k. \tag{2.19}$$

DEFINITION 2.2. The fields ∂_φ are called *evolutionary derivations*; the corresponding section $\varphi \in \mathcal{F}(\pi,\pi)$ is called the *generating section* of the evolutionary derivation.

Note that a special form of coefficients of evolutionary derivations (see (2.15)) guarantees mutual compatibility for evolution of derivatives of the functions u^j:

$$\frac{\partial}{\partial t}\frac{\partial^{|\tau|}u}{\partial x^\tau} = (D_\tau \varphi^j)\left(x,\ldots,\frac{\partial^{|\sigma|}u}{\partial x^\sigma},\ldots\right). \tag{2.20}$$

This is a coordinate expression of the fact that ∂_φ is an automorphism of the Cartan distribution.

It is interesting to note that if one treats evolutionary derivations as vector fields on the "manifold" $\Gamma(\pi)$ of sections of the bundle π, the operator $U_1 \colon \mathcal{F}(\pi) \to \Lambda^1(\pi)$ plays a role similar to that of the external differential $d \colon C^\infty(M) \to \Lambda^1(M)$ in classical differential geometry. In fact, if X is an evolutionary derivation, then one has (see the splitting (2.2)) $X^v = X$, and therefore, by the definition of the operator U_1,

$$X \lrcorner\, U_1(\varphi) = X(\varphi). \tag{2.21}$$

Equation (2.21) is an exact counterpart of the equation

$$Y \lrcorner\, df = Y(f),$$

valid for fields Y and functions f on some finite-dimensional manifold. Therefore, the operator U_1 can be called the *universal evolutionary differential*.

2.5. Jacobi brackets. Above we showed that the mapping

$$\vartheta \colon \mathcal{F}(\pi,\pi) \longrightarrow \operatorname{sym}(\pi) = \mathrm{D}_{\mathcal{C}}(\pi)/\mathcal{C}\mathrm{D}(\pi),$$

putting into correspondence to any section $\varphi \in \mathcal{F}(\pi,\pi)$ the evolutionary derivation ϑ_φ, is a bijection. On the other hand, since the set of evolutionary derivations is identified with the Lie quotient algebra $\operatorname{sym}(\pi)$, the commutator of two evolutionary derivations is an evolutionary derivation again. Therefore, for any two elements $\varphi, \psi \in \mathcal{F}(\pi,\pi)$ one can define their commutator $\{\varphi,\psi\} \in \mathcal{F}(\pi,\pi)$ by setting

$$\{\varphi,\psi\} = \vartheta^{-1}\left([\vartheta_\varphi, \vartheta_\psi]\right), \tag{2.22}$$

or

$$\vartheta_{\{\varphi,\psi\}} = [\vartheta_\varphi, \vartheta_\psi]. \tag{2.23}$$

DEFINITION 2.3. The element $\{\varphi,\psi\}$ defined by (2.22) is called the *higher Jacobi bracket* of the sections φ and ψ.

Since evolutionary derivations are vertical and form a Lie algebra with respect to the commutator of vector fields, the module $\mathcal{F}(\pi,\pi)$ is a Lie $\mathbb{R}$-algebra with respect to the Jacobi bracket. In fact, let us show for example that the bracket $\{\cdot,\cdot\}$ satisfies the Jacobi identity. By (2.23) we have

$$\vartheta_{\{\varphi,\{\psi,\chi\}\}} = [\vartheta_\varphi, \vartheta_{\{\psi,\chi\}}] = [\vartheta_\varphi, [\vartheta_\psi, \vartheta_\chi]] = [[\vartheta_\varphi, \vartheta_\psi], \vartheta_\chi] + [\vartheta_\psi, [\vartheta_\varphi, \vartheta_\chi]]$$
$$= \vartheta_{\{\{\varphi,\psi\},\chi\}} + \vartheta_{\{\psi,\{\varphi,\chi\}\}} = \vartheta_{\{\{\varphi,\psi\},\chi\}+\{\psi,\{\varphi,\chi\}\}},$$

i.e.,

$$\{\varphi,\{\psi,\chi\}\} + \{\psi,\{\chi,\varphi\}\} + \{\chi,\{\varphi,\psi\}\} = 0$$

for any sections $\varphi, \psi, \chi \in \mathcal{F}(\pi,\pi)$. Other properties of Lie algebras can be checked in a similar way.

In local coordinates, the Jacobi bracket of two sections φ and ψ, as follows from (2.15) and (2.23), is of the form

$$\{\varphi,\psi\}^j = \sum_{\sigma,\alpha} \left(D_\sigma(\varphi^\alpha) \frac{\partial \psi^j}{\partial p^\alpha_\sigma} - D_\sigma(\psi^\alpha) \frac{\partial \varphi^j}{\partial p^\alpha_\sigma} \right), \qquad j = 1, \ldots, m. \tag{2.24}$$

2.6. Comparison with Lie fields. In the preceding chapter, we defined Lie fields as derivations $Y \in \mathrm{D}(J^k(\pi))$ preserving the Cartan distribution on $J^k(\pi)$. Since any Lie field can be lifted to any space $J^{k+l}(\pi)$ in a canonical way and this lifting is also a Lie field, we obtain a field $Y^* \in \mathrm{D}(\pi)$. The condition that all the liftings preserve the Cartan distribution on the corresponding spaces of finite jets means that $Y^* \in \mathrm{D}_{\mathcal{C}}(\pi)$, i.e., the infinite lifting of a Lie field is an automorphism of the Cartan distribution on $J^\infty(\pi)$. A characteristic feature of the derivation Y^* is that it preserves filtration degree of elements from $\mathcal{F}(\pi)$: $\deg(Y^*) = 0$. Conversely, let an automorphism $X \in \mathrm{D}_{\mathcal{C}}(\pi)$ possess this property. Then there exists a k_0 such that for all $k \geq k_0$ the restrictions $X_k = X|_{\mathcal{F}_k}$ are derivations of the algebras $\mathcal{F}_k$ to themselves, and consequently these restrictions determine Lie fields on $J^k(\pi)$. For these fields one has

$$X_{k+1} \circ \pi^*_{k+1,k} = \pi^*_{k+1,k} \circ X_k, \tag{2.25}$$

i.e., they are compatible with the projections $\pi_{k+1,k} \colon J^{k+1}(\pi) \to J^k(\pi)$. By the theorem describing Lie fields and proved in the preceding chapter, for any field X_k there exists a uniquely determined Lie field $X_k^\varepsilon \in \mathrm{D}(J^\varepsilon(\pi))$ such that X_k is its

$(k-\varepsilon)$-lifting[6]. Using equations (2.25), we see that X_{k+1} is the lifting of the field X_k to $J^{k+1}(\pi)$ and consequently X is of the form Y^*, where Y is a Lie field. Hence, the following statement is valid.

PROPOSITION 2.6. *An automorphism $X \in \mathrm{D}_\mathcal{C}(\pi)$ is of the form $X = Y^*$, where Y is a Lie field, if and only if $\deg(X) = 0$.*

DEFINITION 2.4. Automorphisms of the Cartan distribution satisfying the assumptions of Proposition 2.6 will be called *Lie fields* on the manifold $J^\infty(\pi)$.

Let X be such a field. Then, by Theorem 2.5, one has the canonical decomposition
$$X = \partial_\varphi + \widehat{X}_M. \tag{2.26}$$
Note that due to coordinate representation of evolutionary derivations (see equation (2.15)), the field ∂_φ raises filtration of elements in $\mathcal{F}(\pi)$ by $\deg(\varphi)$. On the other hand, $\deg(\widehat{X}_M) = 1$ for any field $\widehat{X}_M \in \mathcal{CD}(\pi)$. Therefore, from Proposition 2.6 it follows that if X is a Lie field, then the section φ from the decomposition (2.26) lies in the module $\mathcal{F}_1(\pi, \pi)$, i.e., $\deg(\varphi) = 1$. Recall now that any Lie field on $J^k(\pi)$ is determined by its generating section, and let us compare the expression for Lie fields in terms of generating sections (see §3 of Ch. 3) with the definition of the derivation ∂_φ (see (2.15)). Then it is easily seen that (2.26) can be written as
$$X_\varphi^* = \partial_\varphi - \sum_{i=1}^n \frac{\partial \varphi^j}{\partial p_{1_i}^j} D_i, \tag{2.27}$$
where X_φ^* is the lifting of the Lie field X_φ with the generating section $\varphi \in \mathcal{F}(\pi, \pi)$ to $J^\infty(\pi)$, while $\varphi^1, \ldots, \varphi^m$ are the components of this section. As we know, if $\dim \pi = 1$, then φ is an arbitrary section, while for $\dim \pi > 1$ it is of the form
$$\varphi^j = a_0 + \sum_{i=1}^n p_{1_i}^j a_i, \quad j = 1, \ldots, m, \quad a_0, \ldots, a_n \in \mathcal{F}_0(\pi).$$
Hence, the generating functions of Lie fields defined in Chapter 3 coincide with those considered here.

EXERCISE 2.1. Show that if X_φ and X_ψ are two Lie fields, then
$$[X_\varphi^*, X_\psi^*] = [X_\varphi, X_\psi]^* = \partial_{\{\varphi, \psi\}} - \sum_{i=1}^n \frac{\partial \{\varphi, \psi\}^j}{\partial p_{1_i}^j} D_i.$$

In other words, if φ and ψ are generating sections of Lie fields, then the Jacobi bracket of these two sections defined earlier coincides with the Jacobi bracket introduced in this section. Taking into account that by (2.27) one has $X_\varphi = 0$ if and only if the corresponding evolutionary derivation is trivial, we can formulate the following statement.

PROPOSITION 2.7. *The mapping taking Lie fields to the corresponding evolutionary derivations is a Lie algebra monomorphism. The set of generating sections of Lie fields is a Lie subalgebra in the Lie algebra $\mathcal{F}(\pi, \pi)$ with respect to the Jacobi bracket.*

---

[6]Recall that $\varepsilon = 0$ for $\dim \pi > 1$ and $\varepsilon = 1$ for $\dim \pi = 1$; in the first case X_k^0 is an arbitrary vector field on $J^0(\pi)$ while in the second case X_k^1 is a contact field on $J^1(\pi)$.

Thus, the theory of symmetries of the Cartan distribution on $J^\infty(\pi)$ is a natural generalization of the theory of Lie fields on the manifolds of finite jets.

Let X be a Lie field on $J^\infty(\pi)$. Then, considering one-parameter groups of transformations corresponding to the fields $X_k = X|_{J^k(\pi)}$, we shall see that a one-parameter group of transformations also corresponds to the field X, and it acts on the manifold $J^\infty(\pi)$. In fact, any automorphism of the Cartan distribution on $J^\infty(\pi)$ possessing a one-parameter group of transformations can be represented as a Lie field, possibly in some new bundle [**17**]. Let us explain this, using somewhat informal reasoning.

Let $\pi \colon P \to M$ be a fiber bundle, $X \in D_\mathcal{C}(\pi)$ an integrable vector field, and A_t the corresponding one-parameter group of transformations. Then

$$A_t = \exp(tX) = \mathrm{id} + tX + \cdots + \frac{t^k}{k!}X^k + \cdots \qquad (2.28)$$

From (2.28) it follows that there exists an l such that $X^k(\mathcal{F}_0) \subset \mathcal{F}_l$ for all k. In fact, otherwise the transformation A_t could not transform the manifold $J^0(\pi)$ to any finite manifold $J^k(\pi)$, which would contradict the definition of a smooth map of the manifold $J^\infty(\pi)$.

Consider in the ring $\mathcal{F}_l$ the subring $\mathcal{F}(X)$ generated by the elements of the form $X(\varphi)$, $\varphi \in \mathcal{F}_0$, i.e., we consider the set of sums of the form $\sum X(\varphi_1) \cdot \ldots \cdot X(\varphi_r)$, $\varphi_i \in \mathcal{F}_0$. From the above it follows that the subring $\mathcal{F}(X)$ is invariant with respect to the action of the field X: $X(\mathcal{F}(X)) \subset \mathcal{F}(X)$. Let us pass to the dual viewpoint and consider the maximal $\mathbb{R}$-spectrum of the ring $\mathcal{F}(X)$, i.e., the space whose points are the kernels of homomorphisms $\mathcal{F}(X) \to \mathbb{R}$. Denote this space by P_X. Then locally, in a neighborhood of a generic point, we have the commutative diagram

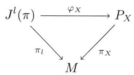

where φ_X is the mapping dual to the embedding $\mathcal{F}(X) \hookrightarrow \mathcal{F}$. By invariance of $\mathcal{F}(X)$ with respect to X, the field X determines on P_X a field $\widetilde{X}_0$. The set of all possible prolongations of the mapping φ_X determines a smooth mapping $\varphi_X^* \colon J^\infty(\pi) \to J^\infty(\pi_X)$ for which the diagram

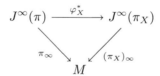

is also commutative. Let $\widetilde{X}$ be the lifting of the field $\widetilde{X}_0$ to $J^\infty(\pi_X)$. Then, by the construction of lifting and the fact that X is an integrable automorphism of the Cartan distribution on $J^\infty(\pi)$, the mapping φ_X^* takes X to the restriction of the field $\widetilde{X}$ to the submanifold $\mathcal{E}_X = \varphi_X^*(J^\infty(\pi)) \subset J^\infty(\pi_X)$. Hence, the structure of the fields X and $\widetilde{X}|_{\mathcal{E}_X}$ is locally the same, and this is what we wanted to show.

2.7. Linearizations. Let us note that the correspondence between evolutionary derivations and generating sections makes it possible, having a pair (φ, ψ), $\varphi \in \mathcal{F}(\pi, \pi)$, $\psi \in \mathcal{F}(\pi)$, to construct a new section $\partial_\varphi(\psi) \in \mathcal{F}(\pi)$. When φ is fixed while ψ is an arbitrary section, we obtain an evolutionary derivation in the algebra

$\mathcal{F}(\pi)$. Passing to the "adjoint" point of view, i.e., by fixing a section ψ and making φ "free", we shall obtain a new operator ℓ_ψ acting from the module $\mathcal{F}(\pi,\pi)$ to the algebra $\mathcal{F}(\pi)$ by the rule

$$\ell_\psi(\varphi) = \partial_\varphi(\psi). \tag{2.29}$$

To understand the meaning of this operator, let us consider equations (2.15) and use them to rewrite the definition (2.29) in the form

$$\ell_\psi(\varphi) = \sum_{j,\sigma} \frac{\partial \psi}{\partial p_\sigma^j} D_\sigma(\varphi^j),$$

or

$$\ell_\psi = \sum_{j,\sigma} \frac{\partial \psi}{\partial p_\sigma^j} D_\sigma^{(j)}, \tag{2.30}$$

where $D_\sigma^{(j)}$ means that the operator D_σ is applied to the j-th component of the corresponding section. Let us study equation (2.30) in more detail.

Let s be a section of the bundle π. Since every field D_i is tangent to all maximal integral manifolds of the Cartan distribution on $J^\infty(\pi)$ and consequently admits restrictions to the graphs of infinite jets, the operator ℓ_ψ, by (2.30), admits such a restriction as well. To be more precise, one has the commutative diagram

$$\begin{array}{ccc} \mathcal{F}(\pi,\pi) & \xrightarrow{\ell_\psi} & \mathcal{F}(\pi) \\ {\scriptstyle j_\infty(s)^*}\downarrow & & \downarrow{\scriptstyle j_\infty(s)^*} \\ \mathcal{F}(\pi) & \xrightarrow{\ell_\psi|_s} & \Gamma(\pi) \end{array} \tag{2.31}$$

where $\ell_\psi|_s$ is the restriction under consideration. Combining (2.31) with (2.30), we see that for any section $\varphi \in \Gamma(\pi)$ the equality

$$\ell_\psi|_s(\varphi) = \sum_{j,\sigma} \left.\frac{\partial \psi}{\partial p_\sigma^j}\right|_s \frac{\partial^{|\sigma|} \varphi^j}{\partial x^\sigma}. \tag{2.32}$$

holds. Formula (2.32) shows that the operator $\ell_\psi|_s$ is the linearization of the operator Δ_ψ corresponding to the function ψ taken at the section $s \in \Gamma(\pi)$. All such linearizations are obtained by restricting the operator ℓ_ψ to the corresponding section.

DEFINITION 2.5. The operator ℓ_ψ is called the *operator of the universal linearization* of the nonlinear differential operator Δ_ψ.

Let us list the main properties of universal linearizations:
1. The operator $\ell_\psi \colon \mathcal{F}(\pi,\pi) \to \mathcal{F}(\pi)$, $\psi \in \mathcal{F}(\pi)$, is linear.
2. It admits restriction to the graphs of infinite jets.
3. The operator ℓ_ψ is dual to evolutionary derivations: $\ell_\psi(\varphi) = \partial_\varphi(\psi)$.

Note also that if ψ is linear with respect to all variables p_σ^j (i.e., the operator Δ_ψ is linear), then

$$\ell_\psi = \widehat{\Delta}_\psi. \tag{2.33}$$

The above constructed operator ℓ_ψ was defined for nonlinear differential operators of the form $\Delta_\psi \colon \Gamma(\pi) \to C^\infty(M)$. Let us generalize the construction of the universal linearization to arbitrary operators $\Delta \colon \Gamma(\pi) \to \Gamma(\pi')$, where both π' and π are locally trivial vector bundles over the manifold M, $\dim(\pi') = m'$. Let

$\psi \in \mathcal{F}(\pi, \pi')$ be the representing section of the operator Δ. Consider an arbitrary element $\varphi \in \mathcal{F}(\pi, \pi)$ and a family of elements $\varphi_t \in \mathcal{F}(\pi, \pi)$ smoothly depending on $t \in \mathbb{R}$, such that $\partial \varphi_t / \partial t|_{t=0} = \varphi$. Let $\nabla_t = \nabla_{\varphi_t} \colon \Gamma(\pi) \to \Gamma(\pi)$ be the corresponding family of operators and $\bar{\varphi}_t = \varphi_{\Delta \circ \nabla_t}$. Let us set

$$\ell_\psi(\varphi) = \left(\frac{\partial}{\partial t} \bar{\varphi}_t\right)\bigg|_{t=0}. \tag{2.34}$$

EXERCISE 2.2. Show that $\ell_\psi(\varphi)$ is well defined by (2.34), i.e., does not depend on the choice of the family φ_t satisfying the condition $\partial \varphi_t / \partial t|_{t=0} = \varphi$.

The operator $\ell_\psi \colon \mathcal{F}(\pi, \pi) \to \mathcal{F}(\pi, \pi')$ is called the *operator of the universal linearization* of the nonlinear differential operator Δ_ψ, and in local coordinates is represented in the form $\ell_\psi = \|\ell_\psi^{\alpha\beta}\|$, where $\|\ell_\psi^{\alpha\beta}\|$ is the $m' \times m$ matrix whose elements are

$$\ell_\psi^{\alpha\beta} = \sum_\sigma \frac{\partial \psi^\alpha}{\partial p_\sigma^\beta} D_\sigma, \qquad \alpha = 1, \ldots, m', \quad \beta = 1, \ldots, m. \tag{2.35}$$

Obviously, in the case $m' = 1$ this operator coincides with the operator introduced above. Moreover, it possesses the above listed properties 1–3, with the only difference that the operator $\partial_\varphi^{\pi'}$ defined by the equality $\partial_\varphi^{\pi'}(\psi) = \ell_\psi(\varphi)$ acts now not in the algebra $\mathcal{F}(\pi)$, but in the module $\mathcal{F}(\pi, \pi')$. If $\psi \in \mathcal{F}(\pi, \pi')$ and $f \in \mathcal{F}(\pi)$, then the operators $\partial_\varphi^{\pi'}$ and ∂_φ are related to each other by the equalities

$$\partial_\varphi^{\pi'}(f\psi) = \partial_\varphi(f)\psi + f \partial_\varphi^{\pi'}(\psi). \tag{2.36}$$

Thus we see that every section $\varphi \in \mathcal{F}(\pi, \pi)$ determines a family of derivations $\partial_\varphi^{\pi'} \colon \mathcal{F}(\pi, \pi') \to \mathcal{F}(\pi, \pi')$, and, as follows from (2.36), every operator $\partial_\varphi^{\pi'}$ is a derivation of the module $\mathcal{F}(\pi, \pi)$ over the derivation ∂_φ of the algebra $\mathcal{F}(\pi)$.

REMARK 2.1. The relation between evolutionary derivations and linearizations becomes clearer if we again use a parallel with classical differential geometry. In fact, if M and M' are smooth finite-dimensional manifolds and $G \colon M \to M'$ is a smooth mapping, then the linearization (i.e., the differential) G_* of the mapping G at a point $x \in M$ is constructed as follows: one takes a curve x_t in M passing through the point x, $x_0 = x$, considers the tangent vector $v = dx_t/dt|_{t=0}$ to this curve at x, and shows that the equality $G_*(v) = dG(x_t)/dt$ determines a tangent vector to M' at the point $G(x)$ depending only on the choice of v. If X is a vector field on M, then, in general, this construction does not allow one to obtain the corresponding vector field $G_*(X)$ on M'. If nevertheless one understands vector fields as sections of the tangent bundle, one can put into correspondence, in a canonical way, to any field $X \in D(M)$ a section $G_*(X)$ of the bundle over M, which is the pullback of the tangent bundle over M' by the mapping G.

In our situation, the role of points of "manifolds" under consideration is played by sections of the bundles π and π' respectively, while a nonlinear differential operator $\Delta = \Delta_\psi$ (or, more exactly, the mapping $\Phi_\Delta \colon J^\infty(\pi) \to J^\infty(\pi')$, see §1.3) plays the role of G. Taking a curve in $\Gamma(\pi)$, we have to consider a family of sections $s(t) \in \Gamma(\pi)$. As we know, "tangent vectors" to such curves are determined by evolutionary derivations in $\mathcal{F}(\pi)$, which play the role of vector fields on $\Gamma(\pi)$, i.e., by sections $\varphi \in \mathcal{F}(\pi, \pi)$. If ∇ is the operator corresponding to the section φ, then

2. THE CARTAN DISTRIBUTION ON $J^\infty(\pi)$

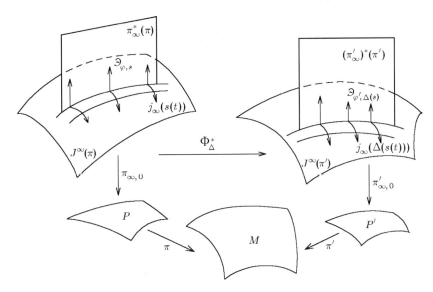

FIGURE 4.6. Action of the linearization on evolutionary vector fields

the curve $s(t)$ is determined by the evolution equation
$$\frac{\partial s}{\partial t} = \nabla(s),$$
while the "tangent vector" to the image of this curve can be found from the equality
$$\frac{\partial}{\partial t}\Delta(s) = j_\infty(s(0))^*(\ell_\psi(\varphi)).$$
Thus, similarly to the finite-dimensional case, the linearization is defined on vector fields, i.e., on evolutionary derivations. The ambiguity can be eliminated if one considers the pullback of the bundle $(\pi'_\infty)^*(\pi')$ (i.e., of a tangent bundle analog) to $J^\infty(\pi)$ by the mapping $G = \Phi_\Delta \colon J^\infty(\pi) \to J^\infty(\pi')$ (defined by all prolongations of the operator Δ); see Figure 4.6. One can easily see that
$$(\Phi_\Delta)^* (\pi'_\infty)^*(\pi') = (\pi'_\infty \circ \Phi_\Delta)^* (\pi') = \pi^*_\infty(\pi').$$

The operators ℓ_ψ and evolutionary derivations of the form $\partial_\varphi^{\pi'}$ make it possible to rewrite some of the relations obtained in this section in a convenient form. First of all, let us note that if $\mathrm{id}\colon \Gamma(\pi) \to \Gamma(\pi)$ is the identity operator, then its linearization $\ell_{\mathrm{id}}\colon \mathcal{F}(\pi,\pi) \to \mathcal{F}(\pi,\pi)$ is the identity as well. Hence, for any section $\varphi \in \mathcal{F}(\pi,\pi)$ we have
$$\partial_\varphi^\pi(\varphi_{\mathrm{id}}) = \ell_{\mathrm{id}}(\varphi) = \varphi. \tag{2.37}$$
Comparing (2.37) with the definition of the Jacobi bracket, we see that
$$[\partial_\varphi^\pi, \partial_\psi^\pi](\varphi_{\mathrm{id}}) = \partial_\varphi^\pi\left(\partial_\psi^\pi(\varphi_{\mathrm{id}})\right) - \partial_\psi^\pi\left(\partial_\varphi^\pi(\varphi_{\mathrm{id}})\right) = \partial_\varphi^\pi(\psi) - \partial_\psi^\pi(\varphi),$$
i.e.,
$$\{\varphi,\psi\} = \partial_\varphi^\pi(\psi) - \partial_\psi^\pi(\varphi). \tag{2.38}$$
Another expression for the Jacobi bracket immediately follows from (2.38) and is of the form
$$\{\varphi,\psi\} = \partial_\varphi^\pi(\psi) - \ell_\varphi(\psi),$$

or
$$\{\varphi, \cdot\} = \partial_\varphi^\pi - \ell_\varphi. \tag{2.39}$$

Comparing equation (2.27) with the coordinate representation of the universal linearization, one can also see that if π is a trivial bundle, then the Lie field X_φ^* on $J^\infty(\pi)$ can be represented in the form

$$X_\varphi^* = \partial_\varphi - \ell_\varphi + \ell_\varphi(\mathbf{1}) = \{\varphi, \cdot\} + \ell_\varphi(\mathbf{1}), \tag{2.40}$$

where $\mathbf{1}$ is the section of the bundle $\pi_\infty^*(\pi)\colon J^\infty(\pi) \times \mathbb{R} \to J^\infty(\pi)$ identically equal to 1 at any point of the manifold $J^\infty(\pi)$.

Evolutionary derivations and universal linearizations play an important role in the theory of higher symmetries for nonlinear differential equation. We now begin to study this theory.

3. Infinitely prolonged equations and the theory of higher symmetries

In the preceding section, studying infinitesimal automorphisms of the Cartan distribution on $J^\infty(\pi)$, we arrived to the concept of evolutionary derivations. These derivations can be informally understood as vector fields on the "manifold" of sections $\Gamma(\pi)$. Our next step consists in extending this approach to arbitrary differential equations $\mathcal{E} \subset J^k(\pi)$. Trying to understand what a vector field on the "manifold" $\mathrm{Sol}(\mathcal{E})$ of solutions to $\mathcal{E}$ is, we shall construct the theory of higher symmetries for nonlinear differential equations. This theory in a natural way generalizes the theory of classical symmetries constructed before.

Since solutions of the equation $\mathcal{E}$ are sections $s \in \Gamma(\pi)$ satisfying the condition $\Gamma_s^k = j_k(s)(M) \subset \mathcal{E}$, it seems natural to assume that at least some of the fields we seek on $\mathrm{Sol}(\mathcal{E})$ are to be obtained by restricting the fields from the enveloping space $\Gamma(\pi)$, i.e., by restricting evolutionary derivations. Nevertheless, if we take an arbitrary evolutionary derivation ∂_φ on $J^\infty(\pi)$ and try restrict it to $\mathcal{E}$ (i.e., to apply it to functions from $C^\infty(\mathcal{E})$), then, in the case $\deg(\varphi) \neq 0$, our efforts will be unsuccessful: the function $\partial_\varphi(\psi)$, $\psi \in C^\infty(\mathcal{E})$, will cease to be an element of the algebra $C^\infty(\mathcal{E})$. Using corrections to ∂_φ by the fields of the form $\widehat{X}_M$ leads to the situation when all "fields" on $\mathrm{Sol}(\mathcal{E})$ obtained in this way are exhausted by the already known classical symmetries of the equation $\mathcal{E}$. The reasons for this are clear: it suffices to remember that evolutionary derivations arose when we passed from manifolds $J^k(\pi)$ of finite jets to the tower

$$\cdots \longleftarrow J^k(\pi) \longleftarrow J^{k+1}(\pi) \longleftarrow \cdots$$

and considered the Cartan distribution on $J^\infty(\pi)$. An analog of this construction for the equation $\mathcal{E}$ is the chain of its *prolongations* $\mathcal{E}^{(l)}$, $l = 0, 1\ldots$, leading to the *infinitely prolonged equation* $\mathcal{E}^\infty \subset J^\infty(\pi)$. Note that the concept of prolongation plays an important role in many aspects of differential equations, such as the theory of formal integrability, singularities of solutions (discontinuities, shock waves), etc.

3.1. Prolongations. Let $\mathcal{E} \subset J^k(\pi)$ be an equation of order k locally given by the conditions

$$\mathcal{E} = \left\{ \theta_k \in J^k(\pi) \mid F_1(\theta_k) = \cdots = F_r(\theta_k) = 0,\ F_1, \cdots, F_r \in \mathcal{F}_k(\pi) \right\}.$$

If $s \in \Gamma(\pi)$ is a solution of the equation $\mathcal{E}$, i.e., if

$$F_\alpha\left(x, \ldots, \frac{\partial^{|\sigma|} s^j}{\partial x^\sigma}, \ldots\right) = 0, \qquad \alpha = 1, \ldots, r, \quad j = 1, \ldots, m, \quad |\sigma| \le k, \qquad (3.1)$$

then s must satisfy all differential consequences of the system $(3.1)^7$. In particular,

$$\frac{\partial F_\alpha}{\partial x_i} + \sum \frac{\partial F_\alpha}{\partial p_\sigma^j} \frac{\partial^{|\sigma|+1} s^j}{\partial x_i \partial x^\sigma} = 0, \qquad \alpha = 1, \ldots, r, \quad j = 1, \ldots, m, \quad |\sigma| \le k, \qquad (3.2)$$

for all $i = 1, \ldots, n$, or

$$(j_{k+1}(s))^* D_i(F_\alpha) = 0, \qquad i = 1, \ldots, n, \quad \alpha = 1, \ldots, r, \qquad (3.3)$$

where the D_i are the operators of total derivatives. Combining systems (3.1) and (3.2), we obtain a differential consequence of (3.1) of order ≤ 1 and thus an equation $\mathcal{E}^{(1)} \subset J^{k+1}(\pi)$ of order $k+1$, which is called the *first prolongation* of $\mathcal{E}$.

To define the concept of the first prolongation in a coordinate-free manner, let us find out what are the conditions for the point $\theta_{k+1} \in J^{k+1}(\pi)$ to belong to the set $\mathcal{E}^{(1)}$. First, it is obvious that the point $\theta_k = \pi_{k+1,k}(\theta_{k+1})$ needs to lie on the equation $\mathcal{E}$. Further, let us represent the point $\theta_{k+1} \in \mathcal{E}^{(1)}$ in the form $\theta_{k+1} = [s]_x^{k+1}$, $s \in \Gamma(\pi)$, and substitute the Taylor expansion of length $k+1$ of the section s into equation (3.1). Expanding the result to the Taylor series and comparing the expression obtained with (3.2), we see that the point θ_{k+1} lies on $\mathcal{E}^{(1)}$ if and only if the corresponding section $s \in \Gamma(\pi)$ satisfies the equation $\mathcal{E}$ at the point θ_k up to infinitesimals of second order. In other words, the point $\theta_{k+1} = [s]_x^{k+1}$ lies on $\mathcal{E}^{(1)}$ if and only if the manifold $j_k(s)(M)$ is tangent to $\mathcal{E}$ at the point $\theta_k = [s]_x^k$. This is the desired invariant definition of the first prolongation. Generalizing it in a natural way, we arrive at the following definition:

DEFINITION 3.1. The set $\mathcal{E}^{(l)} \subset J^{k+l}(\pi)$, consisting of the points $\theta_{k+l} = [s]_x^{k+l}$ such that the graph $j_k(s)(M)$ of the k-th jet of the section $s \in \Gamma(\pi)$ is tangent to the equation $\mathcal{E}$ at the point $\theta_k = [s]_x^k$ with order $\ge l$, is called the *l-th prolongation of the equation* $\mathcal{E} \subset J^k(\pi)$.

Obviously, the conditions describing the l-th prolongation of the equation $\mathcal{E}$ are differential consequences of the conditions (3.1) up to order l, i.e., they are of the form

$$D_\tau(F_\alpha) = 0, \qquad |\tau| \le l, \quad \alpha = 1, \ldots, r, \qquad (3.4)$$

where τ is a multi-index and D_τ is the corresponding composition of total derivatives.

Now let ξ be a bundle of dimension r over M and $\Delta \colon \Gamma(\pi) \to \Gamma(\xi)$ be an operator of order k defining the equation, i.e., possessing the property that $\theta_k \in \mathcal{E}$ if and only if $\varphi_\Delta(\theta_k) = 0$. Comparing the expressions (1.7) obtained in §1 for prolongations of nonlinear differential operators with equations (3.4), we get the following result:

PROPOSITION 3.1. *If the equation $\mathcal{E} \subset J^k(\pi)$ is given by a differential operator $\Delta \colon \Gamma(\pi) \to \Gamma(\xi)$, then its l-th prolongation $\mathcal{E}^{(l)} \subset J^{k+l}(\pi)$ is given by the operator $\Delta_l = j_l \circ \Delta \colon \Gamma(\pi) \to \Gamma(\xi_l)$.*

---

[7]I.e., they satisfy all equations obtained by differentiation of the system at hand.

EXERCISE 3.1. Let the equation $\mathcal{E}$ be such that its l-th prolongation is a smooth submanifold in $J^{k+l}(\pi)$. Show that in this case the equality $(\mathcal{E}^{(l)})^{(t)} = \mathcal{E}^{(l+t)}$ holds for all $t \geq 0$. In particular, this means that in this situation the $(l+1)$-st prolongation can be defined by induction: $\mathcal{E}^{(l+1)} \stackrel{\text{def}}{=} (\mathcal{E}^{(l)})^{(1)}$.

EXERCISE 3.2. Construct examples when $\mathcal{E}^{(1)}$ is not a smooth submanifold in $J^{k+1}(\pi)$.

3.2. Infinitely prolonged equations. Thus, for any $k \geq 0$ we have constructed the set $\mathcal{E}^{(l)} \subset J^{k+l}(\pi)$, the l-th prolongation of the equation $\mathcal{E}$. Since tangency of order $l+1$ leads to tangency with all less orders, the natural mappings $\mathcal{E}^{(l+1)} \to \mathcal{E}^{(l)}$ compatible with the projections of the corresponding jet spaces take place:

$$
\begin{array}{ccc}
J^{k+l+1}(\pi) & \supset & \mathcal{E}^{(l+1)} \\
{\scriptstyle \pi_{k+l+1,k+l}}\downarrow & & \downarrow \\
J^{k+l}(\pi) & \supset & \mathcal{E}^{(l)}
\end{array}
\qquad (3.5)
$$

also denoted by $\pi_{k+l+1,k+l}$. It should be noted that, unlike the projections of jet spaces, the mappings $\pi_{k+l+1,k+l} \colon \mathcal{E}^{(l+1)} \to \mathcal{E}^{(l)}$ need not be surjective.

EXERCISE 3.3. Construct examples of differential equations for which the mapping $\pi_{k+1,k} \colon \mathcal{E}^{(1)} \to \mathcal{E}$ is not surjective.

The chain of mappings

$$
\mathcal{E} = \mathcal{E}^{(0)} \xleftarrow{\pi_{k+1,k}} \mathcal{E}^{(1)} \longleftarrow \cdots \longleftarrow \mathcal{E}^{(l)} \xleftarrow{\pi_{k+l+1,k+l}} \mathcal{E}^{(l+1)} \longleftarrow \cdots \qquad (3.6)
$$

allows one to define the inverse limit $\mathcal{E}^\infty = \operatorname{proj\,lim}_{l \to \infty} \mathcal{E}^{(l)}$ of prolongations of the equation $\mathcal{E}$, which is called the *infinite prolongation* of this equation. The set $\mathcal{E}^\infty$ lies in the manifold $J^\infty(\pi)$ of infinite jets, and its points can be interpreted in a quite obvious manner.

As we know, a point on $\mathcal{E}^{(l)}$ is the Taylor expansion of length $k+l$ of a section of the bundle π satisfying $\mathcal{E}$ up to infinitesimals of order $l+1$. On the other hand, the points of the manifold $J^\infty(\pi)$ are total Taylor series of sections of the bundle π. Therefore, a point $\theta = [s]_x^\infty \in J^\infty(\pi)$ belongs to the set $\mathcal{E}^\infty$ if and only if the Taylor expansion of the section s at the point $x \in M$ satisfies the equation $\mathcal{E}$. In other words, the points of $\mathcal{E}^\infty$ are formal solutions of the equation $\mathcal{E}$. In particular, from this it follows that a necessary condition for solvability of the equation $\mathcal{E}$ at a point $x \in M$ is that set $\mathcal{E}^\infty \cap \pi_\infty^{-1}(x)$ is not empty.

Let us define the algebra of smooth functions on $\mathcal{E}^{(l)}$ as the set of restrictions of smooth functions from the manifold $J^{k+l}(\pi)$:

$$
\mathcal{F}_l(\mathcal{E}) = \left\{ \varphi \colon \mathcal{E}^{(l)} \to \mathbb{R} \mid \exists \bar{\varphi} \in \mathcal{F}_{k+l}(\pi) \colon \bar{\varphi}|_{\mathcal{E}^{(l)}} = \varphi \right\}.
$$

In the case when $\mathcal{E}^{(l)}$ is a smooth submanifold in $J^{k+l}(\pi)$, the equality $\mathcal{F}_l(\mathcal{E}) = C^\infty(\mathcal{E}^{(l)})$ holds. The commutativity of the diagram (3.5) makes it possible, using the chain of mappings (3.6), to construct the chain of commutative algebra homomorphisms

$$
\mathcal{F}_0(\mathcal{E}) = C^\infty(\mathcal{E}) \xrightarrow{\pi_{k+1,k}^*} \mathcal{F}_1(\mathcal{E}) \longrightarrow \cdots \longrightarrow \mathcal{F}_l(\mathcal{E}) \xrightarrow{\pi_{k+l+1,k+l}^*} \mathcal{F}_{l+1}(\mathcal{E}) \longrightarrow \cdots ,
$$

whose direct limit $\mathcal{F}(\mathcal{E}) = \operatorname{inj}\lim_{l\to\infty} \mathcal{F}_l(\mathcal{E})$ is called the *algebra of smooth functions* on the infinite prolongation $\mathcal{E}^\infty$. For any $l \geq 0$, the natural homomorphisms of algebras $\pi^*_{\infty,k+l}\colon \mathcal{F}_l(\mathcal{E}) \to \mathcal{F}(\mathcal{E})$ are defined; since $\operatorname{im}(\pi^*_{\infty,k+l}) \subset \operatorname{im} \pi^*_{\infty,k+l+1}$, the algebra is filtered by the images of these homomorphisms. The image of the homomorphism $\pi^*_{\infty,k+l}$ in $\mathcal{F}(\mathcal{E})$ can be identified with the ring of smooth functions on the set $\mathcal{E}_l = \pi_{\infty,k+l}(\mathcal{E}^\infty) \subset \mathcal{E}^{(l)} \subset J^{k+l}(\pi)$. Therefore, if all the mappings $\mathcal{E}^\infty \to \mathcal{E}^{(l)}$ are surjective, one can consider the algebra $\mathcal{F}(\mathcal{E})$ to be filtered by the subalgebras $\mathcal{F}_l(\mathcal{E})$. As is easily seen, the equations for which the above is valid possess the following important property: any solution of such an equation constructed up to infinitesimals of order l can be continued up to a formal solution.

If $\pi'\colon P' \to M$ is another bundle over M, one can introduce the $\mathcal{F}_l(\mathcal{E})$-modules

$$\mathcal{F}_l(\mathcal{E},\pi') = \{\, \varphi \in \Gamma(\varepsilon_l^*(\pi')) \mid \exists \bar\varphi \in \Gamma(\pi^*_{k+l}(\pi')) : \bar\varphi|_{\mathcal{E}^{(l)}} = \varphi \,\},$$

where $\varepsilon_l \colon \mathcal{E}^{(l)} \hookrightarrow J^{k+l}(\pi)$ is the canonical embedding, and one can construct the filtered $\mathcal{F}(\mathcal{E})$-module $\mathcal{F}(\mathcal{E},\pi') = \operatorname{inj}\lim_{l\to\infty} \mathcal{F}_l(\mathcal{E},\pi')$.

From the definition of the algebras $\mathcal{F}_l(\mathcal{E})$ it follows that for all $l \geq 0$ one has the epimorphisms $\varepsilon_l^*\colon \mathcal{F}_{k+l}(\pi) \to \mathcal{F}_l(\mathcal{E})$. Setting formally $\mathcal{E}^{(l)} \stackrel{\text{def}}{=} \mathcal{E}_l = \pi_{k,k+l}(\mathcal{E})$ for $l < 0$ and defining the algebras $\mathcal{F}_l(\mathcal{E})$ in an appropriate way, we arrive at the commutative diagram

$$\begin{array}{ccccccccc}
\mathcal{F}_{-\infty}(\pi) & \longrightarrow & \cdots & \longrightarrow & \mathcal{F}_{k-1}(\pi) & \longrightarrow & \mathcal{F}_k(\pi) & \longrightarrow & \cdots & \longrightarrow & \mathcal{F}_{k+l}(\pi) & \longrightarrow & \cdots \\
{\scriptstyle \varepsilon^*_{-\infty}}\downarrow & & & & {\scriptstyle \varepsilon^*_{-1}}\downarrow & & {\scriptstyle \varepsilon^*_0}\downarrow & & & & {\scriptstyle \varepsilon^*_l}\downarrow & & \\
\mathcal{F}_{-\infty}(\mathcal{E}) & \longrightarrow & \cdots & \longrightarrow & \mathcal{F}_{-1}(\mathcal{E}) & \longrightarrow & \mathcal{F}_0(\mathcal{E}) & \longrightarrow & \cdots & \longrightarrow & \mathcal{F}_l(\mathcal{E}) & \longrightarrow & \cdots
\end{array} \quad (3.7)$$

where $\mathcal{F}_{-\infty}(\mathcal{E}) \stackrel{\text{def}}{=} C^\infty(\pi_\infty(\mathcal{E}^\infty))$. Let us denote by $I_l(\mathcal{E})$ the kernel of the epimorphism ε_l^*. Then $I_l(\mathcal{E})$ is the ideal of the algebra $\mathcal{F}_{k+l}(\mathcal{E})$ possessing the property that the function φ lies in this ideal if and only if $\varphi(\theta) = 0$ for all points $\theta \in \mathcal{E}^{(l)}$. From commutativity of the diagram (3.7) it follows that for all $l \in \mathbb{Z}$ we have the embedding $I_l(\mathcal{E}) \subset I_{l+1}(\mathcal{E})$. Therefore, the system of ideals $\{I_l(\mathcal{E})\}$ determines the ideal $I(\mathcal{E}) = \bigcup_{l\in\mathbb{Z}} I_l(\mathcal{E})$ of the filtered algebra $\mathcal{F}(\mathcal{E})$, called the *ideal of the equation* $\mathcal{E}$.

Let $X \in \mathrm{D}(\pi)$ be a vector field on $J^\infty(\pi)$ possessing the property that the ideal $I(\mathcal{E})$ is closed with respect to the derivation X: $X(I(\mathcal{E})) \subset I(\mathcal{E})$. Then X generates the derivation $X|_\mathcal{E}$ of the quotient algebra $\mathcal{F}(\pi)/I(\mathcal{E})$, i.e., a vector field on $\mathcal{E}^\infty$. We say in this case that the field X is tangent to the manifold $\mathcal{E}^\infty$, or admits restriction to this manifold.

Note that without loss of generality we may assume that the homomorphism $\varepsilon^*_{-\infty}$ is an isomorphism, i.e., $\mathcal{E}^{(-\infty)} \stackrel{\text{def}}{=} \pi_\infty(\mathcal{E}^\infty) = M$. In fact, if the projection $\pi_\infty|_{\mathcal{E}^\infty} \colon \mathcal{E}^\infty \to M$ is not surjective, we can restrict the bundle π to the submanifold $\pi_\infty(\mathcal{E}^\infty) \subset M$ (or, if necessary, to its nonsingular part) and later on consider all constructions on this restriction. Moreover, by similar reasoning we can assume that $\mathcal{E}^{(-k)} = \pi_{\infty,0}(\mathcal{E}^\infty) = J^0(\pi)$. Thus, in what follows we consider $\mathcal{E}^\infty$ to be surjectively projected to the manifold $J^0(\pi)$.

Let $F_\alpha = 0$, $\alpha = 1,\ldots,r$, be the relations defining the equation $\mathcal{E}$. Then the ideal $I_0(\mathcal{E})$ is generated by the elements $F_\alpha \in \mathcal{F}_k(\pi)$, i.e., any element $\varphi \in I_0(\mathcal{E})$ is of the form $\varphi = \varphi_1 F_1 + \cdots + \varphi_r F_r$, $\varphi_\alpha \in \mathcal{F}_k(\mathcal{E})$. Adding to these generators the elements of the form $D_i F_\alpha$, $i = 1,\ldots,n$, we obtain, as easily follows from (3.2), the ideal $I_1(\mathcal{E})$, etc. Finally, the ideal $I(\mathcal{E})$ is generated by all elements of the form $D_\tau(F_\alpha)$, $|\tau| \geq 0$, $\alpha = 1, \ldots, r$, and $I_l(\mathcal{E}) = I(\mathcal{E}) \cap \mathcal{F}_{k+l}(\pi)$. In particular, the ideal $I_0(\mathcal{E})$ is trivial (and its triviality is equivalent to surjectivity of the mapping $\pi_{\infty,0}|_{\mathcal{E}^\infty}$) if and only if there are no equations of the form $f(x,u) = 0$ in consequences of equations $D_\tau(F_\alpha) = 0$. This means that the equation $\mathcal{E}$ does not contain (even in implicit form) functional relations. If such relations exist, we can locally choose a subsystem of maximal rank among them and express other variables x_i, u^j via the chosen ones. This procedure is a coordinate formulation of the above described procedure of reducing an arbitrary equation $\mathcal{E}$ to an equation for which $\pi_{\infty,0}(\mathcal{E}^\infty) = J^0(\pi)$.

The above description of the ideals $I_l(\mathcal{E})$ shows that the ideal $I(\mathcal{E})$ of the equation $\mathcal{E}$ possesses the following two important properties:

(a) It is an ideal of the filtered algebra, i.e.,
$$I(\mathcal{E}) \cap \mathcal{F}_{k+l}(\mathcal{E}) = I_l(\mathcal{E}), \qquad I(\mathcal{E}) = \bigcup_{l \in \mathbb{Z}} I_l(\mathcal{E}).$$

(b) For any vector field $X \in D(M)$, the ideal $I(\mathcal{E})$ is closed with respect to the lifting $\widehat{X} \in D(\pi)$: $\widehat{X} I(\mathcal{E}) \subset I(\mathcal{E})$.[8]

Conversely, in the algebra $\mathcal{F}(\pi)$ let a filtered differentially closed ideal I be given, which is differentially generated[9] by a finite number of elements $F_1, \ldots, F_r \in \mathcal{F}_k(\pi)$, where F_α are functions of the variables $x_1, \ldots, x_n, \ldots, p_\sigma^j, \ldots$, $|\sigma| \leq k$. Let us define the set $\mathcal{E}_{I,l} \subset J^{k+l}(\pi)$ as the zero manifold of the ideal $I_l = I \cap \mathcal{F}_{k+l}(\pi)$:
$$\mathcal{E}_{I,l} = \left\{ \theta \in J^{k+l}(\pi) \mid \varphi(\theta) = 0 \; \forall \varphi \in I_l \right\}.$$

Then, as is easily seen, $\mathcal{E}_{I,l} = \mathcal{E}_I^{(l)}$, where $l \geq 0$ and $\mathcal{E}_I = \mathcal{E}_{I,0}$, while $I = I(\mathcal{E}_I)$. Thus there exists a one-to-one correspondence between submanifolds of the form $\mathcal{E}^\infty$ in $J^\infty(\pi)$ and finitely generated differentially closed filtered ideals of the algebra $\mathcal{F}(\pi)$. Note a parallel between this correspondence and duality between algebraic manifolds and ideals of commutative algebras known in algebraic geometry (see, e.g., [108]).

3.3. Higher symmetries. Let $\mathcal{E} \subset J^k(\pi)$ be an equation, $\pi_{\infty,0}(\mathcal{E}^\infty) = J^0(\pi)$, and let $I(\mathcal{E}) \subset \mathcal{F}(\pi)$ be its ideal. Then the quotient algebra $\mathcal{F}(\pi)/I(\mathcal{E}) = \mathcal{F}(\mathcal{E})$ is a filtered algebra and is identified with the algebra of smooth functions on the manifold $\mathcal{E}^\infty$. The theory of "differential objects" (vector fields, forms, etc.) is constructed over $\mathcal{E}^\infty$ in exactly the same way as it was done in §1 for the case $J^\infty(\pi)$. For example, a vector field on $\mathcal{E}^\infty$ is a derivation of the algebra $\mathcal{F}(\mathcal{E})$ compatible with filtration, the module $\Lambda^i(\mathcal{E}) = \Lambda^i(\mathcal{E}^\infty)$ of differential i-forms is the direct limit of the modules $\Lambda^i(\mathcal{E}^{(l)})$, etc.

Let us define the *Cartan distribution* $\mathcal{C}(\mathcal{E})$ on $\mathcal{E}^\infty$ by setting[10]
$$\mathcal{C}_\theta(\mathcal{E}) = T_\theta(\mathcal{E}^\infty) \cap \mathcal{C}_\theta, \quad \theta \in \mathcal{E}^\infty,$$
where $\mathcal{C}_\theta$ is the corresponding element of the Cartan distribution on the manifold $J^\infty(\pi)$. By the definition of the infinite prolongation, the distribution $\mathcal{C}(\mathcal{E})$ is nontrivial. From the description of integral manifolds of the Cartan distribution on

---

[8]When this property holds, the ideal $I(\mathcal{E})$ is called *differentially closed*.

[9]We say that an ideal I is differentially generated by elements $F_1, \ldots, F_r$, if it is algebraically generated by the elements $D_\tau F_j$, $|\tau| \geq 0$, $j = 1, \ldots, r$.

[10]We use the same notation for the Cartan distribution on $\mathcal{E}^\infty$ as was used for that on $\mathcal{E}$. It will not cause ambiguity, because everywhere below we deal with infinite prolongations only.

$J^\infty(\pi)$ given in §2 and from the definition of the distribution $\mathcal{C}(\mathcal{E})$ it follows that the maximal integral manifolds of the latter are manifolds of the form $\Gamma_s^\infty = j_\infty(s)(M)$, $s \in \Gamma(\pi)$, lying in $\mathcal{E}^\infty$, and they only. Note that if $\Gamma_s^\infty \subset \mathcal{E}^\infty$, then the manifold $\pi_{\infty,k}(\Gamma_s^\infty) = \Gamma_s^k$ lies in $\mathcal{E}$, i.e., s is a solution of the equation $\mathcal{E}$. Conversely, for any solution s of the equation $\mathcal{E}$ the corresponding manifold Γ_s^∞ lies in $\mathcal{E}^\infty$ and is naturally a maximal integral manifold of the distribution $\mathcal{C}(\mathcal{E})$. Thus maximal integral manifolds of the Cartan distribution on $\mathcal{E}^\infty$ are solutions of the equation $\mathcal{E}$.

REMARK 3.1. It may seem more natural to define the Cartan plane $\mathcal{C}_\theta(\mathcal{E})$ at the point $\theta \in \mathcal{E}^\infty$ as the span of tangent planes to solutions of the equation $\mathcal{E}$ passing through this point. But this approach is seriously deficient: first, to use it, one needs to know solutions of the equation $\mathcal{E}$, and secondly, the theory obtained becomes much worse than the one expounded here. Our approach, if we may say so, consists in constructing the Cartan distribution on $\mathcal{E}^\infty$ using "tangent planes to formal solutions of the equation".

Let us define the module of Cartan forms on $\mathcal{E}^\infty$ as the set $\mathcal{C}\Lambda^1(\mathcal{E}^\infty) \subset \Lambda^1(\mathcal{E}^\infty)$ of one-forms such that at every point $\theta \in \mathcal{E}^\infty$ they are annihilated by vectors of the distribution $\mathcal{C}(\mathcal{E})$.

EXERCISE 3.4. Show that the equality $\mathcal{C}\Lambda^1(\mathcal{E}^\infty) = \mathcal{C}\Lambda^1(\pi)\big|_{\mathcal{E}^\infty}$ holds, i.e., any form $\omega \in \mathcal{C}\Lambda^1(\mathcal{E}^\infty)$ may be represented as a restriction to $\mathcal{E}^\infty$ of a Cartan form from $J^\infty(\pi)$.

From §2 and the above it follows that the ideal $\mathcal{C}\Lambda^1(\mathcal{E}^\infty) \wedge \Lambda^*(\mathcal{E}^\infty)$ is differentially closed with respect to the de Rham differential $d\colon \Lambda^*(\mathcal{E}^\infty) \to \Lambda^*(\mathcal{E}^\infty)$, i.e.,

$$d\left(\mathcal{C}\Lambda^1(\mathcal{E}^\infty) \wedge \Lambda^*(\mathcal{E}^\infty)\right) \subset \mathcal{C}\Lambda^1(\mathcal{E}^\infty) \wedge \Lambda^*(\mathcal{E}^\infty). \tag{3.8}$$

Further, following the already known motivations of the preceding section, let us introduce the sets

$$\mathcal{C}\mathrm{D}(\mathcal{E}^\infty) = \left\{ X \in \mathrm{D}(\mathcal{E}^\infty) \mid X \,\lrcorner\, \omega = 0,\ \forall \omega \in \mathcal{C}\Lambda^1(\mathcal{E}^\infty) \right\}$$

and

$$\mathrm{D}_\mathcal{C}(\mathcal{E}^\infty) = \left\{ X \in \mathrm{D}(\mathcal{E}^\infty) \mid [X, \mathcal{C}\mathrm{D}(\mathcal{E}^\infty)] \subset \mathcal{C}\mathrm{D}(\mathcal{E}^\infty) \right\}.$$

From the definition it follows that the set $\mathrm{D}_\mathcal{C}(\mathcal{E}^\infty)$ is a Lie subalgebra of the Lie algebra of vector fields on $\mathcal{E}^\infty$, while $\mathcal{C}\mathrm{D}(\mathcal{E}^\infty)$ is an ideal in $\mathrm{D}_\mathcal{C}(\mathcal{E}^\infty)$. Literally repeating the reasoning from §2, we shall introduce the Lie $\mathbb{R}$-algebra

$$\mathrm{sym}\,\mathcal{E} = \mathrm{D}_\mathcal{C}(\mathcal{E}^\infty)/\mathcal{C}\mathrm{D}(\mathcal{E}^\infty)$$

of symmetries of the Cartan distribution on $\mathcal{E}^\infty$, and informally identify elements of this algebra with vector fields on the set of maximal integral manifolds of this distribution, i.e., on the "manifold" $\mathrm{Sol}(\mathcal{E})$ of solutions of the equation $\mathcal{E}$.

DEFINITION 3.2. Elements of the Lie algebra $\mathrm{sym}\,\mathcal{E}$ are called *higher (infinitesimal) symmetries* of the equation $\mathcal{E}$.

Our next goal is to describe the algebra $\mathrm{sym}\,\mathcal{E}$. To do this, let us first note the following.

Let $X \in \mathrm{D}(M)$ be a vector field on the manifold M. Then, since the ideal $I(\mathcal{E})$ of the equation $\mathcal{E}$ is differentially closed, i.e., the embedding $\widehat{X}(I(\mathcal{E})) \subset I(\mathcal{E})$

holds, the derivation $\widehat{X}\colon \mathcal{F}(\pi) \to \mathcal{F}(\pi)$ determines a derivation $\widehat{X}|_{\mathcal{E}^\infty}$ of the filtered algebra $\mathcal{F}(\mathcal{E}) = \mathcal{F}(\pi)/I(\mathcal{E})$, i.e., a vector field on $\mathcal{E}^\infty$. In other words, any field of the form $\widehat{X}$, $X \in D(M)$, admits restriction to $\mathcal{E}^\infty$. This is obviously valid for all operators generated by such fields, i.e., for operators of the form

$$\Delta = \sum \varphi_{i_1,\dots,i_h} \widehat{X}_{i_1} \circ \cdots \circ \widehat{X}_{i_h} \colon \mathcal{F}(\pi) \to \mathcal{F}(\pi), \varphi_{i_1,\dots,i_h} \in \mathcal{F}(\pi).$$

In particular, as we can see from the geometric definition of the fields $\widehat{X}$, this is valid for the fields from $\mathcal{C}D(\pi)$. Thus, we have the following homomorphism of Lie algebras:

$$\mathcal{C}D(\pi) \longrightarrow \mathcal{C}D(\mathcal{E}^\infty). \tag{3.9}$$

Below we shall need the following lemma.

LEMMA 3.2. *Assume that the equation* $\mathcal{E} \subset J^k(\pi)$ *is such that* $\mathcal{E}^\infty$ *is surjectively projected to some manifold* $J^{l_0}(\pi)$, $l_0 < k$. *Then for any* $l \leq l_0$ *and any derivation* $X\colon \mathcal{F}_{l-k}(\mathcal{E}) \to \mathcal{F}(\mathcal{E})$ *there exists a derivation* $X'\colon \mathcal{F}_l(\pi) \to \mathcal{F}(\pi)$ *such that the diagram*

$$\begin{array}{ccc} \mathcal{F}_l(\pi) & \xrightarrow{X'} & \mathcal{F}(\pi) \\ \varepsilon^*_{l-k} \downarrow & & \downarrow \varepsilon^* \\ \mathcal{F}_{l-k}(\mathcal{E}) & \xrightarrow{X} & \mathcal{F}(\mathcal{E}) \end{array}$$

is commutative.

PROOF. The lemma's assumptions mean that for $l \leq l_0$ the mapping ε^*_{k-l} is an isomorphism, i.e., $\mathcal{F}_l(\pi) = \mathcal{F}_{l-k}(\mathcal{E})$. Let us (locally) represent X in the form $X = \sum_{i=1}^n \varphi_i \partial/\partial x_i + \sum_{j=1}^m \sum_{|\sigma| \leq l} \varphi_\sigma^j \partial/\partial p_\sigma^j$. In this representation, $\varphi_i, \varphi_\sigma^j$ are functions on some finite prolongation $\mathcal{E}^{(r)}$ of the equation $\mathcal{E}$, and they can be continued to smooth functions on the enveloping manifold $J^{k+r}(\pi)$. □

Let us recall that we reduced the study of arbitrary equations $\mathcal{E} \subset J^k(\pi)$ to equations such that $\mathcal{E}^\infty$ is surjectively projected to $J^0(\pi)$. In this situation, from Lemma 3.2 it follows that the mapping (3.9) is epimorphic, i.e., any field $X \in \mathcal{C}D(\mathcal{E}^\infty)$ is a restriction to $\mathcal{E}^\infty$ of some field $X' \in \mathcal{C}D(\pi)$. Using Proposition 2.2, we see that X is represented in the form $X = \sum_i \varphi_i \widehat{X}_i$ in this case, where $\varphi_i \in \mathcal{F}(\mathcal{E})$, while $X_i \in D(M)$.

Now let $X \in D_\mathcal{C}(\mathcal{E}^\infty)$. Restricting X to $\pi_\infty(\mathcal{E}^\infty) = M$, we obtain the derivation $X_M \colon C^\infty(M) \to \mathcal{F}(\mathcal{E})$, which can be continued to a derivation $X'_M \colon C^\infty(M) \to \mathcal{F}(\pi)$, since the mapping (3.9) is epimorphic. Considering its lifting $\widehat{X}'_M \colon \mathcal{F}(\pi) \to \mathcal{F}(\pi)$, $\widehat{X}'_M \in \mathcal{C}D(\pi)$, and restricting the latter to $\mathcal{E}^\infty$, we obtain the derivation $\mathcal{C}X$, which lies in $\mathcal{C}D(\mathcal{E}^\infty)$.

EXERCISE 3.5. Show that $\mathcal{C}X$ is uniquely determined by the field $X \in D_\mathcal{C}(\mathcal{E}^\infty)$, i.e., is independent of the extension of the derivation X_M up to X'_M.

Let us denote by $D^v_\mathcal{C}(\mathcal{E}^\infty) \subset D_\mathcal{C}(\mathcal{E}^\infty)$ the set of vertical automorphisms of the Cartan distribution on $\mathcal{E}^\infty$, i.e., of elements $X \in D_\mathcal{C}(\mathcal{E}^\infty)$ such that $X|_{C^\infty(M)} = 0$. From the construction of the field $\mathcal{C}X$ it follows that the correspondence $v\colon X \mapsto X^v = X - \mathcal{C}X$ determines the mapping

$$v\colon D_\mathcal{C}(\mathcal{E}^\infty) \longrightarrow D^v_\mathcal{C}(\mathcal{E}^\infty). \tag{3.10}$$

LEMMA 3.3. *The mapping* (3.10) *is a projector, i.e.,* $X^v = X$ *for all* $X \in \mathrm{D}_\mathcal{C}^v(\mathcal{E}^\infty)$.

PROOF. In fact, if X is a vertical field, then the derivation $X_M = X|_{C^\infty(M)}$ is trivial. Hence, from the definition of the field $\mathcal{C}X$ it follows that it is also trivial. $\square$

From the lemma it follows that we have the splitting
$$\mathrm{D}_\mathcal{C}(\mathcal{E}^\infty) = \mathrm{D}_\mathcal{C}^v(\mathcal{E}^\infty) \oplus \ker(v).$$
It is also obvious that $\ker(v) = \mathcal{C}\mathrm{D}(\mathcal{E}^\infty)$. Therefore, the following statement is valid.

PROPOSITION 3.4. *If the projection of the manifold $\mathcal{E}^\infty$ to M is surjective, then the Lie algebra $\mathrm{D}_\mathcal{C}(\mathcal{E}^\infty)$ splits to the semidirect product of the Lie subalgebra $\mathrm{D}_\mathcal{C}^v(\mathcal{E}^\infty)$ of vertical fields and the ideal $\mathcal{C}\mathrm{D}(\mathcal{E}^\infty)$:*
$$\mathrm{D}_\mathcal{C}(\mathcal{E}^\infty) = \mathrm{D}_\mathcal{C}^v(\mathcal{E}^\infty) \,\widetilde{\oplus}\, \mathcal{C}\mathrm{D}(\mathcal{E}^\infty). \tag{3.11}$$

COROLLARY 3.5. *The splitting* (3.11) *induces an isomorphism of Lie algebras*
$$\mathrm{sym}\,\mathcal{E} \simeq \mathrm{D}_\mathcal{C}^v(\mathcal{E}^\infty).$$

3.4. Extrinsic and intrinsic higher symmetries. Let us return to the algebra $\mathrm{sym}\,\mathcal{C}(\pi) = \mathrm{D}_\mathcal{C}(\pi)/\mathcal{C}\mathrm{D}(\pi)$ of symmetries of the Cartan distribution on $J^\infty(\pi)$, and let us note the following. Since elements of a coset $\chi \in \mathrm{sym}\,\mathcal{C}(\pi)$ differ by derivations from $\mathcal{C}\mathrm{D}(\pi)$, then either all of them are tangent to the manifold $\mathcal{E}^\infty$, or, on the contrary, none of them is. In the first case, the element χ generates a symmetry of the equation $\mathcal{E}$ and is called an *extrinsic (higher) symmetry* of this equation. Therefore, as in the case of classical symmetries (§7 of Ch. 3), the problem of comparison of extrinsic and intrinsic approaches to the definition of higher symmetries arises.

Denote the set of extrinsic symmetries by $\mathrm{sym}_e \mathcal{E}$. Obviously, there is a Lie algebra homomorphism
$$\mathrm{sym}_e \mathcal{E} \longrightarrow \mathrm{sym}\,\mathcal{E}, \tag{3.12}$$
determined by the restriction operation. From the results of §2 we know that in every coset $\chi \in \mathrm{sym}\,\mathcal{C}(\pi)$ there is a canonical representative, i.e., the vertical field which is an evolutionary derivation. Therefore, to check whether an element $\chi \in \mathrm{sym}\,\mathcal{C}(\pi)$ is an extrinsic symmetry of the equation $\mathcal{E}$, it suffices to verify whether the corresponding representative is tangent to $\mathcal{E}^\infty$. Using these remarks, we shall now show that the mapping (3.12) is epimorphic. To do this, we shall need the following lemma.

LEMMA 3.6. *Higher symmetries commute with all fields of the form $\widehat{Y}$, $Y \in \mathrm{D}(M)$. More exactly, for any element $X \in \mathrm{D}_\mathcal{C}^v(\mathcal{E}^\infty)$ the equality $[X, \widehat{Y}] = 0$ is valid.*

PROOF. Since the field $\widehat{Y}$ (in fact, its restriction to $\mathcal{E}^\infty$) lies in the set $\mathcal{C}\mathrm{D}(\mathcal{E}^\infty)$, which is an ideal of the Lie algebra $\mathrm{D}_\mathcal{C}(\mathcal{E}^\infty)$, the commutator $[X, \widehat{Y}]$ also lies in $\mathcal{C}\mathrm{D}(\mathcal{E}^\infty)$. Now let $f \in C^\infty(M)$ be a function on the manifold M. Then $\widehat{Y}(f) = Y(f)$ also lies in $C^\infty(M)$. Consequently, by verticality of the field X, the equalities $X(f) = 0$ and $X(Y(f)) = 0$ hold. This means that $[X, \widehat{Y}](f) = X(\widehat{Y}(f)) - \widehat{Y}(X(f)) = 0$, i.e., the field $[X, \widehat{Y}]$ is vertical. Using the decomposition (3.11), we can now state that $[X, \widehat{Y}] = 0$. $\square$

Consider a symmetry $X \in D_{\mathcal{C}}^v(\mathcal{E}^\infty)$ of the equation $\mathcal{E}$. and let us restrict it to the manifold $\pi_{\infty,0}(\mathcal{E}^\infty) = J^0(\pi)$. By Lemma 3.6, the derivation $X_0 \colon \mathcal{F}_0(\pi) \to \mathcal{F}(\mathcal{E})$ thus obtained can be continued to a derivation $X_0' \colon \mathcal{F}_0(\pi) \to \mathcal{F}(\pi)$. Further, from the results obtained in the preceding section it follows that there exists a uniquely defined vertical automorphism $X' \in D_{\mathcal{C}}^v(\pi)$ of the Cartan distribution on $J^\infty(\pi)$ such that $X'|_{\mathcal{F}_0(\pi)} = X_0'$. Moreover, $X'|_{\mathcal{F}_0(\pi)} = X|_{\mathcal{F}_0(\pi)}$, i.e., $X'(\varphi) = X(\varphi)$ for any function $\varphi \in \mathcal{F}_0(\pi)$.

Let $Y_1, \ldots, Y_l$ be arbitrary vector fields on M, and let $\widehat{Y}_* = \widehat{Y}_1 \circ \cdots \circ \widehat{Y}_l$ be the composition of their liftings to $J^\infty(\pi)$. Then, by Lemma 3.6,

$$[X, \widehat{Y}_*] = \sum_{i=1}^{l} \widehat{Y}_1 \circ \cdots \circ [X, \widehat{Y}_i] \circ \cdots \circ \widehat{Y}_l = 0,$$

and it follows that

$$X(\widehat{Y}_*(\varphi)) = \widehat{Y}_*(X(\varphi)) = \widehat{Y}_*(X'(\varphi)) = X'(\widehat{Y}_*(\varphi)). \tag{3.13}$$

Thus we have shown that $X'(\psi) = X(\psi)$ for any function ψ of the form $\psi = \widehat{Y}_*(\varphi)$, $\varphi \in \mathcal{F}_0(\pi)$. Let us locally choose for φ the coordinate functions $p_0^j = u^j$ and for Y_* the composition of total derivatives D_σ. Since $D_\sigma(u^j) = p_\sigma^j$, from (3.13) it follows that $X'(p_\sigma^j) = X(p_\sigma^j|_{\mathcal{E}^\infty})$ for all σ and $j = 1, \ldots, m$. Since any smooth function ψ on $J^\infty(\pi)$ is locally a function in the variables $x_1, \ldots, x_n, p_\sigma^j$, we obtain the equality $X'(\psi) = X(\psi|_{\mathcal{E}^\infty})$ for all $\psi \in \mathcal{F}(\pi)$. In other words, we proved the following theorem.

THEOREM 3.7. *If the equation $\mathcal{E} \subset J^k(\pi)$ is such that the mapping (3.12) is epimorphic, then any intrinsic higher symmetry of the equation $\mathcal{E}$ can be represented as a restriction to $\mathcal{E}^\infty$ of some extrinsic symmetry. More exactly, for any field $X \in D_{\mathcal{C}}^v(\mathcal{E}^\infty)$ there exists a field $X' \in D_{\mathcal{C}}^v(\pi)$ such that $X'|_{\mathcal{E}^\infty} = X$.*

Recall that in the classical theory a similar statement is not valid in general (see §7 of Ch. 3).

3.5. Defining equations for higher symmetries. Now, using Theorem 3.7, we shall give an analytical description of the algebra $\operatorname{sym} \mathcal{E}$ needed in practical computational applications. Let $\mathcal{E} \subset J^k(\pi)$ be an arbitrary equation of order k whose infinite prolongation surjectively projects to $J^0(\pi)$. By the above results, any higher symmetry of the equation $\mathcal{E}$ may be obtained by restricting to $\mathcal{E}^\infty$ some evolutionary derivation $\partial_\varphi \in D_{\mathcal{C}}^v(\pi)$, $\varphi \in \mathcal{F}(\pi, \pi)$. In turn, all evolutionary derivations admitting restriction to $\mathcal{E}^\infty$ are determined by the condition

$$\partial_\varphi(I(\mathcal{E})) \subset I(\mathcal{E}), \tag{3.14}$$

where $I(\mathcal{E}) \subset \mathcal{F}(\pi)$ is the ideal of the equation $\mathcal{E}$. Let $\mathcal{E}$ be given by the relations $F_\alpha = 0$, $F_\alpha \in \mathcal{F}_k(\pi)$, $\alpha = 1, \ldots, r$, and assume that at any point $\theta \in \mathcal{E}$ the differentials $d_\theta F_\alpha$ are linearly independent. This means that the set $\{F_1, \ldots, F_r\}$ is a system of differential generators for the ideal $I(\mathcal{E})$:

$$I(\mathcal{E}) = \left\{ \psi \in \mathcal{F}(\pi) \mid \psi = \sum_{\alpha,\sigma} \psi_{\alpha,\sigma} D_\sigma(F_\alpha),\ \psi_{\alpha,\sigma} \in \mathcal{F}(\pi) \right\}.$$

Hence, condition (3.14) is equivalent to the fact that for arbitrary smooth functions $\psi_{1,\sigma}, \ldots, \psi_{r,\sigma} \in \mathcal{F}(\pi)$ there exist functions $\psi'_{1,\sigma}, \ldots, \psi'_{r,\sigma} \in \mathcal{F}(\pi)$ such that

$$\partial_\varphi \left(\sum_{\alpha,\sigma} \psi_{\alpha,\sigma} D_\sigma(F_\alpha) \right) = \sum_{\alpha,\sigma} \psi'_{\alpha,\sigma} D_\sigma(F_\alpha). \tag{3.15}$$

But ∂_φ is a derivation commuting with the operators of total derivatives, and so

$$\partial_\varphi \left(\sum_{\alpha,\sigma} \psi_{\alpha,\sigma} D_\sigma(F_\alpha) \right) = \sum_{\alpha,\sigma} \partial_\varphi(\psi_{\alpha,\sigma}) D_\sigma(F_\alpha) + \sum_{\alpha,\sigma} \psi_{\alpha,\sigma} D_\sigma(\partial_\varphi(F_\alpha)).$$

Hence, the conditions (3.15) need to be checked for the generators of the ideal $I(\mathcal{E})$ only, i.e., equation (3.14) is equivalent to the system

$$\begin{cases} \partial_\varphi(F_1) = \sum_{\alpha,\sigma} \psi^1_{\alpha,\sigma} D_\sigma(F_\alpha), \\ \partial_\varphi(F_2) = \sum_{\alpha,\sigma} \psi^2_{\alpha,\sigma} D_\sigma(F_\alpha), \\ \qquad \cdots\cdots\cdots\cdots\cdots\cdots\cdots \\ \partial_\varphi(F_r) = \sum_{\alpha,\sigma} \psi^r_{\alpha,\sigma} D_\sigma(F_\alpha), \end{cases} \tag{3.16}$$

where $\varphi \in \mathcal{F}(\pi, \pi)$ is the unknown section and

$$\psi^\beta_{\alpha,\sigma} \in \mathcal{F}(\pi), \qquad \alpha, \beta = 1, \ldots, r, \quad |\sigma| \leq k + \deg(\varphi).$$

Applying the definition of the universal linearization operator given in §2, let us rewrite the system (3.16) in the form

$$\ell_{F_\beta}(\varphi) = \sum_{\alpha,\sigma} \psi^\beta_{\alpha,\sigma} D_\sigma(F_\alpha), \quad \beta = 1, \ldots, r. \tag{3.17}$$

Among solutions of the system (3.17) there are trivial ones corresponding to the kernel of the epimorphism (3.12) and characterized by the condition that the restriction of the derivation ∂_φ to $\mathcal{E}^\infty$ for these solutions leads to the trivial vector field on $\mathcal{E}^\infty$.

EXERCISE 3.6. Show that the set formed by the components φ^j of trivial solutions of the system (3.17) coincides with the ideal $I(\mathcal{E})$ of the equation $\mathcal{E}$.

To eliminate trivial solutions, let us recall that by the representation (2.23) the operator of the universal linearization is expressed in total derivatives and consequently admits restriction to manifolds of the form $\mathcal{E}^\infty$. Set $\ell_F|_{\mathcal{E}^\infty} \stackrel{\text{def}}{=} \ell_F^\mathcal{E}$, and restrict equation (3.17) to $\mathcal{E}^\infty$. Then, since the right-hand side consists of the elements of the ideal $I(\mathcal{E})$, we arrive to the system of equations

$$\ell_{F_\alpha}^\mathcal{E}(\bar{\varphi}) = 0, \qquad \bar{\varphi} = \varphi|_{\mathcal{E}^\infty},$$

whose solutions are in one-to-one correspondence with higher symmetries of the equation $\mathcal{E}$. Moreover, under this correspondence the commutator of two symmetries is taken to the element

$$\{\bar{\varphi}, \bar{\psi}\}_\mathcal{E} \stackrel{\text{def}}{=} \ell_\psi^\mathcal{E}(\bar{\varphi}) - \ell_\varphi^\mathcal{E}(\bar{\psi}) = \{\varphi, \psi\}|_{\mathcal{E}^\infty}, \qquad \bar{\psi} = \psi|_{\mathcal{E}^\infty}.$$

Combining these results, we get

THEOREM 3.8. *If $\mathcal{E} \subset J^k(\pi)$ is an equation such that $\pi_{\infty,0}(\mathcal{E}^\infty) = J^0(\pi)$, and $F_1, \ldots, F_r$ are the generators of the ideal $I(\mathcal{E})$, then the Lie algebra $\operatorname{sym} \mathcal{E}$ is isomorphic to the Lie algebra of solutions of the system of equations*

$$\ell^{\mathcal{E}}_{F_\alpha}(\varphi) = 0, \qquad \alpha = 1, \ldots, r, \quad \varphi \in \mathcal{F}(\mathcal{E}, \pi), \tag{3.18}$$

where the Lie algebra structure is given by the bracket $\{\cdot, \cdot\}_{\mathcal{E}}$.

It is easily seen that if $\Delta \colon \Gamma(\pi) \to \Gamma(\pi')$ is the operator determining the equation $\mathcal{E}$ and chosen in such a way that the corresponding section $F = \varphi_\Delta \in \mathcal{F}(\pi, \pi')$ is transversal to the base of the bundle $\pi^*(\pi')$ (more exactly, such that the images of the zero section and of F are transversal at the points of their intersection), then the system (3.18) can be rewritten in the form

$$\ell^{\mathcal{E}}_F(\varphi) = 0, \qquad \varphi \in \mathcal{F}(\mathcal{E}, \pi), \quad F \in \mathcal{F}(\pi, \pi'). \tag{3.19}$$

Equations (3.18) and (3.19) are called the *defining equations* for higher symmetries.

4. Examples of computation

In this section, we illustrate the techniques of computing higher symmetries for some equations of mathematical physics. Other examples can be found, e.g., in the collection [**137**]. A theoretical base for the computations below is given by Theorem 3.8. Both parts of the theorem are essential for the techniques. Representation of symmetries as solutions of the linear system (3.18) makes it possible to obtain "upper and lower estimates" for the Lie algebra $\operatorname{sym} \mathcal{E}$, while closure with respect to the higher Jacobi bracket $\{\cdot, \cdot\}_{\mathcal{E}}$ allows one to make these estimates more precise and, in some cases, to obtain a complete description of the symmetry algebra. We also rely on the fact that extrinsic and intrinsic symmetries coincide, and use *intrinsic coordinates* for the manifold $\mathcal{E}^\infty$.

EXERCISE 4.1. To obtain an "upper estimate" for the algebra $\operatorname{sym} \mathcal{E}$ it is also useful to apply the *commutator relation*

$$[\bar{\partial}_\varphi - \bar{\ell}_\varphi, \ell^{\mathcal{E}}_F] = \mathcal{D} \circ \ell^{\mathcal{E}}_F, \tag{4.1}$$

where φ is a symmetry, $\bar{\partial}_\varphi$ and $\bar{\ell}_\varphi$ are restrictions of the corresponding operators to $\mathcal{E}^\infty$, and $\mathcal{D}$ is an operator of the form $\mathcal{D} = \sum_\sigma a_\sigma \bar{D}_\sigma$. Prove this equality.

Let us start by specifying general constructions for scalar evolution equations of second order.

4.1. Preparatory remarks.
Consider an equation of the form

$$u_t = \Phi(x, t, u, u_x, u_{xx}). \tag{4.2}$$

Its infinite prolongation $\mathcal{E}^\infty$ is a submanifold in the space $J^\infty(\pi)$ of infinite jets of the trivial one-dimensional bundle over the plane $M = \mathbb{R}^2$ of independent variables $x = x_1$ and $t = x_2$, while a coordinate in the fiber of the bundle π is the dependent variable u. The standard coordinates $p_{(\alpha,\beta)}$, $\alpha, \beta \geq 0$, arise in $J^\infty(\pi)$ uniquely determined by the equalities

$$p_{(\alpha,\beta)}\big|_{j_\infty(s)} = \frac{\partial^{\alpha+\beta} s}{\partial x^\alpha \partial t^\beta}, \tag{4.3}$$

4. EXAMPLES OF COMPUTATION

where $s = s(x,t)$ is an arbitrary section of the bundle π, i.e., a smooth function on $\mathbb{R}^2$. From (4.3), by the definition of the total derivative, it follows that $p_{(\alpha,\beta)} = D_1^\alpha D_2^\beta(u)$, where

$$D_1 = \widehat{\frac{\partial}{\partial x}} = \frac{\partial}{\partial x} + p_{(1,0)}\frac{\partial}{\partial p_{(0,0)}} + \cdots + p_{(\alpha+1,\beta)}\frac{\partial}{\partial p_{(\alpha,\beta)}} + \cdots,$$

$$D_2 = \widehat{\frac{\partial}{\partial t}} = \frac{\partial}{\partial t} + p_{(0,1)}\frac{\partial}{\partial p_{(0,0)}} + \cdots + p_{(\alpha,\beta+1)}\frac{\partial}{\partial p_{(\alpha,\beta)}} + \cdots.$$

Therefore, by (4.2), the equalities

$$p_{(\alpha,\beta+1)} = D_1^\alpha D_2^\beta(\Phi), \quad \alpha,\beta \geq 0,$$

hold on $\mathcal{E}^\infty$, and we can take the functions x,t, and $p_{(\alpha,0)}|_{\mathcal{E}^\infty} \stackrel{\text{def}}{=} p_\alpha$ for intrinsic coordinates on $\mathcal{E}^\infty$. In these coordinates, the restrictions of the total derivatives to $\mathcal{E}^\infty$ are of the form

$$\begin{aligned} D_x &\stackrel{\text{def}}{=} D_1|_{\mathcal{E}^\infty} = \frac{\partial}{\partial x} + \sum_{\alpha \geq 0} p_{\alpha+1}\frac{\partial}{\partial p_\alpha}, \\ D_t &\stackrel{\text{def}}{=} D_2|_{\mathcal{E}^\infty} = \frac{\partial}{\partial t} + \sum_{\alpha \geq 0} D_x^\alpha(\Phi)\frac{\partial}{\partial p_\alpha}, \end{aligned} \quad (4.4)$$

while the system (3.18) determining higher symmetries of the equation $\mathcal{E}$ reduces to the equation

$$D_t\varphi = \Phi_0\varphi + \Phi_1 D_x\varphi + \Phi_2 D_x^2\varphi, \quad (4.5)$$

where $\varphi = \varphi(x,t,p_0,\ldots,p_k)$ is the restriction of the generating function of the symmetry we seek for to $\mathcal{E}^\infty$, while $\Phi_i \stackrel{\text{def}}{=} \partial\Phi/\partial p_i$. The maximal k for which $\varphi_k = \partial\varphi/\partial p_k \neq 0$ will be called the order of the symmetry φ and will be denoted by $\deg\varphi$.

Below we shall need:
(a) To determine for which Φ (inside some chosen class) equation (4.5) possesses solutions of arbitrary high order.
(b) To describe such solutions, if possible.

To obtain answers to these questions, it is more convenient technically to pass from (4.5) to a new system of equations for functions $\varphi_i \stackrel{\text{def}}{=} \partial\varphi/\partial p_i$, where φ is a solution of (4.5). To do this, let us introduce the operators

$$R_\beta^\alpha \stackrel{\text{def}}{=} \begin{cases} \dfrac{\partial}{\partial p_\beta} \circ D_x^\alpha & \text{if } \alpha,\beta \geq 0, \\ 0 & \text{otherwise,} \end{cases}$$

acting in the ring of functions on $\mathcal{E}^\infty$.

LEMMA 4.1. *For any function $\varphi = \varphi(x,t,p_0,\ldots,p_k)$ and for all α,β the equality*

$$R_\beta^\alpha(\varphi) = \sum_{i=0}^k \binom{\alpha}{\alpha-\beta+i} D_x^{\alpha-\beta+i}(\varphi_i) \quad (4.6)$$

holds, where, by definition, $\binom{a}{b} = 0$ if one or both of the numbers a,b is negative.

PROOF. Induction on α. For $\alpha = 0$ the statement is obvious. Let $\alpha > 0$ and assume that for $\alpha - 1$ the identity (4.6) is valid. Note that from (4.4) and from (2.17) it follow that
$$R_\beta^\alpha = R_{\beta-1}^{\alpha-1} + D_x \circ R_\beta^{\alpha-1}.$$
Therefore
$$R_\beta^\alpha(\varphi) = R_{\beta-1}^{\alpha-1}(\varphi) + D_x R_\beta^{\alpha-1}(\varphi)$$
$$= \sum_{i=0}^{k} \binom{\alpha-1}{\alpha-\beta+i} D_x^{\alpha-\beta+i}(\varphi_i) + D_x \sum_{i=0}^{k} \binom{\alpha-1}{\alpha-\beta+i-1} D_x^{\alpha-\beta+i-1}(\varphi_i)$$
$$= \sum_{i=0}^{k} \left[\binom{\alpha-1}{\alpha-\beta+i} + \binom{\alpha-1}{\alpha-\beta+i-1} \right] D_x^{\alpha-\beta+i}(\varphi_i)$$
$$= \sum_{\alpha=0}^{k} \binom{\alpha}{\alpha-\beta+i} D_x^{\alpha-\beta+i}(\varphi_i).$$
$\square$

REMARK 4.1. From (4.6) one can obtain an "asymptotic expansion" of $D_x^\alpha \varphi$ with respect to the variables p_i of higher order, useful in particular computations. Namely, since $\deg D_x^{\alpha-\beta+i}(\varphi) \leq \alpha - \beta + i + k$, the order of the right-hand side in (4.6) is not greater than $\alpha - \beta + 2k$. Therefore, for even α the estimate
$$D_x^{2r} \varphi = \sum_{\beta=r+k+1}^{2r+k} p_\beta \sum_{i=0}^{k} \binom{2r}{2r-\beta+i} D_x^{2r-\beta+i} \varphi_i$$
$$+ \frac{1}{2} p_{r+k}^2 \sum_{i=0}^{k} \binom{2r}{r-k+i} D_x^{r-k+i} \varphi_i + O(r+k-1)$$
is valid, while for odd α one has
$$D_x^{2r+1} \varphi = \sum_{\beta=r+k}^{2r+k+1} p_\beta \sum_{i=0}^{k} \binom{2r+1}{2r-\beta+i+1} D_x^{2r-\beta+i+1} \varphi_i + O(r+k-1),$$
where $O(i)$ is a function on $\mathcal{E}^\infty$ independent of p_β for $\beta > i$.

Let us continue the study of the equation $\ell_F(\varphi) = 0$ and apply the operators $\partial/\partial p_\beta$, $\beta > 2$ to this equation. From (4.5) one has
$$D_t \varphi_\beta + \sum_{\alpha \geq 0} R_\beta^\alpha \varphi = \Phi_0 \varphi_\beta + \Phi_1 R_\beta^1 \varphi + \Phi_2 R_\beta^2 \varphi.$$
Thanks to the lemma and the above agreement on binomial coefficients, the last relation can be rewritten as
$$\ell_F(\varphi_\beta) + \sum_{i=\beta}^{k} \left[\binom{i}{i-\beta} D_x^{i-\beta}(\Phi_0) + \binom{i}{i-\beta+1} D_x^{i-\beta+1}(\Phi_1) \right.$$
$$\left. + \binom{i}{i-\beta+2} D_x^{i-\beta+2}(\Phi_2) \right] \varphi_i + (\beta-1) D_x(\Phi_2) \varphi_{\beta-1} = 2\Phi_2 D_x(\varphi_{\beta-1}).$$
Thus, we have the following statement.

4. EXAMPLES OF COMPUTATION

PROPOSITION 4.2. *If a function $\varphi = \varphi(x,t,p_0,\ldots,p_k)$, $k \geq 3$, is a solution of the equation $\ell_F(\varphi) \equiv D_t(\varphi) - \Phi_0 \varphi - \Phi_1 D_x(\varphi) - \Phi_2 D_x^2(\varphi) = 0$, i.e., is a higher symmetry of order k for the equation $u_t = \Phi(x,t,u,u_x,u_{xx})$, then the functions $\varphi_\beta = \partial\varphi/\partial p_\beta$, $\beta = 2,\ldots,k$, satisfy the following system of equations:*

$$\begin{cases} kD_x(\Phi_2)\varphi_k = 2\Phi_2 D_x(\varphi_k), \\ \ell_F(\varphi_k) + \left[\Phi_0 + \binom{k}{1}D_x(\Phi_1) + \binom{k}{2}D_x^2(\Phi_2)\right]\varphi_k \\ \quad + (k-1)D_x(\Phi_2)\varphi_{k-1} = 2\Phi_2 D_x(\varphi_{k-1}) \\ \cdots\cdots\cdots\cdots\cdots\cdots\cdots\cdots\cdots\cdots\cdots\cdots\cdots\cdots\cdots \\ \ell_F(\varphi_\beta) + \sum_{i=\beta}^{k}\left[\binom{i}{i-\beta}D_x^{i-\beta}(\Phi_0) + \binom{i}{i-\beta+1}D_x^{i-\beta+1}(\Phi_1)\right. \\ \quad \left. + \binom{i}{i-\beta+2}D_x^{i-\beta+2}(\Phi_2)\right]\varphi_i + (\beta-1)D_x(\Phi_2)\varphi_{\beta-1} \\ \quad = 2\Phi_2 D_x(\varphi_{\beta-1}), \\ \cdots\cdots\cdots\cdots\cdots\cdots\cdots\cdots\cdots\cdots\cdots\cdots\cdots\cdots\cdots \\ \ell_F(\varphi_3) + \sum_{i=3}^{k}\left[\binom{i}{i-3}D_x^{i-3}(\Phi_0) + \binom{i}{i-2}D_x^{i-2}(\Phi_1)\right. \\ \quad \left. + \binom{i}{i-1}D_x^{i-1}(\Phi_2)\right]\varphi_i + 2D_x(\Phi_2)\varphi_2 = 2\Phi_2 D_x(\varphi_2). \end{cases} \quad (4.7)$$

System (4.7) is in diagonal form and is easier to study than the initial equation $\ell_F(\varphi) = 0$. Let us illustrate the procedure using one class of evolution equations.

4.2. The Burgers and heat equations. Let us describe equations of the form

$$u_t = u_{xx} + f(u, u_x), \quad (4.8)$$

possessing symmetries of arbitrary high order. We are also going to compute the corresponding symmetry algebras.

In the case of equation (4.8) the system (4.7) acquires the form

$$\begin{cases} D_x(\varphi_k) = 0, \\ 2D_x(\varphi_{k-1}) = \ell_F(\varphi_k) + \left[f_0 + \binom{k}{1}D_x(f_1)\right]\varphi_k, \\ \cdots\cdots\cdots\cdots\cdots\cdots\cdots\cdots\cdots\cdots\cdots\cdots\cdots \\ 2D_x(\varphi_{\beta-1}) = \ell_F(\varphi_\beta) \\ \quad + \sum_{i=\beta}^{k}\left[\binom{i}{i-\beta}D_x^{i-\beta}(f_0) + \binom{i}{i-\beta+1}D_x^{i-\beta+1}(f_1)\right]\varphi_i, \\ \cdots\cdots\cdots\cdots\cdots\cdots\cdots\cdots\cdots\cdots\cdots\cdots\cdots \\ 2D_x(\varphi_2) = \ell_F(\varphi_3) + \sum_{i=3}^{k}\left[\binom{i}{i-3}D_x^{i-3}(f_0) + \binom{i}{i-2}D_x^{i-2}(f_1)\right]\varphi_i. \end{cases} \quad (4.9)$$

Looking at the system (4.9), one can see what can be obstructions for equation (4.8) to possess higher symmetries. Assume that we managed to solve the first i equations of the system (4.9); then the condition for solvability of the next equation is that its right-hand side (which is expressed in terms of the already obtained solutions) belongs to the image of the operator D_x. This condition is determined by the function f. Let us show how this technique works. To simplify computations, we shall make the change of variables $\varphi_\alpha = 2^{\alpha-k}\psi_\alpha$.

From the first equation of the system (4.9) it follows that

$$\psi_k = a_k(t). \tag{4.10}$$

Substituting ψ_k into the second equation, we obtain

$$D_x(\psi_{k-1}) = \ell_F(\psi_k) + \left[f_0 + \binom{k}{1}D_x(f_1)\right]\psi_k = \dot{a}_k + kD_x(f_1)\psi_k,$$

where $\dot{a}_k = da_k/dt$. Thus, we obtain the expression

$$\psi_{k-1} = \dot{a}_k x + k f_1 a_k + a_{k-1}(t). \tag{4.11}$$

Substituting (4.10) and (4.11) into the third equation of (4.9) and making necessary transformations, we obtain

$$\begin{aligned} D_x(\psi_{k-2}) &= \ddot{a}_k x + (k-1)\dot{a}_k\left[f_1 + xD_x(f_1)\right] \\ &\quad + ka_k\left[2D_x(f_0) + (k-2)f_1 D_x(f_1) + (k-2)D_x^2(f_1) + D_t(f_1)\right] \\ &\quad + D_x\left[\tfrac{1}{2}\ddot{a}_k x^2 + (k-1)\dot{a}_k x f_1 + ka_k\left(2f_0 + \tfrac{1}{2}(k-2)f_1^2 + (k-2)D_x(f_1)\right)\right. \\ &\quad\left. + \dot{a}_{k-1}x + (k-1)a_{k-1}f_1\right] + ka_k D_t(f_1). \end{aligned}$$

Thus, if $\deg(\varphi) = k$, then the third equation is solvable if and only if

$$D_t(f_1) \in \operatorname{im} D_x \tag{4.12}$$

(note that this means that f_1 is a *conservation law* for equation (4.8); see Ch. 5).

Let us describe the functions f for which the condition (4.12) holds. We have[11]

$$\begin{aligned} D_t(f_1) &= (p_2 + f)f_{01} + (p_3 + D_x(f))f_{11} \\ &= (p_2 + f)(f_{01} - D_x(f_{11})) + D_x((p_2 + f)f_{11}). \end{aligned}$$

In other words, $D_t(f_1) \in \operatorname{im} D_x$ if and only if

$$(p_2 + f)(f_{01} - D_x(f_{11})) \in \operatorname{im} D_x. \tag{4.13}$$

From (4.4) it is obvious that any element lying in the image of the operator D_x is linear with respect to the variable p_α of highest order; on the other hand, (4.13) is of the form $f_{111}p_2^2 + O(1)$. Therefore, the conditions (4.12) hold if and only if the third derivative f_{111} vanishes. Hence,

$$f = Ap_1^2 + Bp_1 + C, \tag{4.14}$$

where A, B, C are functions of p_0, x, and t. Substituting this into (4.13) and making similar computations, one can see that (4.12) holds if and only if the functions A, B and C in (4.14) satisfy the equations

$$AB_0 = B_{00}, \qquad CB_0 = \text{const}. \tag{4.15}$$

Note now that any equation

$$u_t = u_{xx} + A(u)u_x^2 + B(u)u_x + C(u)$$

---

[11]Below, f_{ij} denotes the partial derivative $\partial f/\partial p_i \partial p_j$, and f_{ijk} is defined similarly.

by the change of variables $u \mapsto \Psi(u)$, where the function Ψ, $\Psi_u \neq 0$, satisfies the differential equation

$$\Psi_{uu} + A(\Psi)\Psi_u^2 = 0,$$

can be transformed to the form

$$u_t = u_{xx} + \widetilde{B}(u)u_x + \widetilde{C}(u).$$

Therefore, without loss of generality, we may put $A = 0$ in (4.15) and take $B = \beta_1 u + \beta_0$, $\beta_0, \beta_1 = \text{const}$, and $\beta_1 C = \text{const}$. We have now two options: $\beta_1 \neq 0$ or $\beta_1 = 0$. In the first case the initial equation transforms to the form

$$u_t = u_{xx} + (\beta_1 u + \beta_0)u_x + \gamma, \qquad \gamma = \text{const}, \quad \beta_1 \neq 0,$$

while in the second case it transforms to

$$u_t = u_{xx} + \beta_0 u_x + C(u). \tag{4.16}$$

By the change of variables

$$x \longmapsto x - \frac{\gamma t^2}{2}, \quad t \longmapsto t, \quad u \longmapsto \frac{u + \gamma t - \beta_0}{\beta_1}$$

the first of these equations transforms to

$$u_t = u_{xx} + uu_x, \tag{4.17}$$

i.e., it is equivalent to the Burgers equation.

If we now come back to system (4.9), we shall see that[12] for (4.17) the fourth equation is solvable, while in the case (4.16) linearity of the function $C(u)$ is necessary. In other words, equation (4.16) has to be of the form

$$u_t = u_{xx} + \beta_0 u_x + \gamma_1 u + \gamma_0.$$

The last equation, by the change of variables

$$u \mapsto u \exp\left[\left(\gamma_1 - \frac{\beta_0^2}{4}\right)t - \frac{\beta_0}{2}x\right] + u_0,$$

where u_0 is an arbitrary solution of the corresponding homogeneous equation, reduces to the form

$$u_t = u_{xx}.$$

Thus, we have proved the following result:

PROPOSITION 4.3. *Any equation*

$$u_t = u_{xx} + f(u, u_x)$$

possessing symmetries of arbitrary high order is equivalent either to the Burgers equation

$$u_t = u_{xx} + uu_x,$$

or to the heat equation

$$u_t = u_{xx}.$$

---

[12]We omit the computations, which are simple but not instructive.

Our next aim is to show that these equations really possess infinite algebras of higher symmetries, and to describe these algebras. In doing this we follow [141]. First of all, we shall need some information on the algebraic structure of the symmetries we are looking for.

From equations (4.10) and (4.11) it follows that any symmetry of order k, if it exists, is of the form

$$\varphi_k[a] = ap_k + \left(\frac{1}{2}\dot{a}x + \frac{k}{2}f_1 a + a'\right) p_{k-1} + O(k-2), \qquad (4.18)$$

where a, a' are functions of t. It is easily seen that any symmetry is uniquely (up to symmetries of lower order) determined by the function a (here $f_1 = p_0$ for the Burgers equation and $f_1 = 0$ for the heat equation). Let $\varphi_l[b]$ also be a function of the form (4.18). Let us compute the Jacobi bracket of functions $\varphi_k[a]$ and $\varphi_l[b]$. For any functions $\varphi, \psi \in \mathcal{F}(\mathcal{E})$, $\deg(\varphi) = k$, $\deg(\psi) = l$, one has (see (2.24))

$$\{\varphi, \psi\}_{\mathcal{E}} = \sum_{i=0}^{l} D_x^i(\varphi)\psi_i - \sum_{j=0}^{k} D_x^j(\psi)\varphi_j.$$

Among the summands on the right-hand side of this equality, $D_x^l(\varphi)\psi_l$ and $-D_x^k(\psi)\varphi_k$ are of the maximal order (equal to $k+l$). But by Remark 4.1

$$D_x^l(\varphi)\psi_l = (\varphi_k p_{k+l} + O(k+l-1))\psi_l, \quad D_x^k(\psi)\varphi_k = (\psi_l p_{k+l} + O(k+l-1))\varphi_k.$$

Therefore, $\deg\{\varphi, \psi\}_{\mathcal{E}} = k+l-1$. In particular, for functions $\varphi_k[a]$ and $\varphi_l[b]$, up to elements of order $k+l-3$, one has

$$\begin{aligned}
\{\varphi_k[a], \varphi_l[b]\}_{\mathcal{E}} &= \{ap_k, bp_l\}_{\mathcal{E}} + \{ap_k, (\tfrac{1}{2}\dot{b}x + \tfrac{l}{2}f_1 b + b')p_{l-1}\}_{\mathcal{E}} \\
&\quad + \{(\tfrac{1}{2}\dot{a}x + \tfrac{k}{2}f_1 a + a')p_{k-1}, bp_l\}_{\mathcal{E}} + O(k+l-3) \\
&= D_x^l(ap_k)b - D_x^k(bp_l)a + D_x^{l-1}(ap_k)(\tfrac{1}{2}\dot{b}x + \tfrac{l}{2}f_1 b + b') \\
&\quad - D_x^k[(\tfrac{1}{2}\dot{b}x + \tfrac{l}{2}f_1 b + b')p_{l-1}]a + D_x^l[(\tfrac{1}{2}\dot{a}x + \tfrac{l}{2}f_1 a + a')p_{k-1}]b \\
&\quad - D_x^{k-1}(bp_l)(\tfrac{1}{2}\dot{a}x + \tfrac{k}{2}f_1 a + a') + O(k+l-3) \\
&= \tfrac{1}{2}(l\dot{a}b - k\dot{b}a)p_{k+l-2} + O(k+l-3).
\end{aligned}$$

Hence, since the algebra of higher symmetries is closed with respect to the Jacobi bracket, and since that the functions $\varphi_k[a]$ and $\varphi_l[b]$ are symmetries, we obtain that the function $\varphi_{k+l-2}[c]$, where

$$c = \tfrac{1}{2}(l\dot{a}b - k\dot{b}a), \qquad (4.19)$$

is also a symmetry of the equation $\mathcal{E}$.

Let us recall that in the preceding chapter we computed classical symmetries of the Burgers equation, and they were of the form

$$\begin{aligned}
\varphi_1^0 &= p_1, \\
\varphi_1^1 &= tp_1 + 1, \\
\varphi_2^0 &= p_2 + p_0 p_1, \\
\varphi_2^1 &= tp_2 + (tp_0 + \tfrac{1}{2}x)p_1 + \tfrac{1}{2}p_0, \\
\varphi_2^2 &= t^2 p_2 + (t^2 p_0 + tx)p_1 + tp_0 + x.
\end{aligned}$$

4. EXAMPLES OF COMPUTATION

Similar computations show that for the heat equation the classical symmetries are

$$\varphi_{-\infty} = \varphi_{-\infty}(x,t),$$
$$\varphi_0^0 = p_0,$$
$$\varphi_1^0 = p_1,$$
$$\varphi_1^1 = tp_1 + \tfrac{1}{2}xp_0,$$
$$\varphi_2^0 = p_2,$$
$$\varphi_2^1 = tp_2 + \tfrac{1}{2}xp_1,$$
$$\varphi_2^2 = t^2 p_2 + txp_1 + (\tfrac{1}{4}x^2 + \tfrac{1}{2}t)p_0,$$

where $\varphi_{-\infty}$ is an arbitrary solution of the heat equation.

Let $\varphi_k[a]$, $k > 2$, be a symmetry; then $\{\varphi_k[a], \varphi_1^0\}_{\mathcal{E}}$ is a symmetry as well and by the above computations it is of the form

$$\{\varphi_k[a], \varphi_1^0\}_{\mathcal{E}} = \tfrac{1}{2}\dot{a}p_{k-1} + O(k-2).$$

Applying the operator $\{\cdot, \varphi_1^0\}_{\mathcal{E}}$ to the function $\varphi_k[a]$ $k-2$ times, we obtain a classical symmetry of the form

$$2^{-k+2}\frac{d^{k-2}a}{dt^{k-2}}p_2 + O(1).$$

But classical symmetries of the equations under consideration have the polynomial coefficient in t of order ≤ 2 at p_2. Therefore, a is also a polynomial in t and its order is not greater than k.

Let us show that any such polynomial determines some symmetry. To do this, note that the equation at hand possesses a symmetry of the form $\varphi_3[t]$. Namely, by direct computations we can find that the Burgers equation has the symmetry

$$\varphi_3^1 = tp_3 + \tfrac{1}{2}(x + 3tp_0)p_2 + \tfrac{3}{2}tp_1^2 + (\tfrac{1}{2}x + \tfrac{3}{4}tp_0)p_0p_1 + \tfrac{1}{4}p_0^2,$$

while

$$\varphi_3^1 = tp_3 + \tfrac{1}{2}xp_2$$

is a symmetry of the heat equation. By (4.19), the symmetry φ_3^1 acts on the functions $\varphi_k[a]$ as follows:

$$\{\varphi_k[a], \varphi_3^1\}_{\mathcal{E}} = \tfrac{1}{2}(3\dot{a}t - ka)p_{k+1} + O(k).$$

In particular, applying the operator $\{\cdot, \varphi_3^1\}_{\mathcal{E}}$ to the function $\varphi_1^0 = p_1$ k times, we obtain a symmetry of the form

$$(-2)^k k! p_{k+1} + O(k),$$

which proves existence of symmetries

$$\varphi_k[1] \stackrel{\text{def}}{=} \varphi_k^0 = p_k + O(k-1), \qquad k = 1, 2, \ldots$$

Finally, consider the symmetry φ_2^2, which acts on functions $\varphi_k[a]$ as follows:

$$\{\varphi_k[a], \varphi_2^2\}_{\mathcal{E}} = t(t\dot{a} - ka)p_k + O(k-1).$$

Consequently, applying the operator $\{\cdot, \varphi_2^2\}_{\mathcal{E}}$ to the symmetry φ_k^0 i times, $i \leq k$, we obtain a symmetry which up to a constant factor equals

$$\varphi_k[t^i] \stackrel{\text{def}}{=} \varphi_k^i = t^i p_k + O(k-1).$$

The above reasoning is equally valid both for the Burgers equation and for the heat equation. Let us make some remarks specific to the latter. First note that any symmetry of the heat equation is linear with respect to all variables $p_0, p_1, \ldots, p_k$, i.e., is of the form

$$\varphi = A(x,t) + \sum_{i=0}^{k} A_i(x,t) p_i,$$

where $\partial A/\partial t = \partial^2 A/\partial x^2$. This fact easily follows either from straightforward analysis of the equation $\ell_F(\varphi) = 0$, or from study of the system (4.7). Since $A(x,t)$ is also a symmetry, the functions φ_k^i can be considered to be linear homogeneous functions of the variables $p_0, \ldots, p_k$. Hence

$$\{p_0, \varphi_k^i\}_{\mathcal{E}} = (\partial_{p_0} - \ell_{p_0})\varphi_k^i = \left(\sum_{\alpha \geq 0} \frac{\partial}{\partial p_\alpha} - 1\right)\varphi_k^i = 0,$$

and the bracket

$$\{\varphi_k^i, \varphi_{-\infty}\}_{\mathcal{E}} = \partial_{\varphi_k^i}(\varphi_{-\infty}) - \ell_{\varphi_k^i}(\varphi_{-\infty}) = -\ell_{\varphi_k^i}(\varphi_{-\infty})$$

$$= -\sum_{\alpha=0}^{k} \frac{\partial \varphi_k^i}{\partial p_\alpha} D_x^\alpha(\varphi_{-\infty}) = -\sum_{\alpha=0}^{k} \frac{\partial \varphi_k^i}{\partial p_\alpha} \frac{\partial^\alpha \varphi_{-\infty}}{\partial x^\alpha}$$

depends on x and t only and thus is a solution of the heat equation. Note also that

$$\{p_0, \varphi_{-\infty}\}_{\mathcal{E}} = \varphi_{-\infty}, \qquad \{\varphi'_{-\infty}, \varphi''_{-\infty}\}_{\mathcal{E}} = 0.$$

As a result of all preceding considerations, we have the following theorem:

THEOREM 4.4. 1. *Any equation of the form* $u_t = u_{xx} + f(u, u_x)$ *possessing symmetries of arbitrary high order is equivalent either to the Burgers equation* $u_t = u_{xx} + u u_x$, *or to the heat equation* $u_t = u_{xx}$.

2. *For any $k > 0$, these equations possess $k+1$ symmetries of order k of the form*

$$\varphi_k^i = t^i p_k + O(k-1), \qquad i = 1, \ldots, k.$$

3. *Symmetries φ_k^i form a Lie $\mathbb{R}$-algebra $A_+(\mathcal{E})$, and*

$$\{\varphi_k^i, \varphi_l^j\}_{\mathcal{E}} = \tfrac{1}{2}(li - kj)\varphi_{k+l-2}^{i+j-1} + \mathcal{S}_{<k+l-2},$$

where $\mathcal{S}_{<k+l-2}$ are symmetries of order $< k+l-2$. The algebra $A_+(\mathcal{E})$ has three generators φ_1^0, φ_2^0, and φ_3^0, where $\varphi_1^0 = p_1$, while

$$\varphi_2^2 = t^2 p_2 + (t^2 p_0 + tx) + p_1 + t p_0 + x,$$
$$\varphi_3^1 = t p_3 + \tfrac{1}{2}(x + 3t p_0)p_2 + \tfrac{3}{2} t p_1^2 + (\tfrac{1}{2}x + \tfrac{3}{4} t p_0)p_0 p_1 + \tfrac{1}{4} p_0^2$$

for the Burgers equation and

$$\varphi_2^2 = t^2 p_2 + txp_1 + (\tfrac{1}{4}x^2 + \tfrac{1}{2}t)p_0,$$
$$\varphi_3^1 = t p_3 + \tfrac{1}{2} x p_2$$

for the heat equation.

4. *In the case of the Burgers equation, the algebra $\operatorname{sym}\mathcal{E}$ of higher symmetries coincides with the algebra $A_+(\mathcal{E})$. For the heat equation, the algebra $\operatorname{sym}\mathcal{E}$ is a semidirect product of $A_+(\mathcal{E})$ with the ideal $A_0(\mathcal{E})$ consisting of functions of the form $ap_0 + \varphi_{-\infty}(x,t)$, where $a = \operatorname{const}$ and $\varphi_{-\infty}$ is an arbitrary solution of the*

heat equation. The functions φ_k^i are linear with respect to all variables p_α, $\alpha = 0, 1, \ldots, k$,

$$\{\varphi_k^i, ap_0 + \varphi_{-\infty}\}_{\mathcal{E}} = -\sum_{\alpha=1}^{k} \frac{\partial \varphi_k^i}{\partial p_\alpha} \frac{\partial^\alpha \varphi_{-\infty}}{\partial x^\alpha}$$

and

$$\{a'p_0 + \varphi'_{-\infty}, a''p_0 + \varphi''_{-\infty}\}_{\mathcal{E}} = a''\varphi'_{-\infty} - a'\varphi''_{-\infty}.$$

REMARK 4.2. All the above symmetries φ_k^i are determined up to symmetries of lower order. The ambiguity arises in commutation relations between the φ_k^i as well (see Theorem 4.4 (3)). For the Burgers equation, this ambiguity can be essentially diminished using the following trick.

Let us assign to the variables x, t, and u *weights* in the following way:

$$\operatorname{gr} x = 1, \qquad \operatorname{gr} t = 2, \qquad \operatorname{gr} u = -1.$$

The Burgers equation becomes homogeneous with respect to this system of weights. Let us also set $\operatorname{gr} p_k = -k - 1$, and for any monomial $\mathcal{M} = x^\alpha t^\beta p_0^{\gamma_0} p_1^{\gamma_1} \cdots p_k^{\gamma_k}$ define its weight as the sum of weights of all factors:

$$\operatorname{gr} \mathcal{M} = \alpha + 2\beta - \sum_{i=0}^{k} \gamma_i(i+1).$$

In the ring $\mathcal{F}(\mathcal{E})$ of smooth functions on the infinitely prolonged Burgers equation, consider the subring $\mathcal{P}(\mathcal{E})$ consisting of functions polynomial with respect to all variables. Then, as it is easily seen, $\mathcal{P}(\mathcal{E})$ is closed with respect to the operators $\ell_F^{\mathcal{E}}$ and $\{\cdot, \cdot\}_{\mathcal{E}}$, and restrictions of these operators to $\mathcal{P}(\mathcal{E})$ are homogeneous with respect to the weighting. Moreover, if $\varphi, \psi \in \mathcal{P}(\mathcal{E})$ are homogeneous polynomials, then

$$\operatorname{gr} \ell_F^{\mathcal{E}}(\varphi) = \operatorname{gr} \varphi - 2, \qquad \operatorname{gr}\{\varphi, \psi\}_{\mathcal{E}} = \operatorname{gr} \varphi + \operatorname{gr} \psi + 1.$$

Consequently, if $\varphi \in \mathcal{P}(\mathcal{E})$ is a solution of the equation $\ell_F^{\mathcal{E}}(\varphi) = 0$, then any homogeneous component of the polynomial φ is also a solution of this equation.

Further, the symmetries φ_1^0, φ_2^2, and φ_3^1 are polynomials and are generators of the Lie algebra $\operatorname{sym} \mathcal{E}$; therefore, $\operatorname{sym} \mathcal{E} \subset \mathcal{P}(\mathcal{E})$. Thus from the above it follows that the functions φ_k^i can be considered to be homogeneous, and $\operatorname{gr} \varphi_k^i = 2i - k - 1$. From this we see that the homogeneity condition *uniquely* determines the classical symmetry of the Burgers equation, as well as the symmetries of the form φ_k^0 and φ_k^k.

Let φ_k^0, φ_l^0 be two homogeneous symmetries. Then, since the order of the symmetry $\{\varphi_k^0, \varphi_l^0\}_{\mathcal{E}}$, if it is not equal to 0, is less than $k + l - 2$, its weight is not greater than $k + l - 3$ and not less than $1 - k - l$. But on the other hand,

$$\operatorname{gr}\{\varphi_k^0, \varphi_l^0\}_{\mathcal{E}} = \operatorname{gr} \varphi_k^0 + \operatorname{gr} \varphi_l^0 = -1 - k - l.$$

This contradiction shows that $\{\varphi_k^0, \varphi_l^0\}_{\mathcal{E}} = 0$, i.e., the symmetries of the form φ_k^0, $k = 1, 2, \ldots$, commute with each other.

Let us consider the action of the operators $\{\varphi_1^0, \cdot\}_{\mathcal{E}}$ and $\{\varphi_2^0, \cdot\}_{\mathcal{E}}$ in more detail. We have

$$\{\varphi_1^0, \cdot\}_{\mathcal{E}} = \partial_{\varphi_1^0} - \ell_{\varphi_1^0} = \sum_{\alpha \geq 0} p_{\alpha+1} \frac{\partial}{\partial p_\alpha} - D_x = -\frac{\partial}{\partial x}$$

and

$$\{\varphi_2^0, \cdot\}_{\mathcal{E}} = \vartheta_{\varphi_2^0} - \ell_{\varphi_2^0} = \sum_{\alpha \geq 0} \left[D^\alpha(p_2 + p_0 p_1) \frac{\partial}{\partial p_\alpha} \right] - p_1 - p_0 D_x - D_x^2 = \ell_F^{\mathcal{E}} - \frac{\partial}{\partial t}.$$

Since the action $\{\varphi_2^0, \cdot\}_{\mathcal{E}}$ is considered on symmetries of the Burgers equation, i.e., on solutions of the equation $\ell_F^{\mathcal{E}}(\varphi) = 0$, we eventually obtain

$$\{\varphi_2^0, \cdot\}_{\mathcal{E}} = \frac{\partial}{\partial t}.$$

Hence, the symmetries of the form φ_k^0 do not depend on x and t. Further, for homogenous components one has obvious equalities

$$\{\varphi_1^0, \varphi_k^1\}_{\mathcal{E}} = -\tfrac{1}{2}\varphi_{k-1}^0, \qquad \{\varphi_2^0, \varphi_k^1\}_{\mathcal{E}} = -\varphi_k^1.$$

Therefore, the symmetries φ_k^1 are linear in x and t. In the same way, by elementary induction it is proved that φ_k^i is a polynomial of i-th degree in t and x.

REMARK 4.3. Let us return to the heat equation and show a simple way to construct its higher symmetries. In this case, the operator ℓ_F is of the form $D_t - D_x^2$ and consequently commutes with the operator D_x. Therefore, if φ is a symmetry, then

$$\ell_F(D_x\varphi) = D_x(\ell_F(\varphi)) = 0,$$

i.e., $D_x\varphi$ is also a symmetry of the heat equation. This fact is a particular case of a more general result.

Let $\mathcal{E} = \{F = 0\}$ be a linear equation and $\Delta = \Delta_F$ be the corresponding linear differential operator. Then (see formula (2.33)) $\ell_F = \widehat{\Delta}$. Let ∂ be another linear operator, for which

$$\Delta \circ \partial = \partial' \circ \Delta,$$

where ∂' is also a linear differential operator. Let us set $\mathcal{R} = \widehat{\partial}$. Then

$$\ell_F \circ \mathcal{R} = \widehat{\Delta} \circ \widehat{\partial} = \widehat{\Delta \circ \partial} = \widehat{\partial' \circ \Delta} = \widehat{\partial'} \circ \ell_F,$$

and thus the operator $\mathcal{R}$ acts on the algebra sym $\mathcal{E}$. Such operators are called *recursion operators* and are widely used for constructing higher symmetries [59]. In particular, applications of recursion operators will be shown in the next example. Nevertheless, for nonlinear equations the situation is more complicated, and we shall discuss it in Chapter 6.

4.3. The plasticity equations. Consider the system of equations $\mathcal{E}$,

$$\begin{cases} \sigma_x = 2k(\theta_x \cos 2\theta + \theta_y \sin 2\theta), \\ \sigma_y = 2k(\theta_x \sin 2\theta - \theta_y \cos 2\theta), \end{cases} \tag{4.20}$$

describing the plane strained state of the medium with von Mises conditions, where σ is hydrostatic pressure, θ is the angle between the x-axis and the first main direction of the stress tensor, and $k \neq 0$ is the plasticity constant. In describing symmetries of this equation[13] we follow the paper [104].

---

[13]In what follows, we do not stress technical details, but pay our main attention to the structure of proofs and to the most interesting specific features. To reconstruct the omitted details is a useful exercise in symmetry computations.

By the change of variables

$$\begin{cases} \sigma = k(\xi + \eta), \\ \theta = \frac{1}{2}(\eta - \xi), \end{cases} \quad \begin{cases} x = u\cos\frac{1}{2}(\eta - \xi) - v\sin\frac{1}{2}(\eta - \xi), \\ y = u\sin\frac{1}{2}(\eta - \xi) + v\cos\frac{1}{2}(\eta - \xi), \end{cases} \quad (4.21)$$

where u, v are new dependent and ξ, η are new independent variables, the system (4.20) is transformed to the system

$$\begin{cases} u_\xi + \frac{1}{2}v = 0, \\ v_\eta + \frac{1}{2}u = 0, \end{cases} \quad (4.22)$$

which will also be denoted by $\mathcal{E}$. Its universal linearization operator is of the form

$$\ell_F = \begin{pmatrix} D_\xi & 1/2 \\ 1/2 & D_\eta \end{pmatrix}.$$

Choose on $\mathcal{E}^\infty$ intrinsic coordinates ξ, η, u_k, v_k in such a way that u_k corresponds to the partial derivative $\partial^k u/\partial \eta^k$ and v_k corresponds to $\partial^k v/\partial \xi^k$. Then the restrictions of the total derivatives to $\mathcal{E}^\infty$ will be written in these coordinates in the form

$$D_\xi = \frac{\partial}{\partial \xi} - \frac{1}{2}v_0 \frac{\partial}{\partial u_0} + \frac{1}{4}\sum_{k \geq 1} u_{k-1} \frac{\partial}{\partial u_k} + \sum_{k \geq 0} v_{k+1} \frac{\partial}{\partial v_k},$$

$$D_\eta = \frac{\partial}{\partial \eta} - \frac{1}{2}u_0 \frac{\partial}{\partial v_0} + \sum_{k \geq 0} u_{k+1} \frac{\partial}{\partial u_k} + \frac{1}{4}\sum_{k \geq 1} v_{k-1} \frac{\partial}{\partial v_k},$$

and if $\Phi = \begin{pmatrix} \varphi \\ \psi \end{pmatrix}$ is a symmetry of order k, then the corresponding defining equations will be

$$\begin{cases} \dfrac{\partial \varphi}{\partial \xi} - \dfrac{1}{2}v_0 \dfrac{\partial \varphi}{\partial u_0} + v_1 \dfrac{\partial \varphi}{\partial v_0} + \displaystyle\sum_{\alpha=1}^k \left(\dfrac{1}{4}u_{\alpha-1} \dfrac{\partial \varphi}{\partial u_\alpha} + v_{\alpha+1} \dfrac{\partial \varphi}{\partial v_\alpha} \right) + \dfrac{1}{2}\psi = 0, \\ \dfrac{\partial \psi}{\partial \eta} + u_1 \dfrac{\partial \psi}{\partial u_0} - \dfrac{1}{2}u_0 \dfrac{\partial \psi}{\partial v_0} + \displaystyle\sum_{\alpha=1}^k \left(u_{\alpha+1} \dfrac{\partial \psi}{\partial u_\alpha} + \dfrac{1}{4}v_{\alpha-1} \dfrac{\partial \psi}{\partial v_\alpha} \right) + \dfrac{1}{2}\varphi = 0. \end{cases} \quad (4.23)$$

Solving system (4.23) for $k \leq 1$, we obtain classical symmetries of equation (4.22); any such a symmetry is a linear combination of the following ones:

$$S_0 = \begin{pmatrix} u_0 \\ v_0 \end{pmatrix}, \quad f_1^0 = \begin{pmatrix} u_1 \\ -\dfrac{u_0}{2} \end{pmatrix}, \quad g_1^0 = \begin{pmatrix} -\dfrac{v_0}{2} \\ v_1 \end{pmatrix}, \quad S_1 = \begin{pmatrix} \eta u_1 + \dfrac{u_0}{3} + \xi \dfrac{v_0}{2} \\ -\xi v_1 - \dfrac{v_0}{2} - \eta \dfrac{u_0}{2} \end{pmatrix},$$

and of the symmetry of the form $H = \begin{pmatrix} f \\ g \end{pmatrix}$, where $u = f$, $v = g$ is an arbitrary solution of equation (4.22).

For "large" k, the solutions of (4.23), if any, have the following "asymptotics":

$$\begin{pmatrix} Au_k - \frac{1}{2}Bv_{k-1} + au_{k-1} + \frac{1}{2}(B' - b)v_{k-2} + (\alpha - \frac{1}{4}\xi A')u_{k-2} + O(k-3) \\ Bv_k - \frac{1}{2}Au_{k-1} + bv_{k-1} + \frac{1}{2}(A' - a)u_{k-2} + (\beta - \frac{1}{4}\eta B')v_{k-2} + O(k-3) \end{pmatrix},$$

where A, a, α are functions of η and B, b, β are functions of ξ, while the prime denotes the derivatives with respect to ξ or η. Let us denote this solution by $\Phi_k(A, B)$.

Following Remark 4.3, let us construct first order recursion operators for equation (4.22). As is easily seen, such operators exist and are of the form

$$\mathcal{R} = \begin{pmatrix} a_{11}D_\xi + b_{11}D_\eta + c_{11} & b_{12}D_\eta + c_{12} \\ a_{21}D_\xi + c_{21} & a_{22}D_\xi + b_{22}D_\eta + c_{22} \end{pmatrix}, \quad (4.24)$$

where $a_{11}, c_{11}, b_{22}, c_{22}$ are arbitrary functions in ξ and η,

$$a_{21} = 2c_{11} + \alpha_1, \quad b_{12} = 2c_{22} + \alpha_2, \quad \alpha_1, \alpha_2 = \text{const},$$

$$a_{22} = \tfrac{1}{2}(\alpha_1 - \alpha_2)\xi + \beta_1, \quad b_{11} = \tfrac{1}{2}(\alpha_2 - \alpha_1)\eta + \beta_2, \quad \beta_1, \beta_2 = \text{const},$$

$$c_{12} = \tfrac{1}{2}(a_{11} - a_{22}), \quad c_{21} = \tfrac{1}{2}(b_{22} - b_{11}).$$

In particular, among the operators (4.24) one has the operator

$$\mathcal{R}_1 = \begin{pmatrix} D_\xi & 0 \\ 0 & D_\xi \end{pmatrix},$$

application of which to the functions of the form $\Phi_k(A, B)$ gives the following expression:

$$\mathcal{R}_1 \Phi_k(A, B) = \Phi_{k+1}(0, B) + \Phi_k(0, B') + \tfrac{1}{4}\Phi_{k-1}(A, 0) + O(k-2).$$

On the other hand, if one acts on the function $\Phi_k(A, B)$ by the symmetry f_1^0, one obtains

$$\{\Phi_k(A, B), f_1^0\}_\mathcal{E} = \frac{\partial}{\partial \eta}\Phi_k(A, B) = \Phi_k(A', 0) - \frac{1}{4}\Phi_{k-2}(0, B') + O(k-3).$$

Therefore,

$$\mathcal{R}_1\{\Phi_k(A, B), f_1^0\}_\mathcal{E} = \tfrac{1}{4}\Phi_{k-1}(A', -B') + O(k-2). \quad (4.25)$$

Now, by induction on k (the base of the induction is the above representation of classical symmetries (4.22), and the induction step consists in using equation (4.25)) it is easily shown that any symmetry of order $k > 0$, if it exists, is a linear combination of the following symmetries:

$$S_k = \Phi_k(\eta^k, (-\xi)^k), \quad f_k^i = \Phi_k(\eta^i, 0), \quad g_k^i = \Phi_k(0, \xi^i), \quad 0 \leq i \leq k.$$

To prove existence of these symmetries, let us note that among the recursion operators of the form (4.24) one has the operator

$$\mathcal{R}_2 = \begin{pmatrix} -\eta D_\xi + \eta D_\eta + \tfrac{1}{2} & \tfrac{1}{2}(\xi - \eta) \\ \tfrac{1}{2}(\xi - \eta) & \xi D_\xi - \xi D_\eta - \tfrac{1}{2} \end{pmatrix}.$$

Application of this operator to the symmetry S_0 k times gives, up to a constant factor, the symmetry S_k. Further, application of the operator $\{\cdot, f_1^0\}_\mathcal{E}$ to the symmetry S_k $k - i$ times leads to the symmetry f_k^i. Finally, applying the operator $\{\cdot, g_1^0\}_\mathcal{E}$ $k - 1$ times to the symmetry S_k, we can prove existence of g_k^i.

The above can be combined in the following result.

THEOREM 4.5. *The algebra of higher symmetries of the system* (4.22) *is a semidirect product of the commutative ideal* $\text{sym}_0 \mathcal{E}$ *consisting of symmetries* $H = \begin{pmatrix} f \\ g \end{pmatrix}$, *where* f, g *is an arbitrary solution of* (4.22), *with the algebra* $\text{sym}_+ \mathcal{E}$. *The latter, as a vector space, is spanned by the elements* $S_k, f_k^i, g_k^i, 0 \leq i < k, k = 0, 1, \ldots,$ *and as a Lie algebra is generated by the elements* $f_1^0, g_1^0, S_0, S_1, \ldots, S_k, \ldots$ *Any symmetry* S_k *is of the form* $S_k = \mathcal{R}_2^k(S_0)$.

4.4. Transformation of symmetries under change of variables.

Let us rewrite our results in terms of the initial equation (4.20). To this end, it is necessary to find out how generating sections and recursion operators transform under change of variables. To do this, let us note the following.

Let $\mathcal{S} = (x_1, \ldots, x_n, u^1, \ldots, u^m, p_1^1, \ldots, p_n^m)$ be a local canonical coordinate system in $J^1(\pi)$, where $p_i^j \stackrel{\text{def}}{=} p_{1_i}^j$. Consider in the same neighborhood another canonical coordinate system $\widetilde{\mathcal{S}} = (\widetilde{x}_1, \ldots, \widetilde{x}_n, \widetilde{u}^1, \ldots, \widetilde{u}^m, \widetilde{p}_1^1, \ldots, \widetilde{p}_n^m)$ compatible with the contact structure in $J^1(\pi)$[14]. The Cartan distribution on $J^1(\pi)$ is determined by the system of Cartan forms

$$\omega^1 = du^1 - \sum_\alpha p_\alpha^1 \, dx_\alpha, \ldots, \omega^m = du^m - \sum_\alpha p_\alpha^m \, dx_\alpha,$$

which will be represented as a column $\Omega = (\omega^1, \ldots, \omega^m)^t$. On the other hand, in the coordinate system $\widetilde{\mathcal{S}}$ the set of forms $\widetilde{\Omega}$ determines, by the above, the same distribution, and consequently the equality

$$\widetilde{\Omega} = \Lambda \Omega \tag{4.26}$$

holds, where $\Lambda = \|\lambda_\beta^\alpha\|$ is a nondegenerate transformation matrix: $\widetilde{\omega}^\alpha = \sum_\beta \lambda_\beta^\alpha \omega^\beta$.

Let X be a Lie field on $J^1(\pi)$. Then its generating section in the coordinate system $\mathcal{S}$, also represented as a column $\varphi = (\varphi^1, \ldots, \varphi^m)^t$, is determined by the equality $\varphi = X \rfloor \Omega = (X \rfloor \omega^1, \ldots, X \rfloor \omega^m)^t$ (see Chapter 3). For the same reason, the generating section of the field X in the new coordinate system is represented in the form $\widetilde{\varphi} = X \rfloor \widetilde{\Omega}$. Hence, using (4.26), we obtain

$$\widetilde{\varphi} = X \rfloor \widetilde{\Omega} = X \rfloor \Lambda \Omega = \Lambda \varphi, \tag{4.27}$$

and it gives us the desired transformation rule for generating sections.

Now let $\mathcal{R}$ be an operator acting in the space of generating sections and represented in the coordinate system $\mathcal{S}$. Then from (4.27) it follows that its representation in the system $\widetilde{\mathcal{S}}$ is of the form

$$\widetilde{\mathcal{R}} = \Lambda \mathcal{R} \Lambda^{-1}. \tag{4.28}$$

Since we are interested in recursion operators of the form $\mathcal{R} = \|\sum_\alpha a_{ij}^\alpha D_\alpha + b_{ij}\|$, we also need to find out how the operators of total derivatives transform under change of coordinates.

Let $D_i = D_{x_i}$ and $\widetilde{D}_i = D_{\widetilde{x}_i}$, $i = 1, \ldots, n$. Since the fields D_i, as well as the fields $\widetilde{D}_i$, form a basis of the Cartan distribution in $J^\infty(\pi)$, the equalities $\widetilde{D}_i = \sum_\alpha \mu_i^\alpha D_\alpha$ must hold, and since $D_i(x_\alpha) = \delta_{i\alpha}$, where $\delta_{i\alpha}$ are the Kronecker symbols, we have $\mu_i^\alpha = \widetilde{D}_i(x_\alpha)$, and consequently

$$\widetilde{D}_i = \sum_{\alpha=1}^n \widetilde{D}_i(x_\alpha) D_\alpha, \qquad i = 1, \ldots, n. \tag{4.29}$$

Let us come back to equation (4.20). For the transformation (4.21), the transformation matrix is of the form

$$\Lambda = \begin{pmatrix} -\sigma_x \cos\theta - \sigma_y \sin\theta & \sigma_x \sin\theta - \sigma_y \cos\theta \\ -\theta_x \cos\theta - \theta_y \sin\theta & \theta_x \sin\theta - \theta_y \cos\theta \end{pmatrix},$$

---

[14]This means that the change of coordinates $\widetilde{x} = \widetilde{x}(x, u)$, $\widetilde{u} = \widetilde{u}(x, u)$, $\widetilde{p} = \widetilde{p}(x, u, p)$ is a Lie transformation (see Chapter 3).

while the total derivative operators are represented as

$$D_\xi = \frac{k}{I}\left[(\frac{\sigma_y}{2k} + \theta_y)D_x - (\frac{\sigma_x}{2k} + \theta_x)D_y\right],$$

$$D_\eta = \frac{k}{I}\left[(\theta_y - \frac{\sigma_y}{2k})D_x - (\theta_x - \frac{\sigma_x}{2k})D_y\right],$$

where $I = \sigma_x\theta_y - \sigma_y\theta_x$. Therefore, to the symmetries H, S_0, f_1^0, and g_1^0 of the transformed equation there correspond the following symmetries of the initial equation:

$$\widetilde{H} = D\begin{pmatrix}(-\sigma_x\cos\theta - \sigma_y\sin\theta)f & (\sigma_x\sin\theta - \sigma_y\cos\theta)g \\ (-\theta_x\cos\theta - \theta_y\sin\theta)f & (\theta_x\sin\theta - \theta_y\cos\theta)g\end{pmatrix},$$

where $f = f(\sigma/2k - \theta, \sigma/2k + \theta)$, $g = g(\sigma/2k - \theta, \sigma/2k + \theta)$, and

$$\widetilde{S}_0 = \begin{pmatrix}-x\sigma_x - y\sigma_y \\ -x\theta_x - y\theta_y\end{pmatrix},$$

$$\widetilde{f}_1^0 = \begin{pmatrix}k + \frac{1}{2}(y\sigma_x - x\sigma_y) \\ -\frac{1}{2} + \frac{1}{2}(y\theta_x - x\theta_y)\end{pmatrix}, \quad \widetilde{g}_1^0 = \begin{pmatrix}k - \frac{1}{2}(y\sigma_x - x\sigma_y) \\ \frac{1}{2} - \frac{1}{2}(y\theta_x - x\theta_y)\end{pmatrix}$$

and to the recursion operator $\mathcal{R}_2$ there corresponds the operator

$$\widetilde{\mathcal{R}}_2 = \begin{pmatrix}r_{11}, & r_{12} \\ r_{21}, & r_{22,}\end{pmatrix}$$

where

$$r_{11} = \frac{\sigma}{k}\Delta + \frac{2k\theta}{I}(\theta_x^2 + \theta_y^2 + c\Delta(d) - d\Delta(c)) - \frac{\Delta(I)}{I},$$

$$r_{12} = 2k^2\theta\Delta + \frac{2k}{I}(c^2 - d^2 - \Delta(I)) + 4k(c\Delta(d) - d\Delta(c)),$$

$$r_{21} = \frac{\theta}{4k}\Delta + \frac{\theta}{I}(c\Delta(d) - d\Delta(c) + c^2 - d^2) + \frac{k}{4},$$

$$r_{22} = \frac{\sigma}{k}\Delta + \frac{2k\theta}{I}(c\Delta(d) - d\Delta(c) - \theta_x^2 - \theta_y^2) - \frac{\Delta(I)}{I}$$

and

$$c = -\theta_x\cos\theta - \theta_y\sin\theta, \quad d = \theta_x\sin\theta - \theta_y\cos\theta, \quad \Delta = \sigma_xD_y - \sigma_yD_x.$$

Applying the operator $\widetilde{\mathcal{R}}_2^k$ to the symmetry $\widetilde{S}_0$, we obtain the symmetries $\widetilde{S}_k$, and applying the operators $\{\cdot, \widetilde{f}_1^0\}_\mathcal{E}$ and $\{\cdot, \widetilde{g}_1^0\}_\mathcal{E}$ $k - i$ times to $\widetilde{S}_k$, we obtain (up to constant factors) the corresponding symmetries $\widetilde{f}_k^i$ and $\widetilde{g}_k^i$. Thus we can efficiently compute any symmetry of equation (4.20) (though the explicit expressions will certainly be quite cumbersome).

4.5. Ordinary differential equations. To finish this section, we shall study higher symmetries of ordinary differential equations. Besides the facts concerning higher symmetries, we also mention some results on classical symmetries of these equations.

Let $M = \mathbb{R}$, and let $\pi \colon \mathbb{R}^m \times \mathbb{R} \to \mathbb{R}$ be a trivial bundle. Consider a determined system of ordinary differential equations $\mathcal{E} \subset J^k(\pi)$, i.e., a system such that $\operatorname{codim}\mathcal{E} = \dim\pi = m$. Since the base of the bundle π is one-dimensional, this means that the dimension of $\mathcal{E}$ coincides with dimension of the space $J^{k-1}(\pi)$, which equals $m(k-1) + 1$. Let us call a point $\theta \in \mathcal{E}$ *generic*, if the projection $\pi_{k,k-1}|_\mathcal{E}$

at this point is of the maximal rank $m(k-1)+1$. We shall confine ourselves to equations such that all their points are generic (otherwise all considerations below will be valid in a neighborhood of such a point). This means that $\mathcal{E}$ is projected to $J^{k-1}(\pi)$ diffeomorphically or, which is the same, that the equation under consideration is solvable with respect to all derivatives $d^k u^1/dx^k, \ldots, d^k u^m/dx^k$ of highest order, where $x = x_1$ is the sole independent variable. Hence, $\mathcal{E}$ can be represented in the form

$$\mathcal{E} = s(J^{k-1}(\pi)), \tag{4.30}$$

where $s = s_0 \cdot J^{k-1}(\pi) \to J^k(\pi)$ is a section of the bundle $\pi_{k,k-1}$. Representation (4.30) allows one to obtain a convenient description of the manifold $\mathcal{E}^\infty$ and of the Cartan distribution on this manifold.

Namely, consider an arbitrary point $\theta \in \mathcal{E}$ and some R-plane L (in our case such a plane is one-dimensional) lying in the space $T_\theta(\mathcal{E})$. Let L' be another R-plane, and let $v \in L$, $v' \in L'$ be vectors such that $\pi_{k,k-1}(v) = \pi_{k,k-1}(v')$. Then $v - v'$ is a $\pi_{k,k-1}$-vertical vector lying in $T_\theta(\mathcal{E})$. But by (4.30), the intersection of $T_\theta(\mathcal{E})$ with the tangent space to the fiber of the bundle $\pi_{k,k-1}$ at θ is trivial. Hence, $v = v'$ and $L = L'$, i.e., the Cartan distribution on $\mathcal{E}$ at every point contains one R-plane at most. Let us show that such a plane always exists. In fact, let us represent the point $\theta \in \mathcal{E}$ as a pair (θ', L_θ), where $\theta' = \pi_{k,k-1}(\theta)$ and L_θ is the R-plane at the point θ' determined by the point θ. Then obviously the R-plane $L_\theta^0 = s_*(L_\theta)$ lies in $T_\theta(\mathcal{E})$. Thus the Cartan distribution on $\mathcal{E}$ coincides with the field of directions $\mathcal{L}^0 = \{ L_\theta^0 \mid \theta \in \mathcal{E} \}$. Note that the section s is an isomorphism between the manifold $J^{k-1}(\pi)$ endowed with the field of directions $\mathcal{L} = \{ L_\theta \mid \theta \in \mathcal{E} \}$ and the pair $(\mathcal{E}, \mathcal{L}^0)$.

Further, a pair (θ, L_θ^0), $\theta \in \mathcal{E}$, determines a point $\theta_1 \in J^{k+1}(\pi)$, and the set of such points coincides with the manifold $\mathcal{E}^{(1)}$. It is obvious that $\mathcal{E}^{(1)}$ can be represented as $s_1(J^{k-1}(\pi))$, where s_1 is a section of the bundle $\pi_{k+1,k-1}$, while the Cartan distribution on $\mathcal{E}^{(1)}$ coincides with the field of directions

$$\mathcal{L}^{(1)} = \{ L_{\theta_1}^1 \mid L_{\theta_1}^1 = (s_1)_*(L_\theta), \theta_1 = s_1(\theta), \theta \in \mathcal{E} \}.$$

Proceeding with this construction, we come to the following statement:

PROPOSITION 4.6. *Let $\mathcal{E} \subset J^k(\pi)$, $\dim(M) = 1$, $\operatorname{codim}(\mathcal{E}) = \dim(\pi)$, be an ordinary differential equation that can be solved with respect to derivatives of highest degree. Then for any $l = 0, 1, \ldots, \infty$ the manifold $\mathcal{E}^{(l)}$ is representable in the form $\mathcal{E}^{(l)} = s_l(J^{k-1}(\pi))$, where $s_l \in \Gamma(\pi_{k+l,k-1})$ and $s_l = \pi_{k+l+1,k+l} \circ s_{l+1}$, while the Cartan distribution on $\mathcal{E}^{(l)}$ coincides with the field of directions*

$$\mathcal{L}^{(l)} = \{ L_{\theta_l}^l \mid L_{\theta_l}^l = (s_l)_*(L_\theta), \theta_l = s_l(\theta), \theta \in \mathcal{E} \}.$$

Thus, all $\mathcal{E}^{(l)}$ ($\mathcal{E}^\infty$ included) as manifolds with distributions are mutually isomorphic and are isomorphic to the manifold $J^{k-1}(\pi)$ endowed with the field of directions $\mathcal{L}$.

Suppose that the equation $\mathcal{E}$ satisfies the assumptions of Proposition 4.6, and consider a higher symmetry $X \in \operatorname{sym} \mathcal{E}$ of this equation. By Proposition 4.6, for any l the field X projects to the vector field X_l on $\mathcal{E}^{(l)}$ and the field preserves the corresponding Cartan distribution. In particular, $X_0 \in \operatorname{D}(\mathcal{E})$ is a classical intrinsic symmetry of the equation $\mathcal{E}$. Taking the set $\operatorname{Sym}_i \mathcal{E}$ of such symmetries, we thus obtain the Lie algebra homomorphism

$$\operatorname{sL}: \operatorname{sym} \mathcal{E} \longrightarrow \operatorname{Sym}_i \mathcal{E}. \tag{4.31}$$

Let us study this homomorphism in more detail, and note the following. Since in the situation under consideration $\mathcal{E}^\infty$ is a finite-dimensional manifold, the symmetry X determines a one-parameter group of transformations $\{A_t\}$ on this manifold. Applying the transformation A_t to a solution $f \in \Gamma(\pi)$ of the equation $\mathcal{E}$, we again obtain a solution $f_t = A_t^*(f)$ of this equation. By the theorem on smooth dependence of solutions of ordinary differential equations on initial values, the set $\mathrm{Sol}(\mathcal{E})$ of solutions of the equation $\mathcal{E}$ is a smooth (possibly, with singularities) manifold, while A_t is a one-parameter group of its diffeomorphisms. Denoting by X^* the corresponding vector field, we obtain the mapping $X \mapsto X^*$ which determines the Lie algebra homomorphism

$$\mathrm{sD}\colon \mathrm{sym}\,\mathcal{E} \longrightarrow \mathrm{D}(\mathrm{Sol}(\mathcal{E})). \tag{4.32}$$

In exactly the same way one can construct the homomorphism

$$\mathrm{LD}\colon \mathrm{Sym}_i(\mathcal{E}) \longrightarrow \mathrm{D}(\mathrm{Sol}(\mathcal{E})). \tag{4.33}$$

Conversely, let $Y \in \mathrm{D}(\mathrm{Sol}(\mathcal{E}))$, and let $\{B_t\}$ be the corresponding one-parameter group of transformations. Consider a point $\theta \in \mathcal{E}^\infty$ and a solution f_θ passing through this point (such a solution exists and is unique). Let us put into correspondence to the point θ the point $B_t'(\theta) = [B_t(f_\theta)]_x^\infty$, $x = \pi_\infty(\theta)$. Then $\{B_t'\}$ is a one-parameter group of transformations of the manifold $\mathcal{E}^\infty$, and these transformations preserve the Cartan distribution. The corresponding vector field is vertical and determines a symmetry of $\mathcal{E}$. Thus we have constructed the homomorphism

$$\mathrm{Ds}\colon \mathrm{D}(\mathrm{Sol}(\mathcal{E})) \longrightarrow \mathrm{sym}\,\mathcal{E}, \tag{4.34}$$

inverse to the homomorphism (4.32) in an obvious way. It is easily seen that the homomorphisms thus constructed are mutually compatible in the following sense.

PROPOSITION 4.7. *Let $\mathcal{E}$ be an equation satisfying the assumptions of Proposition 4.6. Then the above constructed mappings* sD, Ds, sL, *and* LD *are Lie algebra isomorphisms with* $\mathrm{sD} \circ \mathrm{Ds} = \mathrm{id}$ *and* $\mathrm{LD} \circ \mathrm{sL} = \mathrm{sD}$. *In other words, in the situation under consideration the Lie algebras of higher symmetries, of intrinsic symmetries, and of vector fields on the manifold of solutions are isomorphic.*

PROOF. The equalities $\mathrm{sD} \circ \mathrm{Ds} = \mathrm{id}$ and $\mathrm{LD} \circ \mathrm{sL} = \mathrm{sD}$ follow by construction. Hence sL is a monomorphism while LD is an epimorphism. Consequently, to conclude the proof it suffices, for example, to show that the kernel of LD is trivial. From the definition of the homomorphism LD it follows that its kernel consists of symmetries such that all solutions are invariant with respect to them. On the other hand, from the above description of the Cartan distribution on $\mathcal{E}$ it can be seen that these symmetries must lie in the Cartan distribution. But the Cartan distribution on $\mathcal{E}^\infty$ does not contain vertical fields. $\square$

Thus, in the situation considered, the analogy between symmetries and vector field on the manifold $\mathrm{Sol}(\mathcal{E})$ acquires an exact sense.

Note that Proposition 4.6 gives an obvious way to show that the algebras of extrinsic and intrinsic symmetries do not coincide in the case of ordinary differential equations (see §7 of Ch. 3). In fact, intrinsic symmetries are identified with vector fields on $J^{k-1}(\pi)$ preserving the field of directions $\mathcal{L} = \{L_\theta\}$, which is contained in the Cartan distribution on $J^{k-1}(\pi)$ (see Figure 4.7). On the other hand, extrinsic symmetries must preserve both this field of directions and the Cartan distribution

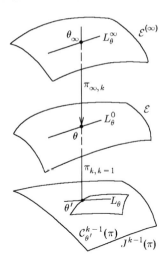

FIGURE 4.7. The field of directions L_θ on $J^{k-1}(\pi)$

on $J^{k-1}(\pi)$ itself. Therefore, if the order of the equation is greater than 1, extrinsic symmetries form a proper subalgebra in the Lie algebra of intrinsic ones.

Let us now pass to coordinate computations and consider an equation of the form

$$L\boldsymbol{f}(x) \equiv \left[\left(\frac{d}{dx}\right)^n + \sum_0^{n-1} A_i(x) \cdot \left(\frac{d}{dx}\right)^i\right]\boldsymbol{f}(x) = \boldsymbol{g}(x), \qquad (4.35)$$

where A_i are $m \times m$ matrices and $\boldsymbol{f}$, $\boldsymbol{g}$ are m-vectors. Below, by a symmetry we mean a higher infinitesimal symmetry.

Equation (4.35) is equivalent to the equation

$$L\boldsymbol{f}(x) = \left[\left(\frac{d}{dx}\right)^n + \sum_0^{n-1} A_i(x) \cdot \left(\frac{d}{dx}\right)^i\right]\boldsymbol{f}(x) = 0. \qquad (4.36)$$

This equivalence is given by the transformation

$$\boldsymbol{f}(x) \mapsto \boldsymbol{f}(x) - \boldsymbol{f}^*(x),$$

where $\boldsymbol{f}^*(x)$ is a solution of (4.35). Equation (4.36) determines a submanifold $\mathcal{E}$ in $J^n(\mathbb{R}, \mathbb{R}^m)$ of the form

$$\mathcal{E} = \left\{ y \in J^n(\mathbb{R}, \mathbb{R}^m) \mid \boldsymbol{p}_n + \sum_{i=0}^{n-1} A_i(x) \cdot \boldsymbol{p}_i = 0 \right\}.$$

Prolongations of $\mathcal{E}$, denoted by $\mathcal{E}^{(N)} \subset J^{n+N}(\mathbb{R}, \mathbb{R}^m)$, are given by the equations

$$\boldsymbol{p}_{n+k} + D^k\left(\sum_{i=0}^{n-1} A_i(x) \cdot \boldsymbol{p}_i\right) = 0, \qquad 0 \leq k \leq N < \infty,$$

where

$$D_x = \frac{\partial}{\partial x} + \sum_{i=0}^{\infty}\sum_{k=1}^{m} p_{i+1}^k \frac{\partial}{\partial p_i^k}$$

is the total derivative along x. The space of solutions of (4.36) is isomorphic to $\mathbb{R}^d$, $d = m \cdot n$, since n initial values for m-vectors $\boldsymbol{f}^{(0)}, \ldots, \boldsymbol{f}^{(n-1)}$ uniquely determine a solution. Let us fix a basis $\{\, \boldsymbol{R}_i \mid 1 \leq i \leq m \cdot n \,\}$ in the space $\ker L$ of solutions of (4.36). Then for any solution $\boldsymbol{f}$ of equation (4.36) we shall have

$$\boldsymbol{f} = \sum_{i=1}^{d} c_i \cdot \boldsymbol{R}_i, \tag{4.37}$$

where $c_i = c_i(\boldsymbol{f}) \in \mathbb{R}$ are constants. The dependence of c on $\boldsymbol{f}$ is explicitly given by the Wronskians:

$$c_i = \frac{W_i(\boldsymbol{f})}{W}. \tag{4.38}$$

Here $W = W(\boldsymbol{R}_1, \ldots, \boldsymbol{R}_i, \ldots, \boldsymbol{R}_d)$,

$$W = \det \begin{pmatrix} \boldsymbol{R}_1 & \ldots & \boldsymbol{R}_i & \ldots & \boldsymbol{R}_d \\ \boldsymbol{R}_1' & \ldots & \boldsymbol{R}_i' & \ldots & \boldsymbol{R}_d' \\ \cdots & \cdots & \cdots & \cdots & \cdots \\ \boldsymbol{R}_1^{(n-1)} & \ldots & \boldsymbol{R}_i^{(n-1)} & \ldots & \boldsymbol{R}_d^{(n-1)} \end{pmatrix} \tag{4.39}$$

and $W_i(f) = W(\boldsymbol{R}_1, \ldots, \boldsymbol{f}(x), \ldots, \boldsymbol{R}_d)$,

$$W_i(\boldsymbol{f}) = \det \begin{pmatrix} \boldsymbol{R}_1 & \ldots & \boldsymbol{f}(x) & \ldots & \boldsymbol{R}_d \\ \boldsymbol{R}_1' & \ldots & \boldsymbol{f}'(x) & \ldots & \boldsymbol{R}_d' \\ \cdots & \cdots & \cdots & \cdots & \cdots \\ \boldsymbol{R}_1^{(n-1)} & \ldots & \boldsymbol{f}^{(n-1)}(x) & \ldots & \boldsymbol{R}_d^{(n-1)} \end{pmatrix}. \tag{4.40}$$

The correspondences $\boldsymbol{f} \mapsto W_i(\boldsymbol{f})$ or $\boldsymbol{f} \mapsto c_i(\boldsymbol{f})$ are linear differential operators of order $n-1$. We shall understand them as functions on the jet space $J^{n-1}(\mathbb{R}, \mathbb{R}^m)$. Let us define

$$\widetilde{W}_i = \det \begin{pmatrix} \boldsymbol{R}_1 & \ldots & \boldsymbol{p}_0 & \ldots & \boldsymbol{R}_d \\ \boldsymbol{R}_1' & \ldots & \boldsymbol{p}_1 & \ldots & \boldsymbol{R}_d' \\ \cdots & \cdots & \cdots & \cdots & \cdots \\ \boldsymbol{R}_1^{(n-1)} & \ldots & \boldsymbol{p}_{n-1} & \ldots & \boldsymbol{R}_d^{(n-1)} \end{pmatrix}. \tag{4.41}$$

Then $W_i(\boldsymbol{f}) = \widetilde{W}\big|_{j_{n-1}(\boldsymbol{f})}$.

The defining equation for symmetries in this case is of the form

$$\left[D^n + \sum_{i=0}^{n-1} A_i(x) \cdot D^i \right] \mathcal{G} \big|_{\mathcal{E}^{(\infty)}} = 0, \tag{4.42}$$

where D is the restriction of the operator D_x to $\mathcal{E}^\infty$.

Any point symmetry is to be of the form

$$\beta(x, \boldsymbol{p}_0) + \xi(x, \boldsymbol{p}_0) \cdot \boldsymbol{p}_1 \tag{4.43}$$

and has to satisfy (4.42). Here x, ξ are scalars, $\boldsymbol{p}_0$, $\boldsymbol{p}_1$ and β are m-vectors.

A coordinate version of Proposition 3.4 is the following statement:

PROPOSITION 4.8. *The complete symmetry algebra* sym $\mathcal{E}$ *is isomorphic to the Lie algebra of vector fields on the space* $\mathrm{Sol}(\mathcal{E})$ *of solutions of* (4.36). *This isomorphism is given by the formula*

$$\sum_{i=1}^{d} F_i(c_1, \ldots, c_d) \cdot \frac{\partial}{\partial c_i} \longmapsto \sum_{i=1}^{d} F_i\left(\frac{W_1}{W}, \ldots, \frac{W_d}{W}\right) \cdot \boldsymbol{R}_i.$$

The following theorem, specifying the structure of the Lie algebra $\operatorname{sym}\mathcal{E}$, is also valid.

THEOREM 4.9. *Let $d = \dim\operatorname{Sol}(\mathcal{E})$. Then the sets of functions*

$$\boldsymbol{p}_0, \quad \boldsymbol{R}_k, \quad \frac{\widetilde{W}_l}{W}\boldsymbol{R}_k, \quad \frac{\widetilde{W}_l}{W}\boldsymbol{p}_0, \qquad l,k = 1,\ldots,d,$$

and

$$\frac{\widetilde{W}_l}{W}\boldsymbol{R}_k, \qquad i,k = 1,\ldots,d,$$

generate in $\operatorname{sym}\mathcal{E}$ *subalgebras isomorphic to* $\mathfrak{gl}(d+1,\mathbb{R})$ *and to* $\mathfrak{gl}(d,\mathbb{R})$ *respectively.*

The following result describes the subalgebra of point symmetries in $\operatorname{sym}\mathcal{E}$[15].

THEOREM 4.10. *Any point symmetry of equation (4.36) is of the form*

$$\mathcal{G} = \left(-\frac{n-1}{2}\cdot\xi'(x) + M\right)\boldsymbol{p}_0 + \xi(x)\cdot\boldsymbol{p}_1 + \boldsymbol{b}(x), \tag{4.44}$$

where ξ is a scalar function, $\boldsymbol{b}$ is an arbitrary solution of (4.36), and M is a constant $m\times m$ matrix commuting with all matrices $A_i(x)$ (coefficients of (4.36)).

COROLLARY 4.11. *The dimension N of the algebra of point symmetries of equation (4.35) satisfies the inequality*

$$m\cdot n + 1 \leq N \leq (m+n)\cdot m + 3.$$

---

[15]The detailed computation can be found in [**99**].

CHAPTER 5

Conservation Laws

This chapter is concerned with the theory of conservation laws for differential equations. We begin with the analysis of the concept itself. It turns out that the most natural and efficient way to study the space of conservation laws is to identify it with a term of the so-called $\mathcal{C}$-*spectral sequence* [**125, 127, 132**]. This spectral sequence arises from the filtration in the de Rham complex on $\mathcal{E}^\infty$ by powers of the ideal of Cartan forms. The theory of the $\mathcal{C}$-spectral sequence, set forth in §2, makes it possible to represent any conservation law by a *generating function*. The space of generating functions is essentially the kernel of a differential operator and so is accessible for computing. In §3, we illustrate the generating function method for computing conservation laws by several examples. In §4, we discuss the connection between symmetries and conservation laws (the Noether theorem) and Hamiltonian formalism on infinite jet spaces.

For the convenience of readers whose primary interests are results and examples we repeat in §§3 and 4 the relevant material from §2 without proofs, so that such readers can jump ahead to §3 immediately after §1.

1. Introduction: What are conservation laws?

Let us take the equation of continuity of fluid dynamics as a prototypical conservation law. If $\rho(x,t)$ is the density of the fluid at the point $x = (x_1, x_2, x_3)$ at the time t and $\boldsymbol{v}(x,t) = (v^1(x,t), v^2(x,t), v^3(x,t))$ is the vector of fluid velocity, then this equation reads

$$\frac{\partial \rho}{\partial t} + \frac{\partial (\rho v^1)}{\partial x_1} + \frac{\partial (\rho v^2)}{\partial x_2} + \frac{\partial (\rho v^3)}{\partial x_3} = 0. \tag{1.1}$$

Using the Gauss theorem, we can rewrite the equation of continuity in the integral form

$$-\frac{d}{dt} \int_V \rho(x,t)\, dx = \int_{\partial V} (\rho \boldsymbol{v} \cdot \boldsymbol{n})\, d\sigma, \tag{1.2}$$

where $V \subset \mathbb{R}^3$ is a space domain, ∂V is its boundary, $\boldsymbol{n}$ is a unit external vector normal to ∂V, and $d\sigma$ is a surface area element. The surface integral equals the flow of the fluid from the region V; thus equation (1.2) says that the time rate of fluid mass decrease within the volume V is equal to the net outflow of the fluid through the boundary ∂V. In particular, when the normal velocity component $\boldsymbol{v} \cdot \boldsymbol{n}$ vanishes, the mass is conserved:

$$\int_V \rho(x,t)\, dx = \text{const.}$$

All the usual conservation laws can be described in the same manner. Let S be the density of some conserved quantity, e.g., the energy density, a component of the linear or angular momentum density, etc. In the example just mentioned, $S = \rho$. Then there exists a flux $\bar{S} = (S_1, \ldots, S_{n-1})$ of S, where $n-1$ is the number of space variables. In our example, $\bar{S} = \rho \boldsymbol{v}$. The equation describing conservation of S and generalizing (1.1) has the form

$$\frac{\partial S}{\partial t} + \sum_{i=1}^{n-1} \frac{\partial S_i}{\partial x_i} = 0, \tag{1.3}$$

or, in the integral form,

$$-\frac{d}{dt}\int_V S\,dx = \int_{\partial V} \bar{S} \cdot \boldsymbol{n}\,d\sigma.$$

A *conserved current* is an n-dimensional vector function $(S_1, \ldots, S_{n-1}, S)$ satisfying (1.3).

Now consider equation (1.3) in terms of the theory of differential equations. Suppose that we study a physical system described by an equation $\mathcal{E} = \{F = 0\} \subset J^k(\pi)$. Then S and S_i can be thought of as functions on $\mathcal{E}^\infty$, and equation (1.3) takes the form

$$\sum_{i=1}^{n} \bar{D}_i(S_i) = 0, \tag{1.4}$$

where $S_n = S$, $\bar{D}_i$ is the restriction of the total derivative D_i to $\mathcal{E}^\infty$, and n is the number of independent variables. This observation makes it possible to define a *conserved current for the equation* $\mathcal{E}$ as a vector function $(S_1, \ldots, S_n)$, $S_i \in \mathcal{F}(\mathcal{E})$, satisfying equation (1.4) on $\mathcal{E}^\infty$.

In the case when one of the independent variables is chosen to be the time $t = x_n$ and the other variables $(x_1, \ldots, x_{n-1})$ are considered as space variables, the component S_n is said to be a *conserved density* (or *charge density*), while the vector function $(S_1, \ldots, S_{n-1})$, as already mentioned, is called the *flux* of S.

There is one very simple method of constructing conserved currents. Take some set of functions $\mathcal{L}_{ij} \in \mathcal{F}(\pi)$ and put

$$S_i = \sum_{j<i} D_j(\mathcal{L}_{ji}) - \sum_{i<j} D_j(\mathcal{L}_{ij}).$$

It is clear that such conserved currents, called *trivial* (or *identically conserved*), are in no way related to the equation under consideration[1]. To get rid of them, let us say that two conserved currents are *equivalent* if they differ by a trivial current, and define *conservation laws* for an equation $\mathcal{E}$ as equivalence classes of conserved currents of this equation. (The term "integral of motion" is also used in the same sense.)

EXERCISE 1.1. Let $\mathcal{E}$ be a system of ordinary differential equations. Check that the notion of a conservation law is identical to the notion of a first integral.

---

[1] This does not necessarily imply that trivial currents are not interesting from the physical point of view. E.g., for gauge theories the conserved currents associated by the Noether theorem with gauge symmetries are trivial and, thus, do not relate to the field equations. Nevertheless, they carry some information on the gauge group and, therefore, are physically relevant (see, e.g., [**10, 11, 56, 90**]).

REMARK 1.1. Unlike ordinary equations, nonlinear partial differential equations do not have (even in a small neighborhood) a complete set of conservation laws[2]. The question of whether there is a remedy remains an interesting and intriguing problem. Here we just mention a conjecture proposed in [**134, 136**]: each regular equation has a complete set of *nonlocal* conservation laws, at least in a sufficiently small neighborhood (see §1.8 of Ch. 6 for the definition of nonlocal conservation laws).

The definition of conservation laws (as it was formulated) is unusable for finding conservation laws for a given differential equation. To this end, we need the means to describe entire classes of conserved currents. One solves this problem by reformulating the definition of conservation laws in terms of *horizontal de Rham cohomology* (i.e., the cohomology of the horizontal de Rham complex, see §1.6 of Ch. 4), which permits the use of homological algebra. In §2, we discuss this cohomological theory. (The reader whose prime interest is in practical aspects of finding conservation laws may skip this section.)

2. The $\mathcal{C}$-spectral sequence

The $\mathcal{C}$-spectral sequence, introduced in [**125, 127**], is of fundamental importance in the theory of conservation laws. For a more detailed discussion of this and related subjects see, e.g., [**4, 5, 8, 16, 21, 24, 32, 64, 81, 82, 91, 117, 118, 119, 120, 126, 132, 138, 139, 148**].

2.1. The definition of the $\mathcal{C}$-spectral sequence. Let $\mathcal{E}^\infty \subset J^\infty(\pi)$ be an infinitely prolonged equation, and let $\Lambda^*(\mathcal{E}) = \sum_{i \geq 0} \Lambda^i(\mathcal{E})$ be the exterior algebra of differential forms on $\mathcal{E}^\infty$. Consider the ideal $\mathcal{C}\Lambda^*(\mathcal{E}) = \sum_{i \geq 0} \mathcal{C}\Lambda^i(\mathcal{E})$ of the algebra $\Lambda^*(\mathcal{E})$ consisting of the Cartan forms (i.e., forms vanishing on the Cartan distribution: $\omega \in \mathcal{C}\Lambda^i(\mathcal{E})$ if and only if $\omega(X_1, \ldots, X_i) = 0$ for all $X_1, \ldots, X_i \in \mathcal{C}D(\mathcal{E})$; see §2.1 of Ch. 4). Denote by $\mathcal{C}^k\Lambda^*(\mathcal{E})$ the k-th power of the ideal $\mathcal{C}\Lambda^*(\mathcal{E})$, i.e., the submodule of $\Lambda^*(\mathcal{E})$ generated by forms $\omega_1 \wedge \cdots \wedge \omega_k$, where $\omega_i \in \mathcal{C}\Lambda^*(\mathcal{E})$. It is obvious that the ideal $\mathcal{C}\Lambda^*(\mathcal{E})$ is stable with respect to the operator d, and, hence, all ideals $\mathcal{C}^k\Lambda^*(\mathcal{E})$ have the same property:

$$d(\mathcal{C}^k\Lambda^*(\mathcal{E})) \subset \mathcal{C}^k\Lambda^*(\mathcal{E}).$$

Thus, we get the filtration

$$\Lambda^*(\mathcal{E}) \supset \mathcal{C}\Lambda^*(\mathcal{E}) \supset \mathcal{C}^2\Lambda^*(\mathcal{E}) \supset \cdots \supset \mathcal{C}^k\Lambda^*(\mathcal{E}) \supset \cdots \qquad (2.1)$$

in the de Rham complex on $\mathcal{E}^\infty$. The spectral sequence[3] $(E_r^{p,q}(\mathcal{E}), d_r^{p,q})$ determined by this filtration is called the *$\mathcal{C}$-spectral sequence* for the equation $\mathcal{E}$.

This filtration is finite in each degree, i.e.,

$$\Lambda^k(\mathcal{E}) \supset \mathcal{C}^1\Lambda^k(\mathcal{E}) \supset \mathcal{C}^2\Lambda^k(\mathcal{E}) \supset \cdots \supset \mathcal{C}^k\Lambda^k(\mathcal{E}) \supset \mathcal{C}^{k+1}\Lambda^k(\mathcal{E}) = 0.$$

Therefore the $\mathcal{C}$-spectral sequence converges to the de Rham cohomology $H^*(\mathcal{E}^\infty)$ of the infinite prolonged equation $\mathcal{E}^\infty$.

As usual, $p + q$ denotes the degree and p is the filtration degree.

---

[2] A set of conservation laws is called *complete*, if the values of the corresponding conserved quantities at a solution of the equation in question completely determine this solution.

[3] For the theory of spectral sequences consult, e.g., [**31, 35, 79, 84**].

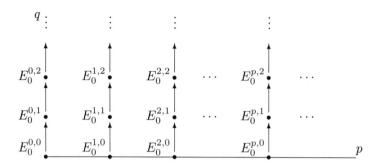

FIGURE 5.1

2.2. The term E_0. Consider the term $E_0(\mathcal{E}) = \sum_{p,q} E_0^{p,q}(\mathcal{E})$ of the $\mathcal{C}$-spectral sequence.

By definition we have

$$E_0^{p,q}(\mathcal{E}) = \mathcal{C}^p \Lambda^{p+q}(\mathcal{E})/\mathcal{C}^{p+1} \Lambda^{p+q}(\mathcal{E}).$$

The differential $d_0^{p,q} \colon E_0^{p,q}(\mathcal{E}) \to E_0^{p,q+1}(\mathcal{E})$ is induced by the exterior differential d. As usual when dealing with spectral sequences, we depict the spaces $E_r^{p,q}$ as integer points on the plane with coordinates (p, q). Then the differential d_0 is represented by arrows pointing upwards (see Figure 5.1).

Thus, $E_0(\mathcal{E})$ is the direct sum of complexes

$$0 \longrightarrow E_0^{p,0}(\mathcal{E}) \longrightarrow E_0^{p,1}(\mathcal{E}) \longrightarrow \cdots \longrightarrow E_0^{p,q}(\mathcal{E}) \longrightarrow E_0^{p,q+1}(\mathcal{E}) \longrightarrow \cdots,$$

which corresponds to columns. The zero column of the term E_0 is the *horizontal de Rham complex* (see §1.6 of Ch. 4):

$$E_0^{0,q}(\mathcal{E}) = \Lambda^q(\mathcal{E})/\mathcal{C}\Lambda^q(\mathcal{E}) = \Lambda_0^q(\mathcal{E}), \qquad d_0^{0,q} = \widehat{d}.$$

The cohomology $\bar{H}^q(\mathcal{E})$ of the horizontal de Rham complex is intimately related to the conservation laws of the equation $\mathcal{E}$. Indeed, let us assign to each conserved current $S = (S_1, \ldots, S_n)$ the horizontal $(n-1)$-form

$$\omega_S = \sum_{i=1}^{n} (-1)^{i-1} S_i \, dx_1 \wedge \cdots \wedge dx_{i-1} \wedge dx_{i+1} \wedge \cdots \wedge dx_n.$$

Then condition (1.4) means that $\widehat{d}\omega_S = 0$, with the current S being trivial if and only if the form ω_S is exact: $\omega_S = \widehat{d}\omega'$.

DEFINITION 2.1. *A conservation law for the equation $\mathcal{E}$ is an $(n-1)$-cohomology class of the horizontal de Rham complex on $\mathcal{E}^\infty$.*

Thus, $\bar{H}^{n-1}(\mathcal{E})$ is the group of conservation laws for the equation $\mathcal{E}$. Sometimes elements of the groups $\bar{H}^q(\mathcal{E})$ for $q < n-1$ are also said to be conservation laws, although for the majority of equations of mathematical physics they are of purely topological nature[4]. The group $\bar{H}^n(\mathcal{E})$ should be interpreted as the set of

---

[4]One encounters physically relevant "conservation laws" belonging to $\bar{H}^q(\mathcal{E})$, $q < n-1$, in gauge theories (see, e.g., [**10, 11, 42**]).

Lagrangians (see §§2.5 and 4.1) in variational problems constrained by the equation $\mathcal{E}$[5].

To complete the description of the term $E_0(\mathcal{E})$, note that from the obvious decomposition $\Lambda^1(\mathcal{E}) = \mathcal{C}\Lambda^1(\mathcal{E}) \oplus \Lambda_0^1(\mathcal{E})$ it follows that $E_0^{p,q}(\mathcal{E}) = \mathcal{C}^p\Lambda^p(\mathcal{E}) \otimes \Lambda_0^q(\mathcal{E})$, with the differential $d_0^{p,q}$ being the composition

$$\mathcal{C}^p\Lambda^p(\mathcal{E}) \otimes \Lambda_0^q(\mathcal{E}) \xrightarrow{d} \mathcal{C}^{p+1}\Lambda^{p+1}(\mathcal{E}) \otimes \Lambda_0^q(\mathcal{E}) \oplus \mathcal{C}^p\Lambda^p(\mathcal{E}) \otimes \Lambda_0^{q+1}(\mathcal{E})$$

$$\xrightarrow{\alpha} \mathcal{C}^p\Lambda^p(\mathcal{E}) \otimes \Lambda_0^{q+1}(\mathcal{E}),$$

where α is the projection onto the second summand. In particular, all nontrivial terms $E_0^{p,q}(\mathcal{E})$ are situated in the domain $0 \leq q \leq n$.

EXERCISE 2.1. Using the direct decomposition $d = \widehat{d} + U_1$ and the properties of the operator U_1 (see §2.1 of Ch. 4), show that

$$d(\mathcal{C}^p\Lambda^p(\mathcal{E}) \otimes \Lambda_0^q(\mathcal{E})) \subset \left(\mathcal{C}^{p+1}\Lambda^{p+1}(\mathcal{E}) \otimes \Lambda_0^q(\mathcal{E})\right) \oplus \left(\mathcal{C}^p\Lambda^p(\mathcal{E}) \otimes \Lambda_0^{q+1}(\mathcal{E})\right).$$

2.3. The term E_1: preparatory results. Now we prove some facts needed to describe the term E_1 of the $\mathcal{C}$-spectral sequence.

Let us begin with the following definition. An $\mathcal{F}(\mathcal{E})$-module P is called *horizontal* if it is isomorphic to the module of sections $\mathcal{F}(\mathcal{E}, \xi)$ for a finite-dimensional vector bundle $\xi \colon N_\xi \to M$. For instance, as shown in §1.6 of Ch. 4, $\Lambda_0^q(\mathcal{E})$ is a horizontal module.

EXERCISE 2.2. Show that $P = \Gamma(\xi) \otimes_{C^\infty(M)} \mathcal{F}(\mathcal{E})$.

Note also that a horizontal module P is filtered by its submodules $P_k = \mathcal{F}_k(\mathcal{E}, \xi) = \Gamma(\xi) \otimes_{C^\infty(M)} \mathcal{F}_k(\mathcal{E})$.

Consider now two vector bundles $\xi \colon N_\xi \to M$ and $\eta \colon N_\eta \to M$, and introduce the notation $P = \mathcal{F}(\mathcal{E}, \xi)$ and $Q = \mathcal{F}(\mathcal{E}, \eta)$. A differential operator $\Delta \colon P \to Q$ is called *$\mathcal{C}$-differential* (or *horizontal*) if it is of the form $\Delta = f_1\widehat{\square}_1 + f_2\widehat{\square}_2 + \cdots + f_r\widehat{\square}_r$, where $\widehat{\square}_i$ are the liftings (see Remark 1.1 of Ch. 4) of some differential operators $\square_i \colon \Gamma(\xi) \to \Gamma(\eta)$, $f_i \in \mathcal{F}(\mathcal{E})$, $i = 1, \ldots, r$. In particular, operators of the form ℓ_φ (universal linearizations) are $\mathcal{C}$-differential. We shall denote the $\mathcal{F}(\mathcal{E})$-module of all $\mathcal{C}$-differential operators acting from P to Q by $\mathcal{C}\text{Diff}(P, Q)$. For its submodules consisting of operators of order $\leq k$, we shall use the notation $\mathcal{C}\text{Diff}_k(P, Q)$.

EXERCISE 2.3. Check that in local coordinates scalar $\mathcal{C}$-differential operators of order $\leq k$ are written as follows:

$$\Delta = \sum_{|\sigma|=0}^{k} a_\sigma \bar{D}_\sigma,$$

where $\bar{D}_\sigma = \bar{D}_1^{\sigma_1} \circ \cdots \circ \bar{D}_n^{\sigma_n}$, $\sigma = (\sigma_1, \ldots, \sigma_n)$, $a_\sigma \in \mathcal{F}(\mathcal{E})$.

EXERCISE 2.4. Show that $\mathcal{C}\text{Diff}(P, Q)$ is a horizontal module.

Let $P_1, P_2, \ldots, P_l, Q$ be horizontal modules. Set

$$\mathcal{C}\text{Diff}(P_1, P_2, \ldots, P_l; Q) = \mathcal{C}\text{Diff}(P_1, \mathcal{C}\text{Diff}(P_2, \ldots, \mathcal{C}\text{Diff}(P_l, Q) \ldots))$$

---

[5]There are a number of nonequivalent formulations of the Lagrangian formalism with constrains. We consider here the case when the sections subject to variation satisfy the "constraint equation" $\mathcal{E}$. This implies, in particular, that the Euler–Lagrange equations are determined by the restriction of the Lagrangian to $\mathcal{E}^\infty$.

and consider the complex

$$0 \longrightarrow \mathcal{C}\mathrm{Diff}(P_1, P_2, \ldots, P_l; \mathcal{F}(\mathcal{E})) \xrightarrow{w} \mathcal{C}\mathrm{Diff}(P_1, P_2, \ldots, P_l; \Lambda_0^1(\mathcal{E})) \xrightarrow{w} \cdots$$
$$\xrightarrow{w} \mathcal{C}\mathrm{Diff}(P_1, P_2, \ldots, P_l; \Lambda_0^n(\mathcal{E})) \longrightarrow 0, \quad (2.2)$$

where $w(\nabla) = \widehat{d} \circ \nabla$, $\nabla \in \mathcal{C}\mathrm{Diff}(P_1, P_2, \ldots, P_l; \Lambda_0^q(\mathcal{E}))$.

THEOREM 2.1. *The cohomology of the complex* (2.2) *is equal to zero at the terms* $\mathcal{C}\mathrm{Diff}(P_1, \ldots, P_l; \Lambda_0^q(\mathcal{E}))$ *for* $q < n$, *while at* $\mathcal{C}\mathrm{Diff}(P_1, \ldots, P_l; \Lambda_0^n(\mathcal{E}))$ *its cohomology is* $\mathcal{C}\mathrm{Diff}(P_1, \ldots, P_{l-1}; \widehat{P}_l)$, *where* $\widehat{P}_l = \mathrm{Hom}_{\mathcal{F}(\mathcal{E})}(P_l, \Lambda_0^n(\mathcal{E}))$.

PROOF. Let $\mathcal{C}\mathrm{Diff}_k(P_1, P_2, \ldots, P_l; Q) \subset \mathcal{C}\mathrm{Diff}(P_1, P_2, \ldots, P_l; Q)$ be the module consisting of operators ∇ such that the operator $\nabla(p_1, p_2, \ldots, p_{l-1}) \colon P_l \to Q$ is of order $\leq k$ for all $p_1 \in P_1$, $p_2 \in P_2$, $\ldots$, $p_{l-1} \in P_{l-1}$ (here and subsequently, $\nabla(p_1, p_2, \ldots, p_{l-1})$ stands for $\nabla(p_1)(p_2) \ldots (p_{l-1})$). Thus,

$$\mathcal{C}\mathrm{Diff}_k(P_1, \ldots, P_l; Q) = \mathcal{C}\mathrm{Diff}(P_1, \ldots, P_{l-1}; \mathcal{C}\mathrm{Diff}_k(P_l, Q))$$

and the embeddings $\mathcal{C}\mathrm{Diff}_k(P_l, Q) \subset \mathcal{C}\mathrm{Diff}_{k+1}(P_l, Q)$ induce the homomorphisms of complexes ($\bullet$ denotes "$P_1, \ldots, P_l$")

$$\begin{array}{ccccc}
\cdots \longrightarrow \mathcal{C}\mathrm{Diff}_{k-n+i}(\bullet; \Lambda_0^i(\mathcal{E})) & \xrightarrow{w} & \mathcal{C}\mathrm{Diff}_{k-n+i+1}(\bullet; \Lambda_0^{i+1}(\mathcal{E})) & \longrightarrow \cdots \\
\uparrow & & \uparrow & & (2.3) \\
\cdots \longrightarrow \mathcal{C}\mathrm{Diff}_{k-n+i-1}(\bullet; \Lambda_0^i(\mathcal{E})) & \xrightarrow{w} & \mathcal{C}\mathrm{Diff}_{k-n+i}(\bullet; \Lambda_0^{i+1}(\mathcal{E})) & \longrightarrow \cdots
\end{array}$$

Hence, there exists the quotient complex

$$0 \longrightarrow \mathcal{C}\mathrm{Diff}(\bullet; S_{k-n}(P_l)) \xrightarrow{\delta} \mathcal{C}\mathrm{Diff}(\bullet; S_{k-n+1}(P_l) \otimes \Lambda_0^1(\mathcal{E})) \xrightarrow{\delta} \cdots$$
$$\xrightarrow{\delta} \mathcal{C}\mathrm{Diff}(\bullet; S_k(P_l) \otimes \Lambda_0^n(\mathcal{E})) \longrightarrow 0, \quad (2.4)$$

where $\bullet$ stands for "$P_1, \ldots, P_{l-1}$" and

$$S_r(P) = \mathcal{C}\mathrm{Diff}_r(P, \mathcal{F}(\mathcal{E}))/\mathcal{C}\mathrm{Diff}_{r-1}(P, \mathcal{F}(\mathcal{E})).$$

The proof will be accomplished if we prove that the complex (2.4) is exact for $k > 0$. Indeed, the cohomologies of complexes (2.3) for $k > 0$ coincide in this case and, hence, the cohomology of (2.2) is equal to the cohomology of the complex

$$0 \longrightarrow \mathcal{C}\mathrm{Diff}_0(P_1, \ldots, P_l; \Lambda_0^n(\mathcal{E})) \longrightarrow 0.$$

The differentials δ of the complex (2.4) are $\mathcal{F}(\mathcal{E})$-linear. Therefore, it suffices to check its exactness at every point $\theta \in \mathcal{E}$. Since the functor $\mathcal{C}\mathrm{Diff}(P, \cdot)$ is exact, we can take $l = 1$. Thus, we need to prove that the complexes

$$0 \longrightarrow S_{k-n}(P)_\theta \longrightarrow S_{k-n+1}(P)_\theta \otimes_\mathbb{R} \Lambda_0^1(\mathcal{E})_\theta \longrightarrow \cdots \longrightarrow S_k(P)_\theta \otimes_\mathbb{R} \Lambda_0^n(\mathcal{E})_\theta \longrightarrow 0$$

are acyclic for $k > 0$.

Let $\{e_\xi\}$ be a basis of $(P_\theta)^*$. Pick the elements $e_\xi \otimes D_\sigma|_\theta$, $|\sigma| = i$, as a basis of the space $S_i(P)_\theta$ and the elements $dx_{i_1} \wedge \cdots \wedge dx_{i_q}|_\theta$, $1 \leq i_1 < \cdots < i_q \leq n$, as a basis of $\Lambda_0^q(\mathcal{E})_\theta$. Then the differential δ takes the form

$$\delta(e_\xi \otimes D_\sigma \otimes dx_{i_1} \wedge \cdots \wedge dx_{i_q}|_\theta) = \sum_{i=1}^n e_\xi \otimes D_{\sigma+1_i} \otimes dx_i \wedge dx_{i_1} \wedge \cdots \wedge dx_{i_q}|_\theta.$$

Consequently, (2.4) is the Koszul complex of the polynomial algebra (see, e.g., [**15**]) and, consequently, is exact for $k > 0$. □

EXERCISE 2.5. Prove that the embedding
$$i_l \colon \mathcal{C}\mathrm{Diff}(P_1, \ldots, P_{l-1}; \widehat{P_l}) \longrightarrow \mathcal{C}\mathrm{Diff}(P_1, \ldots, P_l; \Lambda_0^n(\mathcal{E}))$$
induced by the embedding $\widehat{P_l} = \mathrm{Hom}_{\mathcal{F}(\mathcal{E})}(P_l, \Lambda_0^n(\mathcal{E})) \to \mathcal{C}\mathrm{Diff}(P_l, \Lambda_0^n(\mathcal{E}))$ splits the natural projection
$$\mu_l \colon \mathcal{C}\mathrm{Diff}(P_1, \ldots, P_l; \Lambda_0^n(\mathcal{E})) \longrightarrow \mathcal{C}\mathrm{Diff}(P_1, \ldots, P_l; \Lambda_0^n(\mathcal{E}))/\operatorname{im} w$$
$$= \mathcal{C}\mathrm{Diff}(P_1, \ldots, P_{l-1}; \widehat{P_l}).$$

Thus the module $\mathcal{C}\mathrm{Diff}(P_1, \ldots, P_{l-1}; \widehat{P_l}) = \operatorname{im} i_l$ is a direct summand in the module $\mathcal{C}\mathrm{Diff}(P_1, \ldots, P_l; \Lambda_0^n(\mathcal{E}))$.

Take a $\mathcal{C}$-differential operator $\Delta \colon P \to Q$. It induces the cochain map of complexes (2.2)

$$\begin{array}{ccccccccc}
0 & \longrightarrow & \mathcal{C}\mathrm{Diff}(P, \mathcal{F}(\mathcal{E})) & \xrightarrow{w} & \mathcal{C}\mathrm{Diff}(P, \Lambda_0^1(\mathcal{E})) & \xrightarrow{w} & \cdots & \xrightarrow{w} & \mathcal{C}\mathrm{Diff}(P, \Lambda_0^n(\mathcal{E})) \longrightarrow 0 \\
& & \Delta' \uparrow & & \Delta' \uparrow & & & & \Delta' \uparrow \\
0 & \longrightarrow & \mathcal{C}\mathrm{Diff}(Q, \mathcal{F}(\mathcal{E})) & \xrightarrow{w} & \mathcal{C}\mathrm{Diff}(Q, \Lambda_0^1(\mathcal{E})) & \xrightarrow{w} & \cdots & \xrightarrow{w} & \mathcal{C}\mathrm{Diff}(Q, \Lambda_0^n(\mathcal{E})) \longrightarrow 0
\end{array}$$

where $\Delta'(\nabla) = \nabla \circ \Delta$ for any $\nabla \in \mathcal{C}\mathrm{Diff}(Q, \Lambda_0^q(\mathcal{E}))$. Theorem 2.1 implies that the map Δ' gives rise to the cohomology map $\Delta^* \colon \widehat{Q} \to \widehat{P}$. The operator Δ^* is called *adjoint* to the operator Δ.

EXERCISE 2.6. Show that:

1. If $\Delta = \sum_\sigma a_\sigma D_\sigma$ is a scalar $\mathcal{C}$-differential operator, then
$$\Delta^* = \sum_\sigma (-1)^{|\sigma|} D_\sigma \circ a_\sigma.$$

2. If $\Delta = \|\Delta_{ij}\|$ is a matrix $\mathcal{C}$-differential operator, then $\Delta^* = \|\Delta_{ji}^*\|$.

In particular, this exercise implies that Δ^* is a $\mathcal{C}$-differential operator of the same order as Δ.

EXERCISE 2.7. Prove that for all $\Delta \in \mathcal{C}\mathrm{Diff}(P, Q)$, $p \in P$, and $\widehat{q} \in \widehat{Q}$ there exists a form $\omega_{p, \widehat{q}}(\Delta) \in \Lambda_0^{n-1}(\mathcal{E})$ such that
$$\widehat{q}(\Delta(p)) - (\Delta^*(\widehat{q}))(p) = \widehat{d} \omega_{p, \widehat{q}}(\Delta). \tag{2.5}$$

EXERCISE 2.8. Check that for any two $\mathcal{C}$-differential operators $\Delta_1 \colon P \to Q$ and $\Delta_2 \colon Q \to R$ one has $(\Delta_2 \circ \Delta_1)^* = \Delta_1^* \circ \Delta_2^*$.

EXERCISE 2.9. Consider $X \in \mathcal{C}\mathrm{Diff}_1(\mathcal{F}(\mathcal{E}), \Lambda_0^n(\mathcal{E}))$. Show that $X + X^* \in \Lambda_0^n(\mathcal{E})$, i.e., $X + X^*$ is an operator of zero order.

EXERCISE 2.10. Show that the natural projection
$$\mu_l \colon \mathcal{C}\mathrm{Diff}(P_1, \ldots, P_l; \Lambda_0^n(\mathcal{E})) \longrightarrow \mathcal{C}\mathrm{Diff}(P_1, \ldots, P_l; \Lambda_0^n(\mathcal{E}))/\operatorname{im} w$$
$$= \mathcal{C}\mathrm{Diff}(P_1, \ldots, P_{l-1}; \widehat{P_l}).$$
is given by the formula
$$\mu_l(\nabla)(p_1, \ldots, p_{l-1}) = (\nabla(p_1, \ldots, p_{l-1}))^*(1).$$

Let $\mathcal{C}\mathrm{Diff}_{(l)}(P;Q)$ denote $\mathcal{C}\mathrm{Diff}(\underbrace{P,\ldots,P}_{l\ \text{times}};Q)$, and let $\mathcal{C}\mathrm{Diff}^{\mathrm{alt}}{}_{(l)}(P;Q)$ denote the submodule of $\mathcal{C}\mathrm{Diff}_{(l)}(P;Q)$ consisting of skew-symmetric operators, i.e., of operators $\nabla \in \mathcal{C}\mathrm{Diff}_{(l)}(P;Q)$ such that

$$\nabla(p_1,\ldots,p_i,p_{i+1},\ldots,p_l) = -\nabla(p_1,\ldots,p_{i+1},p_i,\ldots,p_l)$$

for all $p_1,\ldots,p_l \in P$, $i = 1,\ldots,l-1$.

Consider the complex (2.2) for $P_1 = P_2 = \cdots = P_l = P$:

$$0 \longrightarrow \mathcal{C}\mathrm{Diff}_{(l)}(P;\mathcal{F}(\mathcal{E})) \xrightarrow{w} \mathcal{C}\mathrm{Diff}_{(l)}(P;\Lambda^1_0(\mathcal{E})) \xrightarrow{w} \cdots$$
$$\xrightarrow{w} \mathcal{C}\mathrm{Diff}_{(l)}(P;\Lambda^n_0(\mathcal{E})) \longrightarrow 0. \quad (2.6)$$

The permutation group S_l acts in this complex as follows:

$$\tau(\nabla)(p_1,\ldots,p_l) = \nabla(p_{\tau(1)},\ldots,p_{\tau(l)}), \qquad \tau \in S_l.$$

It is obvious that this action commutes with the differential w. Hence from Theorem 2.1 it follows that the skew-symmetric part of complex (2.6), i.e., the subcomplex

$$0 \longrightarrow \mathcal{C}\mathrm{Diff}^{\mathrm{alt}}{}_{(l)}(P;\mathcal{F}(\mathcal{E}))) \xrightarrow{w} \mathcal{C}\mathrm{Diff}^{\mathrm{alt}}{}_{(l)}(P;\Lambda^1_0(\mathcal{E})) \xrightarrow{w} \cdots$$
$$\xrightarrow{w} \mathcal{C}\mathrm{Diff}^{\mathrm{alt}}{}_{(l)}(P;\Lambda^n_0(\mathcal{E})) \longrightarrow 0, \quad (2.7)$$

is exact in all degrees different from n, while its n-th cohomology group is isomorphic to a submodule of $\mathcal{C}\mathrm{Diff}_{(l-1)}(P;\widehat{P})$, which will be denoted by $K_l(P)$.

An explicit description of the module $K_l(P)$ can be obtained in the following manner. First, observe that the embedding (see Exercise 2.5)

$$i_l \colon \mathcal{C}\mathrm{Diff}(P_1,\ldots,P_{l-1};\widehat{P_l}) \longrightarrow \mathcal{C}\mathrm{Diff}(P_1,\ldots,P_l;\Lambda^n_0(\mathcal{E}))$$

commutes with the action of the subgroup $S_{l-1} \subset S_l$ which preserves the l-th index. For this reason

$$K_l(P) \subset \mathcal{C}\mathrm{Diff}^{\mathrm{alt}}{}_{(l-1)}(P;\widehat{P}).$$

Consider the transposition $\tau \in S_l$ interchanging the j-th and the l-th indices, $j < l$, and let us describe its action on an operator $\Delta \in \mathcal{C}\mathrm{Diff}_{(l-1)}(P,\widehat{P})$. Let us fix the elements $p_1,\ldots,p_{j-1},p_{j+1},\ldots,p_{l-1} \in P$ and consider the operator

$$\square(p) = \Delta(p_1,\ldots,p_{j-1},p,p_{j+1},\ldots,p_{l-1}).$$

From (2.5) it follows that

$$\square(p)(p') - \square^*(p')(p) \in \operatorname{im}\widehat{d}$$

for all $p,p' \in P$. Thus the operator $\tau(\Delta)$ has the form $\tau(\Delta)(p_1,\ldots,p_{l-1}) = \square^*(p_j)$. This proves the following.

THEOREM 2.2. *Complex (2.7) is acyclic in the terms* $\mathcal{C}\mathrm{Diff}^{\mathrm{alt}}{}_{(l)}(P;\Lambda^q_0(\mathcal{E}))$ *for* $q < n$. *The cohomology group in the term* $\mathcal{C}\mathrm{Diff}^{\mathrm{alt}}{}_{(l)}(P;\Lambda^n_0(\mathcal{E}))$ *is isomorphic to the module* $K_l(P) \subset \mathcal{C}\mathrm{Diff}^{\mathrm{alt}}{}_{(l-1)}(P;\widehat{P})$ *consisting of all operators* ∇ *satisfying the condition*

$$(\nabla(p_1,\ldots,p_{l-2}))^* = -\nabla(p_1,\ldots,p_{l-2})$$

for all $p_1,\ldots,p_{l-2} \in P$.

2.4. Generalizations. We shall need to generalize Theorems 2.1 and 2.2. To do this, we use the following result:

EXERCISE 2.11. To each $\mathcal{C}$-differential operator $\Delta\colon P_1 \to P_2$ assign the family of operators $\Delta(p_1, p_2^*) \in \mathcal{C}\mathrm{Diff}(\mathcal{F}(\mathcal{E}), \mathcal{F}(\mathcal{E}))$ given by

$$\Delta(p_1, p_2^*)(f) = p_2^*(\Delta(fp_1)), \qquad p_1 \in P_1, \quad p_2^* \in \mathrm{Hom}_{\mathcal{F}(\mathcal{E})}(P_2, \mathcal{F}(\mathcal{E})).$$

Prove that the family $\Delta(p_1, p_2^*)$ uniquely determines the operator Δ and that for any family $\Delta[p_1, p_2^*] \in \mathcal{C}\mathrm{Diff}(\mathcal{F}(\mathcal{E}), \mathcal{F}(\mathcal{E}))$, $p_1 \in P_1$, $p_2^* \in \mathrm{Hom}_{\mathcal{F}(\mathcal{E})}(P_2, \mathcal{F}(\mathcal{E}))$, satisfying

$$\Delta\big[p_1, \sum_i f_i p_2^{*i}\big] = \sum_i f_i \Delta[p_1, p_2^{*i}], \qquad \Delta\big[\sum_i f_i p_1^i, p_2^*\big] = \sum_i \Delta[p_1^i, p_2^*] \circ f_i,$$

there exists an operator $\Delta \in \mathcal{C}\mathrm{Diff}(P_1, P_2)$ such that $\Delta[p_1, p_2^*] = \Delta(p_1, p_2^*)$.

Let Q be a left module over the ring $\mathcal{C}\mathrm{Diff}(\mathcal{F}(\mathcal{E}), \mathcal{F}(\mathcal{E}))$. Since

$$\mathcal{F}(\mathcal{E}) = \mathcal{C}\mathrm{Diff}_0(\mathcal{F}(\mathcal{E}), \mathcal{F}(\mathcal{E})) \subset \mathcal{C}\mathrm{Diff}(\mathcal{F}(\mathcal{E}), \mathcal{F}(\mathcal{E})),$$

the module Q is an $\mathcal{F}(\mathcal{E})$-module as well. Suppose that it is a horizontal $\mathcal{F}(\mathcal{E})$-module. By Exercise 2.11, for any operator $\Delta \in \mathcal{C}\mathrm{Diff}(P_1, P_2)$ there exists a unique operator $\Delta_Q \in \mathcal{C}\mathrm{Diff}(P_1 \otimes Q, P_2 \otimes Q)$ such that

$$\Delta_Q(p_1 \otimes q, p_2^* \otimes q^*) = q^*(\Delta(p_1, p_2^*)q).$$

Note that operators Δ and Δ_Q are of the same order.

It is easy to see that $(\Delta_1 \circ \Delta_2)_Q = (\Delta_1)_Q \circ (\Delta_2)_Q$. Thus any complex of $\mathcal{C}$-differential operators

$$\cdots \longrightarrow P_k \xrightarrow{\Delta_k} P_{k+1} \longrightarrow \cdots$$

can be multiplied by Q:

$$\cdots \longrightarrow P_k \otimes Q \xrightarrow{(\Delta_k)_Q} P_{k+1} \otimes Q \longrightarrow \cdots.$$

In particular, multiplying the complex (2.2) by Q, we get

$$0 \longrightarrow \mathcal{C}\mathrm{Diff}(P_1, \dots, P_l; \mathcal{F}(\mathcal{E})) \otimes Q \longrightarrow \mathcal{C}\mathrm{Diff}(P_1, \dots, P_l; \Lambda_0^1(\mathcal{E})) \otimes Q \longrightarrow \cdots$$
$$\longrightarrow \mathcal{C}\mathrm{Diff}(P_1, \dots, P_l; \Lambda_0^n(\mathcal{E})) \otimes Q \longrightarrow 0. \quad (2.8)$$

Literally repeating the proof of Theorem 2.1, we obtain the following.

THEOREM 2.3. *The cohomology of the complex* (2.8) *in the term*

$$\mathcal{C}\mathrm{Diff}(P_1, P_2, \dots, P_l; \Lambda_0^i(\mathcal{E})) \otimes Q$$

has the form

$$H^i = \begin{cases} 0 & \text{for } i < n, \\ \mathcal{C}\mathrm{Diff}(P_1, P_2, \dots, \widehat{P_l}) \otimes Q & \text{for } i = n. \end{cases}$$

Theorem 2.2 can be generalized in the same way.

EXERCISE 2.12. Show that the set $\mathcal{C}^p\Lambda^p(\mathcal{E})$ can be supplied with a unique left $\mathcal{C}\mathrm{Diff}(\mathcal{F}(\mathcal{E}), \mathcal{F}(\mathcal{E}))$-module structure such that

$$f \cdot \omega = f\omega, \qquad f \in \mathcal{F}(\mathcal{E}),$$
$$X \cdot \omega = L_X(\omega), \qquad X \in \mathcal{CD}(\mathcal{E}).$$

2.5. The term E_1 for $J^\infty(\pi)$. Now consider the term E_1 of the $\mathcal{C}$-spectral sequence for the "empty equation", i.e., for the case $\mathcal{E}^\infty = J^\infty(\pi)$. We shall follow the notation of §2.1, but abbreviate "$J^0(\pi)$" to "π", e.g., $\mathcal{C}^k\Lambda^q(\pi)$ stands for $\mathcal{C}^k\Lambda^q(J^0(\pi))$, etc.

By definition, the first term E_1 of a spectral sequence is the cohomology of its zero term E_0. So the zero column of E_1 consists of the horizontal cohomology groups of the space $J^\infty(\pi)$: $E_1^{0,q}(\pi) = \bar{H}^q(\pi)$. To describe the terms $E_1^{p,q}(\pi)$ for $p > 0$, we need to compute the cohomology of complexes (see §2.3)

$$0 \longrightarrow \mathcal{C}^p\Lambda^p(\pi) \xrightarrow{\hat{d}} \mathcal{C}^p\Lambda^p(\pi) \otimes \Lambda_0^1(\pi) \xrightarrow{\hat{d}} \cdots \xrightarrow{\hat{d}} \mathcal{C}^p\Lambda^p(\pi) \otimes \Lambda_0^1(\pi) \longrightarrow 0. \quad (2.9)$$

Let $\varkappa(\pi) = \mathcal{F}(\pi,\pi)$ be the $\mathcal{F}(\pi)$-module of evolutionary derivations. To each form $\omega \in \mathcal{C}^p\Lambda^p(\pi)$, we assign the operator $\nabla_\omega \in \mathcal{C}\mathrm{Diff}^{\mathrm{alt}}{}_{(p)}(\varkappa(\pi); \mathcal{F}(\pi))$ by setting

$$\nabla_\omega(\chi_1, \ldots, \chi_p) = \partial_{\chi_p} \,\lrcorner\, (\ldots (\partial_{\chi_1} \,\lrcorner\, \omega) \ldots), \quad (2.10)$$

where $\chi_i \in \varkappa(\pi)$.

LEMMA 2.4. *The operation* (2.10) *establishes an isomorphism of the $\mathcal{F}(\mathcal{E})$-modules $\mathcal{C}^p\Lambda^p(\pi)$ and $\mathcal{C}\mathrm{Diff}^{\mathrm{alt}}{}_{(p)}(\varkappa(\pi); \mathcal{F}(\pi))$.*

PROOF. Let us construct the map inverse to the given map $\omega \mapsto \nabla_\omega$. Take $\nabla \in \mathcal{C}\mathrm{Diff}^{\mathrm{alt}}{}_{(p)}(\varkappa(\pi); \mathcal{F}(\mathcal{E}))$. Any vertical tangent vector ξ at the point $\theta \in J^\infty(\pi)$ may be realized in the form $\partial_\chi|_\theta$ for some χ (this can be seen, e.g., from the coordinate expression for ∂_χ (see equality (2.15) of Ch. 4)). Define the form $\omega_\nabla \in \mathcal{C}^p\Lambda^p(\pi)$ by putting

$$\omega_\nabla|_\theta(\xi_1, \ldots, \xi_p) = \nabla(\chi_1, \ldots, \chi_p)(\theta),$$

where $\xi_i = \partial_{\chi_i}|_\theta$. It is clear that the map $\nabla \mapsto \omega_\nabla$ is the desired inverse map. Other details of the proof are left to the reader as an exercise. $\square$

Thus, we can rewrite the complex (2.9) in the form

$$0 \longrightarrow \mathcal{C}\mathrm{Diff}^{\mathrm{alt}}{}_{(p)}(\varkappa(\pi); \mathcal{F}(\pi)) \xrightarrow{w} \mathcal{C}\mathrm{Diff}^{\mathrm{alt}}{}_{(p)}(\varkappa(\pi); \Lambda_0^1(\pi)) \xrightarrow{w} \cdots$$
$$\xrightarrow{w} \mathcal{C}\mathrm{Diff}^{\mathrm{alt}}{}_{(p)}(\varkappa(\pi); \Lambda_0^n(\pi)) \longrightarrow 0. \quad (2.11)$$

EXERCISE 2.13. Prove that the differential w corresponds to the differential $\hat{d}$ in (2.9).

Computing the cohomology of the complex (2.11) by Theorem 2.2, we get the following result:

THEOREM 2.5. *Let π be a smooth vector bundle over a manifold M, $\dim M = n$. Then:*

$$E_1^{0,q}(\pi) = \bar{H}^q(\pi) \qquad \text{for all } q \geq 0;$$
$$E_1^{p,q}(\pi) = 0 \qquad \text{for } p > 0,\ q \neq n;$$
$$E_1^{p,n}(\pi) = K_p(\varkappa(\pi)) \qquad \text{for } p > 0.$$

Since the $\mathcal{C}$-spectral sequence converges to the de Rham cohomology of the manifold $J^\infty(\pi)$ in the case under consideration, Theorem 2.5 has the following obvious corollary:

COROLLARY 2.6. *Under the assumptions of Theorem 2.5, we have*

1. $E_r^{p,q}(\pi) = 0$, for all $1 \leq r \leq \infty$ if $p > 0$, $q \neq n$, or $p = 0$, $q > n$.
2. $E_1^{0,q}(\pi) = E_\infty^{0,q}(\pi) = H^q(J^\infty(\pi)) = H^q(J^0(\pi))$ for $q < n$.
3. $E_2^{p,n}(\pi) = E_\infty^{p,n}(\pi) = H^{p+n}(J^\infty(\pi)) = H^{p+n}(J^0(\pi))$ for $p \geq 0$.

EXERCISE 2.14. Prove the equality

$$H^q(J^\infty(\pi)) = H^q(J^0(\pi)).$$

Let us now consider the differentials $d_1^{p,n}$.

EXERCISE 2.15. a. Show that the operator

$$E_1^{0,n}(\pi) = \bar{H}^n(\pi) \xrightarrow{d_1^{0,n}} E_1^{1,n}(\pi) = \widehat{\varkappa}(\pi)$$

is given by the formula

$$d_1^{0,n}([\omega]) = \ell_\omega^*(1),$$

where $\omega \in \Lambda_0^n(\pi)$ and $[\omega]$ is the horizontal cohomology class of ω. (Note that the expression ℓ_ω makes sense, because ω is a horizontal n-form, i.e., a nonlinear differential operator acting from $\Gamma(\pi)$ to $\Lambda^n(M)$.)

b. Write down the coordinate expression for the operator $d_1^{0,n}$ and check that this operator is the standard Euler operator (i.e., the operator that takes each Lagrangian to the corresponding Euler–Lagrange equation (see §4.1 below)).

Now let us describe the differentials $d_1^{p,n}$, $p > 0$.
Consider an operator $\nabla \in K_p(\varkappa(\pi))$. Define $\square \in \mathcal{C}\text{Diff}_{(p+1)}(\varkappa(\pi), \Lambda_0^n(\pi))$ by putting

$$\square(\chi_1, \ldots, \chi_{p+1}) = \sum_{i=1}^{p+1} (-1)^{i+1} \partial_{\chi_i}(\nabla(\chi_1, \ldots, \widehat{\chi}_i, \ldots, \chi_{p+1}))$$
$$+ \sum_{1 \leq i < j \leq p+1} (-1)^{i+j} \nabla(\{\chi_i, \chi_j\}, \chi_1, \ldots, \widehat{\chi}_i, \ldots, \widehat{\chi}_j, \ldots, \chi_{p+1}). \quad (2.12)$$

EXERCISE 2.16. Prove that

$$d_1^{p,n}(\nabla) = \mu_{p+1}(\square).$$

REMARK 2.1. It is needless to say that this fact follows from the standard formula for the exterior differential. However, it is necessary to prove that one may use this formula even though ∇, as an element of $\mathcal{C}\text{Diff}_{(p)}(\varkappa(\pi); \Lambda_0^n(\pi))$, is not skew-symmetric.

From (2.12) we get

$$\Box(\chi_1,\ldots,\chi_{p+1}) = \sum_{i=1}^{p}(-1)^{i+1}\partial_{\chi_i}(\nabla(\chi_1,\ldots,\widehat{\chi}_i,\ldots,\chi_p))(\chi_{p+1})$$
$$+\sum_{i=1}^{p}(-1)^{i+1}\nabla(\chi_1,\ldots,\widehat{\chi}_i,\ldots,\chi_p,\partial_{\chi_i}(\chi_{p+1}))$$
$$+(-1)^{p}\partial_{\chi_{p+1}}(\nabla(\chi_1,\ldots,\chi_p))$$
$$+\sum_{1\le i<j\le p}(-1)^{i+j}\nabla(\{\chi_i,\chi_j\},\chi_1,\ldots,\widehat{\chi}_i,\ldots,\widehat{\chi}_j,\ldots,\chi_{p+1})$$
$$+\sum_{i=1}^{p}(-1)^{i+p+1}\nabla(\{\chi_i,\chi_{p+1}\},\chi_1,\ldots,\widehat{\chi}_i,\ldots,\chi_p)$$
$$=\sum_{i=1}^{p}(-1)^{i+1}\partial_{\chi_i}(\nabla(\chi_1,\ldots,\widehat{\chi}_i,\ldots,\chi_p))(\chi_{p+1})$$
$$+\sum_{1\le i<j\le p}(-1)^{i+j}\nabla(\{\chi_i,\chi_j\},\chi_1,\ldots,\widehat{\chi}_i,\ldots,\widehat{\chi}_j,\ldots,\chi_{p+1})$$
$$+\sum_{i=1}^{p}(-1)^{i+1}\nabla(\chi_1,\ldots,\widehat{\chi}_i,\ldots,\chi_p,\ell_{\chi_i}(\chi_{p+1}))$$
$$+(-1)^{p}\ell_{\nabla(\chi_1,\ldots,\chi_p)}(\chi_{p+1}).$$

Therefore

$$d_1^{p,n}(\nabla)(\chi_1,\ldots,\chi_p) = \mu_{p+1}(\Box)(\chi_1,\ldots,\chi_p)$$
$$=\sum_{i=1}^{p}(-1)^{i+1}\partial_{\chi_i}(\nabla(\chi_1,\ldots,\widehat{\chi}_i,\ldots,\chi_p))$$
$$+\sum_{i<j}(-1)^{i+j}\nabla(\{\chi_i,\chi_j\},\chi_1,\ldots,\widehat{\chi}_i,\ldots,\widehat{\chi}_j,\ldots,\chi_p) \qquad (2.13)$$
$$+\sum_{i=1}^{p}(-1)^{i+1}\ell^*_{\chi_i}(\nabla(\chi_1,\ldots,\widehat{\chi}_i,\ldots,\chi_p)) + (-1)^{p}\ell^*_{\nabla(\chi_1,\ldots,\chi_p)}(1).$$

Using the obvious equality

$$\ell^*_{\psi(\varphi)}(1) = \ell^*_{\psi}(\varphi) + \ell^*_{\varphi}(\psi), \qquad \varphi \in \varkappa(\pi),\ \psi \in \widehat{\varkappa}(\pi)$$

(cf. Exercise 2.17), let us rewrite the last term of (2.13) in the following way:

$$(-1)^{p}\ell^*_{(\nabla(\chi_1,\ldots,\chi_p))}(1) = \frac{1}{p}\sum_{i=1}^{p}(-1)^{i}\ell^*_{(\nabla(\chi_1,\ldots,\widehat{\chi}_i,\ldots,\chi_p,\chi_i))}(1)$$
$$= \frac{1}{p}\sum_{i=1}^{p}(-1)^{i}(\ell^*_{\nabla(\chi_1,\ldots,\widehat{\chi}_i,\ldots,\chi_p)}(\chi_i) + \ell^*_{\chi_i}(\nabla(\chi_1,\ldots,\widehat{\chi}_i,\ldots,\chi_p))).$$

Finally, we obtain

$$(d_1^{p,n}(\nabla))(\chi_1,\ldots,\chi_p) = \sum_{i=1}^{p}(-1)^{i+1}\partial_{\chi_i}(\nabla(\chi_1,\ldots,\widehat{\chi_i},\ldots,\chi_p))$$

$$+ \sum_{i<j}(-1)^{i+j}\nabla(\{\chi_i,\chi_j\},\chi_1,\ldots,\widehat{\chi_i},\ldots,\widehat{\chi_j},\ldots,\chi_p)$$

$$+ \frac{1}{p}\sum_{i=1}^{p}(-1)^{i+1}((p-1)\ell_{\chi_i}^*(\nabla(\chi_1,\ldots,\widehat{\chi_i},\ldots,\chi_p)) - \ell_{\nabla(\chi_1,\ldots,\widehat{\chi_i},\ldots,\chi_p)}^*(\chi_i)).$$

In particular, for $p = 1$ we have $d_1^{1,n}(\psi)(\varphi) = \partial_{\varphi}(\psi) - \ell_\psi^*(\varphi) = \ell_\psi(\varphi) - \ell_\psi^*(\varphi)$, $\psi \in \widehat{\varkappa}(\pi)$, $\varphi \in \varkappa(\pi)$, that is,

$$d_1^{1,n}(\psi) = \ell_\psi - \ell_\psi^*. \tag{2.14}$$

Note that the horizontal de Rham complex on $J^\infty(\pi)$ can be combined with the complex $(E_1^{p,n}(\pi), d_1^{p,n})$ to give the complex

$$0 \longrightarrow \mathcal{F}(\mathcal{E}) \xrightarrow{\widehat{d}} \Lambda_0^1(\pi) \xrightarrow{\widehat{d}} \cdots \xrightarrow{\widehat{d}} \Lambda_0^n(\pi) \xrightarrow{\mathsf{E}} E_1^{1,n}(\pi) \xrightarrow{d_1^{1,n}} E_1^{2,n}(\pi) \xrightarrow{d_1^{2,n}} \cdots, \tag{2.15}$$

where E is the composition of the natural projection $\Lambda_0^n(\pi) \to \bar{H}^n(\pi)$ and the differential $d_1^{0,n}: \bar{H}^n(\pi) \to E_1^{1,n}(\pi)$[6].

In view of Corollary 2.6, the i-th cohomology group of this complex coincides with $H^i(J^0(\pi))$. The operator E is the Euler operator (see Exercise 2.15b). It takes each Lagrangian density $\omega \in \Lambda_0^n(\pi)$ to the left-hand side of the corresponding Euler–Lagrange equation $\mathsf{E}(\omega) = 0$. Thus the action functional

$$s \mapsto \int_M j_\infty(s)^*(\omega), \qquad s \in \Gamma(\pi),$$

is stationary on a section s if and only if $j_\infty(s)^*(\mathsf{E}(\omega)) = 0$.

The complex (2.15) is often called the (global) *variational complex* of the bundle π. If the cohomology of the space $J^0(\pi)$ is trivial then this complex is exact. This immediately implies a number of consequences. The three most important ones are:

1. $\ker \mathsf{E} = \operatorname{im} \widehat{d}$ (variationally trivial Lagrangians are total divergences).
2. $\widehat{d}\omega = 0$ if and only if ω is of the form $\omega = \widehat{d}\eta$, $\omega \in \Lambda_0^{n-1}(\pi)$ (null total divergences are total curls).
3. $\ell_\psi = \ell_\psi^*$ if and only if ψ is of the form $\psi = \mathsf{E}(\omega)$, $\psi \in \widehat{\varkappa}(\pi)$ (this solves the inverse problem of the calculus of variations).

EXERCISE 2.17. Let P be a horizontal module, $\Delta \in \mathcal{C}\mathrm{Diff}(P, \Lambda_0^n(\pi))$, and $p \in P$. Show that

$$\mathsf{E}(\Delta(p)) = \ell_p^*(\Delta^*(1)) + \ell_{\Delta^*(1)}^*(p).$$

Use this fact to prove the formula

$$\mathsf{E}(\partial_\varphi(\omega)) = \partial_\varphi(\mathsf{E}(\omega)) + \ell_\varphi^*(\mathsf{E}(\omega))$$

for all $\varphi \in \varkappa(\pi)$ and $\omega \in \Lambda_0^n(\pi)$.

---
[6]Below we use the same notation E for the operator $d_1^{0,n}: \bar{H}^n(\pi) \to E_1^{1,n}(\pi)$ as well.

2.6. The term E_1 in the general case. Let us compute the term $E_1(\mathcal{E})$ of the $\mathcal{C}$-spectral sequence of a differential equation $\mathcal{E}$.

Let $I(\mathcal{E}) \subset \mathcal{F}(\mathcal{E})$ be the ideal of the equation $\mathcal{E}$. Recall that if the equation $\mathcal{E} \subset J^k(\pi)$ has the form $\mathcal{E} = \{F = 0\}$, where $F \in P_k = \mathcal{F}_k(\pi, \xi)$, $\xi \colon N_\xi \to M$ is a vector bundle over M, and the projection $\mathcal{E} \to M$ is surjective, then

$$I(\mathcal{E}) = \mathcal{C}\mathrm{Diff}(P, \mathcal{F}(\pi))(F),$$

where $P = \mathcal{F}(\pi, \xi)$ (i.e., the components of the vector function F are differential generators of the ideal $I(\mathcal{E})$). We call such equations *regular*. As we mentioned in Ch. 4, the regularity condition is not restrictive, and equations encountered in mathematical physics, as a rule, satisfy it.

For regular equations, the module $\mathcal{C}\Lambda^1(\mathcal{E})$ can be described as follows. Let $\varkappa = \varkappa(\pi)/(I(\mathcal{E}) \cdot \varkappa(\pi))$ be the restriction of the module $\varkappa(\pi)$ to $\mathcal{E}^\infty$. Consider the submodule L of $\mathcal{C}\mathrm{Diff}(\varkappa, \mathcal{F}(\mathcal{E}))$ consisting of operators of the form $\square \circ \ell_F^\mathcal{E}$, where $\square \in \mathcal{C}\mathrm{Diff}(P, \mathcal{F}(\mathcal{E}))$, $\ell_F^\mathcal{E} = \ell_F|_{\mathcal{E}^\infty}$. Here $P = \mathcal{F}(\mathcal{E}, \xi)$ is the restriction of the module $\mathcal{F}(\pi, \xi)$ to $\mathcal{E}^\infty$. (In the sequel, we shall not distinguish between notation for a module and its restriction to $\mathcal{E}^\infty$.) The next two results are generalizations of Lemma 2.4.

LEMMA 2.7. *To each form $\omega \in \mathcal{C}\Lambda^1(\mathcal{E})$, we assign the $\mathcal{C}$-differential operator $\nabla_\omega \in \mathcal{C}\mathrm{Diff}(\varkappa, \mathcal{F}(\mathcal{E}))$ by setting*

$$\nabla_\omega(\chi) = \omega(\partial_\chi), \qquad \chi \in \varkappa. \tag{2.16}$$

If the equation $\mathcal{E}$ is regular, the correspondence (2.16) establishes an isomorphism of the $\mathcal{F}(\mathcal{E})$-modules $\mathcal{C}\Lambda^1(\mathcal{E})$ and $\mathcal{C}\mathrm{Diff}(\varkappa, \mathcal{F}(\mathcal{E}))/L$.

PROOF. It is obvious that $\mathcal{C}\Lambda^1 = \mathcal{C}\mathrm{Diff}(\varkappa, \mathcal{F}(\mathcal{E}))$, where $\mathcal{C}\Lambda^1$ is the restriction of $\mathcal{C}\Lambda^1(\pi)$ to $\mathcal{E}^\infty$. The kernel L' of the natural projection $\mathcal{C}\Lambda^1 \to \mathcal{C}\Lambda^1(\mathcal{E})$ is generated by the forms $d^v f$, where $f \in I(\mathcal{E})$ and $d^v \colon \mathcal{F}(\pi) \to \mathcal{C}\Lambda^1$ is the composition of the differential $d \colon \mathcal{F}(\pi) \to \Lambda^1(\pi)$ and the projection $\Lambda^1(\pi) \to \mathcal{C}\Lambda^1$. In terms of $\mathcal{C}$-differential operators this means that L' is generated by the operators of the form $\ell_f^\mathcal{E}$, where $f \in I(\mathcal{E})$. Since $\mathcal{E}$ is regular, $f \in I(\mathcal{E})$ can be represented in the form $f = \nabla(F)$, $\nabla \in \mathcal{C}\mathrm{Diff}(P, \mathcal{F}(\pi))$. Thus $\ell_f^\mathcal{E} = \nabla \circ \ell_F^\mathcal{E}$ and, consequently, $L' = L$. $\square$

EXERCISE 2.18. Show that

$$\mathcal{C}^p\Lambda^p(\mathcal{E}) = \mathcal{C}\mathrm{Diff}^{\mathrm{alt}}{}_{(p)}(\varkappa; \mathcal{F}(\mathcal{E}))/L_p,$$

where $L_p = \mathrm{alt}(\mathcal{C}\mathrm{Diff}^{\mathrm{alt}}{}_{(p-1)}(\varkappa; \mathcal{F}(\mathcal{E})) \otimes L)$ and

$$\mathrm{alt} \colon \mathcal{C}\mathrm{Diff}_{(p)}(\varkappa; \mathcal{F}(\mathcal{E})) \to \mathcal{C}\mathrm{Diff}^{\mathrm{alt}}{}_{(p)}(\varkappa; \mathcal{F}(\mathcal{E}))$$

is the alternation operation.

Thus, for a regular equation $\mathcal{E}$ there exists the exact sequence of $\mathcal{F}(\mathcal{E})$-modules

$$\mathcal{C}\mathrm{Diff}(P, \mathcal{F}(\mathcal{E})) \longrightarrow \mathcal{C}\mathrm{Diff}(\varkappa, \mathcal{F}(\mathcal{E})) \longrightarrow \mathcal{C}\Lambda^1(\mathcal{E}) \longrightarrow 0.$$

Generally speaking, the left-hand arrow is not injective: in other words, an operator of the form $\square \circ \ell_F^\mathcal{E}$, $\square \in \mathcal{C}\mathrm{Diff}(P, \mathcal{F}(\mathcal{E}))$, may be trivial for a nontrivial operator $\square$.

A regular equation is called ℓ-*normal*, if the sequence

$$0 \longrightarrow \mathcal{C}\mathrm{Diff}(P, \mathcal{F}(\mathcal{E})) \longrightarrow \mathcal{C}\mathrm{Diff}(\varkappa, \mathcal{F}(\mathcal{E})) \longrightarrow \mathcal{C}\Lambda^1(\mathcal{E}) \longrightarrow 0 \tag{2.17}$$

is exact.

The next proposition gives a way to find out whether a given equation is ℓ-normal.

PROPOSITION 2.8. *Let $\mathcal{E}$ be a regular equation in n independent variables. Assume that in each coordinate neighborhood on $\mathcal{E}^\infty$ one can choose intrinsic coordinates θ such that the functions $D_i(\theta)$, $i = 1, \ldots, n-1$, can be expressed in terms of these coordinates. Then the equation $\mathcal{E}$ is ℓ-normal.*

PROOF. Locally, the equation at hand can be written in the form
$$p^{i_r}_{(0,\ldots,0,k_r)} = f_r, \qquad i_r \in A \subset \{1,\ldots,m\}, \quad r = 1,\ldots,l, \quad k_r > 0,$$
where f_r are functions of the coordinates $x_1, \ldots, x_n$ and p^i_σ, where $\sigma = (\sigma_1, \ldots, \sigma_n)$, $\sigma_n < k_r$, if $i \in A$. We take these coordinates as intrinsic coordinates on $\mathcal{E}^\infty$. Consequently the operator $\ell^\mathcal{E}_F$ has the form $\ell^\mathcal{E}_F = \Delta - \ell_f$, where $f = (f_1, \ldots, f_l)$, $\Delta = \|\Delta_{ij}\|$ is the $l \times m$ matrix such that $\Delta_{r i_r} = \bar{D}^{k_r}_n$, $r = 1, \ldots, l$, and the other elements of Δ are trivial. It follows in the standard way now that if $\square \circ \ell^\mathcal{E}_F = 0$, $\square \in \mathcal{C}\mathrm{Diff}(P, \mathcal{F}(\mathcal{E}))$, then $\square = 0$. □

Thus, if the equation under consideration is not overdetermined, i.e., the number of equations is not greater than the number of unknowns ($l \leq m$), it is almost always ℓ-normal. Nevertheless, let us give an example of a determined but not ℓ-normal equation.

EXAMPLE 2.1. Consider the system
$$\begin{cases} p^3_2 - p^2_3 + u^2 p^3_4 - u^3 p^2_4 = 0, \\ p^1_3 - p^3_1 + u^3 p^1_4 - u^1 p^3_4 = 0, \\ p^2_1 - p^1_2 + u^1 p^2_4 - u^2 p^1_4 = 0, \end{cases} \qquad (2.18)$$
which is the condition for integrability of a three-dimensional distribution in $\mathbb{R}^4$ with coordinates (x_1, x_2, x_3, x_4) given by the form $\omega = dx_4 - \sum_{i=1}^3 u^i \, dx_i$.

EXERCISE 2.19. Find an operator $\square$ such that $\square \circ \ell^\mathcal{E}_F = 0$, where F is the left-hand part of (2.18).
(*The answer:* $\square = (\bar{D}_1 + u^1 \bar{D}_4 - p^1_4, \bar{D}_2 + u^2 \bar{D}_4 - p^2_4, \bar{D}_3 + u^3 \bar{D}_4 - p^3_4)$).

REMARK 2.2. The Maxwell, Yang–Mills, and Einstein equations provide examples of equations which are not ℓ-normal. Roughly speaking, this can be explained as follows. All these equations are invariant under the action of a pseudogroup, i.e., there exists a nontrivial $\mathcal{C}$-differential operator $R \colon Q \to \varkappa$, where Q is a horizontal module, such that $\ell^\mathcal{E}_F \circ R = 0$. Hence $R^* \circ \left(\ell^\mathcal{E}_F\right)^* = 0$ and, since the equations at hand are Euler–Lagrange equations (that is, $\left(\ell^\mathcal{E}_F\right)^* = \ell^\mathcal{E}_F$), we get $R^* \circ \ell^\mathcal{E}_F = 0$. Thus the operator R^* disobeys the condition of ℓ-normality.

In spite of these examples, it is important to stress that the majority of equations of mathematical physics are ℓ-normal.

From now on we consider only ℓ-normal equations[7].

For such an equation $\mathcal{E}$ the sequence (2.17) is exact, and hence there exists the exact sequence of complexes
$$0 \longrightarrow \mathcal{C}\mathrm{Diff}(P, \Lambda^q_0(\mathcal{E})) \longrightarrow \mathcal{C}\mathrm{Diff}(\varkappa, \Lambda^q_0(\mathcal{E})) \longrightarrow \Lambda^q_0(\mathcal{E}) \otimes \mathcal{C}\Lambda^1(\mathcal{E}) \longrightarrow 0.$$

---

[7]Closer examination of the $\mathcal{C}$-spectral sequence makes it possible to compute conservation laws for equations which are not ℓ-normal (see [**120**, **132**]).

Let us multiply each of these complexes by $\mathcal{C}^{p-1}\Lambda^{p-1}(\mathcal{E})$ (see Exercise 2.12):

$$0 \longrightarrow \mathcal{C}\mathrm{Diff}(P,\Lambda_0^q(\mathcal{E})) \otimes \mathcal{C}^{p-1}\Lambda^{p-1}(\mathcal{E}) \longrightarrow \mathcal{C}\mathrm{Diff}(\varkappa,\Lambda_0^q(\mathcal{E})) \otimes \mathcal{C}^{p-1}\Lambda^{p-1}(\mathcal{E})$$
$$\longrightarrow \Lambda_0^q(\mathcal{E}) \otimes \mathcal{C}\Lambda^1(\mathcal{E}) \otimes \mathcal{C}^{p-1}\Lambda^{p-1}(\mathcal{E}) \longrightarrow 0. \quad (2.19)$$

Using the long exact sequence corresponding to (2.19) and Theorem 2.3, we get

PROPOSITION 2.9. *Let $\mathcal{E}$ be an ℓ-normal equation. Then the cohomology of the complex*

$$0 \longrightarrow \mathcal{C}\Lambda^1(\mathcal{E}) \otimes \mathcal{C}^{p-1}\Lambda^{p-1}(\mathcal{E}) \longrightarrow \mathcal{C}\Lambda^1(\mathcal{E}) \otimes \mathcal{C}^{p-1}\Lambda^{p-1}(\mathcal{E}) \otimes \Lambda_0^1(\mathcal{E}) \longrightarrow \cdots$$
$$\longrightarrow \mathcal{C}\Lambda^1(\mathcal{E}) \otimes \mathcal{C}^{p-1}\Lambda^{p-1}(\mathcal{E}) \otimes \Lambda_0^n(\mathcal{E}) \longrightarrow 0 \quad (2.20)$$

is trivial in all degrees different from $n-1$ and n. The cohomology groups in degrees $n-1$ and n are isomorphic to the kernel and cokernel of the operator

$$\left(\ell_F^{\mathcal{E}}\right)^*_{(p-1)} \stackrel{\mathrm{def}}{=} \left(\ell_F^{\mathcal{E}}\right)^*_{\mathcal{C}^{p-1}\Lambda^{p-1}(\mathcal{E})} : \widehat{P} \otimes \mathcal{C}^{p-1}\Lambda^{p-1}(\mathcal{E}) \longrightarrow \widehat{\varkappa} \otimes \mathcal{C}^{p-1}\Lambda^{p-1}(\mathcal{E}),$$

respectively.

EXERCISE 2.20. Let $\nabla_\omega \in \mathcal{C}\mathrm{Diff}^{\mathrm{alt}}_{(p)}(\varkappa;P)$ be the operator that corresponds to a P-valued form $\omega \in \mathcal{C}^p\Lambda^p(\mathcal{E}) \otimes P$ (see Lemma 2.7). Show that if $\Delta \in \mathcal{C}\mathrm{Diff}(P,P_1)$, then the operator

$$\Delta_{\mathcal{C}^p\Lambda^p(\mathcal{E})} \colon \mathcal{C}^p\Lambda^p(\mathcal{E}) \otimes P \longrightarrow \mathcal{C}^p\Lambda^p(\mathcal{E}) \otimes P_1$$

takes ∇_ω to $\Delta \circ \nabla_\omega$.

The complex

$$0 \longrightarrow \mathcal{C}^p\Lambda^p(\mathcal{E}) \longrightarrow \Lambda_0^1(\mathcal{E}) \otimes \mathcal{C}^p\Lambda^p(\mathcal{E}) \longrightarrow \cdots \longrightarrow \Lambda_0^n(\mathcal{E}) \otimes \mathcal{C}^p\Lambda^p(\mathcal{E}) \longrightarrow 0$$

is obviously a direct summand in the complex (2.20). Hence its cohomology is trivial in all degrees different from $n-1$ and n. The cohomology groups in degrees $n-1$ and n are isomorphic to the skew-symmetric parts of the kernel and cokernel of the operator $\left(\ell_F^{\mathcal{E}}\right)^*_{(p-1)}$, respectively.

EXERCISE 2.21. If $\omega \in \ker\left(\ell_F^{\mathcal{E}}\right)^*_{(p-1)} \subset \widehat{P} \otimes \mathcal{C}^{p-1}\Lambda^{p-1}(\mathcal{E})$, then, as follows from the previous exercise, the corresponding operator $\nabla_\omega \in \mathcal{C}\mathrm{Diff}^{\mathrm{alt}}_{(p-1)}(\varkappa;\widehat{P})$ satisfies the equality

$$\left(\ell_F^{\mathcal{E}}\right)^*(\nabla_\omega(\chi_1,\ldots,\chi_{p-1})) = \sum_{i=1}^{p-1} \Delta_i(\chi_1,\ldots,\widehat{\chi}_i,\ldots,\chi_{p-1})(\ell_F^{\mathcal{E}}(\chi_i)),$$

where $\Delta_i \in \mathcal{C}\mathrm{Diff}^{\mathrm{alt}}_{(p-2)}(\varkappa;\mathcal{C}\mathrm{Diff}(P,\widehat{\varkappa}))$. Prove that ω lies in the skew-symmetric part of $\ker\left(\ell_F^{\mathcal{E}}\right)^*_{(p-1)}$ if and only if

$$\nabla_\omega = (-1)^{p-i}\Delta_i^* \mod L_{p-1}\otimes\widehat{P}, \qquad 1 \leq i \leq p-1,$$

where Δ_i^* is the operator adjoint to Δ_i with respect to the last argument (i.e., to that belonging to P) and L_{p-1} is the module defined in Exercise 2.18.

EXERCISE 2.22. Show that $\omega \in \mathrm{coker}\left(\ell_F^{\mathcal{E}}\right)^*_{(p-1)} \subset \widehat{\varkappa} \otimes \mathcal{C}^{p-1}\Lambda^{p-1}(\mathcal{E})$ lies in the skew-symmetric part of $\mathrm{coker}\left(\ell_F^{\mathcal{E}}\right)^*_{(p-1)}$ if and only if

$$\nabla_\omega = -\nabla_\omega^{*i} \mod L_{p-1}\otimes\widehat{\varkappa}, \qquad 1 \leq i \leq p-1,$$

where $\nabla_\omega \in C\mathrm{Diff}^{\mathrm{alt}}_{(p-1)}(\varkappa; \widehat{\varkappa})$ is the operator corresponding to ω and ∇^{*i} denotes the operator adjoint to ∇ with respect to the i-th argument.

EXERCISE 2.23. a. Show that the operator
$$d_1^{0,n-1} \colon E_1^{0,n-1}(\mathcal{E}) = \bar{H}^{n-1}(\mathcal{E}) \longrightarrow E_1^{1,n-1}(\mathcal{E}) = \ker\left(\ell_F^\mathcal{E}\right)^* \subset \widehat{P}$$
has the form
$$d_1^{0,n-1}(h) = \square^*(1),$$
where $h = [\omega] \in \bar{H}^{n-1}(\mathcal{E})$, $\omega \in \Lambda_0^{n-1}(\mathcal{E})$, and $\square \in C\mathrm{Diff}(P, \Lambda_0^n(\mathcal{E}))$ is an operator satisfying $\hat{d}\omega = \square(F)$.

b. Check that the term $E_1^{2,n-1}(\mathcal{E})$ can be described as the quotient set
$$\{\nabla \in C\mathrm{Diff}(\varkappa, \widehat{P}) \mid \left(\ell_F^\mathcal{E}\right)^* \circ \nabla = \nabla^* \circ \ell_F^\mathcal{E}\}/\theta,$$
where $\theta = \{\square \circ \ell_F^\mathcal{E} \mid \square \in C\mathrm{Diff}(P, \widehat{P}), \square^* = \square\}$.

c. Show that the operator $d_1^{1,n-1} \colon E_1^{1,n-1}(\mathcal{E}) = \ker\left(\ell_F^\mathcal{E}\right)^* \to E_1^{2,n-1}(\mathcal{E})$ has the form
$$d_1^{1,n-1}(\psi) = (\ell_\psi^\mathcal{E} + \Delta^*) \mod \theta,$$
where $\Delta \in C\mathrm{Diff}(P, \widehat{\varkappa})$ is an operator satisfying $\ell_F^*(\psi) = \Delta(F)$.

d. Describe the operators $d_1^{p,n}$ and $d_1^{p,n-1}$ for all $p \geq 0$.

Bringing together the above results, we get the following description of the term $E_1(\mathcal{E})$ of the $\mathcal{C}$-spectral sequence:

THE TWO-LINE THEOREM. *Let $\mathcal{E}$ be an ℓ-normal equation. Then:*
1. $E_1^{p,q}(\mathcal{E}) = 0$, *if $p \geq 1$ and $q \neq n-1, n$.*
2. $E_1^{p,n-1}(\mathcal{E})$ *(resp., $E_1^{p,n}(\mathcal{E})$) coincides with the skew-symmetric part (see Exercises 2.21 and 2.22) of the kernel (resp., cokernel) of the operator*
$$\left(\ell_F^\mathcal{E}\right)^*_{(p-1)} \colon \widehat{P} \otimes \mathcal{C}^{p-1}\Lambda^{p-1}(\mathcal{E}) \longrightarrow \widehat{\varkappa} \otimes \mathcal{C}^{p-1}\Lambda^{p-1}(\mathcal{E}).$$

The following result is an obvious consequence of this theorem:

COROLLARY 2.10. *Under the assumptions of the two-line theorem we have:*
1. $E_r^{p,q}(\mathcal{E}) = 0$ *if $p \geq 1$, $q \neq n-1, n$, $1 \leq r \leq \infty$.*
2. $E_3^{p,q}(\mathcal{E}) = E_\infty^{p,q}(\mathcal{E})$.
3. $E_1^{0,q}(\mathcal{E}) = E_\infty^{0,q}(\mathcal{E}) = H^q(\mathcal{E}^\infty)$, $q \leq n-2$.
4. $E_2^{0,n-1}(\mathcal{E}) = E_\infty^{0,n-1}(\mathcal{E}) = H^{n-1}(\mathcal{E}^\infty)$.
5. $E_2^{1,n-1}(\mathcal{E}) = E_\infty^{1,n-1}(\mathcal{E})$.

2.7. Conservation laws and generating functions. Now we apply the results of the previous subsection to the problem of computing conservation laws of an ℓ-normal equation $\mathcal{E}$.

First of all, note that for a formally integrable equation $\mathcal{E}$ the projections $\mathcal{E}^{(k+1)} \to \mathcal{E}^{(k)}$ are affine bundles; therefore $\mathcal{E}^{(k+1)}$ and $\mathcal{E}^{(k)}$ are of the same homotopy type. Hence, $H^*(\mathcal{E}^\infty) = H^*(\mathcal{E})$.

Further, from the two-line theorem we see that the following exact sequence exists:
$$0 \longrightarrow H^{n-1}(\mathcal{E}) \longrightarrow \bar{H}^{n-1}(\mathcal{E}) \xrightarrow{d_1^{0,n-1}} \ker\left(\ell_F^\mathcal{E}\right)^*.$$

Recall that the group $\bar{H}^{n-1}(\mathcal{E})$ was interpreted as the group of conservation laws of the equation $\mathcal{E}$ (see Definition 2.1). Conservation laws $\omega \in H^{n-1}(\mathcal{E}) \subset \bar{H}^{n-1}(\mathcal{E})$ are called *topological* (or *rigid*), since they are determined only by the topology of $\mathcal{E}$. In particular, the conserved quantities do not change under deformations of solutions of $\mathcal{E}$. Thus topological conservation laws[8] are not very interesting for us, and we consider the quotient group $\mathrm{cl}(\mathcal{E}) = \bar{H}^{n-1}(\mathcal{E})/H^{n-1}(\mathcal{E})$, called the group of *proper* conservation laws of the equation $\mathcal{E}$. The two-line theorem immediately implies

THEOREM 2.11. *If $\mathcal{E}$ is an ℓ-normal equation, then*
$$\mathrm{cl}(\mathcal{E}) \subset \ker \left(\ell_F^{\mathcal{E}}\right)^*.$$
If, moreover, $H^n(\mathcal{E}) \subset \bar{H}^n(\mathcal{E})$ (in particular, if $H^n(\mathcal{E}) = 0$), then
$$\mathrm{cl}(\mathcal{E}) = \ker d_1^{1,n-1}.$$

The element $\psi \in \ker \left(\ell_F^{\mathcal{E}}\right)^*$ corresponding to a conservation law $[\omega] \in \mathrm{cl}(\mathcal{E})$ is called its *generating function*.

Theorem 2.11 gives an effective method for computing the group of (proper) conservation laws $\mathrm{cl}(\mathcal{E})$. In §3 we demonstrate this method in action.

EXERCISE 2.24. Show that the generating function of any conservation law can be extended to the entire space $J^\infty(\pi)$ so that the following equality holds:
$$l_F^*(\psi) + l_\psi^*(F) = 0. \tag{2.21}$$
Prove that:

a. Equality (2.21) holds identically, if $\psi = \Box(F)$ and $\Box = -\Box^*$.

b. The solutions to (2.21) of the form $\psi = \Box(F)$ for a self-adjoint operator $\Box$, $\Box = \Box^*$, correspond to topological conservation laws of $\mathcal{E}$.

c. Two different topological conservation laws correspond to the same solution of the above described form if and only if they differ by an element belonging to the image of the natural mapping $H^{n-1}(J^\infty(\pi)) = \bar{H}^{n-1}(\pi) \to \bar{H}^{n-1}(\mathcal{E})$.

To conclude this subsection, note yet another important result. Let $\varphi \in \ker \ell_F^{\mathcal{E}}$ be a symmetry and $[\omega] \in \bar{H}^{n-1}(\mathcal{E})$ be a conservation law of $\mathcal{E}$. Then $[\partial_\varphi(\omega)]$ is obviously a conservation law of $\mathcal{E}$ as well.

EXERCISE 2.25. Prove that if $\psi \in \ker \left(\ell_F^{\mathcal{E}}\right)^*$ is the generating function of a conservation law $[\omega]$ of an ℓ-normal equation $\mathcal{E}$, then the generating function of the conservation law $[\partial_\varphi(\omega)]$ has the form $\partial_\varphi(\psi) + \Delta^*(\psi)$, where the operator $\Delta \in \mathcal{C}\mathrm{Diff}(P, P)$ is defined by $\partial_\varphi(F) = \Delta(F)$.

2.8. Euler–Lagrange equations. Consider a Lagrangian $\mathcal{L} = [\omega] \in \bar{H}^n(\pi)$ and the corresponding Euler–Lagrange equation $\mathcal{E} = \{\mathsf{E}(\mathcal{L}) = 0\}$. Let $\varphi \in \varkappa(\pi)$ be a *Noether symmetry* of $\mathcal{L}$, i.e., $\partial_\varphi(\mathcal{L}) = 0$ on $J^\infty(\pi)$.

EXERCISE 2.26. Check that a Noether symmetry of $\mathcal{L}$ is a symmetry of the corresponding equation $\mathcal{E}$ as well, i.e., $\mathrm{sym}\,\mathcal{L} \subset \mathrm{sym}\,\mathcal{E}$.

EXERCISE 2.27. Show that if $E_2^{0,n}(\mathcal{E}) = 0$, then finding Noether symmetries of the Lagrangian $\mathcal{L} = [\omega]$ amounts to solving the equation
$$\mathsf{E}(\ell_\omega(\varphi)) = \ell_{\mathsf{E}(\mathcal{L})}(\varphi) + \ell_\varphi^*(\mathsf{E}(\mathcal{L})) = 0.$$

---

[8]They are discussed, e.g., in the book [**103**].

(Thus, to compute Noether symmetries of an Euler–Lagrange equation one has no need to know the Lagrangian.)

Let $\partial_\varphi(\omega) = \widehat{d\nu}$, where $\nu \in \Lambda_0^{n-1}(\pi)$. By (2.5), we have

$$\partial_\varphi(\omega) - \widehat{d\nu} = \ell_\omega(\varphi) - \widehat{d\nu} = \ell_\omega^*(1)(\varphi) + \widehat{d}\omega_{\varphi,1}(\ell_\omega) - \widehat{d\nu}$$
$$= \mathsf{E}(\mathcal{L})(\varphi) + \widehat{d}(\omega_{\varphi,1}(\ell_\omega) - \nu) = 0.$$

Set

$$\eta = (\nu - \omega_{\varphi,1}(\ell_\omega))|_{\mathcal{E}^\infty} \in \Lambda_0^{n-1}(\mathcal{E}).$$

Thus, $\widehat{d}\eta|_{\mathcal{E}^\infty} = 0$, i.e., $[\eta] \in \bar{H}^{n-1}(\mathcal{E})$ is a conservation law for $\mathcal{E}$. The map

$$\operatorname{sym}\mathcal{L} \longrightarrow \bar{H}^{n-1}(\mathcal{E}), \qquad \varphi \longmapsto [\eta],$$

is called the *Noether map*.

EXERCISE 2.28. Check that the Noether map is well defined up to the image of the natural homomorphism $H^{n-1}(J^\infty(\pi)) = \bar{H}^{n-1}(\pi) \to \bar{H}^{n-1}(\mathcal{E})$.

REMARK 2.3. The definition of the Noether map implies that it is defined for all Euler–Lagrange equations, not just for ℓ-normal ones.

EXERCISE 2.29. Prove that if the Euler–Lagrange equation $\mathcal{E}$ corresponding to a Lagrangian $\mathcal{L}$ is ℓ-normal, then the Noether map restricted to the set of Noether symmetries of $\mathcal{L}$ is inverse to the differential $d_1^{0,n-1}$:

$$d_1^{0,n-1}([\eta]) = \varphi. \tag{2.22}$$

REMARK 2.4. The Noether map can be understood as a procedure for finding a conserved current for a given generating function.

EXERCISE 2.30. Given an ℓ-normal Euler–Lagrange equation $\mathcal{E}^\infty$, describe the map inverse to the embedding $d_1^{0,n-1}\colon \operatorname{cl}(\mathcal{E}) \to \ker \ell\left(\ell_F^{\mathcal{E}}\right)^* = \ker \ell_F^{\mathcal{E}} = \operatorname{sym}\mathcal{L}$. (In view of the previous exercise, the required inverse map is an extension of the Noether map.)

2.9. Hamiltonian formalism on $J^\infty(\pi)$. Let $A \in \mathcal{C}\mathrm{Diff}(\widehat{\varkappa}(\pi), \varkappa(\pi))$ be a $\mathcal{C}$-differential operator. Define the *Poisson bracket* on $\bar{H}^n(\pi)$ corresponding to the operator A by the formula

$$\{\omega_1, \omega_2\}_A = \langle A(\mathsf{E}(\omega_1)), \mathsf{E}(\omega_2)\rangle,$$

where $\langle\,,\rangle$ denotes the natural paring $\varkappa(\pi) \times \widehat{\varkappa}(\pi) \to \bar{H}^n(\pi)$.

EXERCISE 2.31. a. Consider an operator $A \in \mathcal{C}\mathrm{Diff}(\widehat{\varkappa}(\pi), P)$, where P is a module over $\mathcal{F}(\pi)$. Prove that $A \circ \mathsf{E} = 0$ implies $A = 0$.

b. Consider an operator $A \in \mathcal{C}\mathrm{Diff}_{(l)}(\widehat{\varkappa}(\pi), \Lambda_0^n(\pi))$. Prove that if for all $\omega_1, \ldots, \omega_l \in \bar{H}^n(\pi)$ the element $A(\mathsf{E}(\omega_1), \ldots, \mathsf{E}(\omega_l))$ belongs to the image of the operator $\widehat{d}$, then $\operatorname{im} A \subset \operatorname{im} \widehat{d}$, i.e., $\mu_l(A) = 0$ (see Exercises 2.5 and 2.10). In particular, if for any $\omega \in \bar{H}^n(\pi)$ one has $\partial_\varphi(\omega) = \langle \varphi, \mathsf{E}(\omega)\rangle = 0$, $\varphi \in \varkappa(\pi)$, then $\varphi = 0$. Note that the equality $\partial_\varphi(\omega) = \langle \varphi, \mathsf{E}(\omega)\rangle$ follows from

$$\langle \varphi, \mathsf{E}(\omega)\rangle = \langle \varphi, \ell_\omega^*(1)\rangle = \langle \ell_\omega(\varphi), 1\rangle = \langle \partial_\varphi(\omega), 1\rangle.$$

c. Using the above, check that if the Poisson bracket $\{\omega_1, \omega_2\}_A$ is trivial for arbitrary ω_1 and ω_2, then $A = 0$.

An operator A is called *Hamiltonian*, if the corresponding Poisson bracket defines a Lie algebra structure (over $\mathbb{R}$) on $\bar{H}^n(\pi)$, i.e., if

$$\{\omega_1, \omega_2\}_A = -\{\omega_2, \omega_1\}_A, \qquad (2.23)$$

$$\{\{\omega_1, \omega_2\}_A, \omega_3\}_A + \{\{\omega_2, \omega_3\}_A, \omega_1\}_A + \{\{\omega_3, \omega_1\}_A, \omega_2\}_A = 0. \qquad (2.24)$$

The bracket $\{\,,\,\}_A$ is said to be a *Hamiltonian structure*.

EXERCISE 2.32. Prove that the Poisson bracket corresponding to an operator A is skew-symmetric, i.e., condition (2.23) holds, if and only if the operator A is skew-adjoint, i.e., $A^* = -A$.

Now we derive criteria for a skew-adjoint operator $A \in \mathcal{C}\mathrm{Diff}(\widehat{\varkappa}(\pi), \varkappa(\pi))$ to be Hamiltonian. For this, we use the following observation. Let $P = \mathcal{F}(\pi, \xi)$ and $P' = \mathcal{F}(\pi, \xi')$ be horizontal modules, and let $\Delta \colon P \to P'$ be a $\mathcal{C}$-differential operator. Set $D_\sigma = D_1^{\sigma_1} \circ \cdots \circ D_n^{\sigma_n}$. One has $j_\infty(s)^*(D_\sigma f) = \partial^{|\sigma|}/\partial x_\sigma(j_\infty(s)^* f)$ for any section $s \in \Gamma(\pi)$ (see Chapter 4). Therefore Δ can be considered as a nonlinear operator acting from the sections of π to the set of linear operators $\Gamma(\xi) \to \Gamma(\xi')$. Thus, the universal linearization of this operator, which we denote by ℓ_Δ, belongs to $\mathcal{C}\mathrm{Diff}(\varkappa, \mathcal{C}\mathrm{Diff}(P, P')) = \mathcal{C}\mathrm{Diff}(\varkappa, P; P')$.

For a $\mathcal{C}$-differential operator $A \colon \widehat{\varkappa}(\pi) \to \varkappa(\pi)$ and an element $\psi \in \widehat{\varkappa}(\pi)$, we can define the $\mathcal{C}$-differential operator $\ell_{A,\psi} \colon \varkappa(\pi) \to \varkappa(\pi)$ by

$$\ell_{A,\psi}(\varphi) = (\ell_A(\varphi))(\psi).$$

EXERCISE 2.33. Prove that

$$\ell^*_{A,\psi_1}(\psi_2) = \ell^*_{A^*,\psi_2}(\psi_1).$$

THEOREM 2.12. *Let $A \in \mathcal{C}\mathrm{Diff}(\widehat{\varkappa}(\pi), \varkappa(\pi))$ be a skew-adjoint operator; then the following conditions are equivalent:*

1. *A is a Hamiltonian operator.*
2. *$\langle \ell_A(A(\psi_1))(\psi_2), \psi_3 \rangle + \langle \ell_A(A(\psi_2))(\psi_3), \psi_1 \rangle + \langle \ell_A(A(\psi_3))(\psi_1), \psi_2 \rangle = 0$ for all $\psi_1, \psi_2, \psi_3 \in \widehat{\varkappa}(\pi)$.*
3. *$\ell_{A,\psi_1}(A(\psi_2)) - \ell_{A,\psi_2}(A(\psi_1)) = A(\ell^*_{A,\psi_2}(\psi_1))$ for all $\psi_1, \psi_2 \in \widehat{\varkappa}(\pi)$.*
4. *The expression $\ell_{A,\psi_1}(A(\psi_2)) + \frac{1}{2}A(\ell^*_{A,\psi_1}(\psi_2))$ is symmetric with respect to $\psi_1, \psi_2 \in \widehat{\varkappa}(\pi)$.*
5. *$[\partial_{A(\psi)}, A] = \ell_{A(\psi)} \circ A + A \circ \ell^*_{A(\psi)}$ for all $\psi \in \mathrm{im}\, \mathsf{E} \subset \widehat{\varkappa}(\pi)^9$.*

Moreover, it is sufficient to verify conditions 2–4 for elements $\psi_i \in \mathrm{im}\, \mathsf{E}$ only.

PROOF. Let $\omega_1, \omega_2, \omega_3 \in \bar{H}^n(\pi)$ and $\psi_i = \mathsf{E}(\omega_i)$. The Jacobi identity (2.24) has the form

$$\{\{\omega_1, \omega_2\}_A, \omega_3\}_A + \text{(cyclic)} = -\partial_{A(\psi_3)}\langle A(\psi_1), \psi_2 \rangle + \text{(cyclic)}$$
$$= -\langle \partial_{A(\psi_3)}(A)(\psi_1), \psi_2 \rangle - \langle A(\ell_{\psi_1}(A(\psi_3))), \psi_2 \rangle - \langle A(\psi_1), \ell_{\psi_2}(A(\psi_3)) \rangle + \text{(cyclic)}$$
$$= -\langle \ell_A(A(\psi_3))(\psi_1), \psi_2 \rangle + \langle A(\psi_2), \ell_{\psi_1}(A(\psi_3)) \rangle - \langle A(\psi_1), \ell_{\psi_2}(A(\psi_3)) \rangle + \text{(cyclic)}$$
$$= -\langle \ell_A(A(\psi_3))(\psi_1), \psi_2 \rangle + \text{(cyclic)} = 0, \qquad (2.25)$$

---

[9]Let $P = \mathcal{F}(\pi, \xi)$ and $P' = \mathcal{F}(\pi, \xi')$ be horizontal modules, and let $F \colon P \to P'$ be a map. Then for any $\varphi \in \mathcal{F}(\pi, \pi)$ one can define the commutator $[\partial_\varphi, F] \stackrel{\text{def}}{=} \partial^{\xi'}_\varphi \circ F - F \circ \partial^\xi_\varphi$, since each field ∂_φ defines the family of derivations $\partial^\xi_\varphi \colon \mathcal{F}(\pi, \xi) \to \mathcal{F}(\pi, \xi)$ (see Chapter 4).

where "(cyclic)" denotes terms obtained by the cyclic permutation of indices. It follows from Exercise 2.31 that equality (2.25) holds for all $\psi_i \in \widehat{\varkappa}(\pi)$. Criterion 2 is proved.

Now rewrite the Jacobi identity in the form

$$\langle \ell_{A,\psi_1}(A(\psi_2)), \psi_3 \rangle + \langle A(\psi_1), \ell^*_{A,\psi_3}(\psi_2) \rangle - \langle A(\ell^*_{A,\psi_2}(\psi_1)), \psi_3 \rangle = 0.$$

Using Exercise 2.33, we obtain

$$\langle \ell_{A,\psi_1}(A(\psi_2)), \psi_3 \rangle - \langle \ell_{A,\psi_2}(A(\psi_1)), \psi_3 \rangle - \langle A\left(\ell^*_{A,\psi_2}(\psi_1)\right), \psi_3 \rangle = 0.$$

In view of Exercise 2.31, this implies 3.

Equivalence of 3 and 4 follows from Exercise 2.33.

Finally, 5 is equivalent to 3 by virtue of the following obvious equalities:

$$[\partial_{A(\psi_2)}, A](\psi_1) = \ell_{A,\psi_1}(A(\psi_2)),$$
$$\ell_{A,\psi} \circ A = \ell_{A(\psi)} \circ A - A \circ \ell_\psi \circ A.$$

This concludes the proof. □

Let $A \colon \widehat{\varkappa}(\pi) \to \varkappa(\pi)$ be a Hamiltonian operator. For any $\omega \in \bar{H}^n(\pi)$, the evolutionary derivation $X_\omega = \partial_{A(\mathsf{E}(\omega))}$ is called the *Hamiltonian vector field* corresponding to the Hamiltonian ω. Obviously,

$$X_{\omega_1}(\omega_2) = \langle A\mathsf{E}(\omega_1), \mathsf{E}(\omega_2) \rangle = \{\omega_1, \omega_2\}_A.$$

This yields

$$X_{\{\omega_1,\omega_2\}_A}(\omega) = \{\{\omega_1, \omega_2\}_A, \omega\}_A = \{\omega_1, \{\omega_2, \omega\}_A\}_A - \{\omega_2, \{\omega_1, \omega\}_A\}_A$$
$$= (X_{\omega_1} \circ X_{\omega_2} - X_{\omega_2} \circ X_{\omega_1})(\omega) = [X_{\omega_1}, X_{\omega_2}](\omega)$$

for all $\omega \in \bar{H}^n(\pi)$. Thus,

$$X_{\{\omega_1,\omega_2\}_A} = [X_{\omega_1}, X_{\omega_2}]. \tag{2.26}$$

Similarly to the finite-dimensional case, (2.26) implies a result similar to the Noether theorem.

For each $\mathcal{H} \in \bar{H}^n(\pi)$, the evolution equation

$$u_t = A(\mathsf{E}(\mathcal{H})) \tag{2.27}$$

corresponding to the Hamiltonian $\mathcal{H}$ is called the *Hamiltonian evolution equation*.

THEOREM 2.13. *Hamiltonian operators take generating functions of conservation laws of equation* (2.27) *to symmetries of this equation.*

PROOF. Let A be a Hamiltonian operator, and let

$$\widetilde{\omega}_0(t) + \widetilde{\omega}_1(t) \wedge dt \in \Lambda_0^n(\pi) \oplus \Lambda_0^{n-1}(\pi) \wedge dt$$

be a conserved current of equation (2.27). This means that $\bar{D}_t(\omega_0(t)) = 0$, where $\omega_0(t) \in \bar{H}^n(\pi)$ is the horizontal cohomology class corresponding to the form $\widetilde{\omega}_0(t)$, and $\bar{D}_t$ is the restriction of the total derivative in t to the infinite prolongation of the equation at hand. Furthermore,

$$\bar{D}_t(\omega_0) = \frac{\partial \omega_0}{\partial t} + \partial_{A(\mathsf{E}(\mathcal{H}))}(\omega_0) = \frac{\partial \omega_0}{\partial t} + \{\mathcal{H}, \omega_0\}.$$

This yields

$$\frac{\partial}{\partial t} X_{\omega_0} + [X_\mathcal{H}, X_{\omega_0}] = 0.$$

Hence the field $X_{\omega_0} = \vartheta_{A(\mathsf{E}(\omega_0))}$ is a symmetry of (2.27). It remains to observe that $\mathsf{E}(\omega_0)$ is the generating function of the conservation law under consideration (cf. Exercise 3.2). □

EXERCISE 2.34. a. A skew-symmetric operator
$$B \in E_1^{2,n}(\pi) \subset \mathcal{C}\mathrm{Diff}(\varkappa(\pi), \widehat{\varkappa}(\pi))$$
is called *symplectic* if $d_1^{2,n}(B) = 0$. Prove that the following conditions are equivalent:

1. B is a symplectic operator.
2. $\langle \ell_B(\varphi_1)(\varphi_2), \varphi_3 \rangle + \langle \ell_B(\varphi_2)(\varphi_3), \varphi_1 \rangle + \langle \ell_B(\varphi_3)(\varphi_1), \varphi_2 \rangle = 0$ for all elements $\varphi_1, \varphi_2, \varphi_3 \in \varkappa(\pi)$.
3. $\ell_{B,\varphi_1}(\varphi_2) - \ell_{B,\varphi_2}(\varphi_1) = \ell_{B,\varphi_1}^*(\varphi_2)$ for all $\varphi_1, \varphi_2, \in \varkappa(\pi)$.
4. The expression $\ell_{B,\varphi_1}(\varphi) - \frac{1}{2}\ell_{B,\varphi_1}^*(\varphi_2)$ is symmetric with respect to $\varphi_1, \varphi_2, \in \varkappa(\pi)$.
5. $\ell_B(\varphi) = \ell_{B,\varphi} - \ell_{B,\varphi}^*$, where, as above, the operator $\ell_{B,\varphi} \in \mathcal{C}\mathrm{Diff}(\varkappa(\pi), \widehat{\varkappa}(\pi))$, $\varphi \in \varkappa(\pi)$, is defined by $\ell_{B,\varphi_1}(\varphi_2) = \ell_B(\varphi_2)(\varphi_1)$.

b. An evolution equation $u_t = \varphi$, $\varphi \in \varkappa(\pi)$, is called *Hamiltonian* (with respect to a symplectic operator B) if $B(\varphi) = \mathsf{E}(\mathcal{H})$ for a Hamiltonian $\mathcal{H} \in \Lambda_0^n(\pi)$. Study the relationship between symmetries and conservation laws of such an equation.

3. Computation of conservation laws

In this section, we discuss applications of the $\mathcal{C}$-spectral sequence. To begin with, we show how one computes the space of conservation laws by the method of generating functions based on Theorem 2.11. Generating functions establish an efficient means for describing conservation laws, i.e., conserved currents modulo trivial ones, and allow one to find *all* conservation laws of an equation. After that we present concrete results, to illustrate the technique of computations for conservation laws.

3.1. Basic results. In this subsection, we present the basic results from the theory of conservation laws in coordinate form, suitable for computations. This material can be read independently of §2.

Let $S = (S_1, \ldots, S_n)$ be a conserved current for the equation $\mathcal{E} = \{F = 0\}$ with n independent and m dependent variables. Equality (1.4), which is satisfied on $\mathcal{E}^\infty$, is equivalent to

$$\sum_{i=1}^n D_i(S_i) = \sum_{j=1}^l \square_j(F_j), \qquad (3.1)$$

where l is the number of equations, $\square_j = \sum_\sigma a_\sigma^j D_\sigma$ are scalar differential operators, $F = (F_1, \ldots, F_l)$.

Recall that operators of the form $\sum_\sigma a_\sigma D_\sigma$ are called $\mathcal{C}$-*differential* or *horizontal*. If $\Delta = \sum_\sigma a_\sigma D_\sigma$ is a scalar $\mathcal{C}$-differential operator, $\Delta^* = \sum_\sigma (-1)^{|\sigma|} D_\sigma \circ a_\sigma$ is the operator *(formally) adjoint* to Δ. If $\Delta = \|\Delta_{ij}\|$ is a matrix $\mathcal{C}$-differential operator, then $\Delta^* = \|\Delta_{ji}^*\|$.

The following fundamental result holds (see Theorem 2.11):

THEOREM 3.1. *Suppose that operators $\Box_1, \ldots, \Box_l$ satisfy (3.1). Then the restriction $\psi = (\Box_1^*(1), \ldots, \Box_l^*(1))|_{\mathcal{E}^\infty}$ of the vector function $(\Box_1^*(1), \ldots, \Box_l^*(1))$ to the infinite prolonged equation $\mathcal{E}^\infty$ satisfies the equation*

$$\left(\ell_F^{\mathcal{E}}\right)^*(\psi) = 0.$$

Note that, generally speaking, the vector function ψ is not uniquely defined by the conserved current S.

EXERCISE 3.1. Check that such an ambiguity occurs for equation (2.18).

If any conserved current S defines a unique vector function ψ for the equation at hand, then ψ is the same for all conserved currents equivalent to S and thus characterizes the corresponding conservation law. Such a vector function is called the *generating function* of this conservation law.

EXERCISE 3.2. Prove that for evolution equations

$$u_t = f(x, u, u_x, u_{xx}, \ldots)$$

the generating function corresponding to a conserved current $(S_0, S_1, \ldots, S_n)$, where S_0 is the t-component, has the form $\psi = \ell_{S_0}^*(1)$. In other words, ψ is the left-hand side of the Euler–Lagrange equations for the Lagrangian S_0, which is understood as a function of $x, u, u_x, u_{xx}, \ldots$.

One can prove that for regular and determined ($l = m$) equations, generating functions are uniquely defined by conservation laws. Such equations are called ℓ-*normal* (see §2.6). The majority of equations of mathematical physics are ℓ-normal.

Now the following natural question arises. Given an ℓ-normal equation, is a conservation law uniquely defined by its generating function? To give the answer, note that every de Rham cohomology class $\xi \in H^{n-1}(\mathcal{E})$ can be understood as a conservation law for the equation $\mathcal{E}$. Such *topological* conservation laws are, of course, not very interesting (see, however, [**103**]), since they do not change under continuous deformations of solutions of $\mathcal{E}$. The results of §2.7 imply the following:

THEOREM 3.2. *Let $\mathcal{E}$ be an ℓ-normal equation. Two conservation laws of the equation $\mathcal{E}$ have the same generating function if and only if they differ by a topological conservation law.*

The quotient group of all conservation laws by topological ones is called the group of *proper* conservation laws.

Thus, the problem of finding all the conservation laws for a given ℓ-normal equation reduces to the problem of finding the corresponding generating functions, i.e., the solutions of the following equation:

$$\left(\ell_F^{\mathcal{E}}\right)^*(\psi) = 0. \tag{3.2}$$

Not every solution of (3.2) corresponds to a conservation law. The redundant solutions can be removed by using the following fact.

Let ψ be a solution of (3.2). This means that $\ell_F^*(\psi) = \Delta(F)$, where Δ is an $m \times l$ matrix $\mathcal{C}$-differential operator. The solution ψ corresponds to a conservation law if and only if there exists an $l \times l$ matrix $\mathcal{C}$-differential operator ∇ such that

$$\ell_\psi^{\mathcal{E}} + (\Delta|_{\mathcal{E}^\infty})^* = \nabla|_{\mathcal{E}^\infty} \circ \ell_F^{\mathcal{E}},$$
$$\nabla^* = \nabla.$$

This was proved in §2.6 (see Exercise 2.23c and Corollary 2.10).

To conclude, let us describe the action of symmetries of the equation $\mathcal{E} = \{F = 0\}$ on its conservation laws in terms of generating functions. Let $\varphi \in \operatorname{sym}\mathcal{E}$ be a symmetry of $\mathcal{E}$, and let $\psi \in \ker\left(\ell_F^{\mathcal{E}}\right)^*$ be the generating function of a conservation law of $\mathcal{E}$. Then $\partial_\varphi(F) = \Delta(F)$, where Δ is a $\mathcal{C}$-differential operator. From Exercise 2.25 it follows that ∂_φ acts on the space of generating functions by the formula $\psi \mapsto \partial_\varphi(\psi) + \Delta^*(\psi)$.

EXERCISE 3.3. Work out an analog of the commutator relation (4.1) of Chapter 4 for conservation laws.

3.2. Examples. Now we give several examples to illustrate the above-described algorithm for computing conservation laws[10].

EXAMPLE 3.1. Let us first consider the Burgers equation:
$$F = u_{xx} + uu_x - u_t = 0.$$
The operator $\ell_F^{\mathcal{E}}$, $\mathcal{E} = \{F = 0\}$, has the form $\ell_F^{\mathcal{E}} = D_x^2 + uD_x + p_1 - D_t$ (we use the notation of §4.1 of Ch. 4); therefore,
$$\left(\ell_F^{\mathcal{E}}\right)^* = D_x^2 - D_x \circ u + p_1 + D_t = D_x^2 - uD_x + D_t.$$
Let us look for a generating function ψ in the form $\psi(x, t, u, \ldots, p_k)$, where $k \geq 0$ and $\partial\psi/\partial p_k \neq 0$. Then, using the formulas for D_x and D_t from §4.1 of Ch. 4, we find that
$$D_x^2(\psi) = p_{k+2}\frac{\partial\psi}{\partial p_k} + O(k+1),$$
$$D_t(\psi) = p_{k+2}\frac{\partial\psi}{\partial p_k} + O(k+1).$$
Thus,
$$\left(\ell_F^{\mathcal{E}}\right)^*(\psi) = 2p_{k+2}\frac{\partial\psi}{\partial p_k} + O(k+1),$$
and the equality $\left(\ell_F^{\mathcal{E}}\right)^*(\psi) = 0$ implies that $\partial\psi/\partial p_k = 0$. This contradiction proves that $\psi = \psi(x, t)$ and, hence,
$$\left(\ell_F^{\mathcal{E}}\right)^*(\psi) = \psi_{xx} + \psi_t - u\psi_x = 0.$$
Since ψ does not depend on u, then $\psi_x = 0$. Consequently, $\psi_t = 0$, i.e., $\psi = \text{const}$. It is readily seen that the conserved current $(-u, u_x + u^2/2)$ corresponds to the generating function $\psi = 1$. Thus, we have proved that the group of conservation laws of the Burgers equation is one-dimensional and is generated by the conserved current $(-u, u_x + u^2/2)$.

EXAMPLE 3.2. Consider the n-dimensional quasi-linear isotropic heat equation:
$$u_t = \Delta(u^\alpha) + f(u), \qquad \Delta = \sum_{i=1}^n \frac{\partial^2}{\partial x_i^2}, \qquad \alpha \neq 0.$$

---

[10]The technique of computing conservation laws is much the same as that for symmetries. Since the latter is fully considered in Chapter 4, we shall restrict ourselves to brief descriptions of the computations. For details on Examples 3.2, 3.7, 3.9–3.11 see [**137**].

In this case, equation (3.2) immediately implies that ψ depends on the variables t, x only and satisfies the equation

$$\left(\frac{\partial}{\partial t} + \alpha u^{\alpha-1}\Delta\right)\psi + f'(u)\psi = 0. \tag{3.3}$$

It is easily shown that equation (3.3) has a nontrivial solution only if

$$f(u) = au^\alpha + bu + c.$$

Now, if $\alpha \neq 1$, then $\psi = v(x)e^{-bt}$, with $v = v(x)$ satisfying the equation $\Delta v + av = 0$. If $\alpha = 1$, then $f(u) = bu + c$ and the function $\psi = \psi(x,t)$ satisfies the equation $\psi_t + \Delta\psi + b\psi = 0$.

It can easily be checked that the vector function

$$\left(\psi u, \psi_{x_1}u^\alpha - \alpha\psi u^{\alpha-1}u_{x_1} - c\int \psi\, dx_1, \psi_{x_2}u^\alpha - \alpha\psi u^{\alpha-1}u_{x_2}, \ldots, \right.$$
$$\left.\psi_{x_n}u^\alpha - \alpha\psi u^{\alpha-1}u_{x_n}\right).$$

is a conserved current corresponding to the generating function ψ.

EXAMPLE 3.3. If $\mathcal{E} = \{\Delta = 0\}$ is an arbitrary linear system of equations defined by an operator Δ, then each solution of the adjoint system $\mathcal{E}^* = \{\Delta^* = 0\}$ is the generating function of a conservation law of $\mathcal{E}$ (cf. Example 3.2 for $\alpha = 1$).

EXERCISE 3.4. Do the *linear conservation laws* from Example 3.3 form a complete set of conservation laws (see §1) for an ℓ-normal linear equation?

EXAMPLE 3.4 ([**110**]). The equation

$$u_t - \Delta u_t = \Delta u, \qquad \Delta = \sum_{j=1}^{n}\frac{\partial^2}{\partial x_j^2}, \tag{3.4}$$

which describes the filtration of fluids in crack-porous media, has a complete set of conservation laws. Let us demonstrate this with the example of the following boundary value problem:

$$u(x,0) = \varphi(x), \qquad x \in \bar{V},$$
$$u(x,t) = h(x,t), \qquad x \in \partial V, \quad t \geq 0,$$

where $V \subset \mathbb{R}^n$ is a bounded domain with piecewise smooth boundary ∂V, and $\varphi(x) = h(x,0)$ if $x \in \partial V$. Let us look for generating functions $v(x,t)$ of linear conservation laws of equation (3.4), i.e., solutions of the equation

$$v_t - \Delta v_t + \Delta v = 0 \tag{3.5}$$

satisfying the homogeneous boundary conditions

$$v(x,t)|_{\partial V} = 0, \qquad t \geq 0. \tag{3.6}$$

Using the standard Fourier method of separation of variables, we obtain the following complete and orthogonal in $L^2(V)$ system of functions satisfying (3.5) and (3.6):

$$v_k(x,t) = e^{\lambda_k t}X_k(x), \qquad \lambda_k = \frac{\bar{\lambda}_k}{1+\bar{\lambda}_k}, \qquad k = 1, 2, 3, \ldots,$$

where $\bar{\lambda}_k$ and $X_k(x)$ are eigenvalues and eigenfunctions of the Sturm–Liouville problem

$$\begin{cases} \Delta X + \bar{\lambda} X = 0, \\ X|_{\partial V} = 0, \end{cases}$$

respectively. The conservation law corresponding to the generating function $v(x,t)$ in the integral form reads

$$-\frac{d}{dt} \int_V (v - \Delta v) u \, dx = \int_{\partial V} \left((u_t + u) \frac{\partial v}{\partial n} - \left(\frac{\partial u_t}{\partial n} + \frac{\partial u}{\partial n} \right) v \right) d\sigma, \qquad (3.7)$$

where $\boldsymbol{n}$ is a unit external vector normal to ∂V and $d\sigma$ is a surface area element. Given $v(x,t)$, one can compute the right-hand side of (3.7), so that the value of the integral $\int_V (v - \Delta v) u \, dx$ at the initial moment $t = 0$ uniquely determines its value at each instant of time. This makes it possible to find the Fourier coefficients $c_k(t)$ of $u(x,t)$ with respect to the orthogonal system $\{v_k\}$, since

$$c_k(t) = \frac{1}{\|v_k\|^2} \int_V v_k u \, dx = \frac{1}{(1 + \bar{\lambda}_k) \|v_k\|^2} \int_V (v_k - \Delta v_k) u \, dx.$$

Thus, completeness of the system of functions in the space $L^2(V)$ implies completeness of the set of conservation laws with generating functions $v_k(x,t)$.

EXAMPLE 3.5. It is well known that the Korteweg–de Vries equation

$$u_t = 6uu_x - u_{xxx}$$

possesses an infinite series of conservation laws (see e.g., [**1, 21, 24, 81, 89, 91**]), which starts with the conservation laws of mass, momentum, and energy:

$$u_t + (-3u^2 + u_{xx})_x = 0,$$
$$(u^2)_t + (-4u^3 + 2uu_{xx} - u_x^2)_x = 0,$$
$$(u^3 + u_x^2/2)_t + (-9u^4/2 + 3u^2 u_{xx} - 6uu_x^2 + u_x u_{xxx} - u_{xx}^2/2)_x = 0.$$

This infinite series of conservation laws has been constructed in various ways by many authors. The solution of this problem gave birth to the famous inverse scattering method (which today is dealt with in an extensive literature, see, e.g., the books [**1, 21, 22, 24, 28, 89**] and the references given there). The computation technique described above allows us not only to obtain easily the infinite series of conservation laws, but also to prove that the Korteweg–de Vries equation has no other conservation laws.

EXERCISE 3.5. Prove that an evolution equation $u_t = f(x, u_x, u_{xx}, \dots)$ of *even* order cannot have an infinite number of conservation laws that depend on derivatives of arbitrarily high order.

EXAMPLE 3.6. The nonlinear Schrödinger equation

$$i\psi_t = \Delta \psi + |\psi|^2 \psi, \qquad \Delta = \sum_{j=1}^n \frac{\partial^2}{\partial x_j^2},$$

has two physically obvious conservation laws:

$$(|\psi|^2)_t + i\nabla(\bar{\psi} \nabla \psi - \psi \nabla \bar{\psi}) = 0, \qquad \nabla = \left(\frac{\partial}{\partial x_1}, \dots, \frac{\partial}{\partial x_n} \right);$$

$$(|\nabla \psi|^2 - \frac{1}{2}|\psi|^4)_t + i\nabla((\Delta \psi + |\psi|^2 \psi) \nabla \bar{\psi} - (\Delta \bar{\psi} + |\psi|^2 \bar{\psi}) \nabla \psi) = 0.$$

If the number of space variables equals 1, then, like the Korteweg–de Vries equation, this equation possesses a well-known infinite series of conservation laws. For $n > 1$, there are no other conservation laws of the nonlinear Schrödinger equation.

REMARK 3.1. The explosion of interest in the Korteweg–de Vries equation and the nonlinear Schrödinger equation in the late 60s and early 70s has led to the discovery of many "integrable" equations. Such an equation possesses a complete set of conservation laws and is solvable by the inverse scattering method. The Boussinesq, Kadomtsev–Petviashvili, Harry Dym, sine-Gordon, etc., equations are examples.

EXAMPLE 3.7. Consider the Zakharov equations describing the nonlinear interaction of high-frequency and low-frequency waves:
$$\begin{cases} i\psi_t + \psi_{xx} - n\psi = 0, \\ n_t + u_x = 0, \\ u_t + n_x + (|\psi|^2)_x = 0. \end{cases} \qquad (3.8)$$

Though this system has soliton-like solutions, it is not integrable by the inverse scattering method. The computation of conservation laws by directly solving (3.2) is hardly possible in this case. In fact, one needs the Hamiltonian technique (see §4.2) to prove that (3.8) is not integrable.

Equation (3.8) has eight conservation laws:

1) $(\frac{1}{2i}(\psi\bar{\psi}_x - \bar{\psi}\psi_x) - nu, \frac{1}{2}(|\psi|^2)_{xx} - 2|\psi_x|^2 - n|\psi|^2 - \frac{n^2}{2} - \frac{u^2}{2})$ (conservation of quasi-momentum).

2) $(|\psi_x|^2 + n|\psi|^2 + \frac{n^2}{2} + \frac{u^2}{2}, \frac{1}{i}(\psi_{xx}\bar{\psi}_x - \bar{\psi}_{xx}\psi_x) + \frac{n}{i}(\bar{\psi}\psi_x - \bar{\psi}_x\psi)_x + nu - u|\psi|^2)$ (conservation of quasi-energy).

3) $(|\psi|^2, i(\bar{\psi}_x\psi - \bar{\psi}\psi_x))$ (conservation of quasi-particle number).

4) $(t|\psi|^2 + tn - xu, \frac{t}{i}(\bar{\psi}\psi_x - \psi\bar{\psi}_x) - x|\psi|^2 - xn + tu)$.

5) $(t^2|\psi|^2 + (x^2 + t^2)n - 2txu, \frac{t^2}{i}(\bar{\psi}\psi_x - \psi\bar{\psi}_x) - 2tx|\psi|^2 - 2txn + (x^2 + t^2)u)$.

6) $(tu - xn, tn - xu + t|\psi|^2)$.

7) (n, u).

8) $(u, n + |\psi|^2)$.

EXAMPLE 3.8 ([**104**]). The plasticity equations (see §4.3 of Ch. 4)
$$\begin{cases} u_\xi + \frac{1}{2}v = 0, \\ v_\eta + \frac{1}{2}u = 0, \end{cases}$$

possess an infinite series of conservation laws obtained from the conservation law with the generating function

$$T_0 = \begin{pmatrix} u \\ -v \end{pmatrix}$$

by using the action of the symmetries f_{2k}^i, g_{2k}^i, $0 \leq i \leq 2k$, and the operators $(\mathcal{R}_2^{2k})^*$, $k = 0, 1, 2, \ldots$ (see Theorem 4.5 of Ch. 4). This series, together with the conservation laws corresponding to generating functions $(B_1(\xi, \eta), B_2(\xi, \eta))$ which are the solutions of the system

$$\begin{cases} \dfrac{\partial B_1}{\partial \xi} - \dfrac{1}{2}B_2 = 0, \\ \dfrac{\partial B_2}{\partial \eta} - \dfrac{1}{2}B_1 = 0, \end{cases}$$

generate the whole space of conservation laws of the plasticity equations.

EXERCISE 3.6. Check that $(u\psi_1, v\psi_2)$ is a conserved current corresponding to the generating function (ψ_1, ψ_2).

EXAMPLE 3.9. The Khokhlov–Zabolotskaya equation encountered in the nonlinear acoustics of bounded beams (see §5.3 of Ch. 3) has the form

$$-\frac{\partial^2 u}{\partial q_1 \partial q_2} + \frac{1}{2}\frac{\partial^2 (u^2)}{\partial q_1^2} + \frac{\partial^2 u}{\partial q_3^2} + \frac{\partial^2 u}{\partial q_4^2} = 0.$$

It has a large group of conservation laws, with the corresponding generating functions being of the form

$$\psi = q_1 \mathcal{A}(q_2, q_3, q_4) + \mathcal{B}(q_2, q_3, q_4),$$

where $\mathcal{A}$ and $\mathcal{B}$ are arbitrary solutions of the system

$$\begin{cases} \mathcal{A}_{q_3 q_3} + \mathcal{A}_{q_4 q_4} = 0, \\ \mathcal{B}_{q_3 q_3} + \mathcal{B}_{q_4 q_4} = \mathcal{A}_{q_2}. \end{cases}$$

The conserved current

$$((q_1 \mathcal{A} - \mathcal{B})(u u_{q_1} - u_{q_2}) - \mathcal{A} u^2/2, \mathcal{A} u, (q_1 \mathcal{A} + \mathcal{B}) u_{q_3} - (q_1 \mathcal{A}_{q_3} + \mathcal{B}_{q_3}) u,$$
$$(q_1 \mathcal{A} + \mathcal{B}) u_{q_4} - (q_1 \mathcal{A}_{q_4} + \mathcal{B}_{q_4}) u).$$

corresponds to the generating function ψ.

The two-dimensional analog of the Khokhlov–Zabolotskaya equation is of the form

$$-\frac{\partial^2 u}{\partial q_1 \partial q_2} + \frac{1}{2}\frac{\partial^2 (u^2)}{\partial q_1^2} + \frac{\partial^2 u}{\partial q_3^2} = 0.$$

The generating functions of conservation laws for this equation are

$$\psi = q_1 a(q_2) q_3 + a'(q_2) q_3^3/6 + q_1 b(q_2) + b'(q_2) q_3^2/2 + c(q_2) q_3 + d(q_2), \qquad (3.9)$$

where a, b, c, and d are arbitrary functions of q_2.

For the axially symmetric Khokhlov–Zabolotskaya equation

$$-\frac{\partial^2 u}{\partial q_1 \partial q_2} + \frac{1}{2}\frac{\partial^2 (u^2)}{\partial q_1^2} + \frac{\partial^2 u}{\partial q_3^2} + \frac{1}{q_3}\frac{\partial u}{\partial q_3} = 0$$

the generating functions of conservation laws have the form

$$\begin{aligned}\psi = {}& q_1 a(q_2) q_3 \ln q_3 + \tfrac{1}{4} a'(q_2) q_3^3 (\ln q_3 - 1) \\ & + q_1 b(q_2) q_3 + \tfrac{1}{4} b'(q_2) q_3^3 + c(q_2) q_3 \ln q_3 + d(q_2) q_3,\end{aligned} \qquad (3.10)$$

where a, b, c, and d are again arbitrary functions of q_2.

EXERCISE 3.7. Find the conserved currents corresponding to the generating functions (3.9) and (3.10).

EXAMPLE 3.10. The Navier–Stokes equations of viscous incompressible fluid motion in a three-dimensional domain,

$$\begin{cases} \dfrac{\partial \boldsymbol{u}}{\partial t} + \boldsymbol{u} \cdot \nabla \boldsymbol{u} = -\nabla p + \nu \Delta \boldsymbol{u}, \\ \nabla \cdot \boldsymbol{u} = 0, \end{cases}$$

where $\boldsymbol{u} = (u^1, u^2, u^3)$ is the vector of fluid velocity and p is the pressure, have a 7-dimensional space of conservation laws. The basis generating functions are of the form:

$$\psi_1 = (x_2, -x_1, 0, u^1 x_2 - u^2 x_1),$$
$$\psi_2 = (x_3, 0, -x_1, u^1 x_3 - u^3 x_1),$$
$$\psi_3 = (0, x_3, -x_2, u^2 x_3 - u^3 x_2),$$
$$\psi_4 = (a_1(t), 0, 0, u^1 a_1(t) - a_1'(t) x_1),$$
$$\psi_5 = (0, a_2(t), 0, u^2 a_2(t) - a_2'(t) x_2),$$
$$\psi_6 = (0, 0, a_3(t), u^3 a_3(t) - a_3'(t) x_3),$$
$$\psi_7 = (0, 0, 0, f(t)),$$

where a_1, a_2, a_3, and f are arbitrary functions of t.

EXERCISE 3.8. Find the conserved currents for these generating functions and explain their physical meaning.

EXAMPLE 3.11. The Kadomtsev–Pogutse equations (see §5.4 of Ch. 3), describing nonlinear processes in a high-temperature plasma, are written as follows:

$$\begin{cases} \dfrac{\partial \psi}{\partial t} + [\nabla_\perp \varphi, \nabla_\perp \psi]_z = \dfrac{\partial \varphi}{\partial z}, \\ \dfrac{\partial}{\partial t} \Delta_\perp \varphi + [\nabla_\perp \varphi, \nabla_\perp \Delta_\perp \varphi]_z = \dfrac{\partial}{\partial t} \Delta_\perp \psi + [\nabla_\perp \psi, \nabla_\perp \Delta_\perp \psi]_z, \end{cases}$$

where $\nabla_\perp = (\partial/\partial x, \partial/\partial y)$, $\Delta_\perp = \partial^2/\partial x^2 + \partial^2/\partial y^2$, $[u,v]_z = u_x v_y - u_y v_x$, (x, y, z, t) are the standard coordinates in the space-time, and φ and ψ are the potentials of the velocity and the cross-component of the magnetic field respectively.

Here are the generating functions of conservation laws for this equation depending on the derivatives of order no greater than three[11]:

$$\theta_1 = ((ax + by + c) G(z,t), 0),$$
$$\theta_2 = (\alpha(x^2 + y^2), -4\alpha),$$
$$\theta_3 = (\beta(x^2 + y^2), 4\beta),$$
$$\theta_4 = (\gamma(\varphi - \psi), \gamma(\Delta_\perp \psi - \Delta_\perp \varphi)),$$
$$\theta_5 = (\delta(\varphi + \psi), \delta(\Delta_\perp \varphi + \Delta_\perp \psi)),$$

where $G = G(z,t)$, $\alpha = \alpha(z-t)$, $\beta = \beta(z+t)$, $\gamma = \gamma(z-t)$, $\delta = \delta(z+t)$, and a, b, c are arbitrary constants.

EXERCISE 3.9. Find conserved currents for the functions $\theta_1, \ldots, \theta_5$. What is the physical meaning of the conservation law corresponding to θ_4?

EXERCISE 3.10. Show that the transformation $t \mapsto -z$, $\varphi \mapsto -\psi$, which is a discrete symmetry of the Kadomtsev–Pogutse equations, takes the conservation laws with generating functions θ_2 and θ_4 to the conservation laws with generating functions θ_3 and θ_5 respectively.

---

[11] The Kadomtsev–Pogutse equations supposedly do not have other conservation laws.

4. Symmetries and conservation laws

4.1. The Noether theorem. It is well known that for equations derived from a variational principle the conservation laws are produced by the symmetries of the action. This is the Noether theorem [90]. Prior to the works [125, 127, 132] on the $\mathcal{C}$-spectral sequence this theorem had been the only general method of finding conservation laws, and even now it is widely believed that the existence of conserved quantities is always caused by symmetry properties of the equation at hand. However, the examples of §3 demonstrate that, in general, this is not true. For instance, the Burgers equation has only one conservation law, whereas its symmetry algebra is infinite-dimensional (see Theorem 4.4 of Ch. 4).

On the other hand, symmetries and conservation laws cannot be treated as totally independent concepts, since the generating functions of symmetries and of conservation laws satisfy the mutually adjoint equations (3.19) of Ch. 4 and (3.2). This fact clarifies the nature of the Noether theorem. Recall (see §2.5) that an equation $F = 0$ is derived (locally) from a variational principle if and only if $\ell_F = \ell_F^*$. Therefore, in this case equations (3.19) of Ch. 4 and (3.2) coincide and, consequently, to every conservation law there corresponds a symmetry. It follows from equality (2.22) that this correspondence is inverse to the Noether map that takes each Noether symmetry of the action to a conservation law of the equation.

Recall that a *Lagrangian* (or an *action*, or a *variational functional*) $\mathcal{L} = \int L(x, u, p_\sigma^i) \, dx$, where $dx = dx_1 \wedge \cdots \wedge dx_n$, is a functional on the set of sections of a bundle π and is of the form

$$s \mapsto \int_M L\left(x, s(x), \ldots, \frac{\partial^{|\sigma|} s^i}{\partial x_\sigma}, \ldots\right) dx_1 \wedge \cdots \wedge dx_n, \qquad s \in \Gamma(\pi).$$

The function $L \in \mathcal{F}(\pi)$ (or the form $L\, dx \in \Lambda_0^n(\pi)$) is called the *density* of the Lagrangian. The *Lagrangian* (or *variational*) *derivative* of the Lagrangian $\mathcal{L} = \int L\, dx$ is the function

$$\frac{\delta L}{\delta u^i} = \sum_\sigma (-1)^{|\sigma|} D_\sigma \left(\frac{\partial L}{\partial p_\sigma^i}\right) \in \mathcal{F}(\pi).$$

EXERCISE 4.1. Check that the Lagrangian derivative $\delta L/\delta u^i$ is uniquely determined by the functional $\mathcal{L} = \int L\, dx$ and does not depend on the choice of the density L.

The operator $\mathsf{E}(\mathcal{L}) = (\delta L/\delta u^1, \ldots, \delta L/\delta u^m)$ is called the *Euler operator*, while the equation $\mathsf{E}(\mathcal{L}) = 0$ is called the Euler–Lagrange equation corresponding to the Lagrangian $\mathcal{L}$. In §2.5 we proved that an Euler–Lagrange equation is trivial if and only if the corresponding density L is a total divergence, i.e., if for some functions $P_1, P_2, \ldots, P_n \in \mathcal{F}(\pi)$ one has

$$L = D_1(P_1) + D_2(P_2) + \cdots + D_n(P_n).$$

Furthermore, it was shown in §2.5 that an equation $F = 0$ is an Euler–Lagrange equation if and only if

$$\ell_F = \ell_F^*.$$

An evolutionary derivation ϑ_φ is called a *Noether symmetry* of a Lagrangian $\mathcal{L}$ if $\vartheta_\varphi(\mathcal{L}) = 0$, i.e., if there exist functions $P_1, P_2, \ldots, P_n \in \mathcal{F}(\pi)$ such that

$$\vartheta_\varphi(L) = D_1(P_1) + D_2(P_2) + \cdots + D_n(P_n).$$

If $F = 0$ is an Euler–Lagrange equation corresponding to a Lagrangian $\mathcal{L}$, i.e., $F = \mathsf{E}(\mathcal{L})$, then ∂_φ is a Noether symmetry of $\mathcal{L}$ if and only if

$$\partial_\varphi(F) + \ell_\varphi^*(F) = 0 \tag{4.1}$$

(see Exercise 2.30). In particular, φ is a symmetry of the equation $F = 0$ (the converse statement is *not* true!).

The results of §2.8 imply the following fundamental theorem:

THE NOETHER THEOREM. *Suppose $\mathcal{E} = \{\mathsf{E}(\mathcal{L}) = 0\}$ is an ℓ-normal Euler–Lagrange equation corresponding to a Lagrangian $\mathcal{L}$. Then an evolutionary derivation ∂_φ is a Noether symmetry of $\mathcal{L}$ if and only if φ is a generating function of a conservation law for $\mathcal{E}$[12].*

EXAMPLE 4.1. Consider the sine-Gordon equation

$$u_{xy} = \sin u.$$

It is an Euler–Lagrange equation (check the condition $\ell_F = \ell_F^*$!), the corresponding Lagrangian density being equal to

$$L = \tfrac{1}{2} u_x u_y - \cos u.$$

EXERCISE 4.2. Using (4.1), show that

$$\varphi_1 = u_x,$$
$$\varphi_2 = u_{xxx} + \tfrac{1}{2} u_x^3,$$
$$\varphi_3 = u_{xxxxx} + \tfrac{5}{2} u_x^2 u_{xxx} + \tfrac{5}{2} u_x u_{xx}^2 + \tfrac{3}{8} u_x^5$$

are Noether symmetries of the Lagrangian under consideration.

REMARK 4.1. The symmetries φ_2 and φ_3 are produced from φ_1 by means of the recursion operator $R = D_x^2 + u_x^2 - u_x D_x^{-1} \cdot u_{xx}$: $\varphi_2 = R(\varphi_1)$, $\varphi_3 = R(\varphi_2) = R^2(\varphi_1)$.

By the Noether theorem, φ_1, φ_2, and φ_3 are the generating functions of conservation laws for the sine-Gordon equation.

EXAMPLE 4.2. Consider the problem on the motion of r particles of masses $m_1, \ldots, m_r$ in a potential field. Let $x^j = (x_1^j, x_2^j, x_3^j)$ be the coordinates of the j-th particle. It is well-known that the Newton equations of motion

$$m_j x_{tt}^j = -\operatorname{grad}_j U = \left(-\frac{\partial U}{\partial x_1^j}, -\frac{\partial U}{\partial x_2^j}, -\frac{\partial U}{\partial x_3^j} \right), \qquad j = 1, \ldots, r,$$

where $U(t, x_i^j)$ is the potential energy, are Euler–Lagrange equations with the Lagrangian density

$$L = \sum_{j=1}^{r} \tfrac{1}{2} m_j ((\dot{x}_1^j)^2 + (\dot{x}_2^j)^2 + (\dot{x}_3^j)^2) - U.$$

In this example, we consider the classical Noether symmetries with generating functions

$$\varphi_l^k = b_l^k(t, x_i^j) - a(t, x_i^j) \dot{x}_l^k, \qquad k = 1, \ldots, r, \quad l = 1, 2, 3.$$

---

[12] As noted above, equations encountered in gauge theories, like the Maxwell, Yang–Mills, and Einstein equations, make up an important class of Euler–Lagrange equations which are not ℓ-normal. For such an equation the kernel of the Noether map, i.e., the set of Noether symmetries that produce trivial conservation laws, coincides with the set of gauge symmetries.

Moreover, we restrict ourselves to symmetries satisfying the condition
$$X^{(1)}(L\,dt) = 0,$$
where $X = a(t, x_i^j)\partial/\partial t + \sum_{k,l} b_l^k(t, x_i^j)\partial/\partial x_l^k$ and $X^{(1)}$ is the lifting of X to the space of 1-jets (see Chapter 3).

By the Noether theorem, this symmetry gives rise to the first integral
$$\sum_{i,j} m_j b_i^j \dot{x}_i^j - aE = \text{const},$$
where $E = \sum_{j=1}^{r} m/2((\dot{x}_1^j)^2 + (\dot{x}_2^j)^2 + (\dot{x}_3^j)^2) + U$ is the total energy (check this).

If the potential energy U does not explicitly depend on the time, i.e., $\partial U/\partial t = 0$, then, as is easy to see, $\varphi_i^j = \dot{x}_i^j$ is a Noether symmetry. The resulting first integral is the energy $E = \text{const}$. Here we have a commonly encountered situation: the conservation of energy manifests the symmetry under time translations.

If the function U is invariant under spatial shifts in a fixed direction $\mathbf{l} = (l_1, l_2, l_3) \in \mathbb{R}^3$, then $\varphi_i^j = l_i$ is a Noether symmetry. The corresponding first integral is the momentum
$$\sum_{i,j} m_j l_i \dot{x}_i^j = \text{const}.$$

We see that conservation of momentum follows from invariance under spatial shifts.

Yet another example of this kind: invariance of U under rotations implies conservation of angular momentum. Take, e.g., the z-axis. The corresponding Noether symmetry has the form $\varphi_1^j = -x_2^j$, $\varphi_2^j = x_1^j$, $\varphi_3^j = 0$. It implies conservation of the z-component of angular momentum:
$$\sum_j m_j(x_1^j \dot{x}_2^j - \dot{x}_1^j x_2^j) = \text{const}.$$

REMARK 4.2. From our discussion of the Noether theorem it is readily seen that the connection between symmetries and conservation laws, given by the inverse Noether theorem, can be generalized to equations $\mathcal{E} = \{F = 0\}$ satisfying the condition
$$\left(\ell_F^{\mathcal{E}}\right)^* = \lambda \ell_F^{\mathcal{E}}$$
for a function $\lambda \in \mathcal{F}(\mathcal{E})$. The simplest example of this kind is the equation $u_x = u_y$. It is skew-adjoint: $\left(\ell_F^{\mathcal{E}}\right)^* = -\ell_F^{\mathcal{E}}$.

4.2. Hamiltonian equations. In this subsection we discuss Hamiltonian differential equations. Further details are available, for example, in [**8, 21, 24, 64, 81, 91, 126**].

An $m \times m$ matrix $\mathcal{C}$-differential operator $A = \|A^{ij}\|$ is called *Hamiltonian* if the corresponding Poisson bracket
$$\{\mathcal{L}_1, \mathcal{L}_2\}_A = \int \sum_{i,j} A^{ij}\left(\frac{\delta \mathcal{L}_1}{\delta u^j}\right) \frac{\delta \mathcal{L}_2}{\delta u^i}\, dx,$$
where $\mathcal{L}_1$ and $\mathcal{L}_2$ are Lagrangians, defines a Lie algebra structure on the space of Lagrangians. The skew-symmetry of the Poisson bracket is equivalent to the skew-adjointness of the operator A, i.e., to the equality $A^* = -A$. To check the Jacobi identity one uses the following equivalent criteria (see §2):

1. $\langle \ell_A(A(\psi_1))(\psi_2), \psi_3 \rangle + \langle \ell_A(A(\psi_2))(\psi_3), \psi_1 \rangle + \langle \ell_A(A(\psi_3))(\psi_1), \psi_2 \rangle = 0$ for all $\psi_1, \psi_2, \psi_3 \in \widehat{\varkappa}(\pi)$.
2. $\ell_{A,\psi_1}(A(\psi_2)) - \ell_{A,\psi_2}(A(\psi_1)) = A(\ell^*_{A,\psi_2}(\psi_1))$ for all $\psi_1, \psi_2 \in \widehat{\varkappa}(\pi)$.
3. The expression $\ell_{A,\psi_1}(A(\psi_2)) + \frac{1}{2}A(\ell^*_{A,\psi_1}(\psi_2))$ is symmetric with respect to $\psi_1, \psi_2 \in \widehat{\varkappa}(\pi)$.
4. $[\mathfrak{I}_{A(\psi)}, A] = \ell_{A(\psi)} \circ A + A \circ \ell^*_{A(\psi)}$ for all $\psi \in \mathrm{im}\, \mathsf{E}$.

Here $\psi_i = (\psi_i^1, \ldots, \psi_i^n)$ are vector functions, and $\ell_{A,\psi}$ is an $n \times n$ matrix $\mathcal{C}$-differential operator defined by

$$(\ell_{A,\psi})_\tau^{ij} = \sum_{k,\sigma} \frac{\partial A_\sigma^{ik}}{\partial p_\tau^j} D_\sigma(\psi^k) D_\tau.$$

Moreover, it is sufficient to verify conditions 1–3 for elements $\psi_i \in \mathrm{im}\, \mathsf{E}$ only.

EXERCISE 4.3. a. Write down the Hamiltonian criteria in coordinate form.
b. Show that $\ell^*_{A,\psi_1}(\psi_2) = \ell^*_{A^*,\psi_2}(\psi_1)$ (see Exercise 2.33).

EXAMPLE 4.3. Consider the case of one dependent and one independent variable $n = m = 1$. The simplest Hamiltonian operator of order one is the operator D_x.

EXERCISE 4.4. Describe all Hamiltonian operators of order one in this case.

EXAMPLE 4.4. It is clear that any skew-adjoint operator with coefficients dependent on x only is a Hamiltonian operator.

EXERCISE 4.5. Prove that in the case $n = 1$ an operator of the form $A = 2LD_x + D_x(L)$, where L is a symmetric matrix with elements dependent on x and u^i only, is Hamiltonian.

EXERCISE 4.6. Prove that in the case $n = m = 1$ there is a two-parameter family of Hamiltonian operators of order three

$$\Gamma_{\alpha,\beta} = D_x^3 + (\alpha + \beta u)D_x + \tfrac{1}{2}\beta u_x. \tag{4.2}$$

An evolution differential equation is said to be *Hamiltonian* (with respect to a Hamiltonian operator A) if it has the form

$$u_t^i = A^{ij} \frac{\delta \mathcal{H}}{\delta u^j}$$

for some action functional $\mathcal{H}$, called the *Hamiltonian*. One can prove (see Theorem 2.13) that the operator A takes the generating functions of conservation laws of a Hamiltonian equation to its symmetries.

EXAMPLE 4.5. The Zakharov equations (3.8)

$$\begin{cases} i\psi_t + \psi_{xx} - n\psi = 0, \\ n_t + u_x = 0, \\ u_t + n_x + \left(|\psi|^2\right)_x = 0 \end{cases}$$

are Hamiltonian with respect to the operator

$$A = \begin{pmatrix} 0 & 1/2 & 0 & 0 \\ -1/2 & 0 & 0 & 0 \\ 0 & 0 & 0 & -D_x \\ 0 & 0 & -D_x & 0 \end{pmatrix},$$

with the energy functional
$$\mathcal{H} = \int \left(|\psi_x|^2 + n|\psi|^2 + \frac{n^2}{2} + \frac{u^2}{2} \right) dx$$
being the Hamiltonian.

EXERCISE 4.7. Using Example 3.7, find all Hamiltonian symmetries of the Zakharov equations[13].

EXAMPLE 4.6. The Korteweg–de Vries equation
$$u_t = 6uu_x - u_{xxx}$$
can be written in the Hamiltonian form in two different ways. The first Hamiltonian structure is quite obvious:
$$u_t = D_x(3u^2 - u_{xx}) = D_x \left(\frac{\delta}{\delta u} \left(u^3 + \frac{1}{2}u_x^2 \right) \right) = A_1 \frac{\delta}{\delta u} \mathcal{H}_1,$$
where $A_1 = D_x$ is the Hamiltonian operator and $\mathcal{H}_1 = \int (u^3 + 1/2 u_x^2)\, dx$ is the Hamiltonian.

The second Hamiltonian structure is written as follows:
$$u_t = (D_x^3 - 4uD_x - 2u_x) \left(\frac{\delta}{\delta u} \left(-\frac{u^2}{2} \right) \right) = A_2 \frac{\delta}{\delta u} \mathcal{H}_0.$$
Here $A_2 = D_x^3 - 4uD_x - 2u_x$, $\mathcal{H}_0 = \int(-u^2/2)\, dx$. The Hamiltonian operator A_2 belongs to the family (4.2) with $A_2 = \Gamma_{0,4}$.

We know (see §5.2 of Ch. 3) that the basis of the space of classical symmetries for the Korteweg–de Vries equation has the form

$\varphi_1 = u_x$ (translation along x),

$\varphi_2 = u_t$ (translation along t),

$\varphi_3 = 6tu_x + 1$ (Galilean boost),

$\varphi_4 = xu_x + 3tu_t + 2u$ (scale symmetry).

The first three symmetries are Hamiltonian with respect to the operator $A_1 = D_x$, i.e., they are of the form
$$\varphi_i = A_1 \frac{\delta}{\delta u} \mathcal{Q}_i, \qquad i = 1, 2, 3, \tag{4.3}$$
where
$$\mathcal{Q}_1 = \int \frac{u^2}{2} dx,$$
$$\mathcal{Q}_2 = \int \left(u^3 + \frac{u_x^2}{2} \right) dx,$$
$$\mathcal{Q}_3 = \int (xu + 3tu^2)\, dx.$$

Thus, $\mathcal{Q}_i$ are conserved quantities ($\psi_i = \delta \mathcal{Q}_i / \delta u$ are the corresponding generating functions). The fourth symmetry φ_4 is not of the form (4.3) and does not correspond to a conservation law. Note that the Hamiltonian operator $A_1 = D_x$ has a one-dimensional kernel, so that there exists one more conservation law with the generating function $\psi = 1$ (which is the conservation law of mass, $\mathcal{Q}_0 = \int u\, dx$).

---

[13]In fact, the Zakharov equations have no other symmetries.

Let us turn to the second Hamiltonian operator $A_2 = D_x^3 - 4uD_x - 2u_x$. In this case, the symmetries φ_1, φ_2, and φ_4 are Hamiltonian:

$$\varphi_i = A_2 \frac{\delta}{\delta u} \mathcal{P}_i, \qquad i = 1, 2, 4,$$

where

$$\mathcal{P}_1 = \int -\frac{u}{2} \, dx = -\frac{\mathcal{Q}_0}{2},$$

$$\mathcal{P}_2 = \int -\frac{u^2}{2} \, dx = -\mathcal{Q}_1,$$

$$\mathcal{P}_4 = \int \left(-\frac{xu}{2} - \frac{3tu^2}{2} \right) dx = -\frac{\mathcal{Q}_3}{2}.$$

The symmetry φ_3 is not Hamiltonian.

None of the classical symmetries corresponds to the conservation law of energy $\mathcal{Q}_2$. Therefore this conservation law is produced by a higher symmetry, the corresponding generating function being of the form

$$\varphi_5 = A_2 \frac{\delta}{\delta u} \mathcal{Q}_2 = -u_{xxxxx} + 10uu_{xxx} + 20u_x u_{xx} - 30u^2 u_x.$$

It turns out that the symmetry φ_5 satisfies the Hamiltonian condition (4.3) for the operator A_1, and the corresponding functional

$$\mathcal{Q}_5 = \int \left(-\frac{u_{xx}^2}{2} - 5uu_x^2 - \frac{5u^4}{2} \right) dx$$

is one more conservation law for the Korteweg–de Vries equation. The procedure whereby the conserved quantity $\mathcal{Q}_5$ is obtained from the conserved quantity $\mathcal{Q}_2$ can be applied to $\mathcal{Q}_5$ to give a new conservation law, and so on. In other words, we have

PROPOSITION 4.1. *There exists an infinite series of conserved quantities of the Korteweg–de Vries equation $\mathcal{H}_0, \mathcal{H}_1, \mathcal{H}_2, \ldots$, such that*:

1. $\mathcal{H}_0 = \int -u^2/2 \, dx$.
2. $A_1 \delta/\delta u (\mathcal{H}_i) = A_2 \delta/\delta u (\mathcal{H}_{i-1})$.
3. *The functionals $\mathcal{H}_i$ are mutually in involution with respect to both the Poisson brackets*:

$$\{\mathcal{H}_k, \mathcal{H}_l\}_{A_1} = \{\mathcal{H}_k, \mathcal{H}_l\}_{A_2} = 0 \qquad \text{for all } k, l \geq 0.$$

4. *The symmetries $\varphi_i = A_1 \delta/\delta u(\mathcal{H}_i) = A_2 \delta/\delta u(\mathcal{H}_{i-1})$, $i > 0$, corresponding to the Hamiltonians $\mathcal{H}_i$ commute*:

$$\{\varphi_k, \varphi_l\} = 0 \qquad \text{for all } k, l > 0.$$

PROOF. The proof is left to the reader as an exercise. □

Note that the operator $\mathcal{R} = A_2 \circ A_1^{-1} = D_x^2 - 4u - 2u_x D_x^{-1}$ takes the symmetry φ_k to the symmetry φ_{k+1}.

EXERCISE 4.8. Prove that $\mathcal{R}$ is a recursion operator (see Remark 4.3 of Ch. 4) for the Korteweg–de Vries equation (the Lenard recursion operator).

EXERCISE 4.9. The Harry Dym equation has the form
$$u_t = \left(\frac{1}{\sqrt{u}}\right)_{xxx}.$$
Show that this equation possesses two Hamiltonian structures:
$$u_t = A_1 \frac{\delta}{\delta u}\mathcal{H}_1 = A_2 \frac{\delta}{\delta u}\mathcal{H}_0,$$
where $A_1 = 2uD_x + u_x$, $\mathcal{H}_1 = \int \left(\frac{u_{xx}}{2\sqrt{u^3}} - \frac{5u_x^2}{8\sqrt{u^5}}\right) dx$, $A_2 = D_x^3 = \Gamma_{0,0}$, $\mathcal{H}_0 = \int 2\sqrt{u}\, dx$. Study the symmetries and conservation laws of this equation and prove the analog of Proposition 4.1 for it.

CHAPTER 6

Nonlocal Symmetries

1. Coverings

In preceding chapters we dealt with local objects (symmetries, conservation laws, etc.), i.e., with objects depending on the unknown functions $u^1, \ldots, u^m$ in a differential way. For example, generating functions of symmetries, being elements of the module $\mathcal{F}(\pi, \pi)$, are differential operators in the bundle π. A natural way of generalization is to consider "integro-differential" dependencies, similar to extension of the base field in the theory of algebraic equations. This formal, at first glance, step leads to rather general geometric constructions, which are called *coverings* over infinitely prolonged differential equations [**61, 62, 142**]. Translating the main elements of the theory into the language of coverings, we arrive at the theory of nonlocal symmetries and nonlocal conservation laws for differential equations. It so happens that many well-known constructions (various differential substitutions, Bäcklund transformations, recursion operators, etc.) are elements of the covering theory [**62**].

1.1. First examples. We consider several examples clarifying how nonlocal objects arise.

EXAMPLE 1.1. Consider the Burgers equation $\mathcal{E} = \{u_t = uu_x + u_{xx}\}$. The algebra $\operatorname{sym}\mathcal{E}$ of local symmetries for this equation is identified with the kernel of the operator $\bar{\ell}_F = \bar{D}_x^2 + p_0\bar{D}_x + p_1 - \bar{D}_t$ (see Chapter 4), where

$$\bar{D}_x = \frac{\partial}{\partial x} + \sum_k p_{k+1}\frac{\partial}{\partial p_k}, \qquad \bar{D}_t = \frac{\partial}{\partial t} + \sum_k \bar{D}_x^k(p_0 p_1 + p_2)\frac{\partial}{\partial p_k}.$$

Let us try to find a simplest "integro-differential" symmetry of the Burgers equation, whose generating function depends on the "integral" variable $p_{-1} = \int p_0\,dx = \bar{D}_x^{-1}(p_0)$. Formally, this means that we must extend the manifold $\mathcal{E}^\infty$ up to a manifold $\widetilde{\mathcal{E}}$ by adding a new coordinate to the intrinsic coordinates on $\mathcal{E}^\infty$. The total derivative of this new coordinate function along x should be equal to p_0. The total derivative along t is naturally defined as follows:

$$\bar{D}_t(p_{-1}) = \bar{D}_t(\bar{D}_x^{-1}(p_0)) = \bar{D}_x^{-1}(\bar{D}_t(p_0)) = \bar{D}_x^{-1}(p_2 + p_0 p_1) = p_1 + \tfrac{1}{2}p_0^2 + c,$$

where $c = c(t)$ is the "constant of integration", which may be set equal to zero. In other words, simultaneously with extending the manifold $\mathcal{E}^\infty$ we have to extend the total derivative operators $\bar{D}_x$ and $\bar{D}_t$ by setting

$$\widetilde{D}_x = \bar{D}_x + p_0\frac{\partial}{\partial p_{-1}}, \qquad \widetilde{D}_t = \bar{D}_t + \left(p_1 + \tfrac{1}{2}p_0^2\right)\frac{\partial}{\partial p_{-1}}.$$

Obviously, the integrability condition $[\widetilde{D}_x, \widetilde{D}_t] = 0$ must be preserved.

The operator $\bar{\ell}_F$ is naturally extended up to the operator

$$\widetilde{\ell}_F = \widetilde{D}_x^2 + p_0 \widetilde{D}_x + p_1 - \widetilde{D}_t$$

in this case. It is natural to call a solution of the equation $\widetilde{\ell}_F(\varphi) = 0$ a *nonlocal symmetry* of the Burgers equation depending on p_{-1}.

If the function φ depends on the variables x, t, p_{-1}, and p_0 only, then the equation $\widetilde{\ell}_F(\varphi) = 0$ is easy to solve explicitly, and we obtain the solution

$$\varphi_a = \left(ap_0 - 2\frac{\partial a}{\partial x}\right)e^{-\frac{1}{2}p_{-1}},$$

where $a = a(x,t)$ is an arbitrary solution of the heat equation $a_t = a_{xx}$.

To show possible applications of the above computations, let us find the solutions of the Burgers equation invariant with respect to some symmetry of the form φ_a. By the general theory (see Chapter 4), such solutions are determined by the following system of equations:

$$\begin{cases} u_t = u_{xx} + uu_x, \\ \left(au - 2\dfrac{\partial a}{\partial x}\right)e^{-\frac{1}{2}\int u\,dx} = 0. \end{cases}$$

From this system we obtain, in particular, that $u = 2a_x/a$, i.e., we obtain the Cole–Hopf transformation reducing the Burgers equation to the heat equation.

"Integro-differential" symmetries naturally arise in studies of equation admitting a recursion operator.

EXAMPLE 1.2. Consider the Korteweg–de Vries equation[1]

$$\mathcal{E} = \{u_t = u_{xxx} + uu_x\}.$$

The operator

$$R = \bar{D}_x^2 + \tfrac{2}{3}p_0 + \tfrac{1}{3}p_1\bar{D}_x^{-1}$$

commutes with the universal linearization operator

$$\bar{\ell}_F = \bar{D}_x^3 + p_0\bar{D}_x + p_1 - \bar{D}_t.$$

Consequently, if $\bar{\ell}_F(\varphi) = 0$, then $\bar{\ell}_F(R(\varphi)) = R(\bar{\ell}_F(\varphi)) = 0$. This means that if $\varphi \in \operatorname{sym}\mathcal{E}$ and $R(\varphi) \in \mathcal{F}(\mathcal{E}^\infty)$, then $R(\varphi)$ is also a local symmetry of the equation $\mathcal{E}$. Such operators are called *recursion operators* (cf. Ch. 4; see also, e.g., [**91**]).

The Korteweg–de Vries equation possesses an infinite series of local symmetries $\varphi_k = R^k(p_1)$, where p_1 is the translation along x. It also possesses two other symmetries, Galilean and scale, and the recursion operator takes the former to the latter. These two symmetries can be considered as the starting point for the second series. But one can easily see that by applying the recursion operator to the scale

---

[1] We use here a representation of the KdV equation different from that considered in Chapters 3 and 5. Evidently, both forms are equivalent and reduce to each other by a scale transformation.

symmetry, we obtain the expression containing the term $\bar{D}_x^{-1}(p_0)$.

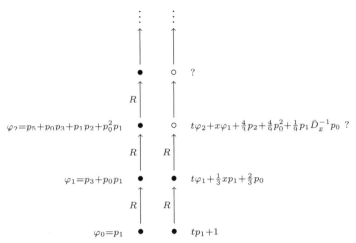

Nevertheless, introducing the additional variable p_{-1} and extending the operators $\bar{D}_x$, $\bar{D}_t$, and $\bar{\ell}_F$ to the operators

$$\widetilde{D}_x = \bar{D}_x + p_0 \frac{\partial}{\partial p_{-1}}, \quad \widetilde{D}_t = \bar{D}_t + \left(p_2 + \frac{1}{2}p_0^2\right)\frac{\partial}{\partial p_{-1}}, \quad \widetilde{\ell}_F = \widetilde{D}_x^3 + p_0 \widetilde{D}_x + p_1 - \widetilde{D}_t,$$

respectively, we shall see that the function $t\varphi_2 + x\varphi_1 + \frac{4}{3}p_2 + \frac{4}{9}p_0^2 + \frac{1}{9}p_1 p_{-1}$ is a solution of the equation $\bar{\ell}_F(\varphi) = 0$, and consequently may be called a nonlocal symmetry of the Korteweg–de Vries equation.

EXERCISE 1.1. Show that the operator $R = \bar{D}_x + \frac{1}{2}p_0 + \frac{1}{2}p_1 \bar{D}_x^{-1}$ is a recursion operator for the Burgers equation.

EXAMPLE 1.3. Consider the potential Korteweg–de Vries equation

$$\mathcal{E} = \{u_t = u_{xxx} + \tfrac{1}{2}u_x^2\}.$$

The procedure of introducing a new variable $p_{-1} = \bar{D}^{-1}(p_0)$ into this equation is more complicated. Namely, let us try to compute $\bar{D}_t(p_{-1})$:

$$\bar{D}_t(p_{-1}) = \bar{D}_t(\bar{D}_x^{-1}(p_0)) = \bar{D}_x^{-1}(\bar{D}_t(p_0)) = \bar{D}_x^{-1}(p_3 + \tfrac{1}{2}p_1^2) = p_2 + \tfrac{1}{2}\bar{D}_x^{-1}(p_1^2).$$

Here we meet the necessity of introducing another variable w whose total derivative along x equals p_1^2. Then

$$\bar{D}_t(w) = \bar{D}_t(\bar{D}_x^{-1}(p_1^2)) = \bar{D}_x^{-1}(\bar{D}_t(p_1^2)) = 2\bar{D}_x^{-1}(p_1(p_4 + p_1 p_2))$$
$$= 2(p_1 p_3 - \tfrac{1}{2}p_2^2 + \tfrac{1}{3}p_1^3).$$

Thus the operators $\bar{D}_x$ and $\bar{D}_t$ can be extended to the operators

$$\widetilde{D}_x = \bar{D}_x + p_0 \frac{\partial}{\partial p_{-1}} + p_1^2 \frac{\partial}{\partial w},$$

$$\widetilde{D}_t = \bar{D}_t + \left(p_2 + \frac{1}{2}w\right)\frac{\partial}{\partial p_{-1}} + 2\left(p_1 p_3 - \frac{1}{2}p_2^2 + \frac{1}{3}p_1^3\right)\frac{\partial}{\partial w}.$$

EXERCISE 1.2. Try to introduce the variable p_{-1} for the equation $\mathcal{E} = \{u_t = u_{xxx} + u_x^3\}$.

The last two examples show that attempts to apply the above constructions in a general situation may meet technical difficulties. In fact, we have no reason, except for a formal parallel with the local picture, to treat "nonlocal" solutions of the equation $\bar{\ell}_F(\varphi) = 0$ as nonlocal symmetries (see, e.g., [50, 30]). Obviously, we need a conceptual analysis of the notion of a nonlocal symmetry itself. Therefore, it seems natural to analyze a definition of nonlocal symmetries by studying geometrical structures on manifolds $\widetilde{\mathcal{E}}$: a similar path has led us to the definition of local symmetries of equations $\mathcal{E}^\infty$.

1.2. Definition of coverings. Let $\mathcal{E}$ be a differential equation in the bundle $\pi\colon E^{n+m} \to M^n$, and $\mathcal{E}^\infty$ its infinite prolongation. Recall that at any point $\theta \in \mathcal{E}^\infty$ the n-dimensional plane $\mathcal{C}_\theta \subset T_\theta(\mathcal{E}^\infty)$ is defined (the Cartan plane). The Cartan distribution $\mathcal{C} = \{\mathcal{C}_\theta\}_{\theta \in \mathcal{E}^\infty}$ on $\mathcal{E}^\infty$ is completely integrable, i.e., it satisfies the assumptions of the classical Frobenius theorem. In local coordinates, the Cartan distribution on $\mathcal{E}^\infty$ is given by the system of n vector fields $\bar{D}_1, \ldots, \bar{D}_n$, where $\bar{D}_i$ is the restriction to $\mathcal{E}^\infty$ of the total derivative operator along the i-th independent variable. The extension procedure of the manifold $\mathcal{E}^\infty$ considered in the previous examples may be formalized by introducing the concept of a covering.

DEFINITION 1.1. We shall say that a *covering* $\tau\colon \widetilde{\mathcal{E}} \to \mathcal{E}^\infty$ of the equation $\mathcal{E}$ is given, if the following objects are fixed:
1. A smooth manifold $\widetilde{\mathcal{E}}$, infinite-dimensional in general[2].
2. An n-dimensional integrable distribution $\widetilde{\mathcal{C}}$ on $\widetilde{\mathcal{E}}$.
3. A regular mapping τ of the manifold $\widetilde{\mathcal{E}}$ onto $\mathcal{E}^\infty$ such that for any point $\theta \in \widetilde{\mathcal{E}}$ the tangent mapping $\tau_{*,\theta}$ is an isomorphism of the plane $\widetilde{\mathcal{C}}_\theta$ to the Cartan plane $\mathcal{C}_{\tau(\theta)}$ of the equation $\mathcal{E}^\infty$ at the point $\tau(\theta)$.

The dimension of the bundle τ is called the dimension of the corresponding covering. Below, by $\mathcal{F}(\widetilde{\mathcal{E}})$ we denote the ring of smooth functions on $\widetilde{\mathcal{E}}$.

From the definition of coverings it follows that the mapping τ takes any n-dimensional integral manifold $\widetilde{\mathcal{U}} \subset \widetilde{\mathcal{E}}$ of the distribution $\widetilde{\mathcal{C}} = \{\widetilde{\mathcal{C}}_\theta\}_{\theta \in \widetilde{\mathcal{E}}}$ to an n-dimensional integral manifold $\mathcal{U} = \tau(\widetilde{\mathcal{U}}) \subset \mathcal{E}^\infty$ of the Cartan distribution on $\mathcal{E}^\infty$, i.e., to a solution of the equation $\mathcal{E}$. Conversely, if $\mathcal{U} \subset \mathcal{E}^\infty$ is a solution of the equation $\mathcal{E}$, then the restriction of the distribution $\widetilde{\mathcal{C}}$ to the inverse image $\widetilde{\mathcal{U}} = \tau^{-1}(\mathcal{U}) \subset \widetilde{\mathcal{E}}$, as is easily seen, is an integrable n-dimensional distribution. If, in particular, $\dim \tau^{-1}(\theta) = N < \infty$, $\theta \in \mathcal{U}$, then the manifold $\widetilde{\mathcal{U}}$ is locally fibered by an N-parameter family of integral manifolds of the distribution $\widetilde{\mathcal{C}}|_{\widetilde{\mathcal{U}}}$.

Thus, to any solution of the equation $\mathcal{E}$ there corresponds a family of integral manifolds lying in $\widetilde{\mathcal{E}}$. The elements of this family are conveniently understood as solutions of $\mathcal{E}$ parametrized by some nonlocal quantities. In Examples 1.1 and 1.2 the integration constant can be taken for such a parameter.

1.3. Coverings in the category of differential equations. To explain the term "covering" for the objects introduced above, we use a parallel between the category of differential equation (called the DE category) and the category of smooth manifolds.

---

[2]All infinite-dimensional manifolds considered below are inverse limits of chains of finite-dimensional ones. Differential geometry for such manifolds is constructed in the same way as it was for the manifolds of the form $\mathcal{E}^\infty$ (see Chapter 4); this allows us to overcome the usual topological difficulties.

In what follows, we take a very simple approach to the DE category suitable for our purposes, though it should be noted that there exists a more thorough approach to the definition of this category (see, for example, [**60**]).

The *objects* of the DE category are (infinite-dimensional) manifolds $\mathcal{O}$ endowed with completely integrable finite-dimensional distributions $\mathcal{P}$. Obviously, infinitely prolonged equations $\mathcal{E}^\infty$ together with the Cartan distribution $\mathcal{C}$ are objects of this category. The *morphisms* in the DE category are smooth mappings $\varphi\colon \mathcal{O} \to \mathcal{O}'$ such that $\varphi_*(\mathcal{P}_\theta) \subset \mathcal{P}'_{\varphi(\theta)}$ at all points $\theta \in \mathcal{O}$, where $\mathcal{P}$ and $\mathcal{P}'$ are the distributions on $\mathcal{O}$ and $\mathcal{O}'$ respectively. The *dimension* of the object $(\mathcal{O},\mathcal{P})$ is the dimension of the distribution $\mathcal{P}$.

Using as an analogy the notion of a covering in the category of smooth manifolds, we can give the following definition:

DEFINITION 1.2. A surjective map $\varphi\colon \mathcal{O} \to \mathcal{O}'$ is called a *covering* in the DE category if it preserves dimensions, i.e., if $\mathrm{Dim}\,\mathcal{O} = \mathrm{Dim}\,\mathcal{O}'$ and for any point $\theta \in \mathcal{O}$ one has $\varphi_*(\mathcal{P}_\theta) = \mathcal{P}'_{\varphi(\theta)}$.

1.4. Examples of coverings. Let us consider some examples of coverings arising in various situations.

EXAMPLE 1.4 (the covering associated to a differential operator). Let us consider a differential operator $\Delta\colon \Gamma(\pi) \to \Gamma(\pi')$ of order k acting from sections of the bundle $\pi\colon E \to M$ to sections of $\pi'\colon E' \to M$. The operator Δ and its prolongations $\Delta_s = j_s \circ \Delta$ determine the family of smooth mappings $\varphi_s = \Phi_{\Delta_s}\colon J^{k+s}(\pi) \to J^s(\pi')$ (see §§1.2 and 1.3 of Ch. 4), which can be included in the following commutative diagram:

$$\begin{array}{ccccccccc}
\cdots & \longrightarrow & J^{k+s+1}(\pi) & \xrightarrow{\pi_{k+s+1,k+s}} & J^{k+s}(\pi) & \longrightarrow & \cdots & \longrightarrow & J^k(\pi) \\
& & \varphi_{s+1}\downarrow & & \varphi_s\downarrow & & & & \varphi_0\downarrow \;\;\searrow^{\pi_k} \\
\cdots & \longrightarrow & J^{s+1}(\pi') & \xrightarrow{\pi'_{s+1,s}} & J^s(\pi') & \longrightarrow & \cdots & \longrightarrow & J^0(\pi') = E' \xrightarrow{\pi'} M
\end{array}$$

It is easily seen that if the operator Δ is regular (i.e., if all the mappings φ_s are surjections), then the system of mappings $\varphi = \{\varphi_s\}$ determines the covering $\varphi\colon J^\infty(\pi) \to J^\infty(\pi')$. We shall say that this is the covering *associated to the operator* Δ.

If an equation $\mathcal{E} \subset J^k(\pi')$ is given, then, using the above construction, one can construct the covering $\varphi\colon \varphi^{-1}(\mathcal{E}^\infty) \to \mathcal{E}^\infty$. Such coverings correspond to differential changes of variables

$$v^i = \varphi^i(x, u^1, \ldots, u^m, \ldots, u^j_\sigma),$$

where v^i and u^j are dependent variables in the bundles π' and π respectively. We shall consider this type of covering in the next two examples.

EXAMPLE 1.5 (Laplace transformation). Consider an equation of the form

$$u_{xy} + Au_x + Bu_y + Cu = 0. \tag{1.1}$$

Recall (see [**92**]) that the functions $h = \partial A/\partial x + AB - C$ and $l = \partial B/\partial y + AB - C$ are called *Laplace invariants* of the equation (1.1).

Assume that $h, l \neq 0$ and consider one-dimensional coverings over the equation $\mathcal{E}^\infty$ determined by the operators

$$\tau_h \colon u = \frac{w_x^1 + w^1}{h} \qquad (x \text{ Laplace transformation}),$$

$$\tau_l \colon u = \frac{w_y^2 + w^2}{l} \qquad (y \text{ Laplace transformation}).$$

It is easily checked that the spaces of these coverings have the form $(\mathcal{E}')^\infty$, where $\mathcal{E}'$ is of the form (1.1) for some new functions A, B, and C. Let us denote the resulting equations by $\mathcal{E}_h$ and $\mathcal{E}_l$ respectively. If the Laplace invariant l of the equation $\mathcal{E}_h$ does not vanish, one can consider the equation $(\mathcal{E}_h)_l$ (or, similarly, $(\mathcal{E}_l)_h$), and this equation, as it happens, is equivalent to $\mathcal{E}$. In such a way, we obtain a two-dimensional covering of the equation $\mathcal{E}^\infty$ over itself. If, at some step of this procedure, we obtain an equation for which either h or l vanishes, we shall be able to construct the formula for a general solution of the equation under consideration (further details can be found in [**92**]).

EXAMPLE 1.6 (the covering associated to a local symmetry). Let an equation $\mathcal{E} \subset J^k(\pi)$ be given and suppose that φ is the generating function of some symmetry of this equation. Then, as was shown in Example 1.4, the covering $\varphi \colon \varphi^{-1}(\mathcal{E}^\infty) \to \mathcal{E}^\infty$ arises. If the equation $\mathcal{E}$ is determined by an operator Δ, i.e., $\mathcal{E} = \{\varphi_\Delta = 0\}$, then the function φ satisfies the equation $\bar{\ell}_\Delta(\varphi) = 0$, where $\bar{\ell}_\Delta$ is the universal linearization operator for Δ restricted to $\mathcal{E}^\infty$. Thus the space of the covering under consideration is determined by the system of equations

$$\begin{cases} \bar{\ell}_\Delta(\varphi) = 0, \\ \varphi^j = u^j, \qquad j = 1, \ldots, \dim \pi. \end{cases}$$

If $\mathcal{E}$ is a linear equation, we may assume that $\bar{\ell}_\Delta = \Delta$. This means that the covering space is the equation $\mathcal{E}^\infty$ itself.

EXAMPLE 1.7 (factorization of differential equations). Let us consider a differential equation $\mathcal{E} \subset J^k(\pi)$, and let F be a finite-dimensional subgroup in the group of its classical symmetries. It is clear that the factorization mapping (see §6 of Ch. 3) $\pi_F \colon \mathcal{E}^\infty \to \mathcal{E}^\infty / F$ is a covering.

For example, the heat equation $u_t = u_{xx}$ admits the one-parameter group of scale symmetries $u \mapsto \varepsilon u$, $\varepsilon \in \mathbb{R}$. The corresponding quotient equation is the Burgers equation. Thus the covering of the Burgers equation by the heat equation arises.

1.5. Coordinates. Consider a coordinate interpretation of the covering concept.

The manifold $\widetilde{\mathcal{E}}$ and the mapping $\tau \colon \widetilde{\mathcal{E}} \to \mathcal{E}^\infty$, by regularity of τ, can be locally realized as the direct product $\mathcal{E}^\infty \times W$, where $W \subseteq \mathbb{R}^N$ is an open set, $0 < N \leq \infty$, and as the natural projection $\mathcal{E}^\infty \times W \to \mathcal{E}^\infty$ respectively. Then the distribution $\widetilde{\mathcal{C}}$ on $\widetilde{\mathcal{E}} = \mathcal{E}^\infty \times W$ can be described by the system of vector fields (see Figure 6.1)

$$\widetilde{D}_i = \bar{D}_i + \sum_{j=1}^N X_{ij} \frac{\partial}{\partial w_j}, \qquad i = 1, \ldots, n, \tag{1.2}$$

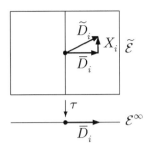

FIGURE 6.1

where $X_i = \sum_{j=1}^{N} X_{ij} \partial/\partial w_j$, $X_{ij} \in \mathcal{F}(\widetilde{\mathcal{E}})$ are τ-vertical fields on $\widetilde{\mathcal{E}}$, and $w_1, w_2, \ldots$ are standard coordinates in $\mathbb{R}^N$. In these terms, the Frobenius condition is equivalent to the equations $[\widetilde{D}_i, \widetilde{D}_j] = 0$, $i, j = 1, \ldots, n$, or, which is the same, to

$$\widetilde{D}_i(X_{jk}) = \widetilde{D}_j(X_{ik}) \tag{1.3}$$

for all $i, j = 1, \ldots, n$, $0 \le k \le N$ (since $[\bar{D}_i, \bar{D}_j] = 0$).

Relations (1.3) constitute a system of differential equations in functions X_{ij} describing all possible N-dimensional coverings over the equation $\mathcal{E}$.

The coordinates w_i will be called *nonlocal*.

1.6. Basic concepts of covering theory. In this subsection, we introduce some concepts of covering theory needed below.

DEFINITION 1.3. Two coverings $\tau_i \colon \widetilde{\mathcal{E}}_i \to \mathcal{E}^\infty$, $i = 1, 2$, are called *equivalent*, if there exists a diffeomorphism $\alpha \colon \widetilde{\mathcal{E}}_1 \to \widetilde{\mathcal{E}}_2$ such that the diagram

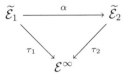

is commutative and $\alpha_*(\widetilde{\mathcal{C}}_y^1) = \widetilde{\mathcal{C}}_{\alpha(y)}^2$ for all points $y \in \widetilde{\mathcal{E}}_1$.

EXERCISE 1.3. Show that if a diffeomorphism α constitutes an equivalence of coverings τ_1 and τ_2 over the equation $\mathcal{E}^\infty$ and the coverings are given by the fields

$$\widetilde{D}_i^{(1)} = \bar{D}_i + \sum_k X_{ik}^{(1)} \frac{\partial}{\partial w_k^{(1)}}, \qquad \widetilde{D}_i^{(2)} = \bar{D}_i + \sum_l X_{ik}^{(2)} \frac{\partial}{\partial w_k^{(2)}},$$

respectively, $i = 1, \ldots, n$, then $\alpha_* \widetilde{D}_i^{(1)} = \widetilde{D}_i^{(2)}$, i.e.,

$$\alpha_* \widetilde{D}_i^{(1)} = \bar{D}_i + \sum_k (\alpha^{-1})^*(\widetilde{D}_i^{(1)}(\alpha_k)) \frac{\partial}{\partial w_k^{(2)}}, \qquad i = 1, \ldots, n, \tag{1.4}$$

where $w_k^{(i)}$ are nonlocal variables in the covering τ_i, $i = 1, 2$, and

$$\alpha(x_i, p_\sigma^j, w_1^{(1)}, w_2^{(1)}, \ldots)$$
$$= (x_i, p_\sigma^j, \alpha_1(x_i, p_\sigma^j, w_1^{(1)}, w_2^{(1)}, \ldots), \alpha_2(x_i, p_\sigma^j, w_1^{(1)}, w_2^{(1)}, \ldots), \ldots).$$

Relations (1.4) can be also rewritten in the form
$$\widetilde{D}_i^{(1)}(\alpha_k) = \alpha^*(X_{ik}^{(2)}).\qquad(1.5)$$

Now let $\mathcal{E}^\infty$ be a differential equation with the Cartan distribution determined by the fields $\bar{D}_1,\ldots,\bar{D}_n$. Consider one particular covering over $\mathcal{E}^\infty$. Let us set $\widetilde{\mathcal{E}} = \mathcal{E}^\infty \times \mathbb{R}^N$ and let $\tau = \mathrm{pr}_N\colon \widetilde{\mathcal{E}} = \mathcal{E}^\infty \times \mathbb{R}^N \to \mathcal{E}^\infty$ be the projection to the first factor. Assume that the distribution on $\widetilde{\mathcal{E}}$ is determined by the fields $\bar{D}_1,\ldots,\bar{D}_n$.

DEFINITION 1.4. A covering $\tau\colon \widetilde{\mathcal{E}} \to \mathcal{E}^\infty$ is called *trivial* if it is equivalent to the covering pr_N for some $0 < N \leq \infty$.

EXERCISE 1.4. Show that if a covering $\tau\colon \widetilde{\mathcal{E}} \to \mathcal{E}^\infty$ is trivial, then the covering space $\widetilde{\mathcal{E}}$ is fibered by integral submanifolds isomorphic to $\mathcal{E}^\infty$.

EXAMPLE 1.8. Let $\mathcal{E} = \{u_t = u_{xx} + uu_x\}$ be the Burgers equation. Consider the one-dimensional covering $\tau\colon \widetilde{\mathcal{E}} \to \mathcal{E}^\infty$, where $\widetilde{\mathcal{E}} = \mathcal{E}^\infty \times \mathbb{R}^1$, τ is the projection to the first factor, v is a nonlocal variable, and the distribution on $\widetilde{\mathcal{E}}$ is given by the fields
$$\widetilde{D}_x = \bar{D}_x + p_1 \frac{\partial}{\partial v}, \qquad \widetilde{D}_t = \bar{D}_t + (p_2 + p_0 p_1)\frac{\partial}{\partial v}.$$

Let us show that this covering is equivalent to the one-dimensional trivial covering of the Burgers equation. Define the mapping $\alpha\colon \mathcal{E}^\infty \times \mathbb{R}^1 \to \mathcal{E}^\infty \times \mathbb{R}^1 = \widetilde{\mathcal{E}}$ by setting $\alpha(x,t,p_k,w) = (x,t,p_k,w+p_0)$. Then α is a fiberwise diffeomorphism, and
$$\alpha_*\bar{D}_x = \bar{D}_x + \bar{D}_x(w+p_0)\frac{\partial}{\partial v} = \bar{D}_x + p_1\frac{\partial}{\partial v} = \widetilde{D}_x,$$
$$\alpha_*\bar{D}_t = \bar{D}_t + \bar{D}_t(w+p_0)\frac{\partial}{\partial v} = \bar{D}_x + (p_2+p_0p_1)\frac{\partial}{\partial v} = \widetilde{D}_t.$$

Thus the mapping α establishes equivalence of the coverings pr_1 and τ.

Let us now describe another construction playing an important role in the theory of coverings. Let $\tau_i\colon \widetilde{\mathcal{E}}_i \to \mathcal{E}^\infty$, $i = 1, 2$, be two coverings over the equation $\mathcal{E}^\infty$. Consider the direct product $\widetilde{\mathcal{E}}_1 \times \widetilde{\mathcal{E}}_2$ and take the subset $\widetilde{\mathcal{E}}_1 \oplus \widetilde{\mathcal{E}}_2$ consisting of the points (y_1, y_2), $y_1 \in \widetilde{\mathcal{E}}_1$, $y_2 \in \widetilde{\mathcal{E}}_2$, such that $\tau_1(y_1) = \tau_2(y_2)$. Then the projection $\tau_1 \oplus \tau_2\colon \widetilde{\mathcal{E}}_1 \oplus \widetilde{\mathcal{E}}_2 \to \mathcal{E}^\infty$ is defined, for which $(\tau_1 \oplus \tau_2)(y_1, y_2) = \tau_1(y_1) = \tau_2(y_2)$. Obviously, we have the commutative diagram

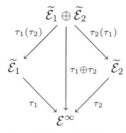

where the mappings $\tau_1(\tau_2)$ and $\tau_2(\tau_1)$ are induced by the projections to the left and right factors respectively. In other words, $\tau_1 \oplus \tau_2\colon \widetilde{\mathcal{E}}_1 \oplus \widetilde{\mathcal{E}}_2 \to \mathcal{E}^\infty$ is the Whitney product of the bundles $\tau_i\colon \widetilde{\mathcal{E}}_i \to \mathcal{E}^\infty$, $i = 1, 2$. Let us now define a distribution $\mathcal{C}^\oplus$ on the manifold $\widetilde{\mathcal{E}}_1 \oplus \widetilde{\mathcal{E}}_2$ by setting $\mathcal{C}^\oplus_{(y_1,y_2)} = (\tau_1(\tau_2))_*^{-1}(\mathcal{C}^{(1)}_{y_1}) \cap (\tau_2(\tau_1))_*^{-1}(\mathcal{C}^{(2)}_{y_2})$, $(y_1, y_2) \in \widetilde{\mathcal{E}}_1 \oplus \widetilde{\mathcal{E}}_2$. Obviously, this is an integrable distribution and $(\tau_1 \oplus \tau_2)_*(\mathcal{C}^{\oplus}_{(y_1,y_2)}) = \mathcal{C}_{(\tau_1 \oplus \tau_2)(y_1,y_2)}$, i.e., the projection $\tau_1 \oplus \tau_2$ and the distribution $\mathcal{C}^{\oplus}$ determine a covering structure over $\mathcal{E}^{\infty}$ on the manifold $\widetilde{\mathcal{E}}_1 \oplus \widetilde{\mathcal{E}}_2$. This covering is called the *Whitney product* of the coverings τ_1 and τ_2. Note also that the projections $\tau_1(\tau_2)$ and $\tau_2(\tau_1)$ are coverings over the objects $\widetilde{\mathcal{E}}_1$ and $\widetilde{\mathcal{E}}_2$ respectively.

A coordinate interpretation of the covering $\tau_1 \oplus \tau_2$ is as follows. Let $\mathbb{R}^{N_\alpha}$ be the fiber of the bundle τ_α; $w_1^{(1)}, w_2^{(1)}, \ldots$ be the coordinates in the fiber of the projection τ_1; $w_1^{(2)}, w_2^{(2)}, \ldots$ be the coordinates in the fiber of the projection τ_2. Let the coverings τ_1 and τ_2 be locally determined by the fields $\widetilde{D}_i^{(1)} = \bar{D}_i + X_i^{(1)}$ and $\widetilde{D}_i^{(2)} = \bar{D}_i + X_i^{(2)}$, $i = 1, \ldots, n$, respectively, $X_i^{(\alpha)} = \sum_k X_{ik}^{(\alpha)} \partial/\partial w_k^{(\alpha)}$, $\alpha = 1, 2$. Then the space $\mathbb{R}^{N_1} \oplus \mathbb{R}^{N_2}$ is the fiber of the bundle $\tau_1 \oplus \tau_2$ with the coordinates $w_1^{(1)}, w_2^{(1)}, \ldots, w_1^{(2)}, w_2^{(2)}, \ldots$, while the distribution on $\widetilde{\mathcal{E}}_1 \oplus \widetilde{\mathcal{E}}_2$ is determined by the fields

$$\widetilde{D}_i^{\oplus} = \bar{D}_i + \sum_k X_{ik}^{(1)} \frac{\partial}{\partial w_k^{(1)}} + \sum_l X_{il}^{(2)} \frac{\partial}{\partial w_l^{(2)}}, \qquad i = 1, \ldots, n.$$

The Whitney product of an arbitrary number of coverings is defined in an obvious way.

The Whitney product construction is convenient when we want to analyze equivalence of coverings. Namely, the following proposition is valid:

PROPOSITION 1.1. *Let τ_1 and τ_2 be two coverings of finite dimension $N = \dim \tau_1 = \dim \tau_2$ over the equation $\mathcal{E}^{\infty}$. These coverings are equivalent if and only if there exists a manifold $\mathcal{X} \subset \widetilde{\mathcal{E}}_1 \oplus \widetilde{\mathcal{E}}_2$, $\operatorname{codim} \mathcal{X} = N$, invariant*[3] *of the distribution $\widetilde{\mathcal{C}}^{\oplus}$ such that the restrictions $\tau_1(\tau_2)|_{\mathcal{X}}$ and $\tau_2(\tau_1)|_{\mathcal{X}}$ are surjective.*

PROOF. The proof consists in a straightforward check of definitions. Indeed, let the coverings τ_1 and τ_2 be equivalent, and let α be a diffeomorphism realizing this equivalence. Set $\mathcal{X} = \{ (y, \alpha(y)) \mid y \in \widetilde{\mathcal{E}}_1 \} \subset \widetilde{\mathcal{E}}_1 \oplus \widetilde{\mathcal{E}}_2$. Obviously, $\operatorname{codim} \mathcal{X} = N$, and $\tau_1(\tau_2)|_{\mathcal{X}}$ and $\tau_2(\tau_1)|_{\mathcal{X}}$ are surjections. Since the mapping α_* takes the distribution $\widetilde{\mathcal{C}}_1$ to $\widetilde{\mathcal{C}}_2$, the submanifold $\mathcal{X}$ is invariant with respect to the distribution $\widetilde{\mathcal{C}}^{\oplus}$.

Conversely, let $\mathcal{X} \subset \widetilde{\mathcal{E}}_1 \oplus \widetilde{\mathcal{E}}_2$ be a submanifold satisfying the assumptions of the proposition. Then the correspondence $y \mapsto \tau_2(\tau_1)\left((\tau_1(\tau_2))^{-1}(y) \cap \mathcal{X}\right)$, $y \in \widetilde{\mathcal{E}}_1$, determines an isomorphism of the bundles τ_1 and τ_2. By invariance of $\mathcal{X}$, this is an equivalence of coverings. □

Let us introduce another notion.

DEFINITION 1.5. A covering $\tau \colon \widetilde{\mathcal{E}} \to \mathcal{E}^{\infty}$ over the equation $\mathcal{E}^{\infty}$ is called *reducible*, if it is equivalent to a covering of the form $\tau_1 \oplus \operatorname{pr}_N$, where τ_1 is some covering over the equation $\mathcal{E}^{\infty}$ and $N > 0$. Otherwise, the covering is called *irreducible*.

A local analog of this definition is obvious.

If $\tau \colon \widetilde{\mathcal{E}} \to \mathcal{E}^{\infty}$ is a reducible covering, then for any point $\theta \in \widetilde{\mathcal{E}}$ there exists an invariant manifold $\mathcal{X}$ of the distribution $\widetilde{\mathcal{C}}$ passing through this point and such that the restriction $\tau|_{\mathcal{X}} \colon \mathcal{X} \to \mathcal{E}^{\infty}$ of the projection τ is surjective and this restriction, together with the restriction of the distribution $\widetilde{\mathcal{C}}$ to $\mathcal{X}$, determines in $\mathcal{X}$ a covering structure over $\mathcal{E}^{\infty}$.

---

[3] A manifold $\mathcal{X} \subset \widetilde{\mathcal{E}}$ is called *invariant* with respect for the distribution $\widetilde{\mathcal{C}}$ on $\widetilde{\mathcal{E}}$, if at any point $x \in \mathcal{X}$ one has the embedding $T_x \mathcal{X} \supset \widetilde{\mathcal{C}}_x$.

Using Proposition 1.1, one can easily prove the following result:

PROPOSITION 1.2. *Let τ_1 and τ_2 be irreducible coverings of finite dimension $N = \dim \tau_1 = \dim \tau_2$ over the equation $\mathcal{E}^\infty$. Then, if the Whitney product $\widetilde{\mathcal{E}}_1 \oplus \widetilde{\mathcal{E}}_2$ of these coverings is reducible and the codimension of invariant manifolds of the distribution $\widetilde{\mathcal{C}}^\oplus$ equals N, then at almost all points $\theta \in \widetilde{\mathcal{E}}$ the covering τ_1 is equivalent to the covering τ_2.*

PROOF. Consider invariant manifolds of the distribution $\widetilde{\mathcal{C}}^\oplus$. Almost all of them project surjectively to $\widetilde{\mathcal{E}}_1$ and $\widetilde{\mathcal{E}}_2$ under the mappings $\tau_1(\tau_2)$ and $\tau_2(\tau_1)$ respectively. In fact, otherwise their images under the projections $\tau_1(\tau_2)$ and $\tau_2(\tau_1)$ would be invariant manifolds of the distributions $\widetilde{\mathcal{C}}_1$ and $\widetilde{\mathcal{C}}_2$ respectively, which contradicts the irreducibility of the coverings τ_1 and τ_2. Choosing one of these manifolds, we arrive at Proposition 1.1, and this concludes the proof. □

COROLLARY 1.3. *If τ_1 and τ_2 are one-dimensional coverings over the equation $\mathcal{E}$ and their Whitney product is reducible, then at almost all points $\theta \in \widetilde{\mathcal{E}}$ the covering τ_1 is equivalent to the covering τ_2.*

In local coordinates, the problem of reducibility is equivalent to studying a certain system of linear differential equations. Namely, the following statement is valid:

PROPOSITION 1.4. *Let $\mathcal{U} \subseteq \mathcal{E}^\infty$ be a domain such that the manifold $\widetilde{\mathcal{U}} = \tau^{-1}(\mathcal{U})$ is representable in the form of the direct product $\mathcal{U} \times \mathbb{R}^N$, $N = 1, 2, \ldots, \infty$, the mapping $\tau|_{\widetilde{\mathcal{U}}} \colon \widetilde{\mathcal{U}} \to \mathcal{U}$ is the projection to the first factor, and the fields $\widetilde{D}_1, \ldots, \widetilde{D}_n$ determine the distribution $\widetilde{\mathcal{C}}$ on $\widetilde{\mathcal{U}}$. Then the covering τ is locally irreducible if and only if the system*

$$\widetilde{D}_1(\varphi) = 0, \ldots, \widetilde{D}_n(\varphi) = 0, \tag{1.6}$$

where $\varphi \in \mathcal{F}(\widetilde{\mathcal{E}})$, possesses constant solutions only.

PROOF. Assume that there exists a solution $\varphi \neq \mathrm{const}$ of (1.6). Then, since the only solutions of the system

$$\bar{D}_1(\varphi) = 0, \ldots, \bar{D}_n(\varphi) = 0$$

are constants, the function φ depends on at least one nonlocal variable. Without loss of generality, we may assume that $\partial \varphi / \partial w_1 \neq 0$ in some neighborhood $\mathcal{U}' \times W$, $\mathcal{U}' \subseteq \mathcal{U}$, $W \subseteq \mathbb{R}^N$. Let us define the diffeomorphism $\alpha \colon \mathcal{U}' \times W \to \alpha(\mathcal{U}' \times W)$ by the formula $\alpha(x_i, p_\sigma^j, w_1, \ldots, w_N) = (x_i, p_\sigma^j, \varphi, w_2, \ldots, w_N)$. Since $\alpha_* \widetilde{D}_i = \bar{D}_i + \sum_{j>1} X_{ij} \partial/\partial w_j$, the covering τ is locally equivalent to the covering $\mathrm{pr}_1 \oplus \tau_1$ for some covering τ_1. In other words, τ is a reducible covering.

Assume now that the covering τ is locally reducible, i.e., is locally equivalent to a covering of the form $\tau_1 \oplus \mathrm{pr}_{N'}$. Let a mapping α be the corresponding equivalence. Then, if f is a smooth function on the space of the covering $\mathrm{pr}_{N'}$, the function $\alpha^*((\mathrm{pr}_{N'}(\tau_1))^*(f)) = f^* \in \mathcal{F}(\widetilde{\mathcal{E}})$ is a solution of system (1.6). Obviously, there exists a function f on the space of the covering $\mathrm{pr}_{N'}$ such that the function $f^* \in \mathcal{F}(\widetilde{\mathcal{E}})$ is not constant. □

1. COVERINGS

1.7. Coverings and connections. Let $\tau\colon \widetilde{\mathcal{E}} \to \mathcal{E}^\infty$ be an arbitrary covering and let $v \in \mathcal{C}_\theta$. Then for any point $\widetilde{\theta} \in \widetilde{\mathcal{E}}$ projecting to θ by the mapping φ, a vector $\widetilde{v} \in \mathcal{C}_{\widetilde{\theta}}$ satisfying $\tau_*\widetilde{v} = v$ is uniquely determined.

Let $\mathcal{E} \subset J^k(\pi)$. Recall that the Cartan connection exists in the bundle $\pi_\infty\colon J^\infty(\pi) \to M$. This connection takes a vector field $X \in D(M)$ to the vector field $\widehat{X} \in D(J^\infty(\pi))$ (see §1.4 of Ch. 4). All vector fields $\widehat{X}$ constructed in such a way are tangent to $\mathcal{E}^\infty$ and span the Cartan distribution on this manifold. Consequently, one can put into correspondence to any field $X \in D(M)$ a uniquely defined field $\widetilde{X} \in D(\widetilde{\mathcal{E}})$ such that $\tau_*(\widetilde{X}) = \widehat{X}$. Besides, it is obvious that if the fields $\widetilde{X}, \widetilde{Y} \in D(\widetilde{\mathcal{E}})$ are the liftings of the fields $X, Y \in D(M)$ respectively, then $[\widetilde{X}, \widetilde{Y}] = \widetilde{[X,Y]}$. Thus the covering $\tau\colon \widetilde{\mathcal{E}} \to \mathcal{E}^\infty$ determines a flat connection in the bundle $\widetilde{\pi} = \pi_\infty \circ \tau\colon \widetilde{\mathcal{E}} \to M$ compatible with the Cartan connection. One can easily see that the converse statement is also valid. More precisely, we have

THEOREM 1.5. *Let $\mathcal{E} \subset J^k(\pi)$ be an equation and $\tau\colon \widetilde{\mathcal{E}} \to \mathcal{E}^\infty$ a fiber bundle. Then the following statements are equivalent:*

1. *A covering structure exists in τ.*
2. *A connection $\nabla^\tau\colon D(M) \to D(\widetilde{\mathcal{E}})$, $\nabla^\tau(X) = \widetilde{X}$, exists in the bundle $\widetilde{\pi} = \pi_\infty \circ \tau\colon \widetilde{\mathcal{E}} \to M$ possessing the following properties:*
   (a) $[\widetilde{X}, \widetilde{Y}] = \widetilde{[X,Y]}$ *for any vector fields $X, Y \in D(M)$, i.e., the connection ∇^τ is flat;*
   (b) $\tau_*(\widetilde{X}) = \widehat{X}$, *i.e., ∇^τ is compatible with the Cartan connection.*

Note that for any point $\widetilde{\theta} \in \widetilde{\mathcal{E}}$ the plane $\widetilde{\mathcal{C}}_{\widetilde{\theta}}$ is spanned by vectors of the form $\widetilde{X}_{\widetilde{\theta}}$, $X \in D(M)$.

In local coordinates, the connection considered above is determined in the covering $\tau\colon \widetilde{\mathcal{E}} \to \mathcal{E}^\infty$ by the correspondence $\bar{D}_i \mapsto \widetilde{D}_i$, $i = 1, \ldots, n$, which makes it possible to lift $\mathcal{C}$-differential operators $\Delta\colon \mathcal{F}(\mathcal{E}^\infty) \to \mathcal{F}(\mathcal{E}^\infty)$ to $\widetilde{\mathcal{E}}$. Namely, if $\Delta = \sum_\sigma a_\sigma \bar{D}_\sigma\colon \mathcal{F}(\mathcal{E}^\infty) \to \mathcal{F}(\mathcal{E}^\infty)$ is a $\mathcal{C}$-differential operator, then it can be extended up to an operator $\widetilde{\Delta}\colon \mathcal{F}(\widetilde{\mathcal{E}}) \to \mathcal{F}(\widetilde{\mathcal{E}})$ by setting

$$\widetilde{\Delta} = \sum_\sigma a_\sigma \widetilde{D}_\sigma.$$

In particular, if $\mathcal{E}^\infty = J^\infty(\pi)$ and $\pi'\colon E' \to M$ is another vector bundle, then for any element $F \in \mathcal{F}(\pi, \pi')$ the lifting $\widetilde{\ell}_F$ of the corresponding universal linearization operator exists. As in Chapter 4, let us set $\widetilde{\partial}_\varphi(F) = \widetilde{\ell}_F(\varphi)$, where φ is a section of the pullback $\pi^*(\pi)$. In coordinates, the field $\widetilde{\partial}_\varphi$ is of the form

$$\widetilde{\partial}_\varphi = \sum_{\sigma,j}{}' \widetilde{D}_\sigma(\varphi^j) \frac{\partial}{\partial p^i_\sigma},$$

where $\varphi_i \in \mathcal{F}(\widetilde{\mathcal{E}})$ and $\sum'$ denotes summation over intrinsic coordinates.

1.8. The horizontal de Rham complex and nonlocal conservation laws. Let a covering $\tau\colon \widetilde{\mathcal{E}} \to \mathcal{E}^\infty$ over the equation $\mathcal{E}^\infty$ be given. Then the horizontal de Rham complex on $\mathcal{E}^\infty$ can also be lifted to $\widetilde{\mathcal{E}}$, since the corresponding differentials are $\mathcal{C}$-differential operators. Locally this is described in the following way. Horizontal k-forms on $\widetilde{\mathcal{E}}$ are represented as

$$\omega = \sum_{i_1<\cdots<i_k} a_{i_1\ldots i_k}\, dx_{i_1} \wedge \cdots \wedge dx_{i_k}, \qquad a_{i_1\ldots i_k} \in \mathcal{F}(\widetilde{\mathcal{E}}). \tag{1.7}$$

The differential $\widetilde{d}$ acts in the following way:

$$\widetilde{d}\omega = \sum_{i_1<\cdots<i_k} \left(\sum_{j=1}^n \widetilde{D}_j(a_{i_1\ldots i_k})\, dx_j\right) \wedge dx_{i_1} \wedge \cdots \wedge dx_{i_k}. \tag{1.8}$$

The cohomology of the horizontal de Rham complex on $\widetilde{\mathcal{E}}$ will be denoted by $\widetilde{H}^k(\widetilde{\mathcal{E}})$. The group $\widetilde{H}^{n-1}(\widetilde{\mathcal{E}})$, similarly to the local situation, will be called the *group of nonlocal conservation laws* of the equation $\mathcal{E}^\infty$.

Note that the horizontal de Rham complex on $\widetilde{\mathcal{E}}$ can be introduced directly, as was done with the horizontal de Rham complex on $\mathcal{E}^\infty$ (i.e., skipping the procedure of lifting $\mathcal{C}$-differential operators to the covering). To do this, let us introduce two notions.

A vector field X on $\widetilde{\mathcal{E}}$ is called *vertical* if $X((\tau \circ \pi_\infty)^* f) = 0$ for any function $f \in C^\infty(M^n)$. A differential form $\omega \in \Lambda^k(\widetilde{\mathcal{E}})$ is called *horizontal* if $\omega(X) = 0$ for any vertical vector field X on $\widetilde{\mathcal{E}}$. Horizontal forms of degree k constitute a submodule in $\Lambda^k(\widetilde{\mathcal{E}})$, which will be denoted by $\Lambda_0^k(\widetilde{\mathcal{E}})$.

EXERCISE 1.5. Show that in local coordinates a form $\omega \in \Lambda_0^k(\widetilde{\mathcal{E}})$ is written down as (1.7).

A differential form $\omega \in \Lambda^k(\widetilde{\mathcal{E}})$ is called a *Cartan form* if $\omega(X) = 0$ for any vector field X lying in the distribution $\widetilde{\mathcal{C}}$. Cartan forms of degree k constitute a submodule in $\Lambda^k(\widetilde{\mathcal{E}})$, which will be denoted by $\mathcal{C}\Lambda^k(\widetilde{\mathcal{E}})$. One has the decomposition

$$\Lambda^k(\widetilde{\mathcal{E}}) = \sum_{\alpha+\beta=k} \Lambda_0^\alpha(\widetilde{\mathcal{E}}) \wedge \mathcal{C}\Lambda^\beta(\widetilde{\mathcal{E}}). \tag{1.9}$$

The de Rham differential $d\colon \Lambda^k(\widetilde{\mathcal{E}}) \to \Lambda^{k+1}(\widetilde{\mathcal{E}})$ acts on the form $\omega \in \Lambda_0^\alpha(\widetilde{\mathcal{E}}) \wedge \mathcal{C}\Lambda^\beta(\widetilde{\mathcal{E}})$ as follows:

$$d\omega = \omega_0 + \omega_C,$$

where $\omega_0 \in \Lambda_0^{\alpha+1}(\widetilde{\mathcal{E}}) \wedge \mathcal{C}\Lambda^\beta(\widetilde{\mathcal{E}})$, $\omega_C \in \Lambda_0^\alpha(\widetilde{\mathcal{E}}) \wedge \mathcal{C}\Lambda_0^{\beta+1}(\widetilde{\mathcal{E}})$. Using the decomposition (1.9), one can introduce the mapping

$$\widehat{d}\colon \Lambda_0^\alpha(\widetilde{\mathcal{E}}) \wedge \mathcal{C}\Lambda^\beta(\widetilde{\mathcal{E}}) \longrightarrow \Lambda_0^{\alpha+1}(\widetilde{\mathcal{E}}) \wedge \mathcal{C}\Lambda^\beta(\widetilde{\mathcal{E}}),$$

by setting $\widehat{d}\omega = \omega_0$.

It is easily checked that $\widehat{d}_h^2 = 0$ and that in local coordinates the operator $\widehat{d}\big|_{\Lambda_0^*(\widetilde{\mathcal{E}})}$ acts by the formula (1.8), i.e., the complex $\{\Lambda_0^*(\widetilde{\mathcal{E}}), \widehat{d}\}$ is the horizontal de Rham complex on $\widetilde{\mathcal{E}}$.

1.9. Covering equations. Consider a covering $\tau\colon \widetilde{\mathcal{E}} \to \mathcal{E}^\infty$. Let us find out when the covering manifold $\widetilde{\mathcal{E}}$ is of the form $(\mathcal{E}')^\infty$ for some equation $\mathcal{E}' \subset J^l(\pi')$. To answer this question, we shall restrict ourselves to the following situation.

Assume that two bundles, $\pi\colon E \to M$ and $\xi\colon W \to M$, over the manifold M are given. Then we have the following commutative diagram:

$$\begin{array}{ccccccc} W^\infty & \longrightarrow & \cdots \longrightarrow & W^1 & \longrightarrow & W^0 & \longrightarrow W \\ {\scriptstyle \pi_\infty^*(\xi)=\tau}\Big\downarrow & & & {\scriptstyle \pi_1^*(\xi)}\Big\downarrow & & {\scriptstyle \pi^*(\xi)}\Big\downarrow & \Big\downarrow{\scriptstyle \xi} \\ J^\infty(\pi) & \longrightarrow & \cdots \longrightarrow & J^1(\pi) & \longrightarrow & E & \longrightarrow M \end{array}$$

where $\pi_i^*(\xi)\colon W^i \to J^k(\pi)$, $i = 1, 2, \ldots, \infty$, denote the pullbacks. Let us denote by τ the projection $\pi_\infty^*(\xi)$ and assume that a covering structure exists in the bundle $\tau\colon W^\infty \to J^\infty(\pi)$. Then, as proved in Theorem 1.5, it is equivalent to existence of a flat connection ∇^τ in the bundle $\pi_\infty \circ \tau$. The coefficients of this connection are functions on $J^\infty(\pi)$, i.e., differential operators acting from $\Gamma(\pi)$ to $C^\infty(M)$. Let $f \in \Gamma(\pi)$ be a section of the bundle π. Then a flat connection is induced in the bundle $j_\infty(f)^*(\tau) \approx \xi$ in a natural way. Denote this connection by $\nabla^\tau(f)$. In other words, the covering structure in τ determines a differential operator, which takes a section of the bundle π to a flat connection in the bundle ξ. Let us denote by $\deg \tau$ the order of this operator and call this number the *order of the covering* τ.

Now consider a point $\theta \in W^k$, $\theta = ([f]_x^k, w_0)$, where $x \in M$, $f \in \Gamma(\pi)$, and $w_0 \in \xi^{-1}(x) \subset W$. Since the bundle $\xi\colon W \to M$ possesses the flat connection $\nabla^\tau(f)$, we can locally consider its integral section g passing through the point w_0. The pair (f, g) determines the local section $\pi \oplus \xi$ of the Whitney product of the bundles π and ξ. Let us set

$$\rho_k(\theta) = [(f,g)]_x^k \in J^k(\pi \oplus \xi).$$

The family $\rho = \{\rho_k\}$ of the mappings $\rho_k\colon W^k \to J^k(\pi \oplus \xi)$ possesses the following properties:

1. The mapping ρ_k is an immersion for any k.
2. For any k, we have the following commutative diagram:

$$\begin{array}{ccc} W^{k+1} & \xrightarrow{\rho_{k+1}} & J^{k+1}(\pi \oplus \xi) \\ \Big\downarrow & & \Big\downarrow{\scriptstyle (\pi\oplus\xi)_{k+1,k}} \\ W^k & \xrightarrow{\rho_k} & J^k(\pi \oplus \xi) \end{array}$$

3. If $k_0 = \max(1, \deg \tau)$ and $\mathcal{E}_\tau = \rho_{k_0}(W^{k_0}) \subset J^{k_0}(\pi \oplus \xi)$, then for all $k > k_0$ the equality $\rho_k(W^k) = (\mathcal{E}_\tau)^{(k-k_0)}$ holds.
4. The mapping $\rho_{\infty,*}$ takes the distribution $\widetilde{\mathcal{C}}$ on W^∞ to the Cartan distribution of the equation $(\mathcal{E}_\tau)^\infty$.

In other words, the space of the covering $\tau\colon W^\infty \to J^\infty(\pi)$, as a manifold with distribution, is isomorphic to the infinitely prolonged equation $(\mathcal{E}_\tau)^\infty$. In the case, when the base of the covering is an equation $\mathcal{E}^\infty \subset J^\infty(\pi)$, we can show by similar reasoning that the space of the covering $\widetilde{\mathcal{E}}$ is isomorphic to the infinitely prolonged equation $(\mathcal{E} \cap \mathcal{E}_\tau)^\infty = \mathcal{E}^\infty \cap (\mathcal{E}_\tau)^\infty$. We shall say that the equation $\mathcal{E}_\tau$ *covers* the equation $\mathcal{E}$. The equation $\mathcal{E}_\tau$ will also be called a *covering equation* for $\mathcal{E}$.

If, in local coordinates, the distribution $\widetilde{\mathcal{C}}$ on $\widetilde{\mathcal{E}}$ is determined by the fields $\widetilde{D}_i = \bar{D}_i + \sum_{j=1}^N X_{ij}(x_\alpha, p_\sigma^\beta, w_\gamma)\partial/\partial w_j$, $i = 1, \ldots, n$, then the covering equation $\mathcal{E}_\tau$

is locally described by the relations

$$\frac{\partial w_j}{\partial x_i} = X_{ij}(x_\alpha, u^\beta_\sigma, w_\gamma), \qquad i = 1, \ldots, n, \quad j = 1, \ldots, N. \tag{1.10}$$

The fact that the connection ∇^τ in the bundle $\tau \colon \widetilde{\mathcal{E}} \to \mathcal{E}^\infty$ is flat means that the compatibility conditions for the system (1.10) are differential consequences of the equation $\mathcal{E}$. To any solution of this equation, there corresponds an N-parameter family of solutions of the covering equation, and the parameters for this family are analogs of integration constants for system (1.10).

EXAMPLE 1.9. Consider the one-dimensional covering of the Korteweg–de Vries equation $\mathcal{E} = \{u_t = uu_x + u_{xxx}\}$ determined by the fields

$$\begin{cases} \widetilde{D}_x = \bar{D}_x + \left(p_0 + \frac{1}{6}w^2\right)\frac{\partial}{\partial w}, \\ \widetilde{D}_t = \bar{D}_t + \left(p_2 + \frac{1}{3}wp_1 + \frac{1}{3}p_0^2 + \frac{1}{18}w^2 p_0\right)\frac{\partial}{\partial w}. \end{cases}$$

Eliminating u from the system

$$\begin{cases} w_x = u + \frac{1}{6}w^2, \\ w_t = u_{xx} + \frac{1}{3}wu_x + \frac{1}{3}u^2 + \frac{1}{18}w^2 u, \end{cases}$$

we see that the covering equation for the Korteweg–de Vries equation is the modified Korteweg–de Vries equation

$$w_t = w_{xxx} - \frac{1}{6}w^2 w_x.$$

The relation between the variables u and w given by the equation $w_x = u + \frac{1}{6}w^2$ is exactly the Miura–Gardner transformation [**86, 88**].

EXERCISE 1.6. Show that if the one-dimensional covering over the Korteweg–de Vries equation is given by the fields

$$\begin{cases} \widetilde{D}_x = \bar{D}_x + p_0 \frac{\partial}{\partial w}, \\ \widetilde{D}_t = \bar{D}_t + \left(p_2 + \frac{1}{2}p_0^2\right)\frac{\partial}{\partial w}, \end{cases}$$

then the covering equation is the potential Korteweg–de Vries equation $w_t = w_{xxx} + \frac{1}{2}w_x^2$.

1.10. Horizontal de Rham cohomology and coverings. In this subsection, we establish a relation between the group $\bar{H}^1(\mathcal{E}^\infty)$ of $(n-1)$-dimensional horizontal de Rham cohomology and special coverings over the equation $\mathcal{E}^\infty$.

First let $\tau \colon \widetilde{\mathcal{E}} \to \mathcal{E}^\infty$ be a one-dimensional covering over the equation $\mathcal{E}^\infty$, $\widetilde{\mathcal{E}} = \mathcal{E}^\infty \times \mathbb{R}$, let τ be the projection to the first factor, and let the distribution on $\widetilde{\mathcal{E}}$ be given by the fields

$$\widetilde{D}_i = \bar{D}_i + X_i \frac{\partial}{\partial w}, \qquad X_i \in \mathcal{F}(\mathcal{E}^\infty). \tag{1.11}$$

Let us put the horizontal form

$$\omega = \sum_{i=1}^n X_i \, dx_i$$

into correspondence with this covering. The conditions (1.3) acquire the form $\bar{D}_i(X_j) = \bar{D}_j(X_i)$ in this case, or $\widehat{d}(\omega) = 0$, i.e., the form ω is closed in the horizontal de Rham complex. Conversely, to any closed horizontal 1-form $\omega = \sum_{i=1}^{n} X_i \, dx_i$ there corresponds a covering over $\mathcal{E}^\infty$ locally described by the fields (1.11).

The following two statements are easily proved:

PROPOSITION 1.6. *Let $\tau\colon \widetilde{\mathcal{E}} \to \mathcal{E}^\infty$ be a one-dimensional covering given by (1.11), and let ω be the corresponding 1-form. The covering τ is trivial if and only if the form ω is exact, i.e., it generates the trivial element in the cohomology group $\bar{H}^1(\mathcal{E}^\infty)$.*

PROPOSITION 1.7. *Let $\tau_i\colon \widetilde{\mathcal{E}}_i \to \mathcal{E}^\infty$, $i = 1, 2$, be two one-dimensional coverings over $\mathcal{E}^\infty$ given by (1.11), and let the forms ω_1 and ω_2 correspond to these coverings. The coverings τ_1 and τ_2 are equivalent if and only if the corresponding cohomology classes $[\omega_1]$ and $[\omega_2]$ coincide.*

EXERCISE 1.7. Prove Propositions 1.6 and 1.7.

Thus, we have a one-to-one correspondence between the elements of the group $\bar{H}^1(\mathcal{E}^\infty)$ and equivalence classes of one-dimensional coverings (1.11) over $\mathcal{E}^\infty$.

Now let $\tau\colon \widetilde{\mathcal{E}} \to \mathcal{E}^\infty$ be an N-dimensional covering over the equation $\mathcal{E}^\infty$, where $\widetilde{\mathcal{E}} = \mathcal{E}^\infty \times \mathbb{R}^N$, τ is the projection to the first factor, and the distribution on $\widetilde{\mathcal{E}}$ is determined by the fields

$$\widetilde{D}_i = \bar{D}_i + \sum_{j=1}^{N} X_{ij} \frac{\partial}{\partial w_j}, \qquad X_{ij} \in \mathcal{F}(\mathcal{E}^\infty), \quad i = 1, \ldots, n.$$

It is easily seen that the horizontal 1-forms

$$\omega_j = \sum_{i=1}^{n} X_{ij} dx_i, \qquad j = 1, \ldots, N,$$

are closed. Therefore, they generate some elements $[\omega_1], \ldots, [\omega_N]$ of the group $\bar{H}^1(\mathcal{E}^\infty)$. Let $V(\tau)$ be the subspace of the linear space $\bar{H}^1(\mathcal{E}^\infty)$ spanned by the classes $[\omega_1], \ldots, [\omega_N]$. The following statement is valid:

PROPOSITION 1.8. *Let $\tau_i\colon \widetilde{\mathcal{E}}_i \to \mathcal{E}^\infty$, $i = 1, 2$, be a locally irreducible N-dimensional, $N < \infty$, covering over the equation $\mathcal{E}^\infty$. Let $\widetilde{\mathcal{E}}_i = \mathcal{E}^\infty \times \mathbb{R}^N$, and let τ_i be the projections to the first factor, while the distribution $\widetilde{\mathcal{E}}_i$ is described by the fields*

$$\widetilde{D}_j^{(i)} = \bar{D}_j + \sum_k X_{jk}^{(i)} \frac{\partial}{\partial w_k^{(i)}}, \qquad X_{jk}^{(i)} \in \mathcal{F}(\mathcal{E}^\infty), \quad j = 1, \ldots, n, \quad i = 1, 2.$$

Then the coverings τ_1 and τ_2 are equivalent if and only if $V(\tau_1) = V(\tau_2)$.

PROOF. Let the coverings τ_1 and τ_2 be equivalent and let a fiberwise diffeomorphism $\alpha\colon \widetilde{\mathcal{E}}_1 \to \widetilde{\mathcal{E}}_2$ establish their equivalence. Then $\alpha_*(\widetilde{D}_j^{(1)}) = \widetilde{D}_j^{(2)}$, i.e., $\alpha^*(X_{jk}^{(2)}) = \widetilde{D}_j^{(1)}(\alpha_k)$, where

$$\alpha(x_i, p_\sigma^j, w_k^{(1)}) = (x_i, p_\sigma^j, \alpha_1(x_i, p_\sigma^j, w_k^{(1)}), \alpha_2(x_i, p_\sigma^j, w_k^{(1)}), \ldots)$$

(see equation (1.5)). Since $X_{jk}^{(i)} \in \mathcal{F}(\mathcal{E}^\infty)$, we have the identities $[\widetilde{D}_j^{(1)}, \partial/\partial w_s^{(1)}] = 0$ and $\widetilde{D}_j^{(1)}(\partial \alpha_k / \partial w_s^{(1)}) = 0$ for any j, k, s. Since the covering τ_1 is irreducible, from

Proposition 1.4 it follows that $\partial\alpha_k/\partial w_s^{(1)} = c_{ks} = \text{const}$, i.e., $\alpha_k = \sum_{s=1}^{N} c_{ks} w_s^{(1)} + \varphi_k(x, p_\sigma^i)$. From this we obtain

$$\alpha^*(X_{jk}^{(2)}) = \widetilde{D}_j^{(1)}(\alpha_k) = \sum_{s=1}^{N} c_{ks} X_{js}^{(1)} + \bar{D}_j(\varphi_k)$$

and

$$\omega_k^{(2)} = \sum_{j=1}^{n} X_{jk}^{(2)} dx_j = \sum_{s=1}^{N} c_{ks} \omega_s^{(1)} + \widehat{d}(\varphi_k).$$

Thus $\omega_k^{(2)} \in V(\tau_1)$ for all $k = 1, \ldots, N$. Similar calculations for the diffeomorphism α^{-1} show that $\omega_k^{(1)} \in V(\tau_2)$ for all $k = 1, \ldots, N$. Consequently, $V(\tau_1) = V(\tau_2)$.

Now let $V(\tau_1) = V(\tau_2) = V$. Since the covering τ_1 is irreducible, we can assume that the forms $\omega_1^{(1)}, \ldots, \omega_N^{(1)}$, determining τ_1, constitute a basis of the space V. Then the representation

$$[\omega_k^{(2)}] = \sum_{s=1}^{N} c_{ks}[\omega_s^{(1)}], \qquad c_{ks} \in \mathbb{R},$$

is valid, or

$$\omega_k^{(2)} = \sum_{s=1}^{N} c_{ks} \omega_s^{(1)} + \widehat{d}(\varphi_k), \qquad \varphi_k \in \mathcal{F}(\mathcal{E}^\infty).$$

It is easily seen that the submanifold in the space $\widetilde{\mathcal{E}}_1 \oplus \widetilde{\mathcal{E}}_2$ determined by the equations

$$w_k^{(2)} - \sum_{s=1}^{N} c_{ks} w_s^{(1)} - \varphi_k = 0, \qquad k = 1, \ldots, N,$$

is an invariant submanifold of the distribution $\widetilde{\mathcal{C}}^\oplus$ and is of codimension N. From Proposition 1.2 it follows that the coverings τ_1 and τ_2 are equivalent. $\square$

EXERCISE 1.8. Prove that a finite-dimensional covering τ of the form under consideration is irreducible if and only if $\dim V(\tau) = \dim \tau$.

REMARK 1.1. From the above it follows that in the case $\dim M = 2$ (i.e., when the equation $\mathcal{E}$ is in two independent variables), there exists a one-to-one correspondence between one-dimensional coverings of the above form and conservation laws for the equation $\mathcal{E}$.

1.11. Bäcklund transformations. The examples considered above show that the theory of coverings is a convenient and adequate language to describe various nonlocal effects arising in studies of differential equations. In particular, it was shown that constructions such as the Cole–Hopf substitution and the Miura–Gardner transformation are common and can be interpreted in terms of coverings in a natural way. In §2 we shall see that the Wahlquist–Estabrook prolongation structures are particular cases of coverings, while in this subsection we give a definition of Bäcklund transformations using the language of coverings.

Let us recall [65, 93] that a Bäcklund transformation between equations $\mathcal{E}$ and $\mathcal{E}'$ is a system of differential relations in unknown functions u and u' possessing the following property: if a function u is a solution of the equation $\mathcal{E}$ and u and u' satisfy the relations at hand, then u' is a solution of $\mathcal{E}'$.

Using the language of coverings, this definition reads as follows [62].

DEFINITION 1.6. A *Bäcklund transformation* between equations $\mathcal{E}$ and $\mathcal{E}'$ is the diagram

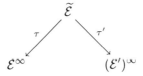

in which the mappings τ and τ' are coverings.

If $\mathcal{E}^\infty = (\mathcal{E}')^\infty$, then the Bäcklund transformation of the equation $\mathcal{E}$ is sometimes called a *Bäcklund autotransformation*.

EXAMPLE 1.10. Consider the following system $\widetilde{\mathcal{E}}$ of equations:
$$\widetilde{w}_x = a \sin w, \quad w_y = \frac{1}{a} \sin \widetilde{w}, \quad a \in \mathbb{R}, \ a \neq 0,$$
in the two-dimensional bundle $\pi \colon \mathbb{R}^2 \times \mathbb{R}^2 \to \mathbb{R}^2$. Then the zero order operator (i.e., a morphism of vector bundles) $\Delta^+ \colon v = w + \widetilde{w}$ determines the covering $\tau^+ \colon J^\infty(\pi) \to J^\infty(\xi)$, where $\xi \colon \mathbb{R} \times \mathbb{R}^2 \to \mathbb{R}^2$ (see §1.4). The restriction of this covering to $\widetilde{\mathcal{E}}^\infty$ determines a covering of the equation $\mathcal{E} = \{v_{xy} = \sin(v)\}$ by the equation $\widetilde{\mathcal{E}}^\infty$. In a similar way, the operator $\Delta^- \colon v = w - \widetilde{w}$ determines another covering $\tau^- \colon \widetilde{\mathcal{E}}^\infty \to \mathcal{E}^\infty$. Thus we obtain the Bäcklund autotransformation of the sine-Gordon equation

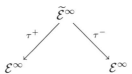

coinciding with the one discovered by Bäcklund [87].

Let us describe a natural way to construct Bäcklund transformations. Consider the diagram

$$\begin{array}{ccc} \widetilde{\mathcal{E}}_1 & \xrightarrow{\varphi} & \widetilde{\mathcal{E}}_2 \\ {\scriptstyle \tau_1}\Big\downarrow & & \Big\downarrow{\scriptstyle \tau_2} \\ \mathcal{E}_1^\infty & & \mathcal{E}_2^\infty \end{array}$$

where the vertical arrows are coverings and φ is an isomorphism. Then obviously $\tau_2 \circ \varphi \colon \widetilde{\mathcal{E}} \to \mathcal{E}_2^\infty$ is a covering, and thus we see that the equations $\mathcal{E}_1^\infty$ and $\mathcal{E}_2^\infty$ are related by the Bäcklund transformation.

The next examples show that some known Bäcklund transformations can be obtained in the this way.

EXAMPLE 1.11. Consider the modified Korteweg–de Vries equation
$$u_t = u_{xxx} + 6u^2 u_x. \tag{1.12}$$

The vector fields $\widetilde{D}_x = \bar{D}_x + X$ and $\widetilde{D}_t = \bar{D}_t + T$, where

$$X = p_0 \frac{\partial}{\partial w} + (p_0 + \alpha \sin(w + \widetilde{w})) \frac{\partial}{\partial \widetilde{w}},$$

$$T = (p_2 + p_0^3) \frac{\partial}{\partial w} + (p_2 + 2\alpha p_1 \cos(w + \widetilde{w}) + 2p_0^3$$
$$+ 2\alpha p_0^2 \sin(w + \widetilde{w}) + 2\alpha^2 p_0 + \alpha^3 \sin(w + \widetilde{w})) \frac{\partial}{\partial \widetilde{w}},$$

determine the two-dimensional covering $\tau\colon \widetilde{\mathcal{E}}^\infty = \mathbb{R}^2 \times \mathcal{E}^\infty \to \mathcal{E}^\infty$. Using the change of coordinates $w \mapsto \widetilde{w}$, $\widetilde{w} \mapsto w$, we obtain a new covering. Consequently, if $u(t,x)$ is a solution of equation (1.12) and the functions $w(t,x)$ and $\widetilde{w}(t,x)$ satisfy the equations

$$u = w_x, \qquad \widetilde{w}_x - w_x = \alpha \sin(w + \widetilde{w}),$$

then the function $\widetilde{u} = \widetilde{w}_x$ is also a solution of equation (1.12). This is a well-known Bäcklund transformations for the modified Korteweg–de Vries equation [**70**].

EXAMPLE 1.12. Consider the Korteweg–de Vries equation

$$u_t = u_{xxx} + 12uu_x$$

There exists a two-dimensional covering over this equation determined by the field X of the form

$$X = p_0 \frac{\partial}{\partial w} + (\alpha - p_0 - (w - \widetilde{w})^2) \frac{\partial}{\partial \widetilde{w}}$$

(we leave the computation of the field T to the reader as an exercise). As in the preceding example, by the change of variables $w \mapsto \widetilde{w}$, $\widetilde{w} \mapsto w$ [**70**] we obtain the Bäcklund transformation

$$u = w_x, \qquad \widetilde{w}_x + w_x = \alpha - (w - \widetilde{w})^2, \qquad \widetilde{u} = \widetilde{w}_x.$$

2. Examples of computations: coverings

In this section, we shall find coverings of a special type for some well-known equations of mathematical physics in two independent variables.

The process of constructing coverings for particular differential equations is a computational procedure similar to that used to compute symmetries. Namely, the covering space $\widetilde{\mathcal{E}}$ of any covering is locally diffeomorphic to the direct product $\mathcal{E}^\infty \times \mathbb{R}^N$, and the mapping $\tau\colon \widetilde{\mathcal{E}} = \mathcal{E}^\infty \times \mathbb{R}^N \to \mathcal{E}^\infty$ is the projection to the first factor. Therefore, a covering is locally determined by a distribution $\widetilde{\mathcal{C}}$, which, in the case of two independent variables, is given by two fields $\widetilde{D}_1 = \bar{D}_1 + X_1$ and $\widetilde{D}_2 = \bar{D}_2 + X_2$, where $X_i = \sum_{j=1}^N X_{ij} \partial/\partial w_j$, $X_{ij} \in \mathcal{F}(\widetilde{\mathcal{E}})$, $i = 1, 2$. Here $w_1, w_2, \ldots, w_N$ are coordinates in the fiber $\mathbb{R}^N$ (nonlocal variables) and $0 < N \leq \infty$. The integrability condition $[\widetilde{D}_1, \widetilde{D}_2] = 0$ for this distribution is of the form

$$[\bar{D}_1, X_2] + [X_1, \bar{D}_2] + [X_1, X_2] = 0. \tag{2.1}$$

This relation is a system of differential equations for the functions $X_{ij} \in \mathcal{F}(\widetilde{\mathcal{E}})$ describing all N-dimensional coverings over the equation $\mathcal{E}^\infty$. Solving this system, we shall find all coverings over the equation under consideration. To establish equivalence of coverings, we use Corollary 1.3.

2.1. Coverings over the Burgers equation.
Let us consider the Burgers equation (see also [**62**])
$$\mathcal{E} = \{u_t = uu_x + u_{xx}\}.$$

Let us choose the functions $x, t, p_1, \ldots, p_k, \ldots$ for intrinsic coordinates on $\mathcal{E}^\infty$ (see §4.1 of Ch. 4).

The total derivatives $\bar{D}_x = \bar{D}_1$ and $\bar{D}_t = \bar{D}_2$ are of the form
$$\bar{D}_x = \frac{\partial}{\partial x} + \sum_k p_{k+1} \frac{\partial}{\partial p_k}, \qquad \bar{D}_t = \frac{\partial}{\partial t} + \sum_k \bar{D}_x^k (p_0 p_1 + p_2) \frac{\partial}{\partial p_k}.$$

Let the distribution on $\widetilde{\mathcal{E}}$ be given by the fields
$$\widetilde{D}_x = \bar{D}_x + X, \qquad \widetilde{D}_t = \bar{D}_t + T,$$
where
$$X = \sum_j X_j \frac{\partial}{\partial w_j}, \qquad T = \sum_k T_k \frac{\partial}{\partial w_k}, \qquad X_j, T_k \in \mathcal{F}(\widetilde{\mathcal{E}}).$$

Then the integrability condition (2.1) acquires the form
$$\frac{\partial T}{\partial x} + \sum_{i \geq 0} p_{i+1} \frac{\partial T}{\partial p_i} - \frac{\partial X}{\partial t} - \sum_{j \geq 0} \bar{D}_x^j (p_0 p_1 + p_2) \frac{\partial X}{\partial p_j} + [X, T] = 0, \qquad (2.2)$$

where the operators $\partial/\partial p_k$ act on the fields X and T component-wise (the coefficients of the fields X and T depend on $x, t, p_1, \ldots, p_k, \ldots, w_1, \ldots, w_N$). We restrict ourselves with the case when all functions X_j, T_k are independent of the variables x, t, and $p_i, i > 1$. In this case, equation (2.2) is written down in the form
$$p_1 \frac{\partial T}{\partial p_0} + p_2 \frac{\partial T}{\partial p_1} - (p_0 p_1 + p_2) \frac{\partial X}{\partial p_0} - (p_1^2 + p_0 p_2 + p_3) \frac{\partial X}{\partial p_1} + [X, T] = 0. \qquad (2.3)$$

The left-hand side of (2.3) is polynomial in p_2 and p_3. Therefore, the coefficients at p_2 and p_3 vanish, from where it follows that the function X is independent of p_1 while $\partial T/\partial p_1 = \partial X/\partial p_0$, i.e.,
$$T = p_1 \frac{\partial X}{\partial p_0} + R, \qquad (2.4)$$
where the coefficients of the field R are independent of p_1. Substituting (2.4) into (2.3), we obtain
$$p_1^2 \frac{\partial^2 X}{\partial p_0^2} + p_1 \left(\frac{\partial R}{\partial p_0} - p_0 \frac{\partial X}{\partial p_0} + \left[X, \frac{\partial X}{\partial p_0} \right] \right) + [X, R] = 0.$$

Since the fields X and R are independent of p_1, the last equation is equivalent to the system
$$\frac{\partial^2 X}{\partial p_0^2} = 0, \qquad \frac{\partial R}{\partial p_0} = p_0 \frac{\partial X}{\partial p_0} + \left[X, \frac{\partial X}{\partial p_0} \right], \qquad [X, R] = 0. \qquad (2.5)$$

From the first equation of this system it follows that
$$X = p_0 A + B, \qquad (2.6)$$
where the coefficients of the fields A and B now depend on the variables w_k only. In other words, A and B are the fields in the fiber $\mathbb{R}^N$ of the covering. Using (2.6),

we see that the second equation of (2.5) acquires the form $\partial R/\partial p_0 = p_0 A + [A, B]$, or

$$R = \tfrac{1}{2} p_0^2 A + p_0[A, B] + C, \tag{2.7}$$

where C is a field on $\mathbb{R}^N$. Finally, substituting (2.6) and (2.7) into the last equation of (2.5), we obtain

$$p_0^2([A, [A, B]] + \tfrac{1}{2}[B, A]) + p_0([A, C] + [B, [A, B]]) + [B, C] = 0,$$

which is equivalent to the relations

$$[A, [A, B]] = \tfrac{1}{2}[A, B], \qquad [B, [B, A]] = [A, C], \qquad [B, C] = 0. \tag{2.8}$$

Thus, we have proved the following result.

THEOREM 2.1. *Any covering of the Burgers equation, where the coefficients of the fields X and T are independent of the variables t, x, and p_k, $k > 1$, are determined by the fields of the form*

$$\widetilde{D}_x = \bar{D}_x + p_0 A + B,$$
$$\widetilde{D}_t = \bar{D}_t + (p_1 + \tfrac{1}{2} p_0^2) A + p_0[A, B] + C,$$

where A, B, and C are arbitrary fields in the fiber of the covering satisfying relations (2.8).

REMARK 2.1. Let us stress here a remarkable fact. Consider the free Lie algebra $\mathcal{G}$ with generators a, b, and c, satisfying the relations

$$[a, [a, b]] = \tfrac{1}{2}[a, b], \qquad [b, [b, a]] = [a, c], \qquad [b, c] = 0.$$

Then the Lie algebra generated by the fields A, B, and C is the image of the algebra $\mathcal{G}$ when it is represented in the Lie algebra of vector fields on the manifold $\mathbb{R}^N$. Thus, Theorem 2.1 shows that coverings over the Burgers equation satisfying the above conditions are uniquely determined by representations of the algebra $\mathcal{G}$ in vector fields. For this reason, we shall call the Lie algebra $\mathcal{G}$ *universal*.

REMARK 2.2. Universal algebras first arose in the context of the so-called *prolongation structures* of Wahlquist and Estabrook [51, 145]. Examples of universal algebras for various evolution equations, as well as some particular representations of these algebras, can be found in [23]. Note that Wahlquist–Estabrook prolongation structures are particular cases of coverings.

Let us now describe one-dimensional coverings of the type under consideration over the Burgers equation. Let w be a nonlocal variable. Then $A = \alpha \partial/\partial w$, $B = \beta \partial/\partial w$, $C = \gamma \partial/\partial w$, where α, β, $\gamma \in C^\infty(\mathbb{R})$, and (2.8) reduces to a system of ordinary differential equations in α, β, and γ.

Consider the case $A = 0$ first. Then system (2.8) reduces to the relation $[B, C] = 0$, or $\beta'\gamma - \beta\gamma' = 0$, where the prime denotes the derivative with respect to w. If the field B does not vanish, the coordinate w in $\mathbb{R}$ can be chosen in such a way that $B = \partial/\partial w$, and it follows that $\gamma = \text{const}$. Note that, choosing another coordinate in the fiber, we pass to a new covering equivalent to the initial one. If $B = 0$, then $C \ne 0$, and this case is considered in a similar way. Thus, the coordinate w can be always be chosen locally in such a way that $\beta, \gamma = \text{const}$. Therefore, the system of fields

$$\widetilde{D}_x = \bar{D}_x + \beta \frac{\partial}{\partial w}, \qquad \widetilde{D}_t = \bar{D}_t + \gamma \frac{\partial}{\partial w}$$

possesses a nontrivial kernel consisting of functions $\psi(w - \beta x - \gamma t)$. In other words, the manifold $\widetilde{\mathcal{E}}$ is fibered by invariant manifolds $\mathcal{X}_\lambda = \{w - \beta x - \gamma t = \lambda\}$, $\lambda \in \mathbb{R}$, isomorphic to the manifold $\mathcal{E}^\infty$, i.e., the covering under consideration is trivial.

Let us pass now to the case $A \neq 0$ and study the structure of the coverings in a neighborhood of a generic point of the field A. Let choose a coordinate w in this neighborhood such that $A = \partial/\partial w$. Then system (2.8) acquires the form

$$\beta'' = \tfrac{1}{2}\beta', \qquad (\beta')^2 - \beta\beta'' = \gamma', \qquad \beta'\gamma = \gamma'\beta$$

and possesses solutions of two types:

$$\beta - \text{const}, \qquad \gamma = \text{const}$$

and

$$\beta = \mu e^{w/2} + \nu, \quad \gamma = -\frac{\nu}{2}(\mu e^{w/2} + \nu), \quad \mu = \text{const} \neq 0, \quad \nu = \text{const}. \tag{2.9}$$

All coverings of this type are pairwise nonequivalent and by the coordinate change $w \mapsto w - \beta x - \gamma t$ reduce to the form

$$\widetilde{D}_x = \bar{D}_x + p_0 \frac{\partial}{\partial w}, \qquad \widetilde{D}_t = \bar{D}_t + \left(p_1 + \frac{1}{2}p_0^2\right)\frac{\partial}{\partial w}.$$

Consider coverings of the second type, and denote by $\tau_{\mu,\nu}$ the covering determined by equation (2.9). Then the coverings τ_{μ_1,ν_1} and τ_{μ_2,ν_2} are equivalent if and only if $\operatorname{sign}\mu_1 = \operatorname{sign}\mu_2$ and $\nu_1 = \nu_2$.

Thus, we have proved the following statement:

PROPOSITION 2.2. *Any nontrivial one-dimensional covering over the Burgers equation* $\mathcal{E} = \{u_t = uu_x + u_{xx}\}$, *where the coefficients of the fields X and T are independent of the variables t, x, and p_k, $k > 1$, are locally equivalent to one of the following*:

$$\tau^0 \colon \widetilde{D}_x = \bar{D}_x + p_0 \frac{\partial}{\partial w},$$

$$\widetilde{D}_t = \bar{D}_t + \left(p_1 + \frac{1}{2}p_0^2\right)\frac{\partial}{\partial w},$$

$$\tau_\nu^+ \colon \widetilde{D}_x = \bar{D}_x + (p_0 + e^{w/2} + \nu)\frac{\partial}{\partial w},$$

$$\widetilde{D}_t = \bar{D}_t + \left(p_1 + \frac{1}{2}p_0^2 + \frac{1}{2}e^{w/2}p_0 - \frac{\nu}{2}e^{w/2} - \frac{\nu^2}{2}\right)\frac{\partial}{\partial w},$$

$$\tau_\nu^- \colon \widetilde{D}_x = \bar{D}_x + (p_0 - e^{w/2} + \nu)\frac{\partial}{\partial w},$$

$$\widetilde{D}_t = \bar{D}_t + \left(p_1 + \frac{1}{2}p_0^2 - \frac{1}{2}e^{w/2}p_0 + \frac{\nu}{2}e^{w/2} - \frac{\nu^2}{2}\right)\frac{\partial}{\partial w},$$

where $\nu = \text{const}$. All these coverings are pairwise inequivalent.

In each of the cases listed above, the manifold $\widetilde{\mathcal{E}}$ is contact equivalent to a manifold of the form $(\mathcal{E}')^\infty$, where $\mathcal{E}'$ is a second order evolution equation. The corresponding change of variables is:

$$u = w_x \text{ for the covering } \tau^0;$$

$$u = w_x - e^{w/2} - \nu \text{ for the coverings } \tau_\nu^+;$$

$$u = w_x + e^{w/2} - \nu \text{ for the coverings } \tau_\nu^-.$$

EXERCISE 2.1. Write down the covering equations in each of the above listed cases.

REMARK 2.3. Similar, but more cumbersome, calculations allow one to describe higher dimensional coverings over the Burgers equation. For example, there exist six 3-parameter families, one 4-parameter, and one 1-parameter family of nontrivial pairwise inequivalent two-dimensional coverings.

2.2. Coverings over the Korteweg–de Vries equation.

Now we consider the Korteweg–de Vries equation
$$\mathcal{E} = \{u_t = uu_x + u_{xxx}\}.$$

Let us take the functions x, t, $p_1, \ldots, p_k, \ldots$ for intrinsic coordinates on $\mathcal{E}^\infty$. The total derivatives on $\mathcal{E}^\infty$ are of the form
$$\bar{D}_x = \frac{\partial}{\partial x} + \sum_k p_{k+1}\frac{\partial}{\partial p_k}, \quad \bar{D}_t = \frac{\partial}{\partial t} + \sum_k \bar{D}_x^k(p_0 p_1 + p_3)\frac{\partial}{\partial p_k}.$$

Let the distribution on $\widetilde{\mathcal{E}}$ be described by the fields
$$\widetilde{D}_x = \bar{D}_x + X, \quad \widetilde{D}_t = \bar{D}_t + T,$$
where
$$X = \sum_j X_j \frac{\partial}{\partial w_j}, \quad T = \sum_k T_k \frac{\partial}{\partial w_k}, \quad X_j, T_k \in \mathcal{F}(\widetilde{\mathcal{E}}).$$

We shall solve equation (2.1) assuming that all the functions X_j, T_k are independent of the variables x, t, and p_i, $i > 2$.

Then computations similar to those in §2.1 lead to the following result:

THEOREM 2.3. *Any covering of the Korteweg–de Vries equation such that the coefficients of the fields X and T are independent of the variables t, x, and p_k, $k > 2$, are determined by the fields of the form*
$$\widetilde{D}_x = \bar{D}_x + p_0^2 A + p_0 B + C,$$
$$\widetilde{D}_t = \bar{D}_t + (2p_0 p_2 - p_1^2 + \tfrac{2}{3}p_0^3 A) + (p_2 + \tfrac{1}{2}p_0^2)B$$
$$+ p_1[B, C] + \tfrac{1}{2}p_0^2[B, [C, B]] + p_0[C, [C, B]] + D,$$
where A, B, C, D are vector fields on the fiber of the covering satisfying
$$[A, B] = [A, C] = [C, D] = 0,$$
$$[B, [B, [B, C]]] = 0, \quad [B, D] + [C, [C, [C, B]]] = 0, \quad (2.10)$$
$$[A, D] + \tfrac{1}{2}[C, B] + \tfrac{3}{2}[B, [C, [C, B]]] = 0.$$

REMARK 2.4. The free Lie algebra with four generators and relations (2.10) is the universal algebra of the Korteweg–de Vries equation. It first arose in studies of prolongation structures [145]. It is known that this algebra is infinite-dimensional [107].

Let us describe one-dimensional coverings over the Korteweg–de Vries equation. It is easily seen that if a covering is nontrivial, then at least one of the fields A or B must not vanish. Assume that $A \neq 0$. Then a nonlocal variable w can be chosen in such a way that $A = \partial/\partial w$. In this case, from (2.10) it easily follows that
$$B = \beta\frac{\partial}{\partial w}, \quad C = \gamma\frac{\partial}{\partial w}, \quad D = \delta\frac{\partial}{\partial w},$$

where β, γ, δ are constants. Thus, any covering over the KdV equation with $A \neq 0$ is equivalent to the covering determined by the fields

$$X = (p_0^2 + \beta p_0 + \gamma)\frac{\partial}{\partial w},$$
$$T = \left(2p_0p_2 - p_1^2 + \frac{2}{3}p_0^3 + \beta\left(p_2 + \frac{1}{2}p_0^2\right) + \delta\right)\frac{\partial}{\partial w}. \tag{2.11}$$

From the results of §1.6 it easily follows that the coverings of the form (2.11) are nontrivial. To check whether they are equivalent to each other, consider the Whitney product of two coverings (2.11),

$$\widetilde{D}_x = \bar{D}_x + (p_0^2 + \beta_1 p_0 + \gamma_1)\frac{\partial}{\partial w_1} + (p_0^2 + \beta_2 p_0 + \gamma_2)\frac{\partial}{\partial w_2},$$
$$\widetilde{D}_t = \bar{D}_t + \left(2p_0p_2 - p_1^2 + \frac{2}{3}p_0^3 + \beta_1\left(p_2 + \frac{1}{2}p_0^2\right) + \delta_1\right)\frac{\partial}{\partial w_1}$$
$$+ \left(2p_0p_2 - p_1^2 + \frac{2}{3}p_0^3 + \beta_2\left(p_2 + \frac{1}{2}p_0^2\right) + \delta_2\right)\frac{\partial}{\partial w_2},$$

and the linear system

$$\widetilde{D}_x(f) = 0, \qquad \widetilde{D}_t(f) = 0,$$

where f is a function on the space of the Whitney product. It is easily seen that this system is equivalent to the following one:

$$\frac{\partial f}{\partial x} + \gamma_1\frac{\partial f}{\partial w_1} + \gamma_2\frac{\partial f}{\partial w_2} = 0, \qquad \frac{\partial f}{\partial t} + \delta_1\frac{\partial f}{\partial w_1} + \delta_2\frac{\partial f}{\partial w_2} = 0,$$
$$\beta_1\frac{\partial f}{\partial w_1} + \beta_2\frac{\partial f}{\partial w_2} = 0, \qquad \frac{\partial f}{\partial w_1} + \frac{\partial f}{\partial w_2} = 0, \tag{2.12}$$

where $f = f(x,t,w_1,w_2)$. Solutions of (2.12) are different from constants if and only if $\beta_1 = \beta_2$. Using (2.12), it is easy to show that the equivalence classes of coverings with $A \neq 0$ are parametrized by β and their representatives can be written as

$$X = (p_0^2 + \beta p_0)\frac{\partial}{\partial w},$$
$$T = \left(2p_0p_2 - p_1^2 + \frac{2}{3}p_0^3 + \beta\left(p_2 + \frac{1}{2}p_0^2\right)\right)\frac{\partial}{\partial w}. \tag{2.13}$$

Assume now that $A = 0$. Then $B \neq 0$, and in a suitable coordinate system the field B can be written in the form $\partial/\partial w$. Then the relations (2.10) become

$$\frac{\partial^3 C}{\partial w^3} = 0, \quad \frac{\partial C}{\partial w} = 3\left[\frac{\partial C}{\partial w}, C\right], \quad \frac{\partial D}{\partial w} = \left[C, \left[C, \frac{\partial C}{\partial w}\right]\right], \quad [C, D] = 0.$$

This system possesses solutions of the two types

$$C = \left(\frac{1}{6}w^2 + \beta w + \gamma\right)\frac{\partial}{\partial w}, \qquad D = \left(\beta^2 - \frac{2}{3}\gamma\right)\frac{\partial}{\partial w}$$

and

$$C = \gamma\frac{\partial}{\partial w}, \qquad D = \delta\frac{\partial}{\partial w},$$

where β, γ, δ are constants. Respectively, we have

$$X = (p_0 + \tfrac{1}{6}w^2 + \beta w + \gamma)\frac{\partial}{\partial w},$$

$$T = (p_2 + p_1(\tfrac{1}{3}w + \beta) + \tfrac{1}{3}p_0^2 + (\tfrac{1}{18}w^2 + \tfrac{1}{3}\beta w + \beta^2 - \tfrac{1}{3}\gamma)p_0$$
$$+ (\beta^2 - \tfrac{2}{3}\gamma)(\tfrac{1}{6}w^2 + \beta w + \gamma))\frac{\partial}{\partial w},$$

and

$$X = (p_0 + \gamma)\frac{\partial}{\partial w}, \qquad T = (p_2 + \tfrac{1}{2}p_0^2 + \delta)\frac{\partial}{\partial w}.$$

In the first case, by the change of variables $w \mapsto w - \tfrac{1}{3}\beta$, we can transform the covering to

$$\begin{aligned}
X &= \left(p_0 + \frac{1}{6}w^2 + \alpha\right)\frac{\partial}{\partial w}, \\
T &= \left(p_2 + \frac{1}{3}wp_1 + \frac{1}{3}p_0^2 + \left(\frac{1}{18}w^2 - \frac{1}{3}\alpha\right)p_0 - \frac{2}{3}\alpha\left(\frac{1}{6}w^2 + \alpha\right)\right)\frac{\partial}{\partial w},
\end{aligned} \qquad (2.14)$$

where α is an arbitrary constant. The coverings belonging to this family are mutually inequivalent.

In the second case, all coverings are equivalent to each other. As a representative of this class, one can take, for example, the covering

$$X = p_0\frac{\partial}{\partial w}, \qquad T = \left(p_2 + \frac{1}{2}p_0^2\right)\frac{\partial}{\partial w}. \qquad (2.15)$$

It is easily seen that all the coverings (2.13)–(2.15) are pairwise inequivalent. Thus, we obtain a complete description of coverings over KdV of the type under consideration. Let us summarize the results.

PROPOSITION 2.4. *Any nontrivial covering of the Korteweg–de Vries equation $\mathcal{E} = \{u_t = uu_x + u_{xxx}\}$ such that the coefficient of the fields X and T are independent of t, x, and p_k, $k > 2$, is locally equivalent to one of the following:*

$$\tau^0 \colon \widetilde{D}_x = \bar{D}_x + p_0\frac{\partial}{\partial w},$$
$$\widetilde{D}_t = \bar{D}_t + \left(p_2 + \frac{1}{2}p_0^2\right)\frac{\partial}{\partial w},$$

$$\tau^1_\alpha \colon \widetilde{D}_x = \bar{D}_x + \left(p_0 + \frac{1}{6}w^2 + \alpha\right)\frac{\partial}{\partial w},$$
$$\widetilde{D}_t = \bar{D}_t + \left(p_2 + \frac{1}{3}wp_1 + \frac{1}{3}p_0^2 + \left(\frac{1}{18}w^2 - \frac{1}{3}\alpha\right)p_0 \right.$$
$$\left. - \frac{2}{3}\alpha\left(\frac{1}{6}w^2 + \alpha\right)\right)\frac{\partial}{\partial w}, \quad \alpha \in \mathbb{R},$$

$$\tau^2_\beta \colon \widetilde{D}_x = \bar{D}_x + \left(p_0^2 + \beta p_0\right)\frac{\partial}{\partial w},$$
$$\widetilde{D}_t = \bar{D}_t + \left(2p_0p_2 - p_1^2 + \frac{2}{3}p_0^3 + \beta\left(p_2 + \frac{1}{2}p_0^2\right)\right)\frac{\partial}{\partial w}, \quad \beta \in \mathbb{R}.$$

All these coverings are pairwise inequivalent.

In each of the listed cases, the manifold $\widetilde{\mathcal{E}}$ is contact diffeomorphic to a manifold of the form $(\mathcal{E}')^\infty$, where $\mathcal{E}'$ is a third order evolution equation. The corresponding change of variables is of the form

$$u = w_x \text{ for the covering } \tau^0;$$
$$u = w_x - \tfrac{1}{6}w^2 - \tfrac{3}{2}\alpha \text{ for the coverings } \tau^1_\alpha;$$
$$u = -\alpha \pm \sqrt{\beta^2 + w_x} \text{ for the coverings } \tau^2_\beta.$$

As we already saw (see §1.9), the covering equation for the covering τ^0 is the potential Korteweg–de Vries equation $w_t = w_{xxx} + \tfrac{1}{2}w_x^2$, while for the covering τ^1_α, $\alpha = 0$, the covering equation is the modified Korteweg–de Vries equation $w_t = w_{xxx} - \tfrac{1}{6}w^2 w_x$.

2.3. Coverings over the equation $u_t = (B(u)u_x)_x$. Let us state, without proof, another result on covering computations [54].

PROPOSITION 2.5. *Any nontrivial one- or two-dimensional covering of the equation $\mathcal{E} = \{u_t = (B(u)u_x)_x\}$, where the coefficients of the fields X and T are independent of the variables t, x, and p_k, $k > 1$, is locally equivalent to one of the following*:

$$\tau^1 \colon \widetilde{D}_x = \bar{D}_x + p_0 \frac{\partial}{\partial w},$$
$$\widetilde{D}_t = \bar{D}_t + B(p_0)p_1 \frac{\partial}{\partial w},$$
$$\tau^2_1 \colon \widetilde{D}_x = \bar{D}_x + p_0 \frac{\partial}{\partial w_1} + w_1 \frac{\partial}{\partial w_2},$$
$$\widetilde{D}_t = \bar{D}_t + B(p_0), p_1 \frac{\partial}{\partial w_1} + \widetilde{B}(p_0) \frac{\partial}{\partial w_2},$$
$$\tau^2_2 \colon \widetilde{D}_x = \bar{D}_x + \frac{\partial}{\partial w_1} - p_0 w_1 \frac{\partial}{\partial w_2},$$
$$\widetilde{D}_t = \bar{D}_t + (\widetilde{B}(p_0) - w_1 B(p_0) p_1) \frac{\partial}{\partial w_2},$$

There are no irreducible coverings of dimension greater than two.

To conclude this section, let us consider two examples, where the equation under consideration is not an evolution one.

2.4. Covering over the f-Gordon equation. Consider the equation $\mathcal{E}_f = \{u_{xy} = f\}$, where $f = f(u)$ is an arbitrary function depending on u. Let us choose the functions x, y, u, p_1, q_1, ..., p_k, q_k, ..., where $x = x_1$, $y = x_2$, $u = p_{(0,0)}$, $p_k = p_{(k,0)}$, $q_k = p_{(0,k)}$ for intrinsic coordinates on $\mathcal{E}_f^\infty$. Consider the coverings $\tau \colon \widetilde{\mathcal{E}}_f = \mathcal{E}_f^\infty \times \mathbb{R}^N \to \mathcal{E}_f^\infty$, $\widetilde{D}_x = \bar{D}_x + X$, $\widetilde{D}_y = \bar{D}_y + Y$, $\bar{D}_x = \bar{D}_1$, $\bar{D}_y = \bar{D}_2$, such that the coefficients of the fields X and Y are independent of the variables x, y and p_i, q_i, $i > 1$. In this case conditions (2.1) become

$$\begin{cases} \dfrac{\partial X}{\partial q_1} = \dfrac{\partial Y}{\partial p_1} = 0, \\[2mm] p_1 \dfrac{\partial Y}{\partial u} + f\left(\dfrac{\partial Y}{\partial q_1} - \dfrac{\partial X}{\partial p_1}\right) - q_1 \dfrac{\partial X}{\partial u} + [X, Y] = 0. \end{cases}$$

Analyzing this system for the case of one-dimensional coverings, we obtain the following result:

PROPOSITION 2.6. *Any nontrivial one-dimensional covering of the equation* $\mathcal{E} = \{u_{xy} = f\}$, $\partial f/\partial u \neq 0$, *where the coefficients of the fields* X *and* Y *are independent of the variables* x, y, *and* p_i, q_i, $i > 1$, *are locally (in a neighborhood of any point where* $\partial^2 X/\partial p_1^2 \neq 0$ *and* $\partial Y^2/\partial q_1^2 \neq 0$), *is equivalent to the following one:*

$$\tau_\alpha^f: \quad \widetilde{D}_x = \bar{D}_x + (\alpha p_1^2 + 2F)\frac{\partial}{\partial w}, \quad \widetilde{D}_y = \bar{D}_y + (q_1^2 + 2\alpha F)\frac{\partial}{\partial w},$$

where $\alpha \in \mathbb{R}$, $F(u) = \int_0^u f(u)\, du$. *These coverings are pairwise inequivalent for different values of the parameter* α.

Consider the equations $\mathcal{E}_f$ and $\mathcal{E}_g$, and let us equate the right-hand sides of the equations determining the covering τ_β^g. As a result, we obtain the system

$$\begin{aligned} \alpha u_x^2 + 2F(u) &= \beta v_x^2 + 2G(v), \\ u_y^2 + 2\alpha F(u) &= v_y^2 + 2\beta G(v). \end{aligned} \quad (2.16)$$

A simple check shows that if $v = v(x, y)$ is a solution of the equation $\mathcal{E}_g$ such that system (2.16) is compatible, then any solution $u = u(x, y)$ of (2.16) with $\alpha u_x + u_y \neq 0$ is also a solution of $\mathcal{E}_f$ for a given v. Thus, (2.16) determines a "partial Bäcklund transformation" between equations $\mathcal{E}_f$ and $\mathcal{E}_g$.

REMARK 2.5. Prolongation structures for the equation $\mathcal{E}_f$ were studied in [**106**]. The result obtained there can be reformulated as follows: the equation $\mathcal{E}_f$ possesses nontrivial one-dimensional coverings such that the coefficients of the fields X and Y are linear in p_1 and q_1 respectively, if and only if the function f satisfies the equation $d^2 f/du^2 = af$, $a = \text{const}$. In particular, if $f = \sin u$, then the manifold $\widetilde{\mathcal{E}}$ is contact diffeomorphic to $(\mathcal{E}_{\sin u})^\infty$, and the corresponding covering is the classical Bäcklund transformation of the sine-Gordon equation. If $f = e^u$, then $\widetilde{\mathcal{E}}_f$ is contact diffeomorphic to the manifold $(\mathcal{E}_0)^\infty$, and the corresponding covering determines the Bäcklund transformation between the Liouville equation and the wave equation.

2.5. Coverings of the equation $u_{xx} + u_{yy} = \varphi(u)$. Consider the equation $\mathcal{E}_\varphi = \{u_{xx} + u_{yy} = \varphi(u)\}$, where φ is a smooth function depending on u. Let us choose for coordinates on $\mathcal{E}^\infty$ the functions x, y, u, $p_1, q_1, \ldots, p_k, q_k, \ldots$, where $p_k = p_{(k,0)}$, $q_k = p_{(k-1,1)}$ (see also [**62**]). Then the total derivatives $\bar{D}_x$ and $\bar{D}_y$ on $\mathcal{E}^\infty$ are of the form

$$\begin{cases} \bar{D}_x = \dfrac{\partial}{\partial x} + p_1 \dfrac{\partial}{\partial u} + \sum_k p_{k+1} \dfrac{\partial}{\partial p_k} + \sum_k q_{k+1} \dfrac{\partial}{\partial q_k}, \\ \bar{D}_y = \dfrac{\partial}{\partial y} + q_1 \dfrac{\partial}{\partial u} + \sum_k q_{k+1} \dfrac{\partial}{\partial p_k} + \sum_k \bar{D}_x^{k-1}(\varphi - p_2) \dfrac{\partial}{\partial q_k}. \end{cases}$$

Let us set $\widetilde{D}_x = \bar{D}_x + X$, $\widetilde{D}_y = \bar{D}_y + Y$ and assume that the coefficients of the fields X and Y depend on the variables $w_1, \ldots, w_s$, u, p_1, q_1 only. In this case, conditions (2.1) become

$$\frac{\partial Y}{\partial p_1} + \frac{\partial X}{\partial q_1} = 0, \qquad \frac{\partial Y}{\partial q_1} - \frac{\partial X}{\partial p_1} = 0,$$
$$p_1 \frac{\partial Y}{\partial u} - q_1 \frac{\partial X}{\partial u} - \varphi \frac{\partial X}{\partial q_1} + [X, Y] = 0. \qquad (2.17)$$

Let us introduce new complex variables $z = p_1 + iq_1$, $H = X + iY$. Then from the first two equations (2.17) it follows that H is an analytic function in z:

$$H(z) = \sum_{k=0}^{\infty} H_k z^k, \qquad H_k = H_k(w, u) = X_k + iY_k, \qquad (2.18)$$

while the third equation in this system acquires the form

$$\bar{z}\frac{\partial H}{\partial u} - z\frac{\partial \bar{H}}{\partial u} + \varphi\left(\frac{\partial H}{\partial z} - \frac{\partial \bar{H}}{\partial z}\right) - [H, \bar{H}] = 0, \qquad (2.19)$$

Substituting (2.18) into (2.19), we obtain

$$\sum_{k=0}^{\infty}\left(\bar{z}z^k \frac{\partial H_k}{\partial u} - z\bar{z}^k \frac{\partial \bar{H}_k}{\partial u} + k\varphi(z^{k-1}H_k - \bar{z}^{k-1}\bar{H}_k)\right) = \sum_{k,j=0}^{\infty} z^k \bar{z}^j [H_k, \bar{H}_j],$$

from which it follows that

$$\varphi(H_1 - \bar{H}_1) = [H_0, \bar{H}_0], \qquad \frac{\partial H_0}{\partial u} - 2\varphi H_2 = [H_0, \bar{H}_1],$$
$$\frac{\partial H_1}{\partial u} - \frac{\partial \bar{H}_1}{\partial u} = [H_1, \bar{H}_1], \qquad (k+1)\varphi H_{k+1} = [H_k, \bar{H}_0], \qquad (2.20)$$
$$\frac{\partial H_k}{\partial u} = [H_k, \bar{H}_1], \qquad [H_k, \bar{H}_j] = 0, \qquad k, j > 1, \quad k \geq j.$$

Let us note now that by a coordinate change in the fiber of the covering, the field $\partial/\partial u + X_1$ may be transformed to $\partial/\partial u$. Then equation (2.20) will take the form

$$2i\varphi Y_1 = [H_0, \bar{H}_0], \qquad 2\varphi H_2 = \left[\frac{\partial}{\partial u} + iY_1, \bar{H}_0\right],$$
$$\frac{\partial Y_1}{\partial u} = 0, \qquad (k+1)\varphi H_{k+1} = [H_k, \bar{H}_0], \qquad (2.21)$$
$$[H_k, \bar{H}_j] = 0, \qquad k, j > 1, \quad k \geq j.$$

There are two options for the field Y_1: (a) $Y_1 = 0$ and (b) $Y_1 \neq 0$. For the first case we have

$$2i[H_0, \bar{H}_0] = 0, \qquad 2\varphi H_2 = \frac{\partial H_0}{\partial u},$$
$$\frac{\partial H_k}{\partial u} = 0, \qquad (k+1)\varphi H_{k+1} = [H_k, \bar{H}_0], \qquad (2.22)$$
$$[H_k, \bar{H}_j] = 0, \qquad k, j > 1, \quad k \geq j.$$

In case (b), after a suitable coordinate change in the fiber, the field Y_1 is transformed to the field $\partial/\partial w$, where $w = w_1$, and system (2.21) acquires the form

$$2i\varphi \frac{\partial}{\partial w} = [H_0, \bar{H}_0], \quad 2\varphi H_2 = i\frac{\partial \bar{H}_0}{\partial \xi},$$
$$\frac{\partial H_k}{\partial \bar{\xi}} = 0, \quad (k+1)\varphi H_{k+1} = [H_k, \bar{H}_0], \quad (2.23)$$
$$[H_k, \bar{H}_j] = 0, \quad k, j > 1, \quad k \geq j,$$

where $\xi = w + iu$. Solving systems (2.22) and (2.23) in the case $\dim \tau = 1$, for an arbitrary function φ we obtain the solution

$$H = (2\mu_2 \Phi + \mu_0 + \bar{\mu}_2 z^2)\frac{\partial}{\partial w}, \quad \Phi = \int_{u_0}^{u} \varphi \, du,$$

where $\mu_0, \mu_2 \in \mathbb{C}$.

For $\varphi = 0$ (and, of course, for equations transforming to this value of φ) we have

$$H = \left(\exp(\gamma \xi) \sum_{k \neq 1} \mu_k z^k + iz\right)\frac{\partial}{\partial w}, \quad \tau \in \mathbb{R},$$

and

$$H = \sum_{k \neq 1} \mu_k z^k \frac{\partial}{\partial w}.$$

If $\varphi = u$, then

$$H = \left(\mu_0 + \mu_1 \xi + iz\right)\frac{\partial}{\partial w}.$$

For

$$\varphi = \frac{1}{2}\left(\frac{1}{\gamma} + \gamma\right)\sinh(2\gamma u), \quad \gamma \in \mathbb{R}, \quad \gamma \neq 0, 1,$$

we have

$$H = \left(\cos(\gamma \xi) + \frac{1}{\gamma}\sin(\gamma \xi) + iz\right)\frac{\partial}{\partial w}$$

and for

$$\varphi = \frac{1}{2}\left(\frac{1}{\gamma} - \gamma\right)\sin(2\gamma u), \quad \gamma \in \mathbb{R}, \quad \gamma \neq 0, 1,$$

the solution is

$$H = \left(\cosh(\gamma \xi) + \frac{1}{\gamma}\sin(\gamma \xi) + iz\right)\frac{\partial}{\partial w}.$$

In the first case (an arbitrary φ) the classes of nontrivial coverings are represented by the following fields:

$$H = \left(2\exp(i\theta)\Phi + \exp(-i\theta)z^2\right)\frac{\partial}{\partial w}, \quad \theta \in [0, 2\pi).$$

We leave it to the reader to analyze the other cases.

3. Nonlocal symmetries

3.1. Definition of nonlocal symmetries. Let $\tau\colon \widetilde{\mathcal{E}} \to \mathcal{E}^\infty$ be a covering over the equation $\mathcal{E}^\infty$. A *nonlocal symmetry* of the equation $\mathcal{E}$ will be called a local symmetry of the object $\widetilde{\mathcal{E}}$. In other words, a nonlocal symmetry of the equation $\mathcal{E}$ is a transformation (finite or infinitesimal) of the object $\widetilde{\mathcal{E}}$ that preserves the distribution $\widetilde{\mathcal{C}}$ on $\widetilde{\mathcal{E}}$. Note that the definition of nonlocal symmetries assumes existence of some covering over the equation $\mathcal{E}^\infty$. To underline this fact, nonlocal symmetries in the covering $\tau\colon \widetilde{\mathcal{E}} \to \mathcal{E}^\infty$ will be called *symmetries of type τ*, or *nonlocal τ-symmetries*. In what follows, we shall consider *infinitesimal* symmetries only and shall call them just "nonlocal symmetries".

Since the general differential-geometric structure of covering manifolds $\widetilde{\mathcal{E}}$ is quite similar to that of infinitely prolonged equations, the definition of nonlocal infinitesimal symmetries corresponds exactly to the definition of higher symmetries for differential equations.

DEFINITION 3.1. The *algebra of nonlocal symmetries* of type τ (or of *nonlocal τ-symmetries*) of the equation $\mathcal{E}$ is the quotient Lie algebra
$$\operatorname{sym}_\tau \mathcal{E} = \mathrm{D}_{\mathcal{C}}(\widetilde{\mathcal{E}})/\mathcal{C}\mathrm{D}(\widetilde{\mathcal{E}}),$$
where
$$\mathcal{C}\mathrm{D}(\widetilde{\mathcal{E}}) = \left\{ \sum_{i=1}^n \varphi_i \widetilde{D}_i \mid \varphi_i \in \mathcal{F}(\widetilde{\mathcal{E}}) \right\},$$
while $\mathrm{D}_{\mathcal{C}}(\widetilde{\mathcal{E}})$ consists of vector fields X on $\widetilde{\mathcal{E}}$ such that $[X, \mathcal{C}\mathrm{D}(\widetilde{\mathcal{E}})] \subset \mathcal{C}\mathrm{D}(\widetilde{\mathcal{E}})$.

EXERCISE 3.1. Prove that when coverings τ_1 and τ_2 are equivalent, then the Lie algebras of nonlocal symmetries $\operatorname{sym}_{\tau_1} \mathcal{E}$ and $\operatorname{sym}_{\tau_2} \mathcal{E}$ are isomorphic.

3.2. How to find nonlocal symmetries? The procedure for finding nonlocal symmetries for a particular differential equation splits in a natural way into two steps: it is necessary first to construct a covering τ over the equation under consideration, and then to find τ-symmetries.

The problem of covering construction was discussed in detail in §2. Let us note here that in searching for nonlocal symmetries, one has to choose the corresponding covering in a special way: in general, it may happen that the algebra $\operatorname{sym}_\tau \mathcal{E}$ diminishes in comparison with $\operatorname{sym} \mathcal{E}$. For example, this is the case when we take the covering over the Korteweg–de Vries equation by the modified Korteweg–de Vries equation (see Example 3.1 below). Besides, some nonlocal symmetries are not interesting in applications (for example, symmetries of trivial coverings).

Consider now the problem of computing the nonlocal symmetry algebra in a given covering $\tau\colon \widetilde{\mathcal{E}} \to \mathcal{E}^\infty$. If the covering object $\widetilde{\mathcal{E}}$ is of the form $\widetilde{\mathcal{E}} = (\mathcal{E}')^\infty$, then the problem reduces to computation of local symmetries for the equation $(\mathcal{E}')^\infty$. When $\widetilde{\mathcal{E}}$ is not represented in this form explicitly, the computational procedure may be based on one of the theorems proved below. Examples of computations of nonlocal symmetries will be given in the next section.

THEOREM 3.1. *The algebra $\operatorname{sym}_\tau \mathcal{E}$ is isomorphic to the Lie algebra of vector fields X on $\widetilde{\mathcal{E}}$ satisfying the following conditions:*
1. *X is a vertical field, i.e., $X(\tau^*(f)) = 0$ for any $f \in C^\infty(M) \subset \mathcal{F}(\mathcal{E}^\infty)$.*
2. *$[X, \widetilde{D}_i] = 0$, $i = 1, \ldots, n$.*

PROOF. Note that the first condition means that in local coordinates the coefficients of the field X at $\partial/\partial x_i$, $i = 1, \ldots, n$, vanish. Therefore, the intersection of the set of vertical fields on $\widetilde{\mathcal{E}}$ with the algebra $\mathcal{CD}(\widetilde{\mathcal{E}})$ is trivial. On the other hand, in every coset $[X] \in \mathrm{sym}_\tau \mathcal{E}$ there exists one and only one vertical representative X^v. Indeed, let X' be an arbitrary representative of the class $[X]$. Then $X^v = X' - \sum_{i=1}^n a_i \widetilde{D}_i$, where a_i is the coefficients of the of the field X' at $\partial/\partial x_i$. $\square$

THEOREM 3.2. *Let the covering* $\tau\colon \widetilde{\mathcal{E}} = \mathcal{E}^\infty \times \mathbb{R}^N \to \mathcal{E}^\infty$ *be locally determined by the fields* $\widetilde{D}_i = \bar{D}_i + \sum_{j=1}^N X_{ij} \partial/\partial w_j$, $i = 1, \ldots, n$, $X_{ij} \in \mathcal{F}(\widetilde{\mathcal{E}})$, *where* $w_1, w_2, \ldots$ *are nonlocal variables. Then any τ-symmetry of the equation* $\mathcal{E} = \{F = 0\}$ *is of the form*

$$\widetilde{\partial}_{\varphi,A} = \widetilde{\partial}_\varphi + \sum_{j=1}^N a_j \frac{\partial}{\partial w_j},$$

where $\varphi = (\varphi_1, \ldots, \varphi_m)$, $A = (a_1, \ldots, a_N)$, $\varphi_i, a_j \in \mathcal{F}(\widetilde{\mathcal{E}})$, *and the functions* φ_i, a_j *satisfy the equations*

$$\widetilde{\ell}_F(\varphi) = 0, \qquad (3.1)$$

$$\widetilde{D}_i(a_j) = \widetilde{\partial}_{\varphi,A}(X_{ij}). \qquad (3.2)$$

PROOF. Let $S \in \mathrm{sym}_\tau \mathcal{E}$. Using Theorem 3.1, let us write down the vertical field S in local coordinates in the form

$$S = \sum_{\sigma,k}{}' b^k_\sigma \frac{\partial}{\partial p^k_\sigma} + \sum_{j=1}^N a_j \frac{\partial}{\partial w_j}, \qquad b^k_\sigma, a_j \in \mathcal{F}(\widetilde{\mathcal{E}}),$$

where the prime denotes summation on all intrinsic coordinates p^k_σ of the equation $\mathcal{E}$. Equating the coefficients of $\partial/\partial p^k_\sigma$ in the commutator $[S, \widetilde{D}_i]$ to zero, we obtain the equations $\widetilde{D}_i(b^k_\sigma) = b^k_{\sigma+1_i}$, if $p^k_{\sigma+1_i}$ is an intrinsic coordinate on $\mathcal{E}^\infty$, or $\widetilde{D}_i(b^k_\sigma) = S(\bar{p}^k_{\sigma+1_i})$ otherwise.

Solving these equations in the same way as for local symmetries, we obtain

$$b^k_\sigma = \widetilde{D}_\sigma(b^k_\varnothing), \qquad \widetilde{\ell}_F(b_\varnothing) = 0, \qquad b_\varnothing = (b^1_\varnothing, \ldots, b^N_\varnothing).$$

In particular, $S = \widetilde{\partial}_{\varphi,A}$, where $\varphi = b_\varnothing$. The equations (3.2) are obtained by equating to zero the coefficients of $\partial/\partial w_j$ in the commutator $[S, \widetilde{D}_i]$. $\square$

Note that for finite N the covering equation is locally of the form $(\mathcal{E}')^\infty$, where $\mathcal{E}' = \{F' = 0\}$ and

$$F' = \left(F, \frac{\partial w_j}{\partial x_i} - X_{ij}\right), \qquad i = 1, \ldots, n, \; j = 1, \ldots, N,$$

and the system of equations on its local symmetry with the generating function $\psi = (\varphi, A)$ coincides exactly with system (3.1)–(3.2).

EXERCISE 3.2. Prove the following identity:

$$[\widetilde{\partial}_{\varphi,A}, \widetilde{\partial}_{\psi,B}] = \widetilde{\partial}_{\chi,C},$$

where $\chi = \widetilde{\partial}_{\varphi,A}(\psi) - \widetilde{\partial}_{\psi,B}(\varphi)$, $C = \widetilde{\partial}_{\varphi,A}(B) - \widetilde{\partial}_{\psi,B}(A)$.

REMARK 3.1. If the covering is zero-dimensional (naturally, in this case $X_{ij} = 0$), i.e., τ is a local isomorphism, then equations (3.2) hold in a trivial way while system (3.1)–(3.2) reduces to the equation $\tilde{\ell}_F(\varphi) = 0$. Thus, the local symmetry theory embeds in the nonlocal one in a natural way. More precisely, the algebra of local symmetries of the equation $\mathcal{E}$ coincides with the algebra of nonlocal ones in the covering id: $\mathcal{E}^\infty \to \mathcal{E}^\infty$.

If a covering $\tau\colon \tilde{\mathcal{E}} \to \mathcal{E}^\infty$ of the equation $\mathcal{E}^\infty$ is given and a function $\varphi \in \mathcal{F}(\tilde{\mathcal{E}})$ satisfies equation (3.1), then, in general, there may be no symmetry of the form $\tilde{\partial}_{\varphi,A}$: the system of equations (3.2) may have no solution for a given φ. In particular, not every local symmetry $\tilde{\partial}_\varphi$, $\varphi \in \mathcal{F}(\mathcal{E}^\infty)$, can be extended to a symmetry $\tilde{\partial}_{\varphi,A}$ in the covering $\tau\colon \tilde{\mathcal{E}} \to \mathcal{E}^\infty$.

EXAMPLE 3.1. Consider the one-dimensional covering $\tau\colon \tilde{\mathcal{E}} = \mathcal{E}^\infty \times \mathbb{R}^1 \to \mathcal{E}^\infty$ over the Korteweg–de Vries equation $\mathcal{E} = \{u_t = u_{xxx} + uu_x\}$ given by the fields

$$\tilde{D}_x = \bar{D}_x + \left(p_0 + \frac{1}{6}w^2\right)\frac{\partial}{\partial w},$$

$$\tilde{D}_t = \bar{D}_t + \left(p_2 + \frac{1}{3}wp_1 + \frac{1}{3}p_0^2 + \frac{1}{18}w^2 p_0\right)\frac{\partial}{\partial w}.$$

As we saw already (see Example 1.9), $\tilde{\mathcal{E}} = (\mathcal{E}')^\infty$, where $\mathcal{E}'$ is the modified Korteweg–de Vries equation

$$w_t = w_{xxx} - \frac{1}{6}w^2 w_x.$$

Consider the local symmetry of the KdV equation with the generating function $\varphi = tp_1 + 1$ (the Galilean symmetry) and try to extend it to a symmetry $\tilde{\partial}_{\varphi,a} = \tilde{\partial}_\varphi + a\partial/\partial w$, $a \in \mathcal{F}(\tilde{\mathcal{E}})$, in the covering τ.

To do this, we have to solve the following system of equations for the function $a \in \mathcal{F}(\tilde{\mathcal{E}})$:

$$\tilde{D}_x(a) = \tilde{\partial}_{\varphi,a}\left(p_0 + \tfrac{1}{6}w^2\right) = \varphi + \tfrac{1}{3}aw,$$

$$\tilde{D}_t(a) = \tilde{\partial}_{\varphi,a}\left(p_2 + \tfrac{1}{3}wp_1 + \tfrac{1}{3}p_0^2 + \tfrac{1}{18}w^2 p_0\right)$$
$$= \bar{D}_x^2(\varphi) + \tfrac{1}{3}w\bar{D}_x(\varphi) + \left(\tfrac{2}{3}p_0 + \tfrac{1}{18}w^2\right)\varphi + a\left(\tfrac{1}{3}p_1 + \tfrac{1}{9}wp_0\right).$$

It is easily seen that this system has no solution. Thus, there is no symmetry $\tilde{\partial}_{\varphi,a}$, where φ is the Galilean symmetry, in the covering under consideration, and the nonlocal symmetry algebra diminishes in comparison with the local one. Nevertheless, one can consider different coverings and hope that in some of them the symmetry algebra will be extended. We consider this problem in §5.

4. Examples of computation: nonlocal symmetries of the Burgers equation

Consider the problem of finding nonlocal symmetries for the Burgers equation in the coverings described by Theorem 2.1 (see [**61**]). Using Theorem 3.1, we shall identify elements of the algebra $\mathrm{sym}_\tau \mathcal{E}$ with the fields S on $\tilde{\mathcal{E}}$ such that $S = P + \Phi$, where

$$P = \sum_{i \geq 0} P_i \frac{\partial}{\partial p_i}, \qquad \Phi = \sum_{j \geq 0} \Phi_j \frac{\partial}{\partial w_j},$$

$P_i, \Phi_j \in \mathcal{F}(\widetilde{\mathcal{E}})$, and $[S, \widetilde{D}_x] = [S, \widetilde{D}_t] = 0$.

Using (2.8), these relations can be transformed into

$$[S, \widetilde{D}_x] = \sum_{i \geq 0}(P_{i+1} - \widetilde{D}_x(P_i))\frac{\partial}{\partial p_i} + P_0 A + [\Phi, \widetilde{D}_x] = 0,$$

$$[S, \widetilde{D}_t] = \sum_{i \geq 0}\left(\sum_{k \geq 0}\frac{\partial}{\partial p_k}\left(\widetilde{D}_x^i(p_0 p_1 + p_2)\right) - \widetilde{D}_t(P_i)\right)\frac{\partial}{\partial p_i}$$
$$+ (P_1 + p_0 P_0)A + P_0[A, B] + [\Phi, \widetilde{D}_t] = 0.$$

The coefficients of the fields $[\Phi, \widetilde{D}_x]$ and $[\Phi, \widetilde{D}_t]$ at $\partial/\partial p_i$ vanish for all $i \geq 0$, i.e., these fields are vertical with respect to the projection τ. Therefore, the first of the above equations is equivalent to

$$P_0 A + [\Phi, \widetilde{D}_x] = 0, \quad P_{i+1} = \widetilde{D}_x(P_i), \quad i = 0, 1, \ldots,$$

which implies that $P_i = \widetilde{D}_x^i(\psi)$, where ψ denotes the function P_0. In other words, $P = \eth_\psi = \sum_{i \geq 0} \widetilde{D}_x^i(\psi) \partial/\partial p_i$. In a similar way, the second equation is equivalent to the equalities

$$(\widetilde{D}_x(\psi) + p_0 \psi)A + \psi[A, B] + [\Phi, \widetilde{D}_t] = 0, \qquad (4.1)$$

$$\eth_\psi(\widetilde{D}_x^i(p_0 p_1 + p_2)) = \widetilde{D}_t(\widetilde{D}_x^i(\psi)), \qquad i \geq 0. \qquad (4.2)$$

Since $[\eth_\psi, \widetilde{D}_x] = 0$, equation (4.2) for $i > 0$ is obtained by applying the operator $\widetilde{D}_x^i$ to equation (4.2) for $i = 0$, i.e.,

$$\eth_\psi(p_0 p_1 + p_2) = \widetilde{D}_t(\psi),$$

or, which is the same,

$$\widetilde{D}_x^2(\psi) + p_0 \widetilde{D}_x(\psi) + p_1 \psi = \widetilde{D}_t(\psi).$$

Thus, we have proved the following result:

PROPOSITION 4.1. *Any nonlocal symmetry of the Burgers equation in the covering (2.8) is of the form $\eth_\psi + \Phi$, where $\Phi = \sum_{j \geq 0} \Phi_j \partial/\partial w_j$, $\psi, \Phi_j \in \mathcal{F}(\widetilde{\mathcal{E}})$, while the function ψ and the field Φ satisfy the following equations*:

$$\psi A = [\widetilde{D}_x, \Phi],$$
$$(\widetilde{D}_x(\psi) + p_0 \psi)A + \psi[A, B] + [\Phi, \widetilde{D}_t] = 0, \qquad (4.3)$$
$$\widetilde{\ell}_F(\psi) = \widetilde{D}_x^2(\psi) + p_0 \widetilde{D}_x(\psi) + p_1 \psi - \widetilde{D}_t(\psi) = 0,$$

where F is the function determining the Burgers equation.

Subsequent analysis of system (4.3) is to be based on particular realizations of the universal algebra for the Burgers equation. Let us start by studying nonlocal symmetries in the coverings described in Proposition 2.4. Consider the covering τ^0. Let $\Phi = \varphi \partial/\partial w$. Then for τ^0 system (4.3) acquires the form

$$\psi = \widetilde{D}_x(\varphi),$$
$$\widetilde{D}_x(\psi) + p_0 \psi = \widetilde{D}_t(\varphi), \qquad (4.4)$$
$$\widetilde{\ell}_F(\psi) = 0.$$

Note that using Theorem 3.2, we would obtain the same system of equations.

It is easily seen that the third equation of system (4.4) is a consequence of the first two. Therefore, the system can be transformed to the following one:

$$\psi = \widetilde{D}_x(\varphi),$$
$$\widetilde{D}_x^2(\varphi) + p_0 \widetilde{D}_x(\varphi) = \widetilde{D}_t(\varphi).$$

Let us now pay attention to the fact that the second equation is of the form $\widetilde{\ell}_H(\varphi) = 0$, where $\widetilde{\ell}_H$ is the universal linearization operator for the equation $\mathcal{E}' = \{v_t = v_{xx} + \frac{1}{2}v_x^2\}$. This agrees with the fact that the equation $\widetilde{\mathcal{E}}$ is contact diffeomorphic to the manifold $(\mathcal{E}')^\infty$. The diffeomorphism is established by the following coordinate change:

$$x = x', \quad w = p'_0, \quad t = t', \quad p_k = p'_{k+1}, \quad k = 0, 1, \ldots,$$

where x', t', p'_k are the standard coordinates on $(\mathcal{E}')^\infty$. From this it follows that $\text{sym}_{\tau^0} \mathcal{E} = \text{sym}\, \mathcal{E}'$. Computing the algebra $\text{sym}\, \mathcal{E}'$ using the scheme described in Chapter 4 and applying the diffeomorphism discussed above, we find that the algebra $\text{sym}\, \mathcal{E}'$ is additively generated by elements of the form

$$\varphi_{-\infty} = \eta(x,t)e^{-w/2}, \quad \eta_{xx} = \eta_t, \quad \varphi^0_{-1} = 1,$$
$$\varphi^i_k = t^i p_k + \tfrac{1}{2}((k+1)t^i p_0 + i x t^{i-1}) p_{k-1} + O(k-2),$$
$$k = 0, 1, \ldots, \quad i = 0, 1, \ldots, k+1,$$

where the function $O(r)$ depends on the variables x, t, $p_0, \ldots, p_r$ only.

We stress the fact that the form of system (4.3) essentially depends on the choice of a coordinate in $\mathbb{R}$, i.e., on the form of a field A representation. If $A = \alpha(w)\partial/\partial w$, while B and C are still equal to zero, the relations (4.3) reduce to the form

$$\psi = \frac{1}{\alpha}\left(tD_x(\varphi) - p_0 \alpha' \varphi\right),$$
$$\widetilde{D}_x^2(\varphi) + (1 - 2\alpha')p_0 \widetilde{D}_x(\varphi) + ((\alpha')^2 - \alpha\alpha'' - \tfrac{1}{2}\alpha')p_0^2 \varphi = \widetilde{D}_t(\varphi).$$

Let us rewrite the last equality in the form $L(\varphi) = 0$, where

$$L = \widetilde{D}_x^2 + (1 - 2\alpha')p_0 \widetilde{D}_x - \widetilde{D}_t + ((\alpha')^2 - \alpha\alpha'' - \tfrac{1}{2}\alpha').$$

The operator L acquires a simpler form if one sets $\alpha = \tfrac{1}{2}w$. Then $L = \widetilde{D}_x^2 - \widetilde{D}_t$.

In this case, $\widetilde{\mathcal{E}} = (\mathcal{E}'')^\infty$, where $\mathcal{E}''$ is the heat equation $w_{xx} = w_t$. Thus, with a suitable choice of a coordinate in the fiber, the covering τ_0 reduces to the form $(\mathcal{E}'')^\infty \to \mathcal{E}^\infty$ and, moreover, the Burgers equation can be represented as a quotient equation of the heat equation $\mathcal{E}''$ over the one-parameter group of its symmetries $w \mapsto \varepsilon w$, where ε is a parameter. Note that such a transformation group is admitted by all linear equations. In the chosen coordinate system on $\widetilde{\mathcal{E}}$ (i.e., when $A = \tfrac{1}{2}w\partial/\partial w$) the generators of the algebra $\text{sym}_{\tau^0}\mathcal{E} = \text{sym}\,\mathcal{E}''$ are of the form

$$\varphi_{-\infty} = \eta(x,t)e^{-w/2}, \quad \eta_{xx} = \eta_t, \quad \varphi^0_{-1} = w,$$
$$\varphi^i_k = w(t^i p_k + \tfrac{1}{2}((k+1)t^i p_0 + i x t^{i-1}) p_{k-1} + O(k-2)),$$
$$k = 0, 1, \ldots, \quad i = 0, 1, \ldots, k+1.$$

Consider the coverings $\tau_\nu^\pm$. System (4.3) for these coverings is easily transformed to
$$\Box(\varphi) = 0,$$
$$\psi = \widetilde{D}_x(\varphi) - \beta'\varphi, \qquad (4.5)$$
where $\beta = \pm e^{w/2} + \nu$ and
$$\Box = \widetilde{D}_x^2 + p_0\widetilde{D}_x - \widetilde{D}_t - \beta'(p_0 + \beta). \qquad (4.6)$$

System (4.5) reduces to the sole equation $\Box(\varphi) = 0$ and is solved in the same way as the equation $\bar{\ell}_F(\varphi) = 0$ determining higher symmetries of the Burgers equation (see Ch. 4). The result shows that the algebras $\text{sym}_{\tau_\nu^\pm}\mathcal{E}$ are isomorphic to $\text{sym}\,\mathcal{E}$. The solutions φ of (4.5) depending on the variables x, t, w, p_0, and p_1 are linear combinations of the following ones:

$$\begin{aligned}
\varphi_1^0 &= p_0 + \beta, \\
\varphi_1^1 &= t(p_0 + \beta) - (\beta')^{-1}, \\
\varphi_2^0 &= p_1 + \tfrac{1}{2}p_0^2 + \beta'p_0 - \tfrac{1}{2}\nu\beta, \\
\varphi_2^1 &= t(p_1 + \tfrac{1}{2}p_0^2) + (t\beta' + \tfrac{1}{2}x)p_0 + \tfrac{1}{2}\beta(x - \nu t) + \tfrac{1}{2}\nu(\beta')^{-1} + 1, \\
\varphi_2^2 &= t^2(p_1 + \tfrac{1}{2}p_0^2) + (t^2\beta' + xt)p_0 + \tfrac{1}{2}\beta(2xt - \nu t^2) + (\nu t - x)(\beta')^{-1} + 2t.
\end{aligned} \qquad (4.7)$$

The corresponding functions ψ are
$$\begin{aligned}
\psi_1^0 &= p_1, \\
\psi_1^1 &= tp_1 + \tfrac{1}{2}(p_0 + \nu)(\beta')^{-1} + 2, \\
\psi_2^0 &= p_2 + p_0p_1, \\
\psi_2^1 &= t(p_2 + p_0p_1) + \tfrac{1}{2}(xp_1 + p_0) - \tfrac{1}{4}\nu(p_0 + \beta)(\beta')^{-1}, \\
\psi_2^2 &= t^2(p_2 + p_0p_1) + t(xp_1 + p_0) - \tfrac{1}{2}\nu(p_0 + \beta)(\beta')^{-1} \\
&\quad + (\tfrac{1}{2}x(p_0 + \beta + 2\beta') - 1(\beta')^{-1}.
\end{aligned}$$

Let S_j^i be a symmetry corresponding to the function φ_j^i. Then from (4.7) it follows that the symmetries S_1^0 and S_2^0 are local, the symmetry S_1^2 is essentially nonlocal for $\nu \neq 0$, and the symmetries S_1^1 and S_2^2 are essentially nonlocal.

The isomorphism between the algebras $\text{sym}_{\tau_\nu^\pm}\mathcal{E}$ and $\text{sym}\,\mathcal{E}$ suggests that the space $\widetilde{\mathcal{E}}_{\tau_\nu^\pm}$ of the covering $\tau_\nu^\pm$ is contact diffeomorphic to the infinite prolongation $\mathcal{E}^\infty$ of the Burgers equation $\mathcal{E}$. To check it, let us try to represent the manifold $\widetilde{\mathcal{E}}_{\tau_\nu^\pm}$ in the form $(\mathcal{E}')^\infty$, where $\mathcal{E}'$ is a differential equation. Assuming x and t to be independent variables in $\mathcal{E}$ and w to be the dependent one, let us introduce in $\widetilde{\mathcal{E}}_{\tau_\nu^\pm}$ the coordinates x, t, $y_0 = w, \ldots, y_i = \widetilde{D}_x^i(w), \ldots$ Then, as is easily seen, $\widetilde{D}_t(w) = y_2 + \tfrac{1}{2}y_1^2 - \beta y_1$, i.e., the desired equation $\mathcal{E}'$ is of the form
$$w_{xx} + \tfrac{1}{2}w_x^2 - \beta w_x = w_t. \qquad (4.8)$$

As expected, the universal linearization operator for equation (4.8) coincides with the operator (4.6).

Further, the change of variables $v = -\beta \mp e^{w/2} - \nu$ transforms equation (4.8) to the Burgers equation $v_t = v_{xx} + vv_x$. Thus, choosing the coordinates on $\widetilde{\mathcal{E}}_{\tau_\nu^\pm}$ in the form
$$x,\ t,\ y_0' = v,\ \ldots,\ y_i' = \widetilde{D}_x^i(v),\ \ldots,$$

we find that $\widetilde{\mathcal{E}}_{\tau_\nu^\pm} \approx \mathcal{E}^\infty$, and so the mapping $\tau_\nu^\pm$ can be considered as a mapping of the Burgers equation to itself. The explicit formula for this mapping is obtained from the relation $v_x = \widetilde{D}_x(v) = -\beta'(u+\beta) = \frac{1}{2}(v+\nu)(u-\nu)$, from which we have

$$u = \frac{2v_x}{v+\nu} + v. \tag{4.9}$$

Thus, we arrive at the following remarkable result. If v is a solution of the Burgers equation, then the function u determined by (4.9) is also a solution of the Burgers equation for all values of the parameter ν.

To conclude this section, we consider one infinite-dimensional covering of the Burgers equation. Let us set

$$A = \frac{\partial}{\partial w_1},$$
$$B = e^{w_1/2}\frac{\partial}{\partial w_2} + w_2\frac{\partial}{\partial w_3} + w_3\frac{\partial}{\partial w_4} + \ldots, \tag{4.10}$$
$$C = e^{w_1/2}\frac{\partial}{\partial w_3} + w_2\frac{\partial}{\partial w_4} + w_3\frac{\partial}{\partial w_5} + \ldots.$$

Obviously, the fields A, B, and C satisfy the relations (2.8) and thus determine an infinite-dimensional covering over the Burgers equation. Let $S = \widetilde{\partial}_\psi + \sum_i \Phi_i \partial/\partial w_i$ be a nonlocal symmetry. Then, since $[A, B] = \frac{1}{2}e^{w_1/2}\partial/\partial w_2$, system (4.3) in the case under consideration is of the form

$$p_1\psi + p_0\widetilde{D}_x(\psi) + \widetilde{D}_x^2(\psi) = \widetilde{D}_t(\psi),$$

$$\psi = \widetilde{D}_x(\Phi_1), \qquad p_0\psi + \widetilde{D}_x(\psi) = \widetilde{D}_t(\Phi_1),$$

$$\Phi_1 = 2e^{-w_1/2}\widetilde{D}_x(\Phi_2), \qquad \psi + \tfrac{1}{2}p_0\Phi_1 = 2e^{-w_1/2}\widetilde{D}_x(\Phi_2),$$

$$\Phi_2 = \widetilde{D}_x(\Phi_3), \qquad \Phi_1 = 2e^{-w_1/2}\widetilde{D}_x(\Phi_3),$$

$$\Phi_3 = \widetilde{D}_x(\Phi_4), \qquad \Phi_2 = \widetilde{D}_t(\Phi_4),$$

$$\cdots\cdots\cdots \qquad\qquad \cdots\cdots\cdots$$

$$\Phi_{k-1} = \widetilde{D}_x(\Phi_k), \qquad \Phi_{k-2} = \widetilde{D}_t(\Phi_k),$$

$$\cdots\cdots\cdots \qquad\qquad \cdots\cdots\cdots$$

It can be easily checked that this system is equivalent to the following one:

$$\psi = \widetilde{D}_x(\Phi_1), \qquad p_1\psi + p_0\widetilde{D}_x(\psi) + \widetilde{D}_x^2(\psi) = \widetilde{D}_t(\psi),$$

$$\Phi_1 = 2e^{-w_1/2}\widetilde{D}_x(\Phi_2), \qquad p_0\widetilde{D}_x(\Phi_1) + \widetilde{D}_x^2(\Phi_1) = \widetilde{D}_t(\Phi_1),$$

$$\Phi_2 = \widetilde{D}_x(\Phi_3), \qquad \widetilde{D}_x^2(\Phi_2) = \widetilde{D}_t(\Phi_2),$$

$$\Phi_3 = \widetilde{D}_x(\Phi_4), \qquad \widetilde{D}_x^2(\Phi_3) = \widetilde{D}_t(\Phi_3), \tag{4.11}$$

$$\cdots\cdots\cdots \qquad\qquad \cdots\cdots\cdots$$

$$\Phi_k = \widetilde{D}_x(\Phi_{k+1}), \qquad \widetilde{D}_x^2(\Phi_k) = \widetilde{D}_t(\Phi_k),$$

$$\cdots\cdots\cdots \qquad\qquad \cdots\cdots\cdots$$

Before solving this system of differential equations, let us introduce a filtration in the algebra $\mathcal{F}(\widetilde{\mathcal{E}})$. Let $k > 0$ and $\varphi \in \mathcal{F}(\widetilde{\mathcal{E}})$. Then we set $\deg \varphi = k$, if $\partial\varphi/\partial p_i = 0$ for all $i > k$; $\deg \varphi = -k$, if $\partial\varphi/\partial w_j = 0$ for all $j < k$; $\deg \varphi = 0$, if $\partial\varphi/\partial p_i = 0$ for all $i \geq 0$. Note further that if an element ψ satisfies all equations of the left column in system (4.11) and if ψ satisfies the k-th equation in the right column, then ψ satisfies all equations in (4.11) for $i < k$.

For subsequent arguments, it is convenient to consider two cases: (a) $\deg \Phi_2 \geq 4$ and (b) $\deg \Phi_2 < 4$. In the first case, since the operator $\widetilde{D}_x$ increases filtration by 1, from the equations of the first column it follows that there exists a number k such that starting from this number the functions Φ_i will depend on the variables $t, x, w_4, w_5, \ldots$ only. Let $(\psi, \Phi_1, \Phi_2, \ldots, \Phi_k)$ be a solution of $(k+1)$ equations of system (4.11), and let k be such that

$$\Phi_k = \Phi_k(t, x, w_4, w_5, \ldots, w_r), \quad \frac{\partial \Phi_k}{\partial w_4} = 0, \quad k \geq 2. \tag{4.12}$$

For functions of the form (4.12) equations (4.11) on Φ_k are reduced to the form

$$\frac{\partial^2 \Phi_k}{\partial x^2} + 2 \sum_{i=4}^{r} w_{i-1} \frac{\partial^2 \Phi_k}{\partial x \partial w_i} + \sum_{i,j=4}^{r} w_{i-1} w_{j-1} \frac{\partial^2 \Phi_k}{\partial w_i \partial w_i} = \frac{\partial \Phi_k}{\partial t}.$$

Using induction on r, one can show that any solution of this equation depends linearly on the variables w_i, i.e.,

$$\Phi_k = \varphi_{k4} w_4 + \varphi_{k5} w_5 + \cdots + \varphi_{kr} w_r + \varphi_{k0},$$

where the function φ_{k0} is a solution of the heat equation $\partial^2 \varphi_{k0}/\partial x^2 = \partial \varphi_{k0}/\partial t$ while the functions φ_{ki}, $i = 4, \ldots, r$, are to satisfy the following system of differential equations:

$$\begin{aligned}
\frac{\partial \varphi_{k4}}{\partial x} &= 0, \\
\frac{\partial^2 \varphi_{k4}}{\partial x^2} + 2 \frac{\partial \varphi_{k5}}{\partial x} &= \frac{\partial \varphi_{k4}}{\partial t}, \\
&\cdots\cdots\cdots\cdots\cdots \\
\frac{\partial^2 \varphi_{kr-1}}{\partial x^2} + 2 \frac{\partial \varphi_{kr}}{\partial x} &= \frac{\partial \varphi_{kr-1}}{\partial t}, \\
\frac{\partial^2 \varphi_{kr}}{\partial x^2} &= \frac{\partial \varphi_{kr}}{\partial t}.
\end{aligned} \tag{4.13}$$

It is easy to show that all solutions of (4.13) are polynomial in x and t, and φ_{k4} is an arbitrary polynomial of degree $r - 4$ depending on t only. To compute nonlocal symmetries in our covering, we need to find solutions of (4.13) defined up to solutions with $\varphi_{k4} = 0$. This space is of dimension $r - 3$ and its basis is determined by the functions $\varphi_{k4} = 1, t, \ldots, t^{r-4}$.

Let $\Phi_k = \varphi_{k4} w_4 + \varphi_{k5} w_5 + \cdots + \varphi_{kr} w_r + \varphi_{k0}$ be a solution of the k-th pair in (4.11). To find the corresponding nonlocal symmetry of the Burgers equation we need to construct Φ_{k+i}, $i = 1, 2, \ldots$, satisfying the $(k+i)$-th pair in (4.11) and such that $\Phi_k = \widetilde{D}_x(\Phi_{k+1})$, $\Phi_{k+1} = \widetilde{D}_x(\Phi_{k+2}), \ldots$. Let us set

$$\Phi_{k+1} = \sum_{i=4}^{r} \sum_{\alpha > 0} (-1)^{\alpha+1} \frac{\partial^{\alpha-1} \varphi_{ki}}{\partial x^{\alpha-1}} w_{i+\alpha}. \tag{4.14}$$

Since, as we said above, all functions φ_{ki} are polynomial in x, the function Φ_{k+1} is well defined by equations (4.11). Calculating $\widetilde{D}_x^2(\Phi_{k+1})$ and $\widetilde{D}_t(\Phi_{k+1})$ and using relations (4.13), it is easy to see that Φ_{k+1} satisfies the $(k+1)$-st pair in (4.11).

The functions $\Phi_{k+2}, \ldots, \Phi_{k+i}, \ldots$ are constructed in the same way, and it is proved that they satisfy the corresponding equations of (4.11).

The above arguments, with insignificant corrections, carry over to the case (b), when $\Phi_2 = \Phi_2(t, x, w_5, \ldots, w_r)$.

Let us denote by Φ_k^i, where $k \geq -2$, $i \geq 0$, the symmetry for which the function Φ_{k+4} is of the form $\Phi_{k+4} = t^i w_4 + \Psi(t, x, w_5, \ldots, w_r)$, and by Φ_k^i, where $k < -2$, $i \geq 0$, the symmetry for which the function Φ_2 is $\Phi_2 = t^i w_{2-k} + \Psi(t, x, w_{2-k+1}, \ldots, w_r)$. Then the following theorem is valid:

THEOREM 4.2. *The Lie algebra of nonlocal symmetries of the Burgers equation in the covering τ determined by the fields (4.10) is additively generated by the functions Φ_k^i, where $k = 0, \pm 1, \pm 2, \ldots$, $i = 0, 1, 2, \ldots$, and by the symmetries for which $\Phi_2 = \Phi_2(x, t)$ is an arbitrary solution of the heat equation $\partial^2 \Phi_2 / \partial x^2 = \partial \Phi_2 / \partial t$.*

Note that local symmetries of the form $\psi_k^i = t^i p_k + O(k-1)$ correspond to the generators Φ_k^i for $k > 0$, $i < k$; no local symmetry corresponds to other generators. For example, the generating function $\psi = (2w_2 - w_3 p_0)e^{-w_1/2}$ corresponds to Φ_{-2}^0; $\psi = 1 - w_2 p_0 e^{-w_1/2}$ corresponds to Φ_{-1}^0; $\psi = tp_1 - 2w_2 p_0 e^{-w_1/2} + 5$ corresponds to Φ_1^1. If Φ_2 is a solution of the heat equation, then $\psi = (2\partial \varphi / \partial t - \partial \varphi / \partial x p_0)e^{-w_1/2}$.

5. The problem of symmetry reconstruction

Let $\tau \colon \widetilde{\mathcal{E}} \to \mathcal{E}^\infty$ be an arbitrary covering over an equation $\mathcal{E}^\infty$, and let $\varphi \in \mathcal{F}(\widetilde{\mathcal{E}}, \pi)$ be a solution of the equation $\widetilde{\ell}_F(\varphi) = 0$, $\mathcal{E} = \{F = 0\}$. As we saw in Chapter 4, if $\varphi \in \mathcal{F}(\mathcal{E}^\infty, \pi)$, then $\bar{\partial}_\varphi$ is a local symmetry of the equation $\mathcal{E}^\infty$. Let us try to find a symmetry τ of the form $\widetilde{\partial}_{\varphi, A} = \bar{\partial}_\varphi + \sum_{i=1}^N a_i \partial / \partial w_i$ for the chosen function $\varphi \in \ker \widetilde{\ell}_F$. This problem will be called the *reconstruction problem* for nonlocal symmetries. As we saw already (see Example 3.1), for an arbitrary covering, this problem has no solution. Nevertheless, below, for an arbitrary differential equation, we shall construct a special covering, where this problem is always solvable. As it happens, additional series of symmetries for equations possessing a recursion operator arise in this covering.

5.1. Universal Abelian covering. Among all coverings (1.2) over the equation $\mathcal{E}^\infty$, where $X_{ij} \in \mathcal{F}(\mathcal{E}^\infty)$, there is a distinguished one: the Whitney product of all coverings determined by basis elements of the space $\bar{H}^1(\mathcal{E}^\infty)$ (see [**52, 53**]). Denote this covering by $\tau_1 \colon \widetilde{\mathcal{E}}^{(1)} \to \mathcal{E}^\infty$. Let us now take the Whitney product of all one-dimensional coverings over $\widetilde{\mathcal{E}}^{(1)}$ determined by basis elements of the space $\widetilde{H}^1(\widetilde{\mathcal{E}}^{(1)})$. We shall obtain the covering $\tau_{2,1} \colon \widetilde{\mathcal{E}}^{(2)} \to \widetilde{\mathcal{E}}^{(1)}$ over $\widetilde{\mathcal{E}}^{(1)}$ and hence the covering $\tau_2 = \tau_1 \circ \tau_{2,1} \colon \widetilde{\mathcal{E}}^{(2)} \to \mathcal{E}^\infty$ over $\mathcal{E}^\infty$. Proceeding with this construction, we shall obtain the tower of coverings[4] over $\mathcal{E}^\infty$

$$\cdots \xrightarrow{\tau_{k+1,k}} \widetilde{\mathcal{E}}^{(k)} \xrightarrow{\tau_{k,k-1}} \widetilde{\mathcal{E}}^{(k-1)} \xrightarrow{\tau_{k-1,k-2}} \cdots \xrightarrow{\tau_{2,1}} \widetilde{\mathcal{E}}^{(1)} \xrightarrow{\tau} \mathcal{E}^\infty. \qquad (5.1)$$

Let us denote the inverse limit of this chain of mappings by $\tau^* \colon \widetilde{\mathcal{E}}^* \to \mathcal{E}^\infty$. We shall call τ^* the *universal Abelian covering* over the equation $\mathcal{E}^\infty$.

---

[4]Infinite-dimensional manifolds arising in this procedure need not be inverse limits of finite-dimensional ones. For example, this is the case when the equation $\widetilde{\mathcal{E}}^{(k)}$, for some $k \geq 0$, possesses a family of conservation laws depending on a functional parameter. Such objects and the corresponding theory, similar to that of Chapter 4, are not considered in this book.

REMARK 5.1. The condition $\widetilde{H}^1(\widetilde{\mathcal{E}}^*) = 0$ is convenient to reformulate in the following way. Let the covering τ^* be locally determined by the fields $\widetilde{D}_1, \ldots, \widetilde{D}_n$. Then, if $A_1, \ldots, A_n \in F(\widetilde{\mathcal{E}}^*)$ and

$$\widetilde{D}_i(A_j) = \widetilde{D}_j(A_i), \tag{5.2}$$

one can state that there exists a function $A \in F(\widetilde{\mathcal{E}}^*)$ such that $A_i = \widetilde{D}_i(A)$ for any i.

5.2. Symmetries in the universal Abelian covering. From now on, till the end of this section, a tilde over an operator will denote the lifting of this operator to the covering τ^*.

THEOREM 5.1. *Let $\tau^* \colon \widetilde{\mathcal{E}}^* \to \mathcal{E}^\infty$ be the universal Abelian covering over the equation $\mathcal{E} = \{F = 0\}$. Then for any vector function $\varphi = (\varphi_1, \ldots, \varphi_m)$, $\varphi_i \in \mathcal{F}(\widetilde{\mathcal{E}}^*)$, satisfying the equation $\widetilde{\ell}_F(\varphi) = 0$, there exists a set of functions $A = (a_{i\alpha})$, $a_{i\alpha} \in \mathcal{F}(\widetilde{\mathcal{E}}^*)$, such that $\Im_{\varphi,A}$ is a nonlocal symmetry of type τ^* for the equation $\mathcal{E}$.*

PROOF. Let us assume that locally the distribution on $\widetilde{\mathcal{E}}^*$ is given by the vector fields $\widetilde{D}_i = \bar{D}_i + \sum_{j,\alpha} X_{ij}^\alpha \partial/\partial w_{j\alpha}$, $X_{ij}^\alpha \in \mathcal{F}(\widetilde{\mathcal{E}}^*)$, where i enumerates independent variables, j enumerates levels of the tower, and α at the j-th level enumerates basis elements in the space $\widetilde{H}^1(\widetilde{\mathcal{E}}^{(j)})$. The functions X_{ij} satisfy the conditions

$$\widetilde{D}_i(X_{jk}^\alpha) = \widetilde{D}_j(X_{ik}^\alpha). \tag{5.3}$$

Note that by construction, X_{ij}^α is a function on $\widetilde{\mathcal{E}}^{(j-1)}$ (we set $\widetilde{\mathcal{E}}^{(0)} = \mathcal{E}^\infty$) for all i, α, i.e., is independent of $w_{k\alpha}$ for $k \geq j$. We are to prove solvability of the equations

$$\widetilde{D}_i(a_{j\alpha}) = \Im_{\varphi,A}(X_{ij}^\alpha) \tag{5.4}$$

for any $\varphi \in \ker \widetilde{\ell}_F$.

By Remark 5.1, it suffices to prove existence of functions $(a_{j\alpha}) = A$ such that the functions $A_i = \Im_{\varphi,A}(X_{ij}^\alpha)$, $1 \leq i \leq n$, for any fixed j and α satisfy (5.2). We shall prove this fact by induction on j.

Let $j = 1$. Since $[\widetilde{D}_i, \Im_{\varphi,A}] = \sum_{j,\alpha}(\widetilde{D}_i(a_{j\alpha}) - \Im_{\varphi,A}(X_{ij}^\alpha))\partial/\partial w_{j\alpha}$, where X_{i1}^α are functions on $\mathcal{E}^\infty$, for any set of functions $A = (a_{j\alpha})$ on $\widetilde{\mathcal{E}}$ one has

$$\widetilde{D}_i(\Im_{\varphi,A}(X_{k1}^\alpha)) = \Im_{\varphi,A}(\widetilde{D}_i(X_{k1}^\alpha)) = \Im_{\varphi,A}(\widetilde{D}_k(X_{i1}^\alpha)) = \widetilde{D}_k(\Im_{\varphi,A}(X_{i1}^\alpha))$$

(the second equality holds by (5.3)), and this implies solvability of (5.4) for $j = 1$.

Suppose now that solvability of equations (5.4) is proved for $j < s$, and let $a_{j\alpha}^0$, $j < s$, be arbitrary solutions. Then for any set $A = (a_{j\alpha})$, where $a_{j\alpha} = a_{j\alpha}^0$ if $j < s$, while the other $a_{j\alpha}$ are arbitrary functions, we have $[\widetilde{D}_i, \Im_{\varphi,A}]|_{\widetilde{\mathcal{E}}^{(s-1)}} = 0$. Since $X_{is}^\alpha \in \mathcal{F}(\widetilde{\mathcal{E}}^{s-1})$, similarly to the case $j = 1$, we obtain equality (5.2) for $A_i = \Im_{\varphi,A}(X_{is}^\alpha)$, $1 \leq i \leq n$. The theorem is proved. $\square$

5.3. Nonlocal symmetries for equations admitting a recursion operator. Consider now the situation when the equation $\mathcal{E} = \{F = 0\}$ possesses a recursion operator R. Recursion operators considered in the literature are not, in general, differential operators. Therefore, the expression $R(\varphi)$, where φ is a symmetry of the given equation, may be undefined. There are various interpretations for this type of expression (see, for example, [**58, 91, 112**]). Here we suggest considering, instead of the operator R, an operator $\widetilde{R}$, which in some cases is defined on the set $\ker \widetilde{\ell}_F$. Let us describe a class of equations where recursion operators possess this property (other examples are given in §5.6).

PROPOSITION 5.2. *Let an evolution equation*
$$\mathcal{E} = \{u_t = D_x(h(x,t,u,u_1,\ldots,u_k))\}$$
possess a recursion operator $R = \sum_{i=-1}^{N} f_i D_x^i$, $f_i \in \mathcal{F}(\mathcal{E}^\infty)$. *Then for any symmetry* $\widetilde{\mathfrak{I}}_{\varphi,A}$ *in the covering* τ^* *there exists a symmetry* $\widetilde{\mathfrak{I}}_{\varphi',A'}$, *where* $\varphi' = R(\varphi)$.

PROOF. Existence of the function φ' easily follows from Remark 5.1. In fact, since $\widetilde{\mathfrak{I}}_{\varphi,A}$ is a symmetry, the function φ satisfies the equation $\widetilde{\ell}_F(\varphi) = 0$, which in the situation under consideration is of the form $\widetilde{D}_t(\varphi) = \widetilde{D}_x(\widetilde{\ell}_h(\varphi))$ and coincides with equations (5.2). Therefore, there exists a function φ' such that $\varphi = \widetilde{D}_x(\varphi')$, or $\varphi' = \widetilde{D}_x^{-1}(\varphi)$. Existence of the symmetry $\widetilde{\mathfrak{I}}_{\varphi',A'}$ is guaranteed by Theorem 5.1. □

5.4. Example: nonlocal symmetries of the Korteweg–de Vries equation. Consider the KdV equation $\mathcal{E} = \{u_t = u_{xxx} + uu_x\}$. By Proposition 5.2, this equation possesses the series of nonlocal symmetries $\widetilde{\mathfrak{I}}_{\psi_n,A}$, where $\psi_n = \widetilde{R}^n(tp_1+1)$, $R = D_x^2 + \frac{2}{3}p_0 + \frac{1}{3}p_1 D_x^{-1}$. In fact, $\psi_n \in \mathcal{F}(\widetilde{\mathcal{E}}^{(1)})$ for any n. This fact is implied by the following statement:

PROPOSITION 5.3. *Let an equation* $\mathcal{E}$ *and its recursion operator* R *satisfy the assumptions of Proposition 5.2. Assume also that there exists a local symmetry* φ *of the equation* $\mathcal{E}$ *evolving by* R *to the infinite series of local symmetries* $\mathcal{S} = \{R^n(\varphi) \mid n = 0, 1, \ldots\}$, *and let the operator* R *be such that* $f_{-1} \in \mathcal{S}$. *Let* $\widetilde{R}^{(1)}$ *be the lifting of* R *to the covering* $\tau_1 \colon \widetilde{\mathcal{E}}^{(1)} \to \mathcal{E}^\infty$, *and let* w_α *be nonlocal variables in the covering* τ_1. *Then, if* $\Phi = \sum \varphi_i w_i + \Psi \in \ker \widetilde{\ell}_F$, *where the number of summands is finite,* $\varphi_i \in \mathcal{S}$, *and* $\Psi \in \mathcal{F}(\mathcal{E}^\infty)$, *then the function* $\widetilde{R}^{(1)}(\Phi)$ *lies in* $\mathcal{F}(\widetilde{\mathcal{E}}^{(1)})$ *and has a similar form.*

PROOF. The subscript "loc" below denotes functions on $\mathcal{E}^\infty$. We shall not need to describe these functions in detail.

One has $\widetilde{R}^{(1)}(\Phi) = \sum w_i R(\varphi_i) + f_{-1}\widetilde{D}_x^{-1}(\Psi - \sum \widetilde{D}_x(w_i)\widetilde{D}_x^{-1}(\varphi_i)) + \Psi_{\mathrm{loc}}$, since by Green's formula (see [**131, 132**]),
$$\widetilde{D}_x^{-1}(w_i \varphi_i) = w_i \widetilde{D}_x^{-1}(\varphi_i) - \widetilde{D}_x^{-1}(\widetilde{D}_x(w_i)\widetilde{D}_x^{-1}(\varphi_i)).$$
Let us set $X_{\mathrm{loc}} = \Psi - \sum \widetilde{D}_x(w_i)\widetilde{D}_x^{-1}(\varphi_i)$ and show that $\bar{D}_t(X_{\mathrm{loc}}) = \bar{D}_x(T_{\mathrm{loc}})$ for some function T_{loc}. Since this means that $X_{\mathrm{loc}}\, dx + T_{\mathrm{loc}}\, dt$ is locally a conserved current density, the statement will be proved. We have
$$\widetilde{D}_x^{-1}(\Phi) = \sum w_i \widetilde{D}_x^{-1}(\varphi_i) + \widetilde{D}_x^{-1}(X_{\mathrm{loc}}) + A_{\mathrm{loc}}.$$
From this and from the equality $\widetilde{D}_t(\Phi) = \widetilde{D}_x(\widetilde{\ell}_g(\Phi))$ it follows that
$$\widetilde{D}_t(\widetilde{D}_x^{-1}(\Phi)) = \widetilde{D}_x^{-1}(\widetilde{D}_t(\Phi)) = \widetilde{\ell}_g(\Phi) = \sum w_i \widetilde{\ell}_g(\varphi_i) + B_{\mathrm{loc}}.$$
On the other hand, since φ_i is a symmetry, we have
$$\widetilde{D}_t(\widetilde{D}_x^{-1}(\Phi)) = \widetilde{D}_t\left(\sum w_i \widetilde{D}_x^{-1}(\varphi_i) + \widetilde{D}_x^{-1}(X_{\mathrm{loc}}) + A_{\mathrm{loc}}\right)$$
$$= \sum w_i \widetilde{D}_t(\widetilde{D}_x^{-1}(\varphi_i)) + \widetilde{D}_t(\widetilde{D}_x^{-1}(X_{\mathrm{loc}})) + C_{\mathrm{loc}}$$
$$= \sum w_i \widetilde{\ell}_g(\varphi_i) + \widetilde{D}_t(\widetilde{D}_x^{-1}(X_{\mathrm{loc}})) + C_{\mathrm{loc}}.$$

Comparing the last two equations, we obtain

$$\widetilde{D}_t(\widetilde{D}_x^{-1}(X_{\text{loc}})) = B_{\text{loc}} - C_{\text{loc}} = T_{\text{loc}}.$$

This completes the proof. □

Let us write down several symmetries of the nonlocal series for the Korteweg–de Vries equation. To this end, let us introduce the nonlocal variables defined by the equalities $\widetilde{D}_x w_0 = p_0$, $\widetilde{D}_x w_1 = \frac{1}{2}p_0^2$, $\widetilde{D}_x w_2 = \frac{1}{6}(p_0^3 - 3p_1^2), \dots$, and set $\varphi_k = R^k p_1$, $k \geq 0$. Then

$$\psi_0 = t\varphi_0 + 1,$$
$$\psi_1 = t\varphi_1 + \tfrac{1}{3}x\varphi_0 + \tfrac{2}{3}p_0,$$
$$\psi_2 = t\varphi_2 + \tfrac{1}{3}x\varphi_1 + \tfrac{4}{3}p_2 + \tfrac{4}{9}p_0^2 + \tfrac{1}{9}\varphi_0 w_0,$$
$$\psi_3 = t\varphi_3 + \tfrac{1}{3}x\varphi_2 + 2\bar{D}_x(\varphi_1) + \tfrac{2}{3}p_0 p_2 + \tfrac{8}{27}p_0^3 + \tfrac{1}{9}\varphi_1 w_0 + \tfrac{1}{9}\varphi_0 w_1.$$

5.5. Master symmetries. Consider in the covering τ^* over the Korteweg–de Vries equation the symmetry $\widetilde{\eth}_{\psi_2,A}$ and take another τ^*-symmetry $\widetilde{\eth}_{\varphi,B}$, where φ is a local symmetry independent of x and t. Taking the commutator of these two symmetries, we obtain a new τ^*-symmetry $\widetilde{\eth}_{\chi,C} = [\widetilde{\eth}_{\psi_2,A}, \widetilde{\eth}_{\varphi,B}]$, where $\chi = \widetilde{\eth}_{\psi_2,A}(\varphi) - \widetilde{\eth}_{\varphi,B}(\psi_2) \in \ker \widetilde{\ell}_F$. By simple calculations, one can see that the function χ is a local symmetry of the Korteweg–de Vries equation independent of the variables x and t. In fact, since φ is a local symmetry and ψ_2 depends only on one nonlocal variable w_0 satisfying the relation $\widetilde{D}_x w_0 = p_0$, one has $\chi = \eth_{\psi_2}(\varphi) - \eth_\varphi(\psi_2) - \tfrac{1}{9}p_1 \widetilde{D}_x^{-1}\varphi$, i.e., the function χ may depend on local variables and on w_0 only.

Further, since local symmetries of the Korteweg–de Vries equation independent of x and t commute, we have in particular $\{\varphi, \varphi_1\} = \{\varphi, \varphi_2\} = 0$, and so

$$\chi = \frac{1}{3}\sum j \bar{D}_x^{j-1}\varphi_1 \frac{\partial \varphi}{\partial p_j} + \widetilde{\eth}_\Psi(\varphi) - \bar{\eth}_\varphi(\Psi) - \frac{1}{9}p_1 \widetilde{D}_x^{-1}\varphi, \qquad (5.5)$$

where $\Psi = \tfrac{4}{3}p_2 + \tfrac{4}{9}p_0^2 + \tfrac{1}{9}\varphi_0 w_0$. From this it follows that χ is independent of x and t. A more detailed analysis shows that χ is independent of w_0. Thus, χ is a local symmetry independent of x and t.

Let $s = 2k + 1$ be the order of the symmetry φ, i.e., $\partial\varphi/\partial p_s = \text{const} \neq 0$ and $\partial\varphi/\partial p_j = 0$ for all $j > s$. Then it is easily seen that the function χ is of the form

$$\chi = \frac{1}{3}sp_{s+2}\frac{\partial \varphi}{\partial p_s} + \text{lower order terms} = c\varphi_{k+1} + \text{lower order symmetries},$$

where $c = \tfrac{1}{3}s\partial\varphi/\partial p_s = \text{const}$.

This means that the operator of commutation with the τ^*-symmetry $\widetilde{\eth}_{\psi_2,A}$ acts on the first component of the generating function of the τ^*-symmetry $\widetilde{\eth}_{\varphi,B}$, up to symmetries of lower order, in the same way as the recursion operator R does. Thus, this operator of commutation plays the role of the recursion operator for the Korteweg–de Vries equation.

Note that a similar situation occurs for the equations considered in the next subsection.

5.6. Examples. In this subsection, we consider examples of equations with recursion operators that formally do not satisfy the assumptions of Proposition 5.2. Nevertheless, additional series of nonlocal symmetries arise in the covering τ^* for these equations.

Consider an equation in two independent variables $x = x_1$, $t = x_2$ possessing a recursion operator of the form

$$R = \sum_{i=0}^{M} f_i D_x^i + a D_x^{-1} \circ b, \qquad f_i, a, b \in \mathcal{F}(\mathcal{E}^\infty).$$

To prove existence of the function $R\varphi \in \mathcal{F}(\mathcal{E}^*)$ for any $\varphi \in \ker \widetilde{\ell}_F$, it suffices to show that the function $b\varphi$ lies in the total derivative image, i.e., $b\varphi = \widetilde{D}_x \psi$ for some function $\psi \in \mathcal{F}(\mathcal{E}^*)$. This fact will be established if we prove existence of a function $T \in \mathcal{F}(\mathcal{E}^*)$ such that $\bar{D}_t(b\varphi) = \bar{D}_x(T)$. It is not difficult to check this condition in each particular case.

EXAMPLE 5.1. Consider the modified Korteweg–de Vries equation

$$\mathcal{E} = \{u_t = u_{xxx} + u^2 u_x\}. \tag{5.6}$$

The recursion operator for this equation is of the form [91]:

$$R = \bar{D}_x^2 + \tfrac{2}{3} p_0^2 + \tfrac{2}{3} p_1 \bar{D}_x^{-1} \circ p_0.$$

Let us show that the operator $\widetilde{R}$ may be applied to any function $\varphi \in \ker \widetilde{\ell}_F$. In fact,

$$\widetilde{D}_t(p_0 \varphi) = \widetilde{D}_x((p_2 + \tfrac{1}{3} p_0^3)\varphi + p_0(\widetilde{D}_x^2(\varphi) + p_0^2 \varphi) - p_1 \widetilde{D}_x(\varphi) - \tfrac{1}{3} p_0^3 \varphi),$$

$$\widetilde{D}_t(\varphi) = \widetilde{D}_x^3(\varphi) + p_0^2 \widetilde{D}_x(\varphi) + 2 p_0 p_1 \varphi.$$

Therefore, equation (5.6) possesses a series of nonlocal symmetries $\partial_{\psi_n, A}$, where

$$\psi_n = \widetilde{R}^n(t(p_3 + p_0^2 p_1) + \tfrac{1}{3} x p_1 + \tfrac{1}{3} p_0).$$

EXAMPLE 5.2. Consider the sine-Gordon equation

$$\mathcal{E} = \{u_{xt} = \sin u\}. \tag{5.7}$$

If φ satisfies the equation $\widetilde{D}_x \widetilde{D}_t(\varphi) = \varphi \cos p_{(0,0)}$, then, since $\widetilde{D}_t(p_{(2,0)} \varphi) = \widetilde{D}_x(p_{(1,0)} \widetilde{D}_t(\varphi))$, the recursion operator $\widetilde{R} = \widetilde{D}_x^2 + p_{(1,0)}^2 - p_{(1,0)} \widetilde{D}_x^{-1} \circ p_{(2,0)}$ (see [91]) is defined on $\ker \widetilde{\ell}_F$. Therefore, equation (5.7) possesses the series of nonlocal symmetries $\partial_{\psi_n, A}$, $\psi_n = \widetilde{R}^n(x p_{(1,0)} - t p_{(0,1)})$.

EXAMPLE 5.3. Consider the nonlinear Schrödinger equation in the form

$$\begin{cases} u_t = i u_{xx} + 2 u u_x v, \\ v_t = -i v_{xx} - 2 u v v_x. \end{cases}$$

This system possesses the following recursion operator [14]

$$R = \begin{pmatrix} \bar{D}_x + i(p_1^1 \bar{D}_x^{-1} \circ p_0^2 \\ \quad - p_0^1 \bar{D}_x^{-1} \circ p_1^2 + p_0^1 p_0^2) & i(p_0^1 \bar{D}_x^{-1} \circ p_1^1 + p_1^1 \bar{D}_x^{-1} \circ p_0^1) \\ i(p_0^2 \bar{D}_x^{-1} \circ p_1^2 + p_1^2 \bar{D}_x^{-1} \circ p_0^2) & -\bar{D}_x + i(p_1^2 \bar{D}_x^{-1} \circ p_0^1 \\ & \quad - p_0^2 \bar{D}_x^{-1} \circ p_1^1 + p_0^1 p_0^2) \end{pmatrix}$$

Here $p_k^i = p_{k0}^i$, $x_1 = x$, $x_2 = t$.

To check that the operator $\widetilde{R}$ generates an infinite series of τ^*-symmetries, it suffices to show that $\widetilde{D}_x^{-1}(p_1^2\varphi - p_1^1\psi)$, $\widetilde{D}_x^{-1}(p_0^2\varphi + p_0^1\psi) \in \mathcal{F}(\widetilde{\mathcal{E}}^*)$ for any vector function $(\varphi, \psi) \in \ker \widetilde{\ell}_F$. But this fact is implied by the following equalities:

$$\widetilde{D}_x^{-1}(p_1^2\varphi - p_1^1\psi) = \widetilde{D}_x(i(p_1^2\widetilde{D}_x\varphi - p_1^2\varphi + p_1^1\widetilde{D}_x\psi - p_1^1\psi) + 2p_0^1p_0^2(p_1^1\psi - p_1^2\varphi)),$$

$$\widetilde{D}_x^{-1}(p_0^2\varphi + p_0^1\psi) = \widetilde{D}_x(i(p_0^2\widetilde{D}_x\varphi - p_1^2\varphi + p_1^1\widetilde{D}_x\psi - p_0^1\psi) - 2p_0^1p_0^2(p_0^1\psi - p_0^2\varphi)).$$

Thus, the operator $\widetilde{R}$ is determined on $\operatorname{sym}_{\tau^*} \mathcal{E}$, and the symmetry

$$\Phi = (tp_2^1 + \tfrac{1}{2}xp_1^1, tp_2^2 + \tfrac{1}{2}xp_1^2 + \tfrac{1}{2}p_0^2)$$

generates a series of symmetries for the system under consideration.

5.7. General problem of nonlocal symmetry reconstruction. Consider the general situation now. Namely, let $\tau \colon \widetilde{\mathcal{E}} \to \mathcal{E}^\infty$ be an arbitrary covering of the equation $\mathcal{E}^\infty$ and $\varphi \in \mathcal{F}(\widetilde{\mathcal{E}}, \pi)$ an arbitrary solution of the equation $\widetilde{\ell}_F(\varphi) = 0$, where $\mathcal{E} = \{F = 0\}$. To find a τ-symmetry of the form $\widetilde{\Im}_{\varphi, A} = \widetilde{\Im}_\varphi + \sum_{i=1}^N a_i \partial/\partial w_i$, we need to find functions $a_1, \ldots, a_N$, $a_j \in \mathcal{F}(\widetilde{\mathcal{E}})$, satisfying the equations

$$\widetilde{D}_i(a_j) = \widetilde{\Im}_{\varphi, A}(X_{ij}). \tag{5.8}$$

As we already saw, this system may have no solution for a given φ. Note that the formal conditions of compatibility for (5.8) are of the form

$$\widetilde{D}_i(Y_{jk}) = \widetilde{D}_j(Y_{ik}), \tag{5.9}$$

where Y_{ij} denotes the right-hand side of (5.8). Note further that the integrability condition for the distribution $\widetilde{\mathcal{C}}$ on $\widetilde{\mathcal{E}}$, i.e., $[\widetilde{D}_i, \widetilde{D}_j] = 0$, $i, j = 1, \ldots, n$, in local coordinates is of the form

$$\widetilde{D}_i(X_{jk} = \widetilde{D}_j(X_{ik}) \tag{5.10}$$

(see (1.3)), where $\widetilde{D}_i = \bar{D}_i + \sum_{j=1}^N X_{ij}\partial/\partial w_j$, $i = 1, \ldots, n$. Conditions (5.9) and (5.10) are of a similar structure. Using this fact, let us construct a new covering $\tau_1 \colon \widetilde{\mathcal{E}}_1 \to \widetilde{\mathcal{E}} \to \mathcal{E}^\infty$ by introducing new nonlocal variables $a_1, \ldots, a_N$ and defining new total derivatives $\widetilde{D}_i^{(1)}$, $i = 1, \ldots, n$, on $\widetilde{\mathcal{E}}_1$ by the formula $\widetilde{D}_i^{(1)} = \widetilde{D}_i + \sum_j Y_{ij}\partial/\partial a_j$. Introducing the covering

$$\tau_1 \colon \widetilde{\mathcal{E}}_1 \longrightarrow \widetilde{\mathcal{E}} \longrightarrow \mathcal{E}^\infty,$$

we can now solve system (5.8), but it does not give us a solution of the initial problem, since the field $\widetilde{\Im}_{\varphi,A} = \widetilde{\Im}_\varphi + \sum_{i=1}^N a_j \partial/\partial w_j$ is not necessary a symmetry in the covering τ_1. This field is to be extended up to a field of the form

$$\widetilde{\Im}_{\varphi, A, B} = \widetilde{\Im}_\varphi + \sum_i a_i \frac{\partial}{\partial w_i} + \sum_j b_j \frac{\partial}{\partial a_j}$$

on the manifold $\widetilde{\mathcal{E}}_1$, where the functions b_j must satisfy a system of equations similar to (5.8). Again, this system may have no solution and we shall have to repeat the construction procedure for a new covering. As a result, we obtain a chain of coverings similar to that which arose when we constructed the universal Abelian covering. In more formal details, this procedure is described in the next two subsections.

5.8. Kiso's construction. Let $\tau\colon \widetilde{\mathcal{E}} \to \mathcal{E}^\infty$ be an arbitrary covering over the equation $\mathcal{E} = \{F = 0\}$ described by the fields (1.2), and let $\varphi \in \ker \widetilde{\ell}_F$, $\varphi = (\varphi_1, \ldots, \varphi_m)$, $\varphi_i \in \mathcal{F}(\widetilde{\mathcal{E}})$. In the paper [55], for evolution equations in one space variable the covering $\tau_\varphi\colon \widetilde{\mathcal{E}}_\varphi \to \mathcal{E}^\infty$ was constructed, where the function φ generates a symmetry. Below we describe a generalization of this construction to arbitrary equations.

Set $\widetilde{\mathcal{E}}_\varphi = \widetilde{\mathcal{E}} \times \mathbb{R}^\infty$, and let w_j^l, $j = 1, \ldots, N$, $l = 0, 1, 2, \ldots$, be coordinates in $\mathbb{R}^\infty$ (new nonlocal variables). The mapping $\tau_\varphi\colon \widetilde{\mathcal{E}}_\varphi \to \mathcal{E}^\infty$ is the composition of the projection to the first factor with τ.

Let us define vector fields $\widetilde{D}_i^\varphi$ on $\widetilde{\mathcal{E}}_\varphi$ by the formula

$$\widetilde{D}_i^\varphi = \bar{D}_i + \sum_{j,l>0} \left(\widetilde{\mathfrak{Z}}_\varphi^{(\varphi)} + S_w\right)^l (X_{ij}) \frac{\partial}{\partial w_j^l}, \qquad i = 1, \ldots, n, \tag{5.11}$$

where[5] $\widetilde{\mathfrak{Z}}_\varphi^{(\varphi)} = \sum'_{\sigma,k} \widetilde{D}_\sigma^\varphi(\varphi_k) \partial/\partial p_\sigma^k$, $S_w = \sum_{j,l} w_j^{l+1} \partial/\partial w_j^l$, $w_j^0 = w_j$.

PROPOSITION 5.4. *Let $\mathcal{E}$ be an equation and $\widetilde{\mathcal{E}}_\varphi = \widetilde{\mathcal{E}} \times \mathbb{R}^\infty$. Then:*

1. $[\widetilde{D}_i^\varphi, \widetilde{D}_k^\varphi] = 0$, $i, k = 1, \ldots, n$, *i.e., the set of fields (5.11) determines a covering structure in $\widetilde{\mathcal{E}}_\varphi$.*

2. *If $\varphi \in \ker \widetilde{\ell}_F$, $\varphi = (\varphi_1, \ldots, \varphi_m)$, $\varphi_i \in \mathcal{F}(\widetilde{\mathcal{E}})$, then $[\widetilde{\mathfrak{Z}}_\varphi^{(\varphi)} + S_w, \widetilde{D}_i^\varphi] = 0$, $i = 1, \ldots, n$, i.e., $\widetilde{\mathfrak{Z}}_\varphi^{(\varphi)} + S_w$ is a symmetry of type τ_φ for the equation $\mathcal{E}^\infty$.*

PROOF. Let us first show that $\widetilde{\ell}_F(\varphi) = 0$ implies $[\widetilde{\mathfrak{Z}}_\varphi^{(\varphi)} + S_w, \widetilde{D}_i^\varphi] = 0$, $i = 1, \ldots, n$. One has

$$[\widetilde{\mathfrak{Z}}_\varphi^{(\varphi)} + S_w, \widetilde{D}_i^\varphi] = \sum'_{\sigma,k} \widetilde{\mathfrak{Z}}_\varphi(\bar{p}_{\sigma+1_i}^k) \frac{\partial}{\partial p_\sigma^k} + \sum_{j,l>0} (\widetilde{\mathfrak{Z}}_\varphi^{(\varphi)} + S_w)^{l+1}(X_{ij}) \frac{\partial}{\partial w_j^l}$$
$$- \sum'_{\sigma,k} \widetilde{D}_i^\varphi(\widetilde{D}_\sigma(\varphi_k)) \frac{\partial}{\partial p_\sigma^k} - \sum_{j,l>0} (\widetilde{\mathfrak{Z}}_\varphi^{(\varphi)} + S_w)^{l+1}(X_{ij}) \frac{\partial}{\partial w_j^l}$$
$$= \sum'_{\sigma,k} (\widetilde{\mathfrak{Z}}_\varphi(\bar{p}_{\sigma+1_i}^k) - \widetilde{D}_{\sigma+1_i}(\varphi_k)) \frac{\partial}{\partial p_\sigma^k} = 0.$$

Now consider the commutator $[\widetilde{D}_i^\varphi, \widetilde{D}_k^\varphi]$. Since $[\bar{D}_i, \bar{D}_k] = 0$ and the coefficients of the vector fields $\bar{D}_j$ are independent of w_j^l, we have

$$[\widetilde{D}_i^\varphi, \widetilde{D}_k^\varphi] = \sum_{j,l>0} \{\widetilde{D}_i^\varphi(\widetilde{\mathfrak{Z}}_\varphi^{(\varphi)} + S_w)^l(X_{kj}) - \widetilde{D}_k^\varphi(\widetilde{\mathfrak{Z}}_\varphi^{(\varphi)} + S_w)^l(X_{ij})\} \frac{\partial}{\partial w_j^l}.$$

By the first statement of our proposition,

$$[\widetilde{D}_i^\varphi, \widetilde{D}_k^\varphi] = \sum_{j,l>0} (\widetilde{\mathfrak{Z}}_\varphi^{(\varphi)} + S_w)^l (\widetilde{D}_i^\varphi(X_{kj}) - \widetilde{D}_k^\varphi(X_{ij})) \frac{\partial}{\partial w_j^l} = 0,$$

since $\widetilde{D}_i^\varphi(X_{kj}) - \widetilde{D}_k^\varphi(X_{ij}) = \widetilde{D}_i(X_{kj}) - \widetilde{D}_k(X_{ij}) = 0$. The proposition is proved. □

---

[5] As above, a prime denotes summation over intrinsic coordinates.

5.9. Construction of the covering τ_S. Below we follow the paper [53].

Let us set $\widetilde{\mathcal{E}}_\tau = \widetilde{\mathcal{E}} \times \mathbb{R}^\infty$, where the coordinates in $\mathbb{R}^\infty$ (nonlocal variables) are the variables[6] v_j^l, $j = 1, \ldots, N$, $l = 1, 2, \ldots p_\sigma^{(k)}$, $k > 0$. The mapping $\tau_S \colon \widetilde{\mathcal{E}}_\tau \to \mathcal{E}^\infty$ is the composition of the projection to the first factor with τ.

Let us define the system of fields $\widetilde{D}_1^\tau, \ldots, \widetilde{D}_n^\tau$ on $\widetilde{\mathcal{E}}_\tau$ as follows:

$$\widetilde{D}_i^\tau = \bar{D}_i^S + \sum_{l \geq 0, j} (S_p + S_v)^l (X_{ij}) \frac{\partial}{\partial v_j^l}, \qquad i = 1, \ldots, n, \tag{5.12}$$

where

$$\bar{D}_i^S = \frac{\partial}{\partial x_i} + \sideset{}{'}\sum_{l \geq 0, \sigma} S_p^l(\bar{p}_{\sigma+1_i}) \frac{\partial}{\partial p_\sigma^{(l)}}, \qquad p_\sigma^{(0)} = p_\sigma, \tag{5.13}$$

$$S_p = \sideset{}{'}\sum_{l \geq 0, \sigma} p_\sigma^{(l+1)} \frac{\partial}{\partial p_\sigma^{(l)}}, \quad S_v = \sum_{l \geq 0, j} v_j^{l+1} \frac{\partial}{\partial v_j^l}, \qquad v_j^0 = w_j. \tag{5.14}$$

PROPOSITION 5.5. 1. $[\widetilde{D}_i^\tau, \widetilde{D}_k^\tau] = 0$, $i, k = 1, \ldots, n$, i.e., the set of fields (5.12) determines a covering structure in $\widetilde{\mathcal{E}}_\tau$.

2. The vector field $S_\tau = S_p + S_v$ is a nonlocal symmetry of type τ_S for the equation $\mathcal{E}^\infty$.

PROOF. From formulas (5.12)–(5.14) it easily follows that $[S_\tau, \widetilde{D}_i^\tau] = 0$, $i = 1, \ldots, n$, i.e., S_τ is a nonlocal symmetry. Let us now compute the commutator $[\widetilde{D}_i^\tau, \widetilde{D}_j^\tau]$:

$$[\widetilde{D}_i^\tau, \widetilde{D}_k^\tau] = \sideset{}{'}\sum \left\{ \widetilde{D}_i^\tau \left(S_p^l(\bar{p}_{\sigma+1_j})\right) - \widetilde{D}_j^\tau \left(S_p^l(\bar{p}_{\sigma+1_i})\right) \right\} \frac{\partial}{\partial p_\sigma^{(l)}}$$

$$+ \sum \left\{ \widetilde{D}_i^\tau \left(S_\tau^l(X_{jk})\right) - \widetilde{D}_j^\tau \left(S_\tau^l(X_{ik})\right) \right\} \frac{\partial}{\partial v_k^l}$$

$$= \sideset{}{'}\sum S_P^l \left\{ \widetilde{D}_i^\tau (\bar{p}_{\sigma+1_j}) - \widetilde{D}_j^\tau (\bar{p}_{\sigma+1_i}) \right\} \frac{\partial}{\partial p_\sigma^{(l)}}$$

$$+ \sum S_\tau^l \left\{ \widetilde{D}_i^\tau (X_{jk}) - \widetilde{D}_j^\tau (X_{ik}) \right\} \frac{\partial}{\partial v_k^l} = 0,$$

since $[\bar{D}_i, \bar{D}_k] = [\widetilde{D}_i, \widetilde{D}_k] = 0$. The proposition is proved. □

The following theorem is a corollary of Propositions 5.4 and 5.5.

THEOREM 5.6. *For any covering $\tau \colon \widetilde{\mathcal{E}} \to \mathcal{E}^\infty$ over the equation $\mathcal{E} = \{F = 0\}$ and for any set Φ of vector functions $\varphi^1, \ldots, \varphi^K$, $\varphi^j = (\varphi_1^j, \ldots, \varphi_m^j)$, $\varphi_i^j \in \mathcal{F}(\widetilde{\mathcal{E}}_\tau)$, satisfying the equation $\widetilde{\ell}_F^\tau(\varphi) = 0$, where $\widetilde{\ell}_F^\tau$ is the lifting of the operator ℓ_F to $\widetilde{\mathcal{E}}_\tau$, there exist a covering $\tau_\Phi \colon \widetilde{\mathcal{E}}_\Phi \to \widetilde{\mathcal{E}}^\infty$ and τ_Φ-symmetries $S_{\varphi^1}, \ldots, S_{\varphi^K}$ such that $S_{\varphi^j}|_{\mathcal{E}^\infty} = \widetilde{\mathfrak{d}}_{\varphi^j}\big|_{\mathcal{E}^\infty}$, $j = 1, \ldots, K$.*

---

[6]To simplify notation, we omit the superscript corresponding to the independent variable number at p_σ. The superscript in parentheses enumerates new nonlocal variables.

5. THE PROBLEM OF SYMMETRY RECONSTRUCTION

5.10. The universal property of the symmetry S_τ. In conclusion, we shall discuss relations between the constructions of §§5.8 and 5.9. Namely, the following statement holds:

PROPOSITION 5.7. *Let $\tau\colon \widetilde{\mathcal{E}} \to \mathcal{E}^\infty$ be a covering over the equation $\mathcal{E} = \{F = 0\}$. Then for any vector function $\varphi \in \ker \widetilde{\ell}_F$, $\varphi = (\varphi_1, \ldots, \varphi_m)$, $\varphi_i \in \mathcal{F}(\widetilde{\mathcal{E}})$, there exists an embedding $i_\varphi \colon \widetilde{\mathcal{E}}_\varphi \to \widetilde{\mathcal{E}}_\tau$ such that the diagram*

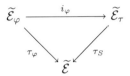

is commutative. Moreover, $\Pi_\varphi = i_\varphi(\widetilde{\mathcal{E}}_\varphi)$ is an integral submanifold of the distribution on $\widetilde{\mathcal{E}}_\tau$ determined by the fields (5.11) and

$$S_\tau\big|_{\Pi_\varphi} = (i_\varphi)_* \left(\widetilde{\partial}_\varphi^{(\varphi)} + S_w\right), \qquad \widetilde{D}_i^\tau\big|_{\Pi_\varphi} = (i_\varphi)_* \widetilde{D}_i^\varphi.$$

PROOF. Let us define the embedding i_φ by the formula $i_\varphi(p_\sigma) = (p_\sigma^{(i)}, v_j^l)$, where $p_\sigma^{(i)} = (\widetilde{\partial}_\varphi + S_w)^i(p_\sigma)$, $v_j^l = w_j^l$. Then the manifold Π_φ is determined by the equations $p_\sigma^{(i)} = (\widetilde{\partial}_\varphi + S_v)^i(p_\sigma)$. Let us show that the field $S_\tau = S_p + S_v$ is tangent to Π_φ. In fact,

$$\left\{S_\tau \left(p_\sigma^{(i)} - \left(\widetilde{\partial}_\varphi + S_v\right)^i (p_\sigma)\right)\right\}\bigg|_{\Pi_\varphi} = \left\{p_\sigma^{(i+1)} - S_\tau \left(\widetilde{\partial}_\varphi + S_v\right)^i (p_\sigma)\right\}\bigg|_{\Pi_\varphi}$$

$$= \left\{\left(\left(\widetilde{\partial}_\varphi + S_v\right)^{i+1} - S_\tau \left(\widetilde{\partial}_\varphi + S_v\right)^i\right)(p_\sigma)\right\}\bigg|_{\Pi_\varphi}$$

$$= \left\{\left(\widetilde{\partial}_\varphi - S_p\right)\left(\widetilde{\partial}_\varphi + S_v\right)^i (p_\sigma)\right\}\bigg|_{\Pi_\varphi}$$

$$= \left\{\left(\sum_{\sigma'}{}' \widetilde{D}_{\sigma'}(\varphi) \frac{\partial}{\partial p_{\sigma'}} - \sum_{\sigma',i}{}' p_{\sigma'}^{(i+1)}\bigg|_{\Pi_\varphi} \frac{\partial}{\partial p_{\sigma'}^{(i)}}\right)\left(\widetilde{\partial}_\varphi + S_v\right)^i (p_\sigma)\right\}\bigg|_{\Pi_\varphi}$$

$$= \left\{\left(\sum_{\sigma'}{}' \widetilde{D}_{\sigma'}(\varphi) \frac{\partial}{\partial p_{\sigma'}} - \sum_{\sigma'}{}' p_{\sigma'}^{(1)}\bigg|_{\Pi_\varphi} \frac{\partial}{\partial p_{\sigma'}}\right)\left(\widetilde{\partial}_\varphi + S_v\right)^i (p_\sigma)\right\}\bigg|_{\Pi_\varphi}$$

$$= \left\{\sum_{\sigma'}{}' \left(\widetilde{D}_{\sigma'}(\varphi) - \left(\widetilde{\partial}_\varphi + S_v\right)^i (p_{\sigma'})\right) \frac{\partial}{\partial p_{\sigma'}} \left(\widetilde{\partial}_\varphi + S_v\right)^i (p_\sigma)\right\}\bigg|_{\Pi_\varphi} = 0,$$

since $\left(\widetilde{\partial}_\varphi + S_v\right)^i (p_\sigma) = \widetilde{\partial}_\varphi(p_\sigma) = \widetilde{D}_\sigma(\varphi)$. Let us prove now that for any function $f \in \mathcal{F}(\widetilde{\mathcal{E}}_\varphi)$ the equality

$$S_\tau\big|_{\Pi_\varphi}\left((i_\varphi^*)^{-1}(f)\right) = (i_\varphi^*)^{-1}\left(\widetilde{\partial}_\varphi^{(\varphi)} + S_w\right)(f)$$

holds. In fact,

$$S_\tau|_{\Pi_\varphi}\left((i_\varphi^*)^{-1}(f)\right) = \left(\sideset{}{'}\sum_{\sigma,i} p_\sigma^{(i+1)}\Big|_{\Pi_\varphi}\frac{\partial}{\partial p_\sigma^i} + S_v\right)\Bigg|_{\Pi_\varphi}\left((i_\varphi^*)^{-1}(f)\right)$$

$$= \left(\sideset{}{'}\sum_\sigma p_\sigma^{(1)}\Big|_{\Pi_\varphi}\frac{\partial}{\partial p_\sigma} + S_v\right)\Bigg|_{\Pi_\varphi}\left((i_\varphi^*)^{-1}(f)\right)$$

$$= \left(\sideset{}{'}\sum_\sigma \widetilde{D}_\sigma(\varphi)\frac{\partial}{\partial p_\sigma} + S_v\right)\Bigg|_{\Pi_\varphi}\left((i_\varphi^*)^{-1}(f)\right) = (i_\varphi^*)^{-1}\left(\widetilde{\Im}_\varphi^{(\varphi)} + S_w\right)(f).$$

From these relations it follows that $S_\tau|_{\Pi_\varphi} = (i_\varphi)_*\left(\widetilde{\Im}_\varphi^{(\varphi)} + S_w\right)$.

Let us show that the fields $\widetilde{D}_i^\tau$ are tangent to Π_φ, $\varphi \in \ker \widetilde{\ell}_F$, i.e., that Π_φ is an invariant submanifold. Indeed,

$$\left\{\widetilde{D}_i^\tau\left(p_\sigma^{(l)} - \left(\widetilde{\Im}_\varphi + S_v\right)^l(p_\sigma)\right)\right\}\Bigg|_{\Pi_\varphi} = S_p^l(\bar{p}_{\sigma+1_i})|_{\Pi_\varphi} - \widetilde{D}_i^\tau(\widetilde{\Im}_\varphi + S_v)^l(p_\sigma)|_{\Pi_\varphi}$$

$$= S_\tau^l(\bar{p}_{\sigma+1_i})|_{\Pi_\varphi} - (i_\varphi^*)^{-1}\left(\widetilde{D}_i^\varphi(\widetilde{\Im}_\varphi + S_v)^l(p_\sigma)\right)$$

$$= S_\tau^l(\bar{p}_{\sigma+1_i})|_{\Pi_\varphi} - (i_\varphi^*)^{-1}\left((\widetilde{\Im}_\varphi + S_v)^l\widetilde{D}_i^\varphi(p_\sigma)\right)$$

$$= S_\tau^l(\bar{p}_{\sigma+1_i})|_{\Pi_\varphi} - (i_\varphi^*)^{-1}\left((\widetilde{\Im}_\varphi + S_v)^l(p_{\sigma+1_i})\right) = 0.$$

Let us prove finally that $\widetilde{D}_i^\tau\Big|_{\Pi_\varphi} = (i_\varphi)_*\widetilde{D}_i^\varphi$. In fact, for an arbitrary function $f \in \mathcal{F}(\widetilde{\mathcal{E}}_\varphi)$ we have

$$\widetilde{D}_i^\tau\Big|_{\Pi_\varphi}((i_\varphi^*)^{-1}(f)) = \left(\bar{D}_i + \sum_{l\geq 0,j} S_\tau^l(X_{ij})|_{\Pi_\varphi}\frac{\partial}{\partial v_j^l}\right)((i_\varphi^*)^{-1}(f))$$

$$= \left(\bar{D}_i + \sum_{l\geq 0,j}(\widetilde{\Im}_\varphi + S_v)^l(X_{ij})|_{\Pi_\varphi}\frac{\partial}{\partial v_j^l}\right)((i_\varphi^*)^{-1}(f)) = (i_\varphi^*)^{-1}\widetilde{D}_i^\varphi(f)$$

The proposition is proved. □

6. Symmetries of integro-differential equations

The above constructed theory of nonlocal symmetries is based on the concept of a covering, i.e., on introduction of nonlocal variables to the space of infinitely prolonged equations. Recall that such nonlocalities are of indefinite integral nature. Now we shall consider the situation when the initial object is nonlocal and the corresponding nonlocalities are definite integrals. Such objects are *integro-differential equations*, and below we construct a symmetry theory for such equations.

6.1. Transformation of integro-differential equations to boundary differential form. Let us represent a system of integro-differential equation in the form

$$G_j(x, u, p_\sigma, I) = 0, \qquad j = 1, \ldots, m_1, \tag{6.1}$$

where $x = (x_1, \ldots, x_n)$ are independent variables, $u = (u^1, \ldots, u^m)$ are dependent ones, $p_\sigma = (p_\sigma^1, \ldots, p_\sigma^m)$ are the derivatives of the latter over the former, and $I =$

$(I_1, \ldots, I_{m_2})$ are integrals. Values of integrals may depend both on the choice of a point x and on the choice of a section u. The most popular integro-differential equations (see [**41**]) contain integrals of three types:

$$\int_a^b f(x, s, u(s+x))\, ds, \tag{6.2}$$

$$\int_0^x f(t, x, s, u(t, s), u(t, x-s), p_\sigma(t, s), p_\sigma(t, x-s))\, ds, \tag{6.3}$$

$$\int_a^b f(t, x, s, u(t, s))\, ds, \tag{6.4}$$

where t, x are the independent variables, u is the dependent one, p_σ is its derivatives, and s is the variable of integration. These, as well as many other types of integrals, are unified in the following way.

Let M be the manifold of independent variables $x = (x_1, \ldots, x_n)$, N the manifold of independent and integration variables $(x, s) = (x_1, \ldots, x_n, s_1, \ldots, s_{n_1})$, $\rho\colon N \to M$ the corresponding fiber bundle (possibly, with singularities), and N_x the fiber of the bundle ρ over the point $x \in M$. Note that the integrand depends on x, s, and $u(x)$, and there exist smooth mappings $h_i\colon N \to M$, $i = 1, \ldots, m_2$, allowing one to rewrite (6.2)–(6.4) in the form

$$\int_{N_x} f(x, s, h_1^*(u), h_1^*(p_\sigma), \ldots, h_l^*(u), h_l^*(p_\sigma))\, ds. \tag{6.5}$$

EXAMPLE 6.1. For the integral (6.2) we have $M = \mathbb{R}$, $N = \{(x, s) \in \mathbb{R}^2 \mid a \leq s \leq b\}$, N_x is the interval $[a, b]$ for any x, $l = 1$, and $h\colon (x, s) \mapsto x + s$.

EXAMPLE 6.2. For the integral (6.3) we have $M = \{(t, x) \in \mathbb{R}^2 \mid x \geq 0\}$, $N = \{(t, x, s) \in \mathbb{R}^3 \mid x \geq 0, 0 \leq s \leq x\}$, $N_{(t,x)}$ is the interval $[0, x]$ (at singular points $\{x = 0\}$, the interval shrinks to a point), $l = 2$, $h_1\colon (t, x, s) \mapsto (t, s)$, and $h_2\colon (t, x, s) \mapsto (t, x - s)$.

EXAMPLE 6.3. For the integral (6.4) in the case $b = \infty$ we have $M = \{(t, x) \in \mathbb{R}^2 \mid x \geq a\}$, $N = \{(t, x, s) \in \mathbb{R}^3 \mid x \geq a, s \geq a\}$, $N_{(t,x)}$ is the ray $\{s \geq a\}$, $l = 1$, and $h\colon (t, x, s) \mapsto (t, s)$. The case of the finite b is considered in a similar way.

To simplify the formulas below, we shall assume that all the integrals under consideration are one-dimensional. The case of multiple integrals can be considered in a similar way: it suffices to represent the integral as an iterated one.

If the integral (6.5) is one-dimensional, then N_x is either an interval or an infinite semi-interval or a straight line. In the case $N_x = [a_x, b_x]$, we shall introduce a nonlocal variable v depending on x, s, such that

$$\frac{\partial v}{\partial s} = f(x, s, h_1^*(u), h_1^*(p_\sigma), \ldots, h_l^*(u), h_l^*(p_\sigma)),$$
$$v\big|_{s=a_x} = 0. \tag{6.6}$$

The value of the integral (6.5) at a point $x \in M$ coincides with $v\big|_{s=b_x}$ in this case. Let us introduce the mappings

$$a\colon M \longrightarrow N, \quad x \longmapsto (x, a_x); \qquad b\colon M \longrightarrow N, \quad x \longmapsto (x, b_x).$$

Then

$$v\big|_{s=a_x} = a^*(v), \qquad v\big|_{s=b_x} = b^*(v).$$

The case $N_x = [a_x, \infty)$ (and, in a similar manner, the cases $(-\infty, b_x]$ and $(-\infty, \infty)$) can be reduced to the case of the interval by passing to a new integration variable. Such a passage is realized by constructing a diffeomorphism $\mu\colon N \to N'$ taking infinite integration intervals $N_x = [a_x, \infty)$ to finite semi-intervals $\mu(N_x) = [\bar{a}_x, b_x)$. Adding to the manifold N' the limit point $\{b_x \mid x \in M\}$ of the semi-intervals $\mu(N_x)$, we shall obtain the manifold $\bar{N}$ and the fiber bundle $\bar{\rho}\colon \bar{N} \to M$, whose restriction to the subset $N' \subset \bar{N}$ coincides with the mapping $\rho \circ \mu^{-1}$, while the fibers of the bundle $\bar{\rho}$ are the intervals $(\bar{\rho})^{-1}(x) = [\bar{a}_x, b_x]$, $x \in M$. Thus, we can restrict ourselves to the case when the sets N_x, $x \in M$, are intervals (see also Remark 6.1 below).

Thus, system (6.1) can be written in the form

$$\begin{aligned} G_j(x, u, p_\sigma, \ldots, b_k^*(v^k), \ldots) &= 0, \qquad j = 1, \ldots, m_1, \\ a_k^*(v^k) &= 0, \qquad k = 1, \ldots, m_2, \\ \frac{\partial v^k}{\partial s_k} &= f_k(x, s_k, \ldots, (h_i^k)^*(u), (h_i^k)^*(p_\sigma), \ldots), \qquad k = 1, \ldots, m_2, \end{aligned} \quad (6.7)$$

where $x = (x_1, \ldots, x_n)$ are coordinates on the manifold M, $u = (u^1, \ldots, u^m)$ is a section of the bundle π over M, $p_\sigma = (p_\sigma^1, \ldots, p_\sigma^m)$ is the vector of its derivatives, v^k is a function on the manifold N_k, b_k, a_k are the embeddings of M into N_k, and the h_i^k, $i = 1, \ldots, l_k$, are the projections of N_k to M, $k = 1, \ldots, m_2$. Note that the variables v^k, $k = 1, \ldots, m_2$, are uniquely determined by the variables u^j, $j = 1, \ldots, m$, by means of the second and the third equations in (6.7). This means that systems (6.1) and (6.7) are equivalent: solutions of one of them can be reconstructed from solutions of the other.

System (6.7) describes sections over different manifolds $M, N_1, \ldots, N_{m_2}$. When the manifolds $N_1, \ldots, N_{m_2}$ are diffeomorphic to each other, all functions and sections in system (6.7) can be represented as functions and sections over the same manifold. In fact, let N be a manifold, $F_k\colon N \to N_k$, $k = 1, \ldots, m_2$, diffeomorphisms, $a_0\colon M \to N$ an embedding (e.g., $a_0 = F_1^{-1} \circ a_1$), and $h_0\colon N \to M$ a smooth mapping left inverse to a_0, i.e., $h_0 \circ a_0 = \mathrm{id}_M$. Let us set

$$\bar{u} = h_0^*(u), \qquad \bar{v}^k = F_k^*(v^k), \qquad k = 1, \ldots, m_2.$$

Then

$$\begin{aligned} a_0^*(\bar{u}) &= a_0^*(h_0^*(u)) = (h_0 \circ a_0)^*(u) = u, \\ v^k &= (F_k^{-1})^*(\bar{v}^k), \quad (h_i^k)^*(u) = (h_i^k)^*(a_0^*(\bar{u})) = (a_0 \circ h_i^k)^*(\bar{u}), \\ b_k^*(v^k) &= b_k^*((F_k^{-1})^*(\bar{v}^k)) = (F_k^{-1} \circ b_k)^*(\bar{v}^k), \end{aligned}$$

etc. Using these equalities and acting by the homomorphism h_0^* on the first and second equations (6.7), we rewrite these equations in the form

$$\begin{aligned} G_j(h_0^*(x), \bar{u}, h_0^*(p_\sigma), \ldots, \bar{b}_k^*(\bar{v}^k), \ldots) &= 0, \\ \bar{a}_k^*(\bar{v}^k) &= 0, \end{aligned} \quad (6.8)$$

where $\bar{b}_k = F_k^{-1} \circ b_k \circ h_0$, $\bar{a}_k = F_k^{-1} \circ a_k \circ h_0$ are mappings from N to N. Let us now act on the third equation in (6.7) by the diffeomorphism F_k^*. Using the equality

$$F_k^*\left(\frac{\partial v^k}{\partial s_k}\right) = (F_k^{-1})_*\left(\frac{\partial}{\partial s_k}\right)(F_k^*(v^k)) = X_k(\bar{v}^k),$$

where $X_k = (F_k^{-1})_*(\partial/\partial s_k)$ is a vector field on N, we obtain the equation
$$X_k(\bar{v}^k) = f_k(F_k^*(x), F_k^*(s_k), (\bar{h}_i^k)^*(\bar{u}), (\bar{h}_i^k)^*(h_0^*(p_\sigma))), \quad (6.9)$$
where $\bar{h}_i^k = a_0 \circ h_i^k \circ F_k$ is a mapping from N to N.

Note that functions of the form $h_0^*(p_\sigma^r)$ can be represented as derivatives of the functions $\bar{u}^r$ if
$$h_{0,*}(Y_i) = \frac{\partial}{\partial x_i}, \quad Y_i \in D(N), \quad i = 1, \ldots, n.$$
In this case
$$h_0^*\left(\frac{\partial u^r}{\partial x_i}\right) = Y_i(\bar{u}^r), \quad h_0^*\left(\frac{\partial^2 u^r}{\partial x_i \partial x_l}\right) = Y_i(Y_l(\bar{u}^r)),$$
etc. Therefore, $h_0^*(p_\sigma^r)$ is a function of x, s, $\bar{u}$, and $\bar{p}_\sigma$. Besides, $\bar{u}$ is a section of the form $h_0^*(u)$, $u \in \Gamma(\pi)$. These and only these sections satisfy the equation
$$(a_0 \circ h_0)^*(\bar{u}) = \bar{u}. \quad (6.10)$$
Indeed,
$$(a_0 \circ h_0)^*(\bar{u}) = h_0^*(a_0^*(\bar{u}))$$
and one can set $u = a_0^*(\bar{u})$. If $\bar{u} = h_0^*(u)$, then
$$(a_0 \circ h_0)^*(\bar{u}) = (h_0 \circ a_0 \circ h_0)^*(u) = h_0^*(u) = \bar{u},$$
since $h_0 \circ a_0 = \mathrm{id}_M$.

Thus, system (6.7) (and consequently, (6.1)) is equivalent to the system of equations (6.8)–(6.10), $j = 1, \ldots, m_1$, $k = 1, \ldots, m_2$. The functions $\bar{u}^j$, $\bar{v}^k$ in these equations are functions on N, while $\bar{b}_k$, $\bar{a}_k$, $\bar{h}_i^k$, $\bar{h}_0 = a_0 \circ h_0$ are mappings from N to N. Coming back to the previous notation, i.e., denoting by u the section $(\bar{u}^1, \ldots, \bar{u}^m, \bar{v}^1, \ldots, \bar{v}^{m_2})$ over N and by $x = (x_1, \ldots, x_{n+m_2})$ coordinates on N, this system can be written as a system of equations of the form
$$G(x, u, p_\sigma, g^*(u), g^*(p_\sigma)) = 0, \quad (6.11)$$
where $g = g_1, \ldots, g_l$ are mappings from N to N. We shall call equations of the form (6.11) *boundary differential equations*. The reason for this name is that in examples, as a rule, the image of g_k, $k = 1, \ldots, m_2$, lies in the boundary of N, while the value of $g_k^*(u)$ is determined by the values of u on the boundary of N. The other term for these equations is *functional differential equations* (see, for example, [**41**]).

For prolongation of equations of the form (6.11), we can use both derivations and the action of the homomorphism g^*, where g is a mapping from N to N. It is easy to see that the new (prolonged) equation will contain sections of the form $g^*(g_i^*(u)) = (g_i \circ g)^*(u)$, where $g_1, \ldots, g_l$ are the mappings in (6.11). Therefore, besides g, we must consider all compositions $g_i \circ g$, $i = 1, \ldots, l$. Taking for g the mappings in (6.11) and their prolongations only, we obtain the semigroup containing $g_1, \ldots, g_l$, the identity mapping $g_0 = \mathrm{id}_N$ and various compositions of the mappings $g_1, \ldots, g_l$. We call this semigroup the *semigroup of the boundary differential system* (6.11) and denote it by $\mathcal{G}$.

EXAMPLE 6.4. Let a system $\mathcal{E}_0$ of integro-differential equations contain one integral of type (6.3) and one integral of type (6.4) with $a = 0$, $b = \infty$. Let us find the semigroup of mappings for the corresponding system $\mathcal{E}_1$ of boundary differential equations. The manifold of independent variables for the system $\mathcal{E}_0$

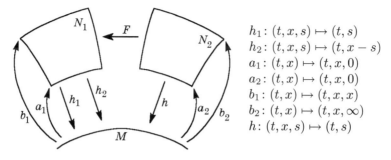

FIGURE 6.2

is $M = \{(t,x) \mid x \geq 0\}$ (see Examples 6.2 and 6.3). The manifold N_1 for the integral (6.3) is $N_1 = \{(t,x,s) \mid x \geq 0, 0 \leq s \leq x\}$, while for the integral (6.4) it is $N_2 = \{(t,x,s) \mid x \geq 0, s \geq 0\}$ ($a = 0$, $b = \infty$). One can see the mappings defined on these manifolds in Figure 6.2.

REMARK 6.1. The above scheme for transforming an integro-differential equation to a system of boundary differential equations needs a reduction of an improper integral of the form (6.4) with $b = \infty$ to a definite integral with finite limits of integration. This is done by construction of the diffeomorphism $\mu \colon N_2 \to N_2'$ and of the manifold $\bar{N}$ obtained from N_2' by adding some limit points (see p. 268). Such an approach makes solution essentially more difficult. Instead, we can assume that the sets (t,x,∞), (t,∞,s), (t,∞,∞), $x \geq 0$, $s \geq 0$, are added to the manifold N_2, and that a smooth structure on the set obtained is introduced by defining the ring of smooth functions on this set. Namely, a function f in the variables x, t, s is called *smooth* if there exist the limits

$$f(t,x,\infty) = \lim_{s \to \infty} f(t,x,s),$$
$$f(t,\infty,s) = \lim_{x \to \infty} f(t,x,s),$$
$$f(t,\infty,\infty) = \lim_{\substack{x \to \infty \\ s \to \infty}} f(t,x,s),$$

and they are equal to the values of the function f at the added point. The manifold thus obtained is diffeomorphic to $\bar{N}_2$ and will also be denoted by N_2. We introduce similar smooth structures on M and N_1. Such an approach allows us to define uniquely the mappings b_2, g_3 (see below) and various compositions containing g_3.

The manifolds N_1 and N_2 are diffeomorphic, and $F \colon N_2 \to N_1$, $F(t,x,s) = (t, x+s, s)$ is their diffeomorphism. Let us choose for the manifold of independent variables of the desired system of boundary differential equations the manifold $N = N_2$ and take for the embedding the mapping $a_0 = F^{-1} \circ a_1 \colon M \to N$. Then the mapping $h_0 = h_2 \circ F \colon N \to M$ can be taken for the left inverse mapping. Consequently, the mappings

$g_1 = \bar{a}_1 = \bar{h}_2 = F^{-1} \circ a_1 \circ h_2 \circ F = \bar{a}_2 = a_2 \circ h_2 \circ F, \quad g_1 \colon (t,x,s) \longmapsto (t,x,0);$

$g_2 = \bar{b}_1 = F^{-1} \circ b_1 \circ h_2 \circ F, \quad g_2 \colon (t,x,s) \longmapsto (t,0,x);$

$g_3 = \bar{b}_2 = b_2 \circ h_2 \circ F, \quad g_3 \colon (t,x,s) \longmapsto (t,x,\infty);$

$g_4 = \bar{h}_1 = F^{-1} \circ a_1 \circ h_1 \circ F = \bar{h} = F^{-1} \circ a_1 \circ h, \quad g_4 \colon (t,x,s) \longmapsto (t,s,0)$

generate a semigroup $\mathcal{G}$ consisting of eleven elements $g_0 = \mathrm{id}_M$, g_1, g_2, g_3, g_4, g_2^2, $g_3 \circ g_2$, $g_2 \circ g_4$, $g_3 \circ g_4$, $g_4 \circ g_3$, $g_3 \circ g_4 \circ g_3$.

EXAMPLE 6.5. Let us find the system of boundary differential equations for the *Smoluchowski coagulation equation* [**111, 144**]:

$$\frac{\partial u(t,x)}{\partial t} = \frac{1}{2} \int_0^x K(x-s,s) u(t,x-s) u(t,s) \, ds - u(t,x) \int_0^\infty K(x,s) u(t,s) \, ds, \tag{6.12}$$

where $K(x,s)$ is a fixed function satisfying the condition $K(s,x) = K(x,s)$ for all $x \geq 0$ and $s \geq 0$. Let us apply the above general procedure to this particular equation. Equation (6.12) contains one integral of type (6.3) and one integral of type (6.4). The manifold N and the semigroup are described in Example 6.4. For any solution $u(t,x)$ of equation (6.12), let us introduce the function $v^1(t,x,s)$ on N_1 and the function $v^2(t,x,s)$ on N_2:

$$\frac{\partial v^1}{\partial s} = K(x-s,s) h_1^*(u) h_2^*(u), \tag{6.13}$$

$$a_1^*(v^1) = 0, \tag{6.14}$$

$$\frac{\partial v^2}{\partial s} = K(x,s) h^*(u),$$

$$a_2^*(v^2) = 0 \tag{6.15}$$

(the mappings h, h_1, h_2, a_1, b_1, a_2, b_2, F are defined in Example 6.4). Then the coagulation equation can be written in the form

$$\frac{\partial u}{\partial t} = \frac{1}{2} b_1^*(v^1) - u b_2^*(v^2). \tag{6.16}$$

Let us introduce the notation $u^1 = (h_2 \circ F)^*(u)$, $u^2 = F^*(v^1)$, $u^3 = v^2$. Then, acting on equations (6.14), (6.15) and (6.16) by the homomorphism $(h_2 \circ F)^*$ and on equation (6.13) by the isomorphism F^*, and using the equality $(F^{-1})_*(\partial/\partial s) = \partial/\partial s - \partial/\partial x$, we obtain the system of boundary differential equations

$$\frac{\partial u^1}{\partial t} = \frac{1}{2} g_2^*(u^2) - u^1 g_3^*(u^3),$$

$$\frac{\partial u^2}{\partial s} - \frac{\partial u^2}{\partial x} = K g_4^*(u^1) u^1, \tag{6.17}$$

$$\frac{\partial u^3}{\partial s} = K g_4^*(u^1),$$

$$g_1^*(u^1) = u^1, \quad g_1^*(u^2) = 0, \quad g_1^*(u^3) = 0,$$

equivalent to equation (6.12).

Let us give a geometric interpretation of boundary differential equations, using a parallel with differential equations.

6.2. Spaces of $(k,\mathcal{G})$-jets. Let us generalize the concept of jet spaces in such a way that boundary differential equations could be interpreted as submanifolds of these spaces. Let $\pi\colon E \to M$ be a smooth locally trivial bundle over the manifold M with boundary, $k = 1, 2, \ldots, \infty$, let $\pi_k\colon J^k(\pi) \to M$ be the corresponding bundle of k-jets, and let $\mathcal{G}$ be a finite set of smooth mappings from M to M containing the identity mapping id_M. Let us denote by $\pi_k^{\mathcal{G}}$ the Whitney product of the pullbacks

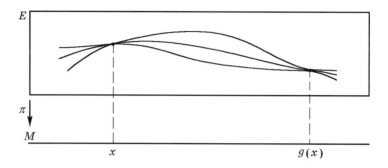

FIGURE 6.3

$g^*(\pi_k)$: $\pi_k^{\mathcal{G}} = \bigoplus_{g \in \mathcal{G}} g^*(\pi_k)$. We denote by $J^k(\pi; \mathcal{G})$ the total space of the bundle $\pi_k^{\mathcal{G}}$. The set $J^k(\pi; \mathcal{G})$ is a finite-dimensional smooth manifold if k is finite, and an infinite-dimensional one if k is infinite.

Every point of $J^k(\pi; \mathcal{G})$ over $x \in M$ is a set of k-jets $\theta_k^g \in J^k(\pi)$, $g \in \mathcal{G}$, satisfying the condition $\pi_k(\theta_k^g) = g(x)$. For any $g \in \mathcal{G}$ there exists a smooth section s_g of the bundle π such that $[s_g]_{g(x)}^k = \theta_k^g$. The set of k-jets $\{[s_g]_{g(x)}^k\}_{g \in \mathcal{G}}$ will be called the $(k, \mathcal{G})$-*jet of the family of sections* $\{s_g\}$ at the point x. Any family of sections $\{s_g\}_{g \in \mathcal{G}}$ determines the section $j_k(\{s_g\})$ of the bundle $\pi_k^{\mathcal{G}}$:

$$j_k(\{s_g\})(x) = \{[s_g]_{g(x)}^k\}_{g \in \mathcal{G}}. \tag{6.18}$$

If all sections s_g, $g \in \mathcal{G}$, coincide with the section s, then the corresponding set of jets will be called the $(k, \mathcal{G})$-*jet of the section* s at the point x and will be denoted by $[s]_x^{(k,\mathcal{G})}$. The section $x \mapsto [s]_x^{(k,\mathcal{G})}$ will be denoted by $j_k(s)$, and the subset of $(k, \mathcal{G})$-jets of sections of the bundle π will be denoted by $J^k(\pi; \mathcal{G})_0$. Geometrically, the $(k, \mathcal{G})$-jet of a section s is interpreted as the class of sections of the bundle π tangent to the section s with order $\geq k$ at all points $g(x)$, $g \in \mathcal{G}$ (see Figure 6.3). We shall call $J^k(\pi; \mathcal{G})$ the *manifold (space) of $(k, \mathcal{G})$-jets*; the bundle $\pi_k^{\mathcal{G}}$ will be called the *bundle of $(k, \mathcal{G})$-jets*.

Let us describe the set $J^k(\pi; \mathcal{G})_0$. If a point $x \in M$ is such that all the points $g(x)$, $g \in \mathcal{G}$, are different, then for any set of sections $\{s_g\}_{g \in \mathcal{G}}$ there exists a smooth section s tangent to the section s_g at the point $g(x)$ with order $\geq k$ for any $g \in \mathcal{G}$. Consequently, the fiber $(\pi_k^{\mathcal{G}})^{-1}(x)$ lies in $J^k(\pi; \mathcal{G})_0$. If for a point $x \in M$ two mappings $g_1, g_2 \in \mathcal{G}$, $g_1 \neq g_2$, exist such that $g_1(x) = g_2(x)$, then the $(k, \mathcal{G})$-jet of the family of sections $\{s_g\}_{g \in \mathcal{G}}$ is the $(k, \mathcal{G})$-jet of a section if and only if

$$[s_{g_1}]_{g_1(x)}^k = [s_{g_2}]_{g_2(x)}^k.$$

A submanifold $\mathcal{E}$ of the manifold $J^k(\pi; \mathcal{G})$ will be called a *boundary differential equation* of order k in the bundle π with respect to the set of mappings $\mathcal{G}$ (or simply, an *equation*). A *solution* of the equation $\mathcal{E} \subset J^k(\pi; \mathcal{G})$ is a section $s \in \Gamma(\pi)$ such that $j_k(s)(M) \subset \mathcal{E}$.

REMARK 6.2. Submanifolds of the form $j_k(s)(M)$ are contained in $J^k(\pi; \mathcal{G})_0$. It might seem that in the boundary differential case the set $J^k(\pi; \mathcal{G})_0$ must be taken for an analog of the space $J^k(\pi)$. But since in the sequel we shall use the language of differential geometry and the set $J^k(\pi; \mathcal{G})_0$ is not a manifold, we extend $J^k(\pi; \mathcal{G})_0$ to the manifold $J^k(\pi; \mathcal{G})$ and take the latter for an analog of $J^k(\pi)$.

Let us introduce the following canonical coordinate system on the manifold $J^k(\pi; \mathcal{G})$. Let $x_1, \ldots, x_n$ be coordinates in a neighborhood $\mathcal{U} \subset M$, $u^1, \ldots, u^m$ coordinates in the fiber of the bundle $\pi|_{\mathcal{U}}$, and x_i, p_σ^j, $i = 1, \ldots, n$, $j = 1, \ldots, m$, $|\sigma| \leq k$, the corresponding coordinates in $J^k(\pi)$. Then, denoting by $p_{\sigma g}^j$ the corresponding coordinates in the fiber of the bundle $g^*(\pi_k)$, we obtain the coordinate system
$$x_i, \ p_{\sigma g}^j, \quad i = 1, \ldots, n, \quad j = 1, \ldots, m, \quad |\sigma| \leq k, \quad g \in \mathcal{G}, \tag{6.19}$$
on $J^k(\pi; \mathcal{G})$. The p-coordinates of the $(k, \mathcal{G})$-jet of the family of sections $\{s_g\}_{g \in \mathcal{G}}$ at the point x are equal to
$$p_{\sigma g}^j = \frac{\partial^{|\sigma|} s_g^j}{\partial x^\sigma}(g(x)),$$
where $s_g^1, \ldots, s_g^m$ are the components of the section s_g. We have the same coordinate system (6.19) on the manifold $J^\infty(\pi; \mathcal{G})$, but with no restriction imposed on $|\sigma|$.

EXERCISE 6.1. Describe the set $J^\infty(\pi; \mathcal{G})_0$ in canonical coordinates.

As in the case of "usual" jets, for $k > l$ and for $k = \infty$ the bundle
$$\pi_{k,l}^{\mathcal{G}} \colon J^k(\pi; \mathcal{G}) \longrightarrow J^l(\pi; \mathcal{G}) \tag{6.20}$$
is defined. Namely,
$$\pi_{k,l}^{\mathcal{G}}(\{\theta_k^g\}_{g \in \mathcal{G}}) = \{\pi_{k,l}(\theta_k^g)\}_{g \in \mathcal{G}}. \tag{6.21}$$
The following equality holds:
$$\pi_{k,l}^{\mathcal{G}} \circ j_k(\{s_g\}) = j_l(\{s_g\}), \tag{6.22}$$
where $\{s_g\}_{g \in \mathcal{G}}$ is an arbitrary family of sections of the bundle π.

EXERCISE 6.2. Show that the space on infinite jets $J^\infty(\pi; \mathcal{G})$ is the inverse limit of the tower of finite jets generated by the mapping $\pi_{l+1,l}^{\mathcal{G}}$, $l \geq 0$.

If $\mathcal{G}_1$ is a subset in $\mathcal{G}$, then the formula
$$\pi_k^{\mathcal{G}, \mathcal{G}_1}(\{\theta_k^g\}_{g \in \mathcal{G}}) = \{\theta_k^g\}_{g \in \mathcal{G}_1} \tag{6.23}$$
determines the bundle
$$\pi_k^{\mathcal{G}, \mathcal{G}_1} \colon J^k(\pi; \mathcal{G}) \longrightarrow J^k(\pi; \mathcal{G}_1), \tag{6.24}$$
and the following equalities hold:
$$\pi_k^{\mathcal{G}_1} \circ \pi_k^{\mathcal{G}, \mathcal{G}_1} = \pi_k^{\mathcal{G}}, \tag{6.25}$$
$$\pi_{k,l}^{\mathcal{G}_1} \circ \pi_k^{\mathcal{G}, \mathcal{G}_1} = \pi_l^{\mathcal{G}, \mathcal{G}_1} \circ \pi_{k,l}^{\mathcal{G}}, \tag{6.26}$$
$$\pi_k^{\mathcal{G}, \mathcal{G}_1} \circ j_k(\{s_g\}) = j_k(\{s_g\}_{g \in \mathcal{G}_1}), \tag{6.27}$$
where $\{s_g\}$ is an arbitrary family of sections of the bundle π, $g \in \mathcal{G}$, and $\{s_g\}_{g \in \mathcal{G}_1}$ denotes the sections s_g such that $g \in \mathcal{G}_1$.

REMARK 6.3. If the set $\mathcal{G}$ consists of the identity mapping only, then the space $J^k(\pi; \{\mathrm{id}_M\})$ coincides with the "usual" jet space $J^k(\pi)$ defined in Chapters 3 and 4. If the set $\mathcal{G}$ contains other mappings besides the identity, then $J^0(\pi; \mathcal{G}) \neq E = J^0(\pi)$. Note that for any k the bundle
$$\pi_0^{\mathcal{G}, \{\mathrm{id}_M\}} \circ \pi_{k,0}^{\mathcal{G}} \colon J^k(\pi; \mathcal{G}) \longrightarrow J^0(\pi; \mathcal{G}) \longrightarrow E$$
is defined.

In the case when the set $\mathcal{G}$ is infinite, the manifold of $(k,\mathcal{G})$-jets is defined in the following way. Let $\{\mathcal{G}_\alpha\}$ be all finite subsets of the set $\mathcal{G}$ containing id_M. Then for any pair $\mathcal{G}_\alpha \subset \mathcal{G}_\beta$ the mappings $\pi_k^{\mathcal{G}_\beta,\mathcal{G}_\alpha}\colon J^k(\pi;\mathcal{G}_\beta) \to J^k(\pi;\mathcal{G}_\alpha)$ are defined, and $\pi_k^{\mathcal{G}_\beta,\mathcal{G}_\alpha} \circ \pi_k^{\mathcal{G}_\gamma,\mathcal{G}_\beta} = \pi_k^{\mathcal{G}_\gamma,\mathcal{G}_\alpha}$ if $\mathcal{G}_\alpha \subset \mathcal{G}_\beta \subset \mathcal{G}_\gamma$. Let us define $J^k(\pi;\mathcal{G})$ as the inverse limit of the manifolds $J^k(\pi;\mathcal{G}_\alpha)$ over the system of mappings $\{\pi_k^{\mathcal{G}_\beta,\mathcal{G}_\alpha}\}_{\mathcal{G}_\alpha \subset \mathcal{G}_\beta}$. In other words, a point θ_k of the set $J^k(\pi;\mathcal{G})$ is a family of points $\theta_k^{\mathcal{G}_\alpha} \in J^k(\pi;\mathcal{G}_\alpha)$ such that (a) $\mathcal{G}_\alpha$ is an arbitrary finite subset of $\mathcal{G}$, and (b) if $\mathcal{G}_\alpha \subset \mathcal{G}_\beta$, then

$$\pi_k^{\mathcal{G}_\beta,\mathcal{G}_\alpha}(\theta_k^{\mathcal{G}_\beta}) = \theta_k^{\mathcal{G}_\alpha}. \tag{6.28}$$

In this case, the mapping $\pi_k^{\mathcal{G}}\colon J^k(\pi;\mathcal{G}) \to M$ is defined by the formula $\pi_k^{\mathcal{G}}(\theta_k) = \pi_k^{\mathcal{G}_\gamma}(\theta_k^{\mathcal{G}_\gamma})$, where $\theta_k^{\mathcal{G}_\gamma}$ is an arbitrary element of the set $\theta_k = \{\theta_k^{\mathcal{G}_\alpha}\}_{\mathcal{G}_\alpha \subset \mathcal{G}}$. From (6.25) and (6.28) it follows that the point $\pi_k^{\mathcal{G}_\gamma}(\theta_k^{\mathcal{G}_\gamma}) \in M$ is independent of the choice of elements $\theta_k^{\mathcal{G}_\gamma}$ from the set θ_k.

The points $\theta_k^{\mathcal{G}_\alpha} \in J^k(\pi;\mathcal{G}_\alpha)$, in the case of a finite set $\mathcal{G}_\alpha$, are themselves sets of k-jets $\{\theta_k^g\}_{g\in\mathcal{G}_\alpha}$ while condition (6.28) means only that the set $\{\theta_k^g\}_{g\in\mathcal{G}_\alpha}$ is a part of the set $\{\theta_k^g\}_{g\in\mathcal{G}_\beta}$ provided $\mathcal{G}_\alpha \subset \mathcal{G}_\beta$. Therefore a point of $J^k(\pi;\mathcal{G})$, in the case of an infinite set $\mathcal{G}$, may be understood as an infinite set of k-jets $\{\theta_k^g\}_{g\in\mathcal{G}}$, or more exactly, as $\{[s_g]_{g(x)}^k\}_{g\in\mathcal{G}}$. Formulas (6.21), (6.23), and (6.18) define the maps (6.20), (6.24), and $j_k(\{s_g\})\colon M \to J^k(\pi;\mathcal{G})$ in the case of infinite $\mathcal{G}$ as well; and, as in the finite case, equalities (6.22), (6.25), (6.26), (6.27) hold, as well as the equality $\pi_k^{\mathcal{G}} \circ j_k(\{s_g\}) = \mathrm{id}_M$, etc. (see Chapter 4).

Similar to the case $J^\infty(\pi)$, the set $J^k(\pi;\mathcal{G})$, when $\mathcal{G}$ is infinite and k is either finite or infinite, is endowed with a structure of infinite-dimensional manifold. Coordinates in $J^k(\pi;\mathcal{G})$ are given by the set of functions (6.19) with infinite set $\mathcal{G}$. The concepts of smooth functions, smooth mappings, tangent vectors, vector fields, differential forms, etc., are introduced on $J^k(\pi;\mathcal{G})$ in the same way as in the case of $J^\infty(\pi)$ (see Chapter 4).

Let us denote by $\mathcal{F}(\pi;\mathcal{G})$ and by $\mathcal{F}_k(\pi;\mathcal{G})$ the algebras of smooth functions on $J^\infty(\pi;\mathcal{G})$ and on $J^k(\pi;\mathcal{G})$, respectively, and denote by $\Lambda^*(\pi;\mathcal{G})$ the module of differential forms on $J^\infty(\pi;\mathcal{G})$. The monomorphisms $(\pi_{\infty,k}^{\mathcal{G}})^*$ and $(\pi_\infty^{\mathcal{G},\mathcal{G}'})^*$, where $\mathcal{G}' \subset \mathcal{G}$, embed the algebras $\mathcal{F}_k(\pi;\mathcal{G})$ and $\mathcal{F}(\pi;\mathcal{G}')$ in the algebra $\mathcal{F}(\pi;\mathcal{G})$. The images of these monomorphisms will be also denoted by $\mathcal{F}_k(\pi;\mathcal{G})$ and $\mathcal{F}(\pi;\mathcal{G}')$. In particular, $\mathcal{F}(\pi) = \mathcal{F}(\pi;\{\mathrm{id}_M\})$ is a subalgebra of the algebra $\mathcal{F}(\pi;\mathcal{G})$. The above embeddings allow us to define two filtrations in the algebra $\mathcal{F}(\pi;\mathcal{G})$ (as well as in the module $\Lambda^*(\pi;\mathcal{G})$): filtration with respect to k, similar to filtration in $\mathcal{F}(\pi)$ (see Chapter 4) and filtration corresponding to the embeddings of finite subsets $\mathcal{G}_\alpha$ of the set $\mathcal{G}$. The second filtration puts into correspondence to a function φ from $\mathcal{F}(\pi;\mathcal{G})$ the finite subset $\mathcal{G}_\alpha$ of the set $\mathcal{G}$, if φ is a function on $J^k(\pi;\mathcal{G}_\alpha)$. The first filtration possesses the same properties as filtration in the algebra $\mathcal{F}(\pi)$ (see (1.3) of Chapter 4). For the second filtration the properties (1.3) of Chapter 4 should be rewritten as follows: $\deg\varphi$ is to be understood as the corresponding subset $\mathcal{G}_\alpha \subset \mathcal{G}$, while the inequality $\leq$ must be changed to the embedding $\subseteq$.

To simplify the subsequent exposition, we shall not use the notion of filtration, though this exposition is completely in parallel with that of Chapter 4. For example, a smooth mapping G from $J^\infty(\pi;\mathcal{G})$ to $J^\infty(\xi;\mathcal{G}')$, where $\xi\colon Q \to M$ is a vector bundle and $\mathcal{G}'$ is another set of mappings from M to M, is a mapping of sets such that for any function $\varphi \in \mathcal{F}(\xi;\mathcal{G}')$ the element $G^*(\varphi) = \varphi \circ G$ is a smooth

function on the space $J^\infty(\pi; \mathcal{G})$, and for any number k and any finite subset of the set $\mathcal{G}_\alpha \subset \mathcal{G}'$ there exist a number l and a finite subset $\mathcal{G}_\beta \subset \mathcal{G}$ such that
$$G^*(\mathcal{F}_k(\xi; \mathcal{G}_\alpha)) \subset \mathcal{F}_l(\pi; \mathcal{G}_\beta).$$

A tangent vector X_θ to the manifold $J^\infty(\pi; \mathcal{G})$ at a point θ is defined as the set $\{X_x, X_{\theta_k^{\mathcal{G}_\alpha}}\}$ of tangent vectors to finite-dimensional manifolds M and $J^k(\pi; \mathcal{G}_\alpha)$ at points $x = \pi_\infty^{\mathcal{G}}(\theta)$ and $\theta_k^{\mathcal{G}_\alpha} = \pi_k^{\mathcal{G},\mathcal{G}_\alpha}(\pi_{\infty,k}^{\mathcal{G}}(\theta))$, respectively, such that
$$(\pi_{k+1,k}^{\mathcal{G}_\alpha})_*(X_{\theta_{k+1}^{\mathcal{G}_\alpha}}) = X_{\theta_k^{\mathcal{G}_\alpha}}, \quad (\pi_k^{\mathcal{G}_\beta,\mathcal{G}_\alpha})_*(X_{\theta_k^{\mathcal{G}_\beta}}) = X_{\theta_k^{\mathcal{G}_\alpha}}, \quad (\pi_k^{\mathcal{G}_\alpha})_*(X_{\theta_k^{\mathcal{G}_\alpha}}) = X_x.$$

A vector field on the manifold $J^\infty(\pi; \mathcal{G})$ is, by definition, a derivation X of the algebra $\mathcal{F}(\pi; \mathcal{G})$ such that for any number k and any finite subset $\mathcal{G}_\alpha \subset \mathcal{G}$ there exist a number l and a finite subset $\mathcal{G}_\beta \subset \mathcal{G}$ such that
$$X(\mathcal{F}_k(\pi; \mathcal{G}_\alpha)) \subset \mathcal{F}_l(\pi; \mathcal{G}_\beta).$$

The set of vector fields on $J^\infty(\pi; \mathcal{G})$ will be denoted by $\mathrm{D}(\pi; \mathcal{G})$. The *inner product* of a vector field X on $J^\infty(\pi; \mathcal{G})$ with a form $\omega \in \Lambda^i(\pi; \mathcal{G})$ gives us the form $X \,\lrcorner\, \omega \in \Lambda^{i-1}(\pi; \mathcal{G})$, whose value at a point $\theta = \{x, \theta_k^{\mathcal{G}_\alpha}\}$ equals $X_{\theta_l^{\mathcal{G}_\beta}} \,\lrcorner\, \omega_{\theta_l^{\mathcal{G}_\beta}}$, if ω is a form on $J^l(\pi; \mathcal{G}_\beta)$.

EXERCISE 6.3. Define the de Rham differential $d\omega$ and the Lie derivative $X(\omega)$ in the case $\omega \in \Lambda^*(\pi; \mathcal{G}_1)$, $X \in \mathrm{D}(\pi; \mathcal{G}_2)$, $\mathcal{G}_1 \neq \mathcal{G}_2$.

6.3. Boundary differential operators. Let π and π' be smooth locally trivial vector bundles over the manifold M, and let $\mathcal{G}$ be a finite set of smooth mappings from M to M containing the identity mapping id_M. A mapping $\Delta \colon \Gamma(\pi) \to \Gamma(\pi')$ is called a *boundary differential operator* of order k with respect to the set $\mathcal{G}$, if for any section $s \in \Gamma(\pi)$ the value of the section $\Delta(s)$ at an arbitrary point $x \in M$ is determined by the k-jets of the section s at the points $g(x)$, $g \in \mathcal{G}$. Similarly to the purely differential case (see Chapter 4), any section φ of the bundle $(\pi_k^{\mathcal{G}})^*(\pi')$ determines a boundary differential operator $\Delta_\varphi \colon \Gamma(\pi) \to \Gamma(\pi')$ by the formula
$$\Delta_\varphi(s) = \varphi \circ j_k(s). \tag{6.29}$$

In coordinates $x_1, \ldots, x_n, \ldots, p_{\sigma g}^j, \ldots$ on $J^k(\pi; \mathcal{G})$ this correspondence takes the following form. If $\varphi = \{\varphi^i(x_1, \ldots, x_n, \ldots, p_{\sigma g}^j, \ldots)\}$ and $s = \{s^j(x_1, \ldots, x_n)\}$, then
$$\Delta_\varphi(s) = (\varphi^1(\ldots, x_i, \ldots, g^*(\frac{\partial^{|\sigma|} s^j}{\partial x^\sigma}), \ldots), \ldots, \varphi^{m'}(\ldots, x_i, \ldots, g^*(\frac{\partial^{|\sigma|} s^j}{\partial x^\sigma}), \ldots)),$$
where $m' = \dim \pi'$.

A particular feature of the boundary differential case is that there exist boundary differential operators of order k such that no section of the bundle $(\pi_k^{\mathcal{G}})^*(\pi')$ correspond to these operators. The reason is that, using formula (6.29) and having the function Δ_φ, we can reconstruct the values of the section φ at the $(k, \mathcal{G})$-jets of sections of the bundle π only, i.e., at points of the subset $J^k(\pi; \mathcal{G})_0$. Therefore, (6.29) determines the boundary differential operator Δ_φ not only in the case when φ is a section the bundle $(\pi_k^{\mathcal{G}})^*(\pi')$, but also in the case when φ is a section of the restriction of $(\pi_k^{\mathcal{G}})^*(\pi')$ to the subset $J^k(\pi; \mathcal{G})_0$.

EXAMPLE 6.6. Let $M = \mathbb{R}$, and let $\pi = \pi' \colon \mathbb{R}^2 \to \mathbb{R}$ be a trivial bundle. Let the set $\mathcal{G}$ consist of two mappings: the identity mapping $g_0 = \mathrm{id}_\mathbb{R}$ and the projection g of the manifold $\mathbb{R}$ to the point $0 \in \mathbb{R}$. We have the coordinate system x, p_{0g_0},

p_{0g} on $J^0(\pi;\mathcal{G})$. In these coordinates, points of the set $J^0(\pi;\mathcal{G}) \smallsetminus J^0(\pi;\mathcal{G})_0$ form the plane $\{x = 0\}$ without the line $\{x = 0, p_{0g_0} = p_{0g}\}$. The boundary differential operator

$$\Delta \colon s \longmapsto \begin{cases} \exp\left(-\dfrac{1}{x^2 + (s(x) - s(0))^2}\right), & x > 0, \\ 0, & x \leq 0, \end{cases}$$

where $s \in \Gamma(\pi)$, determines a smooth function on $J^0(\pi;\mathcal{G})_0$ that cannot be extended to all of $J^0(\pi;\mathcal{G})$.

EXERCISE 6.4. Prove this fact.

In what follows, only operators of the form Δ_φ, where φ is a section of the bundle $(\pi_k^{\mathcal{G}})^*(\pi')$, will be called *boundary differential operators*. Not only boundary differential operators satisfy this condition, but also rather "exotic" operators (see Example 6.6), and thus this restriction is not too essential. We shall denote by $\mathcal{F}_k(\pi, \pi'; \mathcal{G})$ the set of sections of the bundle $(\pi_k^{\mathcal{G}})^*(\pi')$. Similarly to the differential case, for any k there exist the embeddings

$$\mathcal{F}_k(\pi, \pi'; \mathcal{G}) \subset \mathcal{F}_{k+1}(\pi, \pi'; \mathcal{G}).$$

Therefore, we can set

$$\mathcal{F}(\pi, \pi'; \mathcal{G}) = \bigcup_{k=0}^{\infty} \mathcal{F}_k(\pi, \pi'; \mathcal{G}).$$

Let us show now that the composition of two boundary differential operators of the form Δ_φ is again a boundary differential operator of the same form. First we prove this fact for the composition $j_{k'}^{\mathcal{G}'} \circ j_k^{\mathcal{G}}$, where $j_k^{\mathcal{G}}$ denotes the operator taking an arbitrary section s of the bundle π to the section $j_k(s)$ of the bundle $\pi_k^{\mathcal{G}}$. In the composition $j_{k'}^{\mathcal{G}'} \circ j_k^{\mathcal{G}}$, the operator $j_{k'}^{\mathcal{G}'}$ takes sections of the bundle $\pi_k^{\mathcal{G}}$ to sections of the bundle $(\pi_k^{\mathcal{G}})_{k'}^{\mathcal{G}'}$. Denote by $\mathcal{G} \circ \mathcal{G}'$ the set of mappings of the manifold M of the form $g \circ g'$, $g \in \mathcal{G}$, $g' \in \mathcal{G}'$. Let us construct the mapping

$$\Phi_{k,k'}^{\mathcal{G},\mathcal{G}'} \colon J^{k+k'}(\pi; \mathcal{G} \circ \mathcal{G}') \longrightarrow J^{k'}(\pi_k^{\mathcal{G}}; \mathcal{G}')$$

by setting

$$\Phi_{k,k'}^{\mathcal{G},\mathcal{G}'}(\{[s_g]_{g(x)}^{k+k'}\}_{g \in \mathcal{G} \circ \mathcal{G}'}) = \{[j_k(\{s_g\}_{g \in \mathcal{G}})]_{g'(x)}^{k'}\}_{g' \in \mathcal{G}'}.$$

Then for any section $s \in \Gamma(\pi)$ one has

$$\Phi_{k,k'}^{\mathcal{G},\mathcal{G}'} \circ j_{k+k'}^{\mathcal{G} \circ \mathcal{G}'}(s) = (j_{k'}^{\mathcal{G}'} \circ j_k^{\mathcal{G}})(s), \tag{6.30}$$

i.e., $j_{k'}^{\mathcal{G}'} \circ j_k^{\mathcal{G}} = \Delta_\varphi$, where φ is a section of the bundle $(\pi_{k+k'}^{\mathcal{G} \circ \mathcal{G}'})^*((\pi_k^{\mathcal{G}})_{k'}^{\mathcal{G}'})$ determined by the mapping $\Phi_{k,k'}^{\mathcal{G},\mathcal{G}'}$. Therefore, the composition $j_{k'}^{\mathcal{G}'} \circ j_k^{\mathcal{G}}$ is a boundary differential operator of order $k + k'$ with respect to the set $\mathcal{G} \circ \mathcal{G}'$.

For any morphism Φ of the bundles π and π' let us define its lifting

$$\Phi^{(k)} \colon J^k(\pi; \mathcal{G}) \to J^k(\pi'; \mathcal{G})$$

by the formula

$$\Phi^{(k)}(\{[s_g]_{g(x)}^k\}_{g \in \mathcal{G}}) = \{[\Phi \circ s_g]_{g(x)}^k\}_{g \in \mathcal{G}}.$$

Then, obviously, for any section $s \in \Gamma(\pi)$ the equality

$$\Phi^{(k)} \circ j_k(s) = j_k(\Phi \circ s) \tag{6.31}$$

holds.

Now let $\Delta_\varphi \colon \Gamma(\pi) \to \Gamma(\pi')$ and $\Delta_{\varphi'} \colon \Gamma(\pi') \to \Gamma(\pi'')$ be two boundary differential operators of orders k and k' with respect to the sets of mappings $\mathcal{G}$ and $\mathcal{G}'$ respectively. Then, using consecutively the equalities (6.29), (6.31), and (6.30), we obtain

$$\Delta_{\varphi'}(\Delta_\varphi(s)) = \Delta_{\varphi'}(\varphi \circ j_k^{\mathcal{G}}(s)) = \varphi' \circ j_{k'}^{\mathcal{G}'}(\varphi \circ j_k^{\mathcal{G}}(s))$$
$$= \varphi' \circ \varphi^{(k)} \circ j_{k'}^{\mathcal{G}'}(j_k^{\mathcal{G}}(s)) = \varphi' \circ \varphi^{(k)} \circ \Phi_{k,k'}^{\mathcal{G},\mathcal{G}'} \circ j_{k+k'}^{\mathcal{G} \circ \mathcal{G}'}(s).$$

Therefore, the mapping

$$\Phi = \varphi' \circ \varphi^{(k)} \circ \Phi_{k,k'}^{\mathcal{G},\mathcal{G}'} \colon J^{k+k'}(\pi; \mathcal{G} \circ \mathcal{G}') \longrightarrow E'',$$

where E'' is the space of the bundle π'', is the section of the bundle $(\pi_{k+k'}^{\mathcal{G} \circ \mathcal{G}'})^*(\pi'')$ determining the operator $\Delta_{\varphi'} \circ \Delta_\varphi$. In other words, the composition $\Delta_{\varphi'} \circ \Delta_\varphi = \Delta_\Phi$ is a boundary differential operator of order $k+k'$ with respect to the set of mappings $\mathcal{G} \circ \mathcal{G}'$.

If $\Delta \colon \Gamma(\pi) \to \Gamma(\pi')$ is a boundary differential operator of order k with respect to the set of mappings $\mathcal{G}$ and $\mathcal{G}'$ is another set of mappings of the manifold M, then the composition

$$\Delta_l^{\mathcal{G}'} = j_l^{\mathcal{G}'} \circ \Delta \colon \Gamma(\pi) \to \Gamma(\pi'{}_l^{\mathcal{G}'})$$

is again a boundary differential operator of order $k + l$ with respect to the set of mappings $\mathcal{G} \circ \mathcal{G}'$. The operator $\Delta_l^{\mathcal{G}'}$ will be called the $(l, \mathcal{G}')$-*prolongation* of the boundary differential operator Δ.

EXERCISE 6.5. Let $\mathcal{G}$ and $\mathcal{G}'$ be sets of mappings of the manifold M containing the identity mapping, and let $\mathcal{G}$ be finite. Show that the set of mappings

$$\{\varphi_{\Delta_l^{\mathcal{G}'}} \colon J^{k+l}(\pi; \mathcal{G} \circ \mathcal{G}') \longrightarrow J^l(\pi'; \mathcal{G}')\}_{l \geq 0}$$

defines a smooth mapping of $J^\infty(\pi; \mathcal{G} \circ \mathcal{G}')$ to $J^\infty(\pi'; \mathcal{G}')$.

Now let $\nabla \colon \Gamma(\xi) \to \Gamma(\xi')$ be a linear boundary differential operator with respect to the set of mappings $\mathcal{G}'$, π a bundle over the same base M, and $\mathcal{G}$ a set of mappings of M. Similarly to the differential case, let us define the *lifting*

$$\widehat{\nabla} \colon \mathcal{F}(\pi, \xi; \mathcal{G}) \to \mathcal{F}(\pi, \xi'; \mathcal{G} \circ \mathcal{G}')$$

of the operator ∇. Let θ be a point of the space $J^\infty(\pi; \mathcal{G})$, let $\{s_g\}_{g \in \mathcal{G}}$ be a family of sections whose infinite jet at the point x is θ, and let φ be a section of the bundle $(\pi_k^{\mathcal{G}})^*(\xi)$. We set, by definition,

$$\widehat{\nabla}(\varphi)(\theta) = \nabla(j_k(\{s_g\})^*(\varphi))(x).$$

EXERCISE 6.6. Prove the identity $\widehat{\nabla}(\varphi) = \varphi_{\nabla \circ \Delta}$ at points of $J^\infty(\pi; \mathcal{G})_0$ where $\Delta = \Delta_\varphi$. Construct an example in which this equality does not hold at other points of the space $J^\infty(\pi; \mathcal{G})$. Thus, we cannot use the "global" definition of the lifting given in Chapter 4.

A vector field X on the manifold M and the homomorphism g_1^*, where g_1 is a mapping from M to M, are linear boundary differential operators acting from $C^\infty(M)$ to $C^\infty(M)$ with respect to the sets of mappings $\{\mathrm{id}_M\}$ and $\{\mathrm{id}_M, g_1\}$. Therefore, the liftings $\widehat{X}$ and $\widehat{g_1^*}$ are defined.

EXERCISE 6.7. Prove that $\widehat{X}$ is a vector field on $J^\infty(\pi; \mathcal{G})$ satisfying the equalities (1.12) of Chapter 4 for $\varphi \in \mathcal{F}(\pi; \mathcal{G})$.

A smooth mapping $g_1 \colon M \to M$ lifts to the smooth mapping
$$\widehat{g_1} \colon J^k(\pi; \mathcal{G} \circ \mathcal{G}_{g_1}) \to J^k(\pi; \mathcal{G}),$$
where $\mathcal{G}_{g_1} = \{\mathrm{id}_M, g_1\}$, $k = 0, 1, \ldots, \infty$. Namely, the mapping $\widehat{g_1}$ takes the point $\theta_k = \{\theta_k^g\}_{g \in \mathcal{G} \circ \mathcal{G}_{g_1}}$ to the point $\widetilde{\theta}_k = \{\widetilde{\theta}_k^g\}_{g \in \mathcal{G}}$, where $\widetilde{\theta}_k^g = \theta_k^{g \circ g_1}$. If $\pi_k^{\mathcal{G}}(\theta) = x$, then θ_k^g is a k-jet at the point $g(x)$ and $\widetilde{\theta}_k^g$ is a k-jet at the point $g(x_1)$, where $x_1 = g_1(x)$. Therefore, the diagram

$$\begin{array}{ccc} J^k(\pi; \mathcal{G} \circ \mathcal{G}_{g_1}) & \xrightarrow{\widehat{g_1}} & J^k(\pi; \mathcal{G}) \\ \pi_k^{\mathcal{G} \circ \mathcal{G}_{g_1}} \downarrow & & \downarrow \pi_k^{\mathcal{G}} \\ M & \xrightarrow{g_1} & M \end{array} \qquad (6.32)$$

is commutative, as well as the diagram

$$\begin{array}{ccc} J^k(\pi; \mathcal{G} \circ \mathcal{G}_{g_1}) & \xrightarrow{\widehat{g_1}} & J^k(\pi; \mathcal{G}) \\ \pi_k^{\mathcal{G} \circ \mathcal{G}_{g_1}, \mathcal{G}' \circ \mathcal{G}_{g_1}} \downarrow & & \downarrow \pi_k^{\mathcal{G}, \mathcal{G}'} \\ J^k(\pi; \mathcal{G}' \circ \mathcal{G}_{g_1}) & \xrightarrow{\widehat{g_1}} & J^k(\pi; \mathcal{G}') \end{array} \qquad (6.33)$$

where $\mathcal{G}' \subset \mathcal{G}$. The following formula is also valid:
$$\widehat{g_1} \circ j_k(\{s_{\widetilde{g}}\}) = j_k(\{\widetilde{s}_g\}) \circ g_1, \qquad (6.34)$$
where $\widetilde{g} \in \mathcal{G} \circ \mathcal{G}_{g_1}$, $g \in \mathcal{G}$, $\widetilde{s}_g = s_{g \circ g_1}$, since
$$\widehat{g_1}(j_k(\{s_{\widetilde{g}}\})(x)) = \widehat{g_1}(\{[s_{\widetilde{g}}]_{\widetilde{g}(x)}^k\}_{\widetilde{g} \in \mathcal{G} \circ \mathcal{G}_{g_1}})$$
$$= \{[s_{g \circ g_1}]_{g(g_1(x))}^k\}_{g \in \mathcal{G}} = j_k(\{s_{g \circ g_1}\})(g_1(x)).$$

The mapping $\widehat{g_1}$ is smooth for $k = 0, 1, \ldots, \infty$, since for any number $l \geq 0$ and any finite subset $\mathcal{G}_\alpha \subset \mathcal{G}$ one has
$$(\widehat{g_1})^*(\mathcal{F}_l(\pi; \mathcal{G}_\alpha)) \subset \mathcal{F}_l(\pi; \mathcal{G}_\alpha \circ \mathcal{G}_{g_1}).$$

EXERCISE 6.8. Prove that $\widehat{g}^* = \widehat{g^*}$ when $k = \infty$.

Note that the equality $\mathcal{G} \circ \mathcal{G}_{g_1} = \mathcal{G}$ means invariance of the set of mappings $\mathcal{G}$ with respect to the right action of g_1, i.e., $g \circ g_1 \in \mathcal{G}$ for any $g \in \mathcal{G}$. If the set $\mathcal{G}$ is invariant with respect to the right action of g_1 and g_2, then
$$\widehat{g_1 \circ g_2} = \widehat{g_1} \circ \widehat{g_2}. \qquad (6.35)$$

In fact, $\widehat{g_2}(\{\theta_k^g\}_{g \in \mathcal{G}}) = \{\widetilde{\theta}_k^g\}_{g \in \mathcal{G}}$, where $\widetilde{\theta}_k^g = \theta_k^{g \circ g_2}$, and
$$\widehat{g_1}(\{\widetilde{\theta}_k^g\}_{g \in \mathcal{G}}) = \{\widetilde{\theta}_k^{g \circ g_1}\}_{g \in \mathcal{G}} = \{\theta_k^{g \circ g_1 \circ g_2}\}_{g \in \mathcal{G}} = \widehat{g_1 \circ g_2}(\{\theta_k^g\}_{g \in \mathcal{G}}).$$

Equality (6.35) is valid, for example, if g_1 and g_2 are elements of a semigroup $\mathcal{G}$.

Let us describe coordinate representation of liftings of vector fields and mappings and coordinate representation of $(l, \mathcal{G}')$-prolongations of boundary differential operators. Let $x_1, \ldots, x_n, \ldots, p^j_{\sigma g}, \ldots$ be coordinates in $J^\infty(\pi; \mathcal{G})$, $X = \partial/\partial x_i$, $\varphi = \varphi(x_1, \ldots, x_n, \ldots, p^j_{\sigma g}, \ldots) \in \mathcal{F}(\pi; \mathcal{G})$, and $s = (s^1, \ldots, s^m) \in \Gamma(\pi)$. Then

$$\left(\frac{\partial}{\partial x_i} \circ \Delta_\varphi\right)(s) = \frac{\partial}{\partial x_i}\left(\varphi\left(x_1, \ldots, x_n, \ldots, g^*\left(\frac{\partial^{|\sigma|} s^j}{\partial x^\sigma}\right), \ldots\right)\right)$$

$$= \frac{\partial \varphi}{\partial x_i} + \sum_{j,\sigma,g} \frac{\partial}{\partial x_i}\left(g^*\left(\frac{\partial^{|\sigma|} s^j}{\partial x^\sigma}\right)\right) \frac{\partial \varphi}{\partial p^j_{\sigma g}}.$$

Therefore, by the equality $\partial/\partial x_i \circ g^* = \sum_l \partial g^*(x_l)/\partial x_i g^* \circ \partial/\partial x_l$, we have

$$\widehat{\frac{\partial}{\partial x_i}} = \frac{\partial}{\partial x_i} + \sum_{|\sigma|,j,g,l} \frac{\partial g^*(x_l)}{\partial x_i} p^j_{\sigma+1_l,g} \frac{\partial}{\partial p^j_{\sigma g}}. \tag{6.36}$$

Similarly to the differential case, we shall denote the operator $\widehat{\partial/\partial x_i}$ by D_i and call it the *total derivative* along the variable x_i.

The mapping $\widehat{g_1}$, where g_1 is a mapping from M to M, obviously acts on the coordinate functions in the following way: $\widehat{g_1}^*(x_i) = g_1^*(x_i)$, $\widehat{g_1}^*(p^j_{\sigma g}) = p^j_{\sigma g_2}$, where $g_2 = g \circ g_1$.

Finally, let $\Delta: \Gamma(\pi) \to \Gamma(\pi')$ be a boundary differential operator, let $x_1, \ldots, x_n$, $v^1, \ldots, v^{m'}, \ldots, q^j_{\sigma g'}, \ldots$ be coordinates in $J^\infty(\pi'; \mathcal{G}')$, and let the operator Δ be determined by the relations

$$v^{j'} = \varphi^{j'}(x_1, \ldots, x_n, \ldots, p^j_{\sigma g}, \ldots), \qquad j' = 1, \ldots, m'.$$

Then the $(l, \mathcal{G}')$-prolongation of the operator Δ is described by the relations

$$\begin{aligned} v^{j'} &= \varphi^{j'}, \\ q^{j'}_{\tau g'} &= g'^*(D_\tau \varphi^{j'}), \qquad j' = 1, \ldots, m', \quad |\tau| = 0, \ldots, l, \quad g' \in \mathcal{G}', \end{aligned} \tag{6.37}$$

where $D_\tau = D_1^{l_1} \circ \cdots \circ D_n^{l_n}$ for $\tau = (l_1, \ldots, l_n)$.

6.4. The Cartan distribution on $J^\infty(\pi; \mathcal{G})$. If $\mathcal{G}$ is a finite set, then, in parallel with the differential case, we denote by $L_{\theta_{k+1}}$ the tangent plane to the graph of the section $j_k(\{s_g\})$ at a point $\theta_k \in J^k(\pi; \mathcal{G})$, where $\{s_g\}_{g \in \mathcal{G}}$ is a family of sections of the bundle π such that its $(k+1, \mathcal{G})$-jet is $\theta_{k+1} \in J^{k+1}(\pi; \mathcal{G})$ and $\theta_k = \pi^\mathcal{G}_{k+1,k}(\theta_{k+1})$. In the case of finite $\mathcal{G}$ and k, the *Cartan distribution* $\mathcal{C}^k$ on $J^k(\pi; \mathcal{G})$ at the point θ_k is defined as the span of the spaces $L_{\theta_{k+1}}$, where $\theta_{k+1} \in (\pi^\mathcal{G}_{k+1,k})^{-1}(\theta_k)$.

From (6.22) and (6.27) it follows that the projections $\pi^\mathcal{G}_{k,k-1}$ and $\pi^{\mathcal{G},\mathcal{G}_1}_k$, where $\mathcal{G}_1$ is a subset in $\mathcal{G}$, take the graph of the section $j_k(\{s_g\})$ to a graph of a similar type. For tangent maps, this fact implies the equalities

$$(\pi^\mathcal{G}_{k,k-1})_*(L_{\theta_{k+1}}) = L_{\theta_k}, \qquad (\pi^{\mathcal{G},\mathcal{G}_1}_k)_*(L_{\theta_{k+1}}) = L_{\widetilde{\theta}_{k+1}}, \tag{6.38}$$

where $\theta_k = \pi^\mathcal{G}_{k+1,k}(\theta_{k+1})$, $\widetilde{\theta}_{k+1} = \pi^{\mathcal{G},\mathcal{G}_1}_{k+1}(\theta_{k+1})$. Consequently,

$$(\pi^\mathcal{G}_{k,k-1})_*(\mathcal{C}^k_{\theta_k}) = L_{\theta_k} \subset \mathcal{C}^{k-1}_{\theta_{k-1}}, \tag{6.39}$$

$$(\pi^{\mathcal{G},\mathcal{G}_1}_k)_*(\mathcal{C}^k_{\theta_k}) = \mathcal{C}^k_{\widetilde{\theta}_k}, \tag{6.40}$$

where $\theta_{k-1} = \pi_{k,k-1}^{\mathcal{G}}(\theta_k)$, $\widetilde{\theta}_k = \pi_k^{\mathcal{G},\mathcal{G}_1}(\theta_k)$.

Formula (6.39) means that the distribution $\mathcal{C}^k$ on $J^k(\pi;\mathcal{G})$ determines the distribution $\mathcal{C} = \mathcal{C}(\pi;\mathcal{G})$ on $J^\infty(\pi;\mathcal{G})$ in the case of a finite set $\mathcal{G}$. If $\mathcal{G}$ is an infinite set, then (6.40) allows us to define the distributions $\mathcal{C}^k$ on $J^k(\pi;\mathcal{G})$ and $\mathcal{C} = \mathcal{C}(\pi;\mathcal{G})$ on $J^\infty(\pi;\mathcal{G})$. Namely, a tangent vector $X_\theta = \{X_x, X_{\theta_k^{\mathcal{G}_\alpha}}\}$ to the manifold $J^\infty(\pi;\mathcal{G})$ at a point $\theta = \{x, \theta_k^{\mathcal{G}_\alpha}\}$ lies, by definition, in the plane $\mathcal{C}_\theta$, if for any number $k \geq 0$ and any finite subset $\mathcal{G}_\alpha \subset \mathcal{G}$ the vector $X_{\theta_k^{\mathcal{G}_\alpha}}$ lies in the Cartan distribution $\mathcal{C}^k$ on $J^k(\pi;\mathcal{G}_\alpha)$ at the point $\theta_k^{\mathcal{G}_\alpha}$. We shall call the distribution $\mathcal{C}_\theta$, $\theta \in J^\infty(\pi;\mathcal{G})$, the *Cartan distribution* on $J^\infty(\pi;\mathcal{G})$. Let us describe the Cartan distribution on $J^\infty(\pi;\mathcal{G})$ from various points of view.

PROPOSITION 6.1. *At an arbitrary point* $\theta = \{[s_g]_{g(x)}^\infty\}_{g \in \mathcal{G}} \in J^\infty(\pi;\mathcal{G})$, *the Cartan plane* $\mathcal{C}_\theta$ *is the tangent plane to the graph of the section* $j_\infty(\{s_g\})$ *at the point* θ. *Namely, a vector* $X_\theta = \{X_x, X_{\theta_k^{\mathcal{G}_\alpha}}\}$ *lies in the plane* $\mathcal{C}_\theta$ *if and only if for any number* $k \geq 0$ *and any finite subset* $\mathcal{G}_\alpha \subset \mathcal{G}$ *the vector* $X_{\theta_k^{\mathcal{G}_\alpha}}$ *is tangent to the submanifold* $j_k(\{s_g\})(M) \subset J^k(\pi;\mathcal{G}_\alpha)$.

PROOF. If the vector $X_\theta = \{X_x, X_{\theta_k^{\mathcal{G}_\alpha}}\}$ at the point $\theta = \{x, \theta_k^{\mathcal{G}_\alpha}\}$ belongs to the plane $\mathcal{C}_\theta$, then, by the definition of the Cartan distribution, for any number $k \geq 0$ and a finite subset $\mathcal{G}_\alpha \subset \mathcal{G}$ the vector $X_{\theta_{k+1}^{\mathcal{G}_\alpha}}$ lies in the plane $\mathcal{C}_{\theta_{k+1}^{\mathcal{G}_\alpha}}^{k+1}$. Therefore, from (6.39) we obtain

$$X_{\theta_k^{\mathcal{G}_\alpha}} = (\pi_{k+1,k}^{\mathcal{G}_\alpha})_*(X_{\theta_{k+1}^{\mathcal{G}_\alpha}}) \in L_{\theta_k^{\mathcal{G}_\alpha}} = T_{\theta_k^{\mathcal{G}_\alpha}}(j_k(\{s_g\})(M)),$$

where $\theta = \{[s_g]_{g(x)}^\infty\}_{g \in \mathcal{G}}$ and $\theta_{k+1}^{\mathcal{G}_\alpha} = \{[s_g]_{g(x)}^{k+1}\}_{g \in \mathcal{G}_\alpha}$.

Conversely, if the vector $X_{\theta_k^{\mathcal{G}_\alpha}}$ is tangent to the submanifold $j_k(\{s_g\})(M)$, then it lies in the Cartan distribution on $J^k(\pi;\mathcal{G}_\alpha)$. Since this is true for all $k \geq 0$ and $\mathcal{G}_\alpha \subset \mathcal{G}$, we have $X_\theta = \{X_x, X_{\theta_k^{\mathcal{G}_\alpha}}\} \in \mathcal{C}_\theta$. $\square$

PROPOSITION 6.2. *The Cartan distribution on* $J^\infty(\pi;\mathcal{G})$ *determines a connection in the bundle* $\pi_\infty^{\mathcal{G}}$: *at any point* $\theta \in J^\infty(\pi;\mathcal{G})$ *the tangent mapping* $(\pi_\infty^{\mathcal{G}})_{*,\theta}$ *isomorphically takes the Cartan plane* $\mathcal{C}_\theta$ *to the tangent space* $T_x(M)$, *where* $x = \pi_\infty^{\mathcal{G}}(\theta)$.

PROOF. At any point $\theta = \{x, \theta_k^{\mathcal{G}_\alpha}\} \in J^\infty(\pi;\mathcal{G})$ we shall construct the inverse to the mapping $(\pi_\infty^{\mathcal{G}})_{*,\theta} \colon \mathcal{C}_\theta \to T_x(M)$. For any number $k \geq 0$ and a finite subset $\mathcal{G}_\alpha \subset \mathcal{G}$, the tangent mapping $(\pi_k^{\mathcal{G}_\alpha})_{*,\theta_k^{\mathcal{G}_\alpha}}$ isomorphically maps the plane $L_{\theta_{k+1}^{\mathcal{G}_\alpha}}$ to the space $T_x(M)$. Therefore, any vector $X_x \in T_x(M)$ uniquely determines a set of vectors $X_{\theta_k^{\mathcal{G}_\alpha}} \in L_{\theta_{k+1}^{\mathcal{G}_\alpha}}$, $k \geq 0$, $\mathcal{G}_\alpha \subset \mathcal{G}$, such that $(\pi_k^{\mathcal{G}_\alpha})_*(X_{\theta_k^{\mathcal{G}_\alpha}}) = X_x$. From uniqueness of this set and from (6.38) it follows that for any number $k \geq 0$ and a finite subset $\mathcal{G}_\beta \subset \mathcal{G}_\alpha$ of the set $\mathcal{G}$ the equalities

$$(\pi_{k,k-1}^{\mathcal{G}_\alpha})_*(X_{\theta_k^{\mathcal{G}_\alpha}}) = X_{\theta_{k-1}^{\mathcal{G}_\alpha}}, \qquad (\pi_k^{\mathcal{G}_\alpha,\mathcal{G}_\beta})_*(X_{\theta_k^{\mathcal{G}_\alpha}}) = X_{\theta_k^{\mathcal{G}_\beta}}$$

hold. Therefore, the set of vectors $\{X_x, X_{\theta_k^{\mathcal{G}_\alpha}}\}$ determines a tangent vector X_θ lying in the Cartan plane $\mathcal{C}_\theta$. From uniqueness of vectors $\{X_{\theta_k^{\mathcal{G}_\alpha}}\}$ it also follows that the above constructed mapping $X_x \mapsto X_\theta$ is inverse to the mapping $(\pi_\infty^{\mathcal{G}})_*|_{\mathcal{C}_\theta}$. $\square$

Let us now prove generalizations of Propositions 2.1–2.3 of Chapter 4.

PROPOSITION 6.3. *The set $CD(\pi; \mathcal{G})$ of vector fields lying in the Cartan distribution is spanned by the liftings to $J^\infty(\pi; \mathcal{G})$ of fields on the manifold M:*

$$CD(\pi; \mathcal{G}) = \{\varphi_1 \widehat{X_1} + \cdots + \varphi_l \widehat{X_l} \mid \varphi_i \in \mathcal{F}(\pi; \mathcal{G}),\ X_i \in D(M),\ l = 1, 2, \ldots\}.$$

PROOF. Let us prove that any vector of the form $\widehat{X}_\theta$, where $\theta = \{x, \theta_k^{\mathcal{G}_\alpha}\} \in J^\infty(\pi; \mathcal{G})$, $X \in D(M)$, lies in the plane $\mathcal{C}_\theta$. This is equivalent to the fact that any projection $\widehat{X}_{\theta_k^{\mathcal{G}_\alpha}}$ of the vector $\widehat{X}_\theta$ to the finite-dimensional space of jets $J^k(\pi; \mathcal{G}_\alpha)$ lies in the corresponding plane $L_{\theta_{k+1}^{\mathcal{G}_\alpha}}$. Consider an arbitrary function φ on $J^k(\pi; \mathcal{G}_\alpha)$ constant on the graph of the section $j_k(\{s_g\})$ whose $(k+1, \mathcal{G}_\alpha)$-jet equals $\theta_{k+1}^{\mathcal{G}_\alpha}$, where $\{s_g\}_{g \in \mathcal{G}_\alpha}$ is a family of sections of the bundle π. Then

$$\widehat{X}_\theta(\varphi) = \widehat{X}_{\theta_k^{\mathcal{G}_\alpha}}(\varphi) = X(j_k(\{s_g\})^*(\varphi))(x) = X(\text{const})(x) = 0.$$

Since this is true for any function φ of the chosen type, the vector $\widehat{X}_{\theta_k^{\mathcal{G}_\alpha}}$ is tangent to the graph of the section $j_k(\{s_g\})$ and consequently lies in $L_{\theta_{k+1}^{\mathcal{G}_\alpha}}$. Hence, $\widehat{X}_\theta \in \mathcal{C}_\theta$ and $\widehat{X} \in CD(\pi; \mathcal{G})$.

Consider now an arbitrary field X from $CD(\pi; \mathcal{G})$. By the definition of vector fields, there exists a number k such that $X(C^\infty(M)) \subset \mathcal{F}_k(\pi; \mathcal{G})$. Similarly to the differential case (see Chapter 4), the restriction X_M of the field X to $C^\infty(M)$ can be locally represented in the form $X_M = \varphi_1 X_1 + \cdots + \varphi_l X_l$, where $X_1, \ldots, X_l \in D(M)$, $\varphi_1, \ldots, \varphi_l \in \mathcal{F}_k(\pi; \mathcal{G})$. Let us show that the field X locally coincides with the field

$$\widehat{X}_M = \varphi_1 \widehat{X_1} + \cdots + \varphi_l \widehat{X_l} \in CD(\pi; \mathcal{G}).$$

To this end, consider an arbitrary point $\theta = \{x, \theta_k^{\mathcal{G}_\alpha}\} \in J^\infty(\pi; \mathcal{G})$, where the field $\widehat{X}_M$ is defined. The vectors X_θ and $\widehat{X}_{M,\theta}$ lie in $\mathcal{C}_\theta$ and, by the construction of the field $\widehat{X}_M$, their projections to M by the mapping $\pi_\infty^{\mathcal{G}}$ coincide. From this and from Proposition 6.2 it follows that $X_\theta = \widehat{X}_{M,\theta}$. Standard arguments using partition of unity [**115**, **146**] give global representation of the field $\varphi_1 \widehat{X_1} + \cdots + \varphi_l \widehat{X_l}$. □

PROPOSITION 6.4. *A field $X \in D(\pi; \mathcal{G})$ lies in the Cartan distribution on $J^\infty(\pi; \mathcal{G})$ if and only if, for any function $\varphi \in \mathcal{F}(\pi)$ and any mapping $g \in \mathcal{G}$,*

$$X \,\lrcorner\, (\widehat{g} \circ \pi_\infty^{\mathcal{G}, \mathcal{G}_g})^*(U_1(\varphi)) = 0,$$

where $U_1(\varphi)$ is the Cartan form on $J^\infty(\pi)$ and $\mathcal{G}_g$ is the set consisting of the two mappings id_M and g.

The forms $(\widehat{g} \circ \pi_\infty^{\mathcal{G}, \mathcal{G}_g})^*(U_1(\varphi))$ will be called the *Cartan forms* on $J^\infty(\pi; \mathcal{G})$. The module generated by these forms is called the *Cartan module* and is denoted by $C\Lambda^1(\pi; \mathcal{G})$.

PROOF. From (6.34) and (6.27) it follows that the mappings $\widehat{g}$ and $\pi_\infty^{\mathcal{G}, \mathcal{G}_g}$ preserve the Cartan distribution. Therefore, if $X_\theta \in \mathcal{C}_\theta$, $\theta \in J^\infty(\pi; \mathcal{G})$, then

$$X_\theta \,\lrcorner\, (\widehat{g} \circ \pi_\infty^{\mathcal{G}, \mathcal{G}_g})^*(U_1(\varphi)) = (\widehat{g} \circ \pi_\infty^{\mathcal{G}, \mathcal{G}_g})_*(X_\theta) \,\lrcorner\, U_1(\varphi) = 0.$$

Since the bundle $\pi_\infty^{\mathcal{G}}$ is the Whitney product of the bundles $g^*(\pi_\infty)$, $g \in \mathcal{G}$, the fiber $(\pi_\infty^{\mathcal{G}})^{-1}(x)$ is the direct product of the fibers $(g^*(\pi_\infty))^{-1}(x)$, $g \in \mathcal{G}$, for any point $x \in M$. On the other hand, by the definition of the bundle $g^*(\pi_\infty)$, the fiber $(g^*(\pi_\infty))^{-1}(x)$ coincides with the fiber $\pi_\infty^{-1}(g(x))$. Thus we obtain the equality

$$(\pi_\infty^{\mathcal{G}})^{-1}(x) = \prod_{g \in \mathcal{G}} \pi_\infty^{-1}(g(x)).$$

The mapping $(\widehat{g} \circ \pi_\infty^{\mathcal{G},\mathcal{G}_g})|_{(\pi_\infty^{\mathcal{G}})^{-1}(x)}$ is the projection to the component $\pi_\infty^{-1}(g(x))$ of this direct product. In fact, any point θ of the fiber $(\pi_\infty^{\mathcal{G}})^{-1}(x)$ is the set $\{\theta^g\}_{g \in \mathcal{G}}$ of infinite jets at the corresponding points $g(x)$, $g \in \mathcal{G}$, i.e., $\theta^g \in \pi_\infty^{-1}(g(x))$. The mapping $\pi_\infty^{\mathcal{G},\mathcal{G}_g}$ takes the point θ to the set consisting of two jets $\{\theta^{g_0}, \theta^g\}$, where $g_0 = \mathrm{id}_M$, and the mapping $\widehat{g}$ takes this set to the jet θ^g.

Thus, the linear space of vertical vectors of the bundle $\pi_\infty^{\mathcal{G}}$ at the point $\theta \in J^\infty(\pi;\mathcal{G})$ is the direct sum of the linear spaces of vertical vectors of the bundle π_∞ at the points $\theta^g \in J^\infty(\pi)$ over all $g \in \mathcal{G}$:

$$V_\theta(J^\infty(\pi;\mathcal{G})) \simeq \bigoplus_{g \in \mathcal{G}} V_{\theta^g}(J^\infty(\pi)),$$

where V_θ denotes the space of vertical fields at the point θ of the corresponding jet bundle.

From Proposition 6.2 it follows that the tangent space to the manifold $J^\infty(\pi;\mathcal{G})$ at the point θ is the direct sum of the space of vertical vectors and the Cartan plane $\mathcal{C}_\theta$. We obtain the splitting

$$T_\theta(J^\infty(\pi;\mathcal{G})) \simeq \left(\bigoplus_{g \in \mathcal{G}} V_{\theta^g}(J^\infty(\pi))\right) \oplus \mathcal{C}_\theta, \qquad \theta = \{\theta^g\}_{g \in \mathcal{G}}. \tag{6.41}$$

The tangent mapping $(\widehat{g} \circ \pi_\infty^{\mathcal{G},\mathcal{G}_g})_*$ isomorphically maps the component $\mathcal{C}_\theta$ to the Cartan plane $\mathcal{C}_{\theta^g} \subset T_{\theta^g}(J^\infty(\pi))$; the component $V_{\theta^g}(J^\infty(\pi))$ is mapped identically to itself, and all other components are taken to zero.

Consider now an arbitrary tangent vector X_θ to the manifold $J^\infty(\pi;\mathcal{G})$ at the point θ and its splitting with respect to (6.41): $X_\theta = \sum_{g \in \mathcal{G}} X_{\theta^g} + Y_\theta$, where $X_{\theta^g} \in V_{\theta^g}(J^\infty(\pi))$, $Y_\theta \in \mathcal{C}_\theta$. If the vector X_θ does not lie in the Cartan distribution, then there exists at least one mapping $g_1 \in \mathcal{G}$ such that the vertical vector $X_{\theta^{g_1}}$ does not vanish. Then from Proposition 2.1 of Chapter 4 it follows that there exists a function $\varphi \in \mathcal{F}(\pi)$ such that $X_{\theta^{g_1}} \lrcorner U_1(\varphi) \neq 0$. Hence we have

$$X_\theta \lrcorner (\widehat{g_1} \circ \pi_\infty^{\mathcal{G},\mathcal{G}_{g_1}})^*(U_1(\varphi))$$
$$= \left(\sum_{g \in \mathcal{G}} X_{\theta^g}\right) \lrcorner (\widehat{g_1} \circ \pi_\infty^{\mathcal{G},\mathcal{G}_{g_1}})^*(U_1(\varphi)) + Y_\theta \lrcorner (\widehat{g_1} \circ \pi_\infty^{\mathcal{G},\mathcal{G}_{g_1}})^*(U_1(\varphi))$$
$$= (\widehat{g_1} \circ \pi_\infty^{\mathcal{G},\mathcal{G}_{g_1}})_* \left(\sum_{g \in \mathcal{G}} X_{\theta^g}\right) \lrcorner U_1(\varphi) + 0 = X_{\theta^{g_1}} \lrcorner U_1(\varphi) \neq 0,$$

i.e., X_θ does not satisfy our identity in this case. □

PROPOSITION 6.5. *Any maximal integral manifold of the Cartan distribution on $J^\infty(\pi;\mathcal{G})$ is locally the graph of infinite jets for a family $\{s_g\}_{g \in \mathcal{G}}$ of sections of the bundle π.*

PROOF. The proof is similar to that of Proposition 2.3 of Chapter 4. Namely, let $\mathcal{R}^\infty$ be a maximal integral manifold of the Cartan distribution on $J^\infty(\pi;\mathcal{G})$.

The projections
$$\pi^{\mathcal{G}}_{\infty,k}|_{\mathcal{R}^\infty}: \mathcal{R}^\infty \longrightarrow \mathcal{R}^k \stackrel{\text{def}}{=} \pi^{\mathcal{G}}_{\infty,k}(\mathcal{R}^\infty) \subset J^k(\pi;\mathcal{G}), \qquad k=0,1,\ldots,$$
and
$$\pi^{\mathcal{G}}_{\infty}|_{\mathcal{R}^\infty}: \mathcal{R}^\infty \longrightarrow \mathcal{R} \stackrel{\text{def}}{=} \pi^{\mathcal{G}}_{\infty}(\mathcal{R}^\infty) \subset M$$
are local diffeomorphisms, since the Cartan planes $\mathcal{C}_\theta$, $\theta \in J^\infty(\pi;\mathcal{G})$ (and consequently, the tangent planes to $\mathcal{R}^\infty$) do not contain nontrivial vertical vectors (see Proposition 6.2). Therefore, the projection $\pi^{\mathcal{G}}_0|_{\mathcal{R}^0}: \mathcal{R}^0 \to \mathcal{R}$ is also a local diffeomorphism. Hence, there exists a mapping $s': \mathcal{R} \to \mathcal{R}^0$ such that $\pi^{\mathcal{G}}_0|_{\mathcal{R}^0} \circ s'$ is the identity. Let us extend the mapping s', as in the differential case (see Chapter 4), to a section $\tilde{s}$ of the bundle $\pi^{\mathcal{G}}_0$. Since $\pi^{\mathcal{G}}_0$ is the Whitney product of the bundles $g^*(\pi_0) = g^*(\pi)$, $g \in \mathcal{G}$, the section $\tilde{s}$ of the bundle $\pi^{\mathcal{G}}_0$ is the set $\{s_g\}_{g \in \mathcal{G}}$ of sections of the bundle π. The manifolds $j_k(\{s_g\})(M) \subset J^k(\pi;\mathcal{G})$ for $k=0,1,\ldots,\infty$ are integral manifolds of the Cartan distribution and contain the manifolds $\mathcal{R}^k$ for the corresponding k. But since $\mathcal{R}^\infty$ is a maximal integral manifold, we obtain that at least locally the manifolds $\mathcal{R}^\infty$ and $j_\infty(\{s_g\})(M)$ coincide. $\square$

From this proposition it follows that besides the manifolds $j_\infty(s)(M)$, which correspond to solutions of equations $\mathcal{E} \subset J^k(\pi;\mathcal{G})$, other maximal integral manifolds of the Cartan distribution on $J^\infty(\pi;\mathcal{G})$ may exist. The following proposition allows us to distinguish the needed ones among all the other maximal integral manifolds.

PROPOSITION 6.6. *Let the set $\mathcal{G}$ be a semigroup of mappings of the manifold M, and let N be the graph of a section of the bundle $\pi^{\mathcal{G}}_\infty$. Then N is a graph of the form $j_\infty(s)$, $s \in \Gamma(\pi)$, if and only if N is a maximal integral manifold of the Cartan distribution on $J^\infty(\pi;\mathcal{G})$ invariant with respect to any mapping $\hat{g}$, $g \in \mathcal{G}$, i.e., such that $\hat{g}(N) \subset N$.*

PROOF. From the definition of the Cartan distribution on $J^\infty(\pi;\mathcal{G})$ it follows that the manifold $N = j_\infty(s)(M)$ is a maximal integral manifold of this distribution. Besides, if $\theta = [s]_x^{(\infty,\mathcal{G})} \in N$, then $\hat{g}(\theta) = [s]_{g(x)}^{(\infty,\mathcal{G})} \in N$.

Now let N be a maximal integral manifold invariant with respect to $\mathcal{G}$. Since N is the graph of a section of the bundle $\pi^{\mathcal{G}}_\infty$, we have that $\mathcal{R} = \pi^{\mathcal{G}}_\infty(N) = M$ and s' is a section of the bundle $\pi^{\mathcal{G}}_0$ defined at all points of M (see the proof of the previous proposition). Therefore, N is the graph of a section of the form $j_\infty(\{s_g\})$, where $\{s_g\}_{g \in \mathcal{G}}$ is a family of sections of the bundle π determined by the section s'.

Consider an arbitrary mapping $g_1 \in \mathcal{G}$, an arbitrary point x of the manifold M, and the corresponding point $\theta = j_\infty(\{s_g\})(x) = \{[s_g]_{g(x)}^\infty\}_{g \in \mathcal{G}}$ of the manifold N. By the assumption, the point $\hat{g}_1(\theta)$, which, by the definition of $\hat{g}_1$, is the infinite jet at the point $x_1 = g_1(x)$, lies in the manifold $N = j_\infty(\{s_g\})(M)$. Therefore, $\hat{g}_1(\theta) = j_\infty(\{s_g\})(x_1)$. Hence $\{[s_{g \circ g_1}]_{g(x_1)}^\infty\}_{g \in \mathcal{G}} = \{[s_g]_{g(x_1)}^\infty\}_{g \in \mathcal{G}}$, and consequently
$$[s_{g \circ g_1}]_{(g \circ g_1)(x)}^\infty = [s_g]_{(g \circ g_1)(x)}^\infty$$
for any $g \in \mathcal{G}$. Substituting $g = \text{id}_M$ into this equality and denoting s_{id_M} by s, we obtain the equality $[s_{g_1}]_{g_1(x)}^\infty = [s]_{g_1(x)}^\infty$ for any $g_1 \in \mathcal{G}$. Consequently,
$$\theta = \{[s]_{g(x)}^\infty\}_{g \in \mathcal{G}} = j_\infty(s)(x).$$
Since this is valid for any $x \in M$, we have $N = j_\infty(s)(M)$. $\square$

Everywhere below we assume that $\mathcal{G}$ is a semigroup of mappings of the manifold M.

6.5. $\mathcal{G}$-invariant symmetries of the Cartan distribution on $J^\infty(\pi;\mathcal{G})$.
Let us describe infinitesimal transformations of the space $J^\infty(\pi;\mathcal{G})$ preserving maximal integral manifolds of the form $j_\infty(s)(M)$. Note that any manifold $j_\infty(s)(M)$ lies in $J^\infty(\pi;\mathcal{G})_0$. Therefore, in general, such a transformation is a transformation of the set $J^\infty(\pi;\mathcal{G})_0$. But for the reasons set forth in Remark 6.2 we shall assume that our transformation X is extended to the entire space $J^\infty(\pi;\mathcal{G})$. From Proposition 6.1 it follows that tangent planes to manifolds of the form $j_\infty(s)(M)$ are the Cartan planes at points of the set $J^\infty(\pi;\mathcal{G})_0$. Consequently (see Chapter 4), at points of the set $J^\infty(\pi;\mathcal{G})_0$ we have the embedding

$$X(\mathcal{C}\Lambda^1(\pi;\mathcal{G})) \subset \mathcal{C}\Lambda^1(\pi;\mathcal{G}). \tag{6.42}$$

On the other hand, a field lying in $\mathcal{CD}(\pi;\mathcal{G})$ is tangent to all integral manifolds of the distribution $\mathcal{C}(\pi;\mathcal{G})$. Therefore, it is tangent to manifolds of the form $j_\infty(s)(M)$ and hence is the desired but "trivial" transformation (cf. Chapter 4). Therefore it suffices to find vertical infinitesimal transformations only, that preserve the class of manifolds of the form $j_\infty(s)(M)$.

Let X be such a vertical field, let θ be an arbitrary point of the manifold $N = j_\infty(s)(M)$, let $\mathcal{G}$ be a semigroup of mappings of the manifold M, and let $g \in \mathcal{G}$. Assume also that the field X possesses a one-parameter group of diffeomorphisms $\{A_t\}$. Any manifold $N_t = A_t(N)$ is of the form $j_\infty(s_t)(M)$, $s_t \in \Gamma(\pi)$, and hence $\widehat{g}(N_t) \subset N_t$ (see Proposition 6.6). Hence, $\widehat{g}(A_t(\theta)) \in N_t$. Since $\widehat{g}(\theta) \in \widehat{g}(N) \subset N$, the point $A_t(\widehat{g}(\theta))$ also lies in N_t. Moreover, since X is a vertical field, the mapping A_t preserves every fiber of the bundle $\pi_\infty^\mathcal{G}$. From this and from the commutative diagram (6.32) we have

$$\pi_\infty^\mathcal{G}(A_t(\widehat{g}(\theta))) = \pi_\infty^\mathcal{G}(\widehat{g}(\theta)) = g(\pi_\infty^\mathcal{G}(\theta)),$$

as well as

$$\pi_\infty^\mathcal{G}(\widehat{g}(A_t(\theta))) = g(\pi_\infty^\mathcal{G}(A_t(\theta))) = g(\pi_\infty^\mathcal{G}(\theta)).$$

Therefore, the points $A_t(\widehat{g}(\theta))$ and $\widehat{g}(A_t(\theta))$ are projected to the same point under the mapping $\pi_\infty^\mathcal{G}$ and lie on the graph N_t of a section of the bundle $\pi_\infty^\mathcal{G}$. Hence, these points coincide. But since N is an arbitrary manifold of the form $j_\infty(s)(M)$, we obtain that θ is an arbitrary point of the set $J^\infty(\pi;\mathcal{G})_0$. Consequently, at points of the set $J^\infty(\pi;\mathcal{G})_0$ we have the equality $A_t \circ \widehat{g} = \widehat{g} \circ A_t$ for sufficiently small t, while for any function $\varphi \in \mathcal{F}(\pi;\mathcal{G})$ the equality $\widehat{g}^*(A_t^*(\varphi)) = A_t^*(\widehat{g}^*(\varphi))$ holds. Differentiating the last equality with respect to t at $t = 0$, we obtain the formula $\widehat{g}^*(X(\varphi)) = X(\widehat{g}^*(\varphi))$, which is valid at points of the set $J^\infty(\pi;\mathcal{G})_0$ for any function $\varphi \in \mathcal{F}(\pi;\mathcal{G})$.

Thus, a vertical vector field X preserving the class of submanifolds of the form $j_\infty(s)(M)$, $s \in \Gamma(\pi)$, and possessing the one-parameter group of diffeomorphisms should satisfy condition (6.42) and commute with any homomorphism $\widehat{g}^*$, $g \in \mathcal{G}$. These conditions are to be satisfied at points of the set $J^\infty(\pi;\mathcal{G})_0$ only. But if the closure of the set $J^\infty(\pi;\mathcal{G})_0$ coincides with the entire space $J^\infty(\pi;\mathcal{G})$ (and in all examples under consideration the situation is exactly this), then the field X satisfies these conditions at all point of the space $J^\infty(\pi;\mathcal{G})$. This fact explains why we must consider vertical fields X on $J^\infty(\pi;\mathcal{G})$ satisfying conditions (6.42) and such that $\widehat{g}^* \circ X = X \circ \widehat{g}^*$ for any $g \in \mathcal{G}$. We shall call such fields $\mathcal{G}$-*invariant* (*infinitesimal*) *symmetries* of the Cartan distribution on $J^\infty(\pi;\mathcal{G})$. Denote by $\mathrm{sym}^{\mathcal{G}}\,\mathcal{C}(\pi)$ the set of such fields.

PROPOSITION 6.7. *Any vector field* $X \in \mathrm{sym}^{\mathcal{G}}\,\mathcal{C}(\pi)$ *is uniquely determined by its restriction to the subalgebra* $\mathcal{F}_0(\pi) \subset \mathcal{F}(\pi;\mathcal{G})$.

PROOF. An automorphism X of the Cartan distribution on $J^\infty(\pi;\mathcal{G})$ vanishing on the subalgebra $\mathcal{F}_0(\pi) \subset \mathcal{F}(\pi;\mathcal{G})$ equals zero on the subalgebra $\mathcal{F}(\pi) \subset \mathcal{F}(\pi;\mathcal{G})$ as well. This fact is proved in exactly the same way as Proposition 2.4 of Chapter 4 was proved. The fact that X is a derivation of the algebra $\mathcal{F}(\pi;\mathcal{G})$, but not of $\mathcal{F}(\pi)$, is inessential.

Let $X \in \mathrm{sym}^{\mathcal{G}}\,\mathcal{C}(\pi)$ and $X|_{\mathcal{F}(\pi)} = 0$. Consider in M an arbitrary coordinate neighborhood and the corresponding local coordinates $x_1,\ldots,x_n,\ldots,p^j_{\sigma g},\ldots$ in $J^\infty(\pi;\mathcal{G})$. If $g_0 = \mathrm{id}_M$, then $p^j_{\sigma g_0} \in \mathcal{F}(\pi)$ and $X(p^j_{\sigma g_0}) = 0$. If $g \neq g_0$, then

$$X(p^j_{\sigma g}) = X(\widehat{g}^*(p^j_{\sigma g_0})) = \widehat{g}^*(X(p^j_{\sigma g_0})) = 0,$$

and consequently $X = 0$. □

Let $X \in \mathrm{sym}^{\mathcal{G}}\,\mathcal{C}(\pi)$. Then, as in the differential case (see Chapter 4), the restriction $X_0 = X|_{\mathcal{F}_0(\pi)}$ is a derivation of the algebra $\mathcal{F}_0(\pi)$ with values in the algebra $\mathcal{F}_k(\pi;\mathcal{G}_\alpha)$ for some number $k \geq 0$ and a finite subset $\mathcal{G}_\alpha \subset \mathcal{G}$. Therefore, at any point $\theta_k \in J_k(\pi;\mathcal{G}_\alpha)$ the vector X_{0,θ_k} is identified with a tangent vector to the fiber $\pi^{-1}(x)$ of the bundle π, where $x = \pi_k^{\mathcal{G}_\alpha}(\theta_k)$. Since π is a linear bundle, tangent vectors to its fibers are identified with points of these fibers. Hence, the vector X_{0,θ_k} identifies with a point of the fiber $\pi^{-1}(x)$ while the derivation X_0 identifies with a section of the bundle $(\pi_k^{\mathcal{G}_\alpha})^*(\pi)$. Thus we obtain the mapping

$$\mathrm{sym}^{\mathcal{G}}\,\mathcal{C}(\pi) \longrightarrow \mathcal{F}(\pi,\pi;\mathcal{G}). \tag{6.43}$$

Injectivity of this mapping is implied by Proposition 6.7. Let us prove surjectivity.

Let φ be a section of the bundle $(\pi_k^{\mathcal{G}_\alpha})^*(\pi)$. Consider a coordinate neighborhood $\mathcal{U} \subset M$ and the corresponding neighborhood $\mathcal{U}^\infty = (\pi_\infty^{\mathcal{G}})^{-1}(\mathcal{U}) \subset J^\infty(\pi;\mathcal{G})$. Let us define a vector field in the neighborhood $\mathcal{U}^\infty$ by setting

$$\eth_{\varphi,\mathcal{U}} = \sum_{j,\sigma,g} \widehat{g}^*(D_\sigma(\varphi^j)) \frac{\partial}{\partial p^j_{\sigma g}},$$

where φ^j is the j-th component of the restriction of the section φ to the neighborhood $\mathcal{U}^\infty$, and D_σ is the composition of total derivatives corresponding to the multi-index σ. Let us show that the derivation $\eth_{\varphi,\mathcal{U}}$ in the neighborhood $\mathcal{U}^\infty$ is a $\mathcal{G}$-symmetry of the Cartan distribution on $J^\infty(\pi;\mathcal{G})$.

The field $\eth_{\varphi,\mathcal{U}}$ is vertical. Let us prove that it commutes with any homomorphism $\widehat{g}_1^*$, $g_1 \in \mathcal{G}$. The function $\widehat{g}_1^*(x_i) = g_1^*(x_i)$ is a function on M for any $i = 1,\ldots,n$, and consequently

$$\eth_{\varphi,\mathcal{U}}(\widehat{g}_1^*(x_i)) = 0 = \widehat{g}_1^*(\eth_{\varphi,\mathcal{U}}(x_i)).$$

For the function $p^j_{\sigma g}$, using formula (6.35), we obtain

$$\eth_{\varphi,\mathcal{U}}(\widehat{g}_1^*(p^j_{\sigma g})) = \eth_{\varphi,\mathcal{U}}(p^j_{\sigma,g \circ g_1}) = (\widehat{g \circ g_1})^*(D_\sigma(\varphi^j))$$
$$= (\widehat{g}_1^* \circ \widehat{g}^*)(D_\sigma(\varphi^j)) = \widehat{g}_1^*(\eth_{\varphi,\mathcal{U}}(p^j_{\sigma g})),$$

i.e., the equality $\eth_{\varphi,\mathcal{U}}(\widehat{g}_1^*(\psi)) = \widehat{g}_1^*(\eth_{\varphi,\mathcal{U}}(\psi))$ holds for any coordinate function ψ on $\mathcal{U}^\infty$ and consequently for any function on $\mathcal{U}^\infty$.

Consider the Cartan form
$$\omega = (\widehat{g_1} \circ \pi_\infty^{\mathcal{G},\mathcal{G}_{g_1}})^*(U_1(\psi)),$$
where $g_1 \in \mathcal{G}$, $\psi \in \mathcal{F}(\pi)$. From the commutative diagram (6.33), where $k = \infty$, $\mathcal{G}' = \{\mathrm{id}_M\}$, and $\mathcal{G} \circ \mathcal{G}_{g_1} = \mathcal{G}$, due to the fact that $\mathcal{G}$ is a semigroup and $g_1 \in \mathcal{G}$, we obtain
$$\omega = (\pi_\infty^{\mathcal{G},\{\mathrm{id}_M\}} \circ \widehat{g_1})^*(U_1(\psi)).$$
Since $(\pi_\infty^{\mathcal{G},\{\mathrm{id}_M\}})^*$ is an embedding of the module $\Lambda^1(\pi)$ in $\Lambda^1(\pi;\mathcal{G})$, we have $\omega = \widehat{g_1}^*(U_1(\psi))$. Thus, we obtain
$$\eth_{\varphi,\mathcal{U}}(\omega) = \eth_{\varphi,\mathcal{U}}(\widehat{g_1}^*(U_1(\psi))) = \widehat{g_1}^*(\eth_{\varphi,\mathcal{U}}(U_1(\psi)))$$
$$= \widehat{g_1}^*\left(d\eth_{\varphi,\mathcal{U}}(\psi) - \sum_i \eth_{\varphi,\mathcal{U}}(D_i(\psi))\, dx_i\right).$$
Using calculations similar to those in the differential case (see Chapter 4) and the fact that $\psi \in \mathcal{F}(\pi)$, it is easy to prove the equalities
$$\eth_{\varphi,\mathcal{U}}(D_i(\psi)) = D_i(\eth_{\varphi,\mathcal{U}}(\psi)), \qquad i = 1,\ldots,n.$$
Therefore,
$$\eth_{\varphi,\mathcal{U}}(\omega) = \widehat{g_1}^*\left(d\eth_{\varphi,\mathcal{U}}(\psi) - \sum_i D_i(\eth_{\varphi,\mathcal{U}}(\psi))\, dx_i\right).$$
The form $\Omega = d\eth_{\varphi,\mathcal{U}}(\psi) - \sum_i D_i(\eth_{\varphi,\mathcal{U}}(\psi))\, dx_i$ is an element of the Cartan module, since elements of this module, and they only, vanish on the Cartan distribution (this is easily derived from Proposition 6.4) and $D_j \lrcorner \Omega = 0$, $j = 1, \ldots, n$. The mapping $\widehat{g_1}$ preserves the Cartan distribution (see formula (6.34)), and therefore the induced mapping $\widehat{g_1}^*$ preserves the Cartan module. This means that
$$\eth_{\varphi,\mathcal{U}}(\omega) = \widehat{g_1}^*(\Omega) \in \mathcal{C}\Lambda^1(\pi;\mathcal{G}) \quad\text{and}\quad \eth_{\varphi,\mathcal{U}}(\mathcal{C}\Lambda^1(\pi;\mathcal{G})) \subset \mathcal{C}\Lambda^1(\pi;\mathcal{G}).$$

Since the restriction of the field $\eth_{\varphi,\mathcal{U}}$ to the subalgebra $\mathcal{F}_0(\pi|_\mathcal{U})$ is the derivation $\varphi^1 \partial/\partial p_0^1 + \cdots + \varphi^m \partial/\partial p_0^m$, the field $\eth_{\varphi,\mathcal{U}}$ is taken to the section $\varphi|_\mathcal{U}$ by the mapping (6.43). Now let $\mathcal{U}, \mathcal{U}' \subset M$ be two coordinate neighborhoods in M. Then the restrictions of the fields $\eth_{\varphi,\mathcal{U}}$ and $\eth_{\varphi,\mathcal{U}'}$ to the algebra $\mathcal{F}_0(\pi|_{\mathcal{U}\cap\mathcal{U}'})$ coincide. Thus, by Proposition 6.7, we see that these field coincide in the neighborhood $(\pi_\infty^\mathcal{G})^{-1}(\mathcal{U}\cap\mathcal{U}')$. Thus, as in the differential case, any section φ of the bundle $(\pi_k^\mathcal{G})^*(\pi)$ determines the field $\eth_\varphi \in \mathrm{sym}^\mathcal{G} \mathcal{C}(\pi)$. Therefore, we have proved the following result:

THEOREM 6.8. *Any $\mathcal{G}$-symmetry of the Cartan distribution on $J^\infty(\pi;\mathcal{G})$ is of the form $\eth_\varphi$, where $\varphi \in \mathcal{F}(\pi,\pi;\mathcal{G})$. The algebra $\mathrm{sym}^\mathcal{G} \mathcal{C}(\pi)$ is identified with the module $\mathcal{F}(\pi,\pi;\mathcal{G})$.*

As in the differential case, the derivations $\eth_\varphi$, $\varphi \in \mathcal{F}(\pi,\pi;\mathcal{G})$, are called *evolutionary derivations*. They determine evolution of sections of the bundle π governed by the evolution equations
$$\frac{\partial u^j}{\partial t} = \varphi^j\left(x,\ldots,g^*\left(\frac{\partial^{|\sigma|} u^l}{\partial x^\sigma}\right),\ldots\right), \qquad j = 1,\ldots,m.$$

The field $\eth_\varphi$, $\varphi \in \mathcal{F}(\pi,\pi;\mathcal{G}')$, may be understood both as a field on $J^\infty(\pi;\mathcal{G}')$, and as a field on any jet space $J^\infty(\pi;\mathcal{G})$, where $\mathcal{G}$ is a semigroup of mappings such that $\mathcal{G}' \subset \mathcal{G}$. If it is necessary to stress that the field ∂_φ acts on $J^\infty(\pi;\mathcal{G})$, we shall use the notation $\partial_\varphi^\mathcal{G}$.

PROPOSITION 6.9. *Assume that $\mathcal{G}'$ is a subsemigroup in the semigroup $\mathcal{G}$, and let $\varphi \in \mathcal{F}(\pi,\pi;\mathcal{G}')$. Then the field $\partial_\varphi^\mathcal{G}$ projects to the filed $\partial_\varphi^{\mathcal{G}'}$ under the mapping $\pi_\infty^{\mathcal{G},\mathcal{G}'}$.*

PROOF. The induced mapping $(\pi_\infty^{\mathcal{G},\mathcal{G}'})^*$ embeds the algebra $\mathcal{F}(\pi;\mathcal{G}')$ in $\mathcal{F}(\pi;\mathcal{G})$. From the definition of the fields $\partial_\varphi^\mathcal{G}$ and $\partial_\varphi^{\mathcal{G}'}$ it follows that they coincide on the algebra $\mathcal{F}(\pi;\mathcal{G}')$. Hence, at any point $\theta \in J^\infty(\pi;\mathcal{G})$, for an arbitrary function $f \in \mathcal{F}(\pi;\mathcal{G}')$ we have

$$((\pi_\infty^{\mathcal{G},\mathcal{G}'})_*(\partial_\varphi^\mathcal{G}|_\theta))(f) = \partial_\varphi^\mathcal{G}|_\theta((\pi_\infty^{\mathcal{G},\mathcal{G}'})^*(f)) = \partial_\varphi^{\mathcal{G}'}|_{\theta'}(f),$$

where $\theta' = \pi_\infty^{\mathcal{G},\mathcal{G}'}(\theta) \in J^\infty(\pi;\mathcal{G}')$. Consequently, $(\pi_\infty^{\mathcal{G},\mathcal{G}'})_*(\partial_\varphi^\mathcal{G}|_\theta) = \partial_\varphi^{\mathcal{G}'}|_{\theta'}$. □

From Proposition 6.9 it follows, in particular, that $\mathcal{G}$-invariant symmetries of the Cartan distribution on $J^\infty(\pi;\mathcal{G})$ correspond to classical symmetries of the Cartan distribution on $J^\infty(\pi)$ for any semigroup $\mathcal{G}$. These symmetries will be called *classical symmetries* of the Cartan distribution on $J^\infty(\pi;\mathcal{G})$.

The *Jacobi bracket* $\{\varphi,\psi\}$ of two sections $\varphi,\psi \in \mathcal{F}(\pi,\pi;\mathcal{G})$ is defined by the formula

$$\partial_{\{\varphi,\psi\}} = [\partial_\varphi, \partial_\psi].$$

The equality $\{\varphi,\psi\} = \partial_\varphi(\psi) - \partial_\psi(\varphi)$ holds for the Jacobi bracket, which in local coordinates can be rewritten as

$$\{\varphi,\psi\}^j = \sum_{\alpha,\sigma,g} \left(\widehat{g}^*(D_\sigma(\varphi^\alpha))\frac{\partial \psi^j}{\partial p_{\sigma g}^\alpha} - \widehat{g}^*(D_\sigma(\psi^\alpha))\frac{\partial \varphi^j}{\partial p_{\sigma g}^\alpha}\right), \quad j=1,\ldots,m.$$

EXERCISE 6.9. Prove the last two equalities.

The formula

$$l_\psi(\varphi) = \partial_\varphi(\psi)$$

defines the *universal linearization operator* l_ψ of the nonlinear boundary differential operator Δ_ψ, $\psi \in \mathcal{F}(\pi;\mathcal{G})$. The operator l_ψ acts from $\mathcal{F}(\pi,\pi;\mathcal{G})$ to $\mathcal{F}(\pi;\mathcal{G})$ and its properties are similar to those in the differential case (see Chapter 4). In addition, the formula

$$l_{\widehat{g}^*(\psi)} = \widehat{g}^* \circ l_\psi$$

is easily proved.

EXERCISE 6.10. State and prove the above mentioned properties of the operator l_ψ.

6.6. Higher symmetries of boundary differential equations. Let us first introduce the notion of infinite prolongation for boundary differential equations. Let $\mathcal{E} \subset J^k(\pi;\mathcal{G})$ be an equation of order k, and let $\mathcal{G}$ be a semigroup. The *l-prolongation* of the equation $\mathcal{E}$ is the set $\mathcal{E}^{(l)} \subset J^{k+l}(\pi;\mathcal{G})$ consisting of points $\theta_{k+l} = \{[s_g]_{g(x)}^{k+l}\}_{g \in \mathcal{G}}$ such that all points $\widehat{g}(\theta_k)$, $g \in \mathcal{G}$, where $\theta_k = \pi_{k+l,k}^\mathcal{G}(\theta_{k+l})$, lie in $\mathcal{E}$ while the graph of the section $j_k(\{s_g\})$ is tangent to the equation $\mathcal{E}$ with order $\geq l$ at all points $\widehat{g}(\theta_k)$, $g \in \mathcal{G}$.

REMARK 6.4. The set $\mathcal{E}^{(0)}$ is always contained in $\mathcal{E}$, though, contrary to the differential case, it may not coincide with $\mathcal{E}$. Besides, the equation $\mathcal{E}$ may not be invariant with respect to the mappings $\widehat{g}$, $g \in \mathcal{G}$, but any prolongation $\mathcal{E}^{(l)}$ of this equation, including $\mathcal{E}^{(0)}$, is invariants with respect to these mappings.

If the equation $\mathcal{E}$ is determined by the system
$$G_j\left(x, \ldots, g^*\left(\frac{\partial^{|\sigma|} s^l}{\partial x^\sigma}\right), \ldots\right) = 0, \qquad j = 1, \ldots, r,$$
where s^l, $l = 1, \ldots, m$, are the components of the section $s \in \Gamma(\pi)$, $|\sigma| \leq k$, $g \in \mathcal{G}$, then its zero prolongation $\mathcal{E}^{(0)}$ is described by the system
$$G_j\left(g_1{}^*(x), \ldots, (g \circ g_1)^*\left(\frac{\partial^{|\sigma|} s^l}{\partial x^\sigma}\right), \ldots\right) = 0, \qquad j = 1, \ldots, r, \quad g_1 \in \mathcal{G},$$
while the l-prolongation $\mathcal{E}^{(l)}$ is given by the equations
$$\widehat{g_1}^*(D_\sigma(G_j)) = 0, \qquad j = 1, \ldots, r, \quad g_1 \in \mathcal{G}, \quad |\sigma| \leq l,$$
where D_σ is the composition of total derivatives corresponding to the multi-index σ. Thus, to obtain all consequences of the equation $\mathcal{E} \subset J^k(\pi; \mathcal{G})$ we must use both the total derivative operators D_i, $i = 1, \ldots, n$, and the action of the homomorphisms $\widehat{g_1}^*$, $g_1 \in \mathcal{G}$.

EXERCISE 6.11. Show that if the equation $\mathcal{E} \subset J^k(\pi; \mathcal{G})$ is given by a boundary differential operator $\Delta\colon \Gamma(\pi) \to \Gamma(\xi)$, then the l-prolongation $\mathcal{E}^{(l)} \subset J^{k+l}(\pi; \mathcal{G})$ corresponds to the $(l, \mathcal{G})$-prolongation $\Delta_l^{\mathcal{G}} = j_l^{\mathcal{G}} \circ \Delta\colon \Gamma(\pi) \to \Gamma(\xi_l^{\mathcal{G}})$ of this operator.

Similarly to the differential case (see Chapter 4), the set $\mathcal{E}^{(l+1)}$ projects to the set $\mathcal{E}^{(l)}$ by the mapping $\pi_{k+l+1,k+l}^{\mathcal{G}}$ for any l. The inverse limit of the tower of equations $\mathcal{E}^{(l)}$, $l \geq 0$, together with the projections $\pi_{k+l+1,k+l}^{\mathcal{G}}$, is called the *infinite prolongation* of the equation $\mathcal{E}$ and is denoted by $\mathcal{E}^\infty$. A point $\theta = [s]_x^{(\infty, \mathcal{G})} \in J^\infty(\pi; \mathcal{G})_0$ belongs to the set $\mathcal{E}^\infty$ if and only if the Taylor series of the section s at the points $g(x) \in M$, $g \in \mathcal{G}$, satisfy the equation $\mathcal{E}$. Therefore, points of the set $\mathcal{E}^\infty \cap J^\infty(\pi; \mathcal{G})_0$ are naturally called *formal solutions* of the boundary differential equation $\mathcal{E}$.

The set of restrictions to $\mathcal{E}^{(l)}$ of smooth functions on the enveloping space $J^{k+l}(\pi; \mathcal{G})$ will be denoted by $\mathcal{F}_l(\mathcal{E})$. For any $l \geq 0$ the embedding $\mathcal{F}_l(\mathcal{E}) \subset \mathcal{F}_{l+1}(\mathcal{E})$ is defined. Elements of the set $\mathcal{F}(\mathcal{E}) = \bigcup_{l=0}^\infty \mathcal{F}_l(\mathcal{E})$ will be called smooth functions on the equation $\mathcal{E}^\infty$. The set $\mathcal{F}(\mathcal{E})$ is an algebra, because the sets $\mathcal{F}_l(\mathcal{E})$ are algebras.

If $\theta = \{[s_g]_{g(x)}^\infty\}_{g \in \mathcal{G}} \in \mathcal{E}^\infty$, then the graph of the section $j_\infty(\{s_g\})$ is tangent to the set $\mathcal{E}^\infty$ at the point θ. Hence, by Proposition 6.1, the Cartan plane $\mathcal{C}_\theta$ at the point $\theta \in \mathcal{E}^\infty$ lies in the tangent space $T_\theta(\mathcal{E}^\infty)$. The distribution $\mathcal{C}_\theta(\mathcal{E}) = \mathcal{C}_\theta$, $\theta \in \mathcal{E}^\infty$, is called the *Cartan distribution* on $\mathcal{E}^\infty$. From this definition it follows that integral manifolds of the distribution $\mathcal{C}(\mathcal{E})$ are integral manifolds of the Cartan distribution on $J^\infty(\pi; \mathcal{G})$. Besides, since any prolongation of the equation $\mathcal{E}$ is invariant with respect to any mapping $\widehat{g}$, $g \in \mathcal{G}$, this condition is also satisfied by the infinite prolongation $\mathcal{E}^\infty$. From this and from Proposition 6.6 we obtain the following statement:

PROPOSITION 6.10. *The graph of a section $\pi_\infty^{\mathcal{G}}\big|_{\mathcal{E}^\infty}$ is a maximal integral manifold of the Cartan distribution on $\mathcal{E}^\infty$ invariant with respect to any mapping $\widehat{g}$,*

$g \in \mathcal{G}$, if and only if it is of the form $j_\infty(s)(M)$, where $s \in \Gamma(\pi)$ is a solution of the equation $\mathcal{E}$.

The restriction of the mapping $\widehat{g}$, $g \in \mathcal{G}$, to $\mathcal{E}^\infty$ will be also denoted by $\widehat{g}$.

Since the planes of the Cartan distribution on $J^\infty(\pi; \mathcal{G})$ at points of $\mathcal{E}^\infty$ coincide with the planes of this distribution on $\mathcal{E}^\infty$, any field lying in the Cartan distribution on $J^\infty(\pi; \mathcal{G})$ is tangent to the equation $\mathcal{E}^\infty$. The set of restrictions of such fields to $\mathcal{E}^\infty$ will be denoted by $\mathcal{C}\mathrm{D}(\mathcal{E}^\infty)$.

As in the differential case, let us define the module of Cartan forms on $\mathcal{E}^\infty$ to be the set $\mathcal{C}\Lambda^1(\mathcal{E}^\infty) \subset \Lambda^1(\mathcal{E}^\infty)$ of 1-forms annihilating by vectors of the distribution $\mathcal{C}(\mathcal{E})$ at any point $\theta \in \mathcal{E}^\infty$. A vector field X on $\mathcal{E}^\infty$ will be called a *higher (infinitesimal) symmetry* of the equation $\mathcal{E} \subset J^k(\pi; \mathcal{G})$ if it satisfies the conditions

$$X(\mathcal{C}\Lambda^1(\mathcal{E}^\infty)) \subset \mathcal{C}\Lambda^1(\mathcal{E}^\infty), \qquad \widehat{g}^* \circ X = X \circ \widehat{g}^*$$

for any $g \in \mathcal{G}$. The set of such fields forms a Lie algebra, denoted by $\mathrm{sym}(\mathcal{E})$.

The restriction to $\mathcal{E}^\infty$ of a $\mathcal{G}$-invariant symmetry of the Cartan distribution on $J^\infty(\pi; \mathcal{G})$ tangent to $\mathcal{E}^\infty$ is a symmetry of this equation. Such symmetries will be called *extrinsic*.

THEOREM 6.11. *If the equation $\mathcal{E} \subset J^k(\pi; \mathcal{G})$ is such that*

$$(\pi_0^{\mathcal{G},\{\mathrm{id}_M\}} \circ \pi_{\infty,0}^{\mathcal{G}})(\mathcal{E}^\infty) = J^0(\pi), \tag{6.44}$$

then any higher symmetry of the equation $\mathcal{E}$ is a restriction to $\mathcal{E}^\infty$ of some extrinsic symmetry of this equation.

PROOF. The proof of this fact is similar to the proof of Theorem 3.7 of Ch. 4. Namely, if X is a higher symmetry, then the restriction of X to $\mathcal{F}_0(\pi)$ extends to a derivation $X'_0 \colon \mathcal{F}_0(\pi) \to \mathcal{F}(\pi; \mathcal{G})$. The derivation X'_0 determines an evolutionary derivation ∂_φ such that $\partial_\varphi|_{\mathcal{F}_0(\pi)} = X'_0$ (cf. the arguments after the proof of Proposition 6.7). The restriction of the field ∂_φ to $\mathcal{E}^\infty$ coincides with X. Indeed, any coordinate function on $J^\infty(\pi; \mathcal{G})$ is obtained from elements of the algebra $\mathcal{F}_0(\pi)$ by action of fields $\widehat{Y}$, $Y \in D(M)$, and of homomorphisms $\widehat{g}^*$, $g \in \mathcal{G}$. Coordinate functions on $\mathcal{E}^\infty$ are obtained by restriction to $\mathcal{E}^\infty$ of coordinate functions from $J^\infty(\pi; \mathcal{G})$. Since the fields ∂_φ and X commute both with $\widehat{Y}$ and with $\widehat{g}^*$ while $\partial_\varphi|_{\mathcal{F}_0(\pi)} = X|_{\mathcal{F}_0(\pi)}$, the fields ∂_φ and X coincide at any function on $\mathcal{E}^\infty$. $\square$

Let us derive defining equations for higher symmetries. Let the equation $\mathcal{E}$ be given by the relations

$$G_j = 0, \qquad G_j \in \mathcal{F}_k(\pi; \mathcal{G}), \qquad j = 1, \ldots, r. \tag{6.45}$$

From Theorems 6.8 and 6.11 it follows that any higher symmetry of the equation $\mathcal{E}$ can be obtained by restriction to $\mathcal{E}^\infty$ of some evolutionary derivation ∂_φ, $\varphi \in \mathcal{F}(\pi, \pi; \mathcal{G})$, tangent to the equation $\mathcal{E}^\infty$. As in the differential case (see Chapter 4), the condition of tangency of the field ∂_φ to the equation $\mathcal{E}^\infty$ is written in the form

$$\partial_\varphi(G_j) = \sum_{l,\sigma,g} \psi'_{l,\sigma,g} \widehat{g}^*(D_\sigma(G_l)), \qquad j = 1, \ldots, r.$$

The right-hand sides of these equalities vanish on $\mathcal{E}^\infty$. The left-hand sides, restricted to $\mathcal{E}^\infty$, are of the form

$$\partial_\varphi(G_j)|_{\mathcal{E}^\infty} = l_{G_j}(\varphi)|_{\mathcal{E}^\infty} = l_{G_j}^{\mathcal{E}}(\bar\varphi), \qquad j = 1, \ldots, r,$$

where $\bar{\varphi} = \varphi|_{\mathcal{E}^\infty}$, $l_G^\mathcal{E} = l_G|_{\mathcal{E}^\infty}$. Denote by $\mathcal{F}(\mathcal{E}, \pi; \mathcal{G})$ the set of restrictions to $\mathcal{E}^\infty$ of elements of the module $\mathcal{F}(\pi, \pi; \mathcal{G})$, and for elements $\bar{\varphi}, \bar{\psi} \in \mathcal{F}(\mathcal{E}, \pi; \mathcal{G})$ define the bracket

$$\{\bar{\varphi}, \bar{\psi}\}_\mathcal{E} \stackrel{\text{def}}{=} l_{\bar{\psi}}^\mathcal{E}(\varphi) - l_{\bar{\varphi}}^\mathcal{E}(\psi) = \{\varphi, \psi\}|_{\mathcal{E}^\infty},$$

where $\bar{\varphi} = \varphi|_{\mathcal{E}^\infty}$, $\bar{\psi} = \psi|_{\mathcal{E}^\infty}$.

THEOREM 6.12. *If an equation $\mathcal{E} \subset J^k(\pi; \mathcal{G})$ is given by the relations (6.45) and satisfies (6.44), then the Lie algebra* $\operatorname{sym}(\mathcal{E})$ *is isomorphic to the algebra of solutions of the system*

$$l_{G_j}^\mathcal{E}(\varphi) = 0, \qquad j = 1, \ldots, r, \quad \varphi \in \mathcal{F}(\mathcal{E}, \pi; \mathcal{G}), \tag{6.46}$$

where the Lie algebra structure is determined by the bracket $\{\cdot, \cdot\}_\mathcal{E}$.

EXERCISE 6.12. Write down equations (6.46) for the system of boundary differential equations (6.17).

6.7. Examples. The generating sections φ of symmetries for integro-differential equations contain many more variables than the corresponding sections in the differential case. For this reason, to solve equation (6.46) without a computer is rather difficult.

When one analyzes methods of solution of equations (6.46) (see examples in Chapters 3 and 6), one sees that the main trick is differentiation of both sides of equations with respect to some variable. Most often, such a variable is contained in the equation, but either all components of the section φ are independent of this variable, or only a few of them depend on it. After differentiating, we obtain a new equation in components of φ. If the new equation is sufficiently simple and contains a derivative of the component φ^i with respect to a variable v, then, solving this equation, we derive the dependence of φ^i on v. Substituting the expression for φ into (6.46) and differentiating the result with respect to another variable (w, for example), we obtain new equations, and repeat the procedure until we reach the final solution of (6.46).

The first of the examples below was computed using a program written by means of the MAPLE V software. This program produces and analyzes the system of equations (6.46), seeks for suitable variables for differentiation, differentiates the corresponding equations with respect to these variables, and chooses the simplest of the resulting equations. The latter were solved by hand, and the corresponding procedures are described below. When solving the equations, we used the following lemma, which may be useful in solving other equations.

LEMMA 6.13. *Let M be a smooth manifold. Then*

1. *If a submanifold N of the manifold M is given by the equations*

$$f_i = 0, \qquad i = 1, \ldots, k,$$

where $f_i \in C^\infty(M)$ and the covectors $df_i|_x$, $i = 1, \ldots, k$, are linearly independent at all points $x \in N$, then any function f on M vanishing on N can be represented in the form

$$f = \sum_{i=1}^{k} f_i a_i, \tag{6.47}$$

where a_i, $i = 1, \ldots, k$, are smooth functions on M.

2. Let a second submanifold $N_1 \subset M$ be given by the equations

$$g_j = 0, \quad j = 1, \ldots, l, \quad g_j \in C^\infty(M),$$

and let the covectors $df_i|_x$, $dg_j|_x$, $i = 1, \ldots, k$, $j = 1, \ldots, l$, be linearly independent at all points $x \in N_1$. Then any smooth function f on M, vanishing both on N and on N_1, can be represented in the form

$$f = \sum_{i,j} f_i g_j b_{ij}, \tag{6.48}$$

where $b_{ij} \in C^\infty(M)$, $i = 1, \ldots, k$, $j = 1, \ldots, l$.

PROOF. The result is a consequence of [80, Lemma 2.1, Ch.3]. □

EXAMPLE 6.7. Let us find symmetries of equation (6.12). This equation is equivalent to the system of boundary differential equations (6.17). Let us use the following notation. Let $\mathcal{G}$ be the semigroup of mappings considered in Example 6.4. If v is a coordinate in the fiber of the jet space with the semigroup $\mathcal{G}$, then the coordinates corresponding to the derivatives of v over x, s, and t will be denoted by v_1, v_2, and v_3 respectively. For example, the coordinate u_{13}^2 corresponds to the derivative $\partial^2 u^2/\partial x \partial t$. By $g_{[ab]}$, where the letters a, b can be any of the symbols 0, x, s, or ∞, we denote the mapping belonging to the semigroup $\mathcal{G}$ and taking the point with coordinates t, x, s to the point with coordinates (t, a, b). For example, $g_1 = g_{[x0]}$, $g_2 = g_{[0x]}$, etc. (See Example 6.4.) By $v_{[ab]}$, where v is a function on the jet space, we denote the function $g_{[ab]}^*(v)$. For example, $g_2^*(u^2) = u_{[0,x]}^2$, $g_1^*(\partial u^1/\partial x) = u_{1[x0]}^1$, etc. We shall write just v instead of $v_{[xs]}$. Note that if the restriction of a function v to the section $j_k(h)$ is $f(t, x, s)$, then the restriction of the function $v_{[ab]}$ to $j_k(h)$ is $f(t, a, b)$.

EXERCISE 6.13. Describe e, f via a, b, c, d, if $g_{[ef]} = g_{[ab]} \circ g_{[cd]}$.

Using the above notation, let us rewrite system (6.17) in the form

$$u_3^1 = \tfrac{1}{2} u_{[0x]}^2 - u^1 u_{[x\infty]}^3, \quad u_2^2 = u_1^2 + K u_{[s0]}^1 u^1,$$
$$u_2^3 = K u_{[s0]}^1, \quad u_{[x0]}^1 = u^1, \quad u_{[x0]}^2 = 0, \quad u_{[x0]}^3 = 0. \tag{6.49}$$

The linearization of the system (6.49) is of the form

$$D_3(\varphi^1) = \tfrac{1}{2} \varphi_{[0x]}^2 - u^1 \varphi_{[x\infty]}^3 - u_{[x\infty]}^3 \varphi^1, \tag{6.50}$$
$$D_2(\varphi^2) = D_1(\varphi^2) + K u_{[s0]}^1 \varphi^1 + K u^1 \varphi_{[s0]}^1, \tag{6.51}$$
$$D_2(\varphi^3) = K \varphi_{[s0]}^1, \tag{6.52}$$
$$\varphi_{[x0]}^1 = \varphi^1, \tag{6.53}$$
$$\varphi_{[x0]}^2 = 0, \tag{6.54}$$
$$\varphi_{[x0]}^3 = 0, \tag{6.55}$$

where D_1, D_2, D_3 are the total derivatives with respect to x, s, t respectively, $\varphi_{[0x]}^2 = g_{[0x]}^*(\varphi^2)$, etc.

We shall solve system (6.50)–(6.55) assuming that the desired functions φ^1, φ^2, φ^3 depend on coordinates in $\pi^{\mathcal{G}}_{\infty,1}(\mathcal{E}^\infty) \subset J^1(\pi;\mathcal{G})$, i.e., on the variables

$$
\begin{aligned}
&x,\ s,\ t,\ u^1,\ u^2,\ u^3,\ u^1_1,\ u^2_1,\ u^3_1,\ u^3_1,\ u^3_3,\ u^1_{[00]},\ u^1_{[\infty 0]},\ u^1_{[s0]},\ u^2_{[0\infty]},\\
&u^2_{[0s]},\ u^2_{[0x]},\ u^2_{[\infty\infty]},\ u^2_{[s\infty]},\ u^2_{[x\infty]},\ u^3_{[0\infty]},\ u^3_{[0s]},\ u^3_{[0x]},\ u^3_{[\infty\infty]},\\
&u^3_{[s\infty]},\ u^3_{[x\infty]},\ u^1_{1[00]},\ u^1_{1[\infty 0]},\ u^1_{1[s0]},\ u^2_{1[0\infty]},\ u^2_{1[0s]},\ u^2_{1[0x]},\\
&u^2_{1[\infty\infty]},\ u^2_{1[s\infty]},\ u^2_{1[x\infty]},\ u^3_{1[0\infty]},\ u^3_{1[0s]},\ u^3_{1[0x]},\ u^3_{1[\infty\infty]},\\
&u^2_{3[s\infty]},\ u^2_{3[x\infty]},\ u^3_{1[0\infty]},\ u^3_{1[0s]},\ u^3_{1[0x]},\ u^3_{1[\infty\infty]},\ u^3_{1[s\infty]},\\
&u^3_{1[x\infty]},\ u^3_{3[0\infty]},\ u^3_{3[0s]},\ u^3_{3[0x]},\ u^3_{3[\infty\infty]},\ u^3_{3[s\infty]},\ u^3_{3[x\infty]}.
\end{aligned}
\quad (6.56)
$$

From (6.53) it follows that the function φ^1 depends only on the variables (6.56) by the action of the homomorphism $g_{[x,0]}{}^*$. These variables are

$$
\begin{aligned}
&x,\ t,\ u^1_{[00]},\ u^1_{[\infty 0]},\ u^1,\ u^2_{[0\infty]},\ u^2_{[0x]},\ u^2_{[\infty\infty]},\ u^2_{[x\infty]},\ u^3_{[0\infty]},\\
&u^3_{[0x]},\ u^3_{[\infty\infty]},\ u^3_{[x\infty]},\ u^1_{1[00]},\ u^1_{1[\infty 0]},\ u^1_1,\ u^2_{1[0\infty]},\ u^2_{1[0x]},\\
&u^2_{1[\infty\infty]},\ u^2_{1[x\infty]},\ u^2_{3[0\infty]},\ u^2_{3[0x]},\ u^2_{3[\infty\infty]},\ u^2_{3[x\infty]},\ u^3_{1[0\infty]},\\
&u^3_{1[0x]},\ u^3_{1[\infty\infty]},\ u^3_{1[x\infty]},\ u^3_{3[0\infty]},\ u^3_{3[0x]},\ u^3_{3[\infty\infty]},\ u^3_{3[x\infty]}.
\end{aligned}
$$

The above-mentioned differentiating program was applied to equations (6.50)–(6.52). The result is a new set of equations, etc. The equations essential for subsequent exposition will be written down and analyzed.

Step 1. Choose the variables corresponding to second order derivatives. Differentiation with respect to these variables shows that φ^1 is independent of

$$
\begin{aligned}
&u^2_{1[0\infty]},\ u^2_{1[0x]},\ u^2_{1[\infty\infty]},\ u^2_{1[x\infty]},\ u^2_{3[0\infty]},\ u^2_{3[0x]},\ u^2_{3[\infty\infty]},\ u^2_{3[x\infty]},\\
&u^3_{1[0\infty]},\ u^3_{1[0x]},\ u^3_{1[\infty\infty]},\ u^3_{1[x\infty]},\ u^3_{3[0\infty]},\ u^3_{3[0x]},\ u^3_{3[\infty\infty]},\ u^3_{3[x\infty]}.
\end{aligned}
$$

The function φ^2 is independent of the variables

$$
\begin{aligned}
&u^3_1,\ u^3_3,\ u^1_{1[s0]},\ u^1_{1[x0]},\ u^2_{1[0s]},\ u^2_{1[0x]},\ u^2_{1[s\infty]},\ u^2_{1[x\infty]},\ u^2_{3[0s]},\ u^2_{3[0x]},\\
&u^2_{3[s\infty]},\ u^2_{3[x\infty]},\ u^3_{1[s\infty]},\ u^3_{1[x\infty]},\ u^3_{3[s\infty]},\ u^3_{3[x\infty]},
\end{aligned}
$$

and φ^3 is independent of

$$
u^2_1,\ u^2_3,\ u^1_{1[s0]},\ u^2_{1[0s]},\ u^2_{1[s\infty]},\ u^2_{3[0s]},\ u^2_{3[s\infty]},\ u^3_{1[s\infty]},\ u^3_{3[s\infty]}.
$$

Step 2. Differentiation of (6.51) with respect to the variables u^3_1, $u^2_{1[0s]}$, $u^2_{1[0x]}$, $u^2_{1[s\infty]}$, $u^2_{1[x\infty]}$, $u^3_{1[s\infty]}$, $u^3_{1[x\infty]}$ and of (6.52) with respect to u^2_1, $u^2_{1[0s]}$, $u^2_{1[s\infty]}$, $u^3_{1[s\infty]}$ proves that φ^2 is independent of the variables

$$
u^3,\ u^2_{[0s]},\ u^2_{[0x]},\ u^2_{[s\infty]},\ u^2_{[x\infty]},\ u^3_{[s\infty]},\ u^3_{[x\infty]}
$$

and φ^3 is independent of

$$
u^2,\ u^2_{[0s]},\ u^2_{[s\infty]},\ u^3_{[s\infty]}.
$$

Step 3. Let us now differentiate (6.50) with respect to $u^2_{3[x\infty]}$, (6.51) with respect to $u^2_{[0x]}$, $u^3_{[x\infty]}$, u^1_1, $u^2_{[0s]}$, $u^1_{1[s0]}$, and (6.52) with respect to $u^2_{[0s]}$, $u^1_{1[s0]}$. Then we obtain the following equations:

$$\frac{\partial \varphi^1}{\partial u^2_{[x\infty]}} = 0, \qquad (6.57)$$

$$\frac{\partial \varphi^2}{\partial u^2_3} K u^1_{[s0]} - \frac{\partial \varphi^2}{\partial u^3_{3[0x]}} K_{[0x]} = 2K u^1_{[s0]} \frac{\partial \varphi^1}{\partial u^2_{[0x]}}, \qquad (6.58)$$

$$-\frac{\partial \varphi^2}{\partial u^2_3} K u^1_{[s0]} u^1 + \frac{\partial \varphi^2}{\partial u^3_{3[0x]}} K_{[0x]} u^1 = K u^1_{[s0]} \frac{\partial \varphi^1}{\partial u^3_{[x\infty]}}, \qquad (6.59)$$

$$-\frac{\partial \varphi^2}{\partial u^1} + \frac{\partial \varphi^2}{\partial u^2_1} K u^1_{[s0]} - K u^1_{[s0]} \frac{\partial \varphi^1}{\partial u^1_1} = 0, \qquad (6.60)$$

$$\frac{\partial \varphi^2}{\partial u^3_{3[0s]}} K_{[0s]} + \frac{\partial \varphi^2}{\partial u^2_3} K u^1 = 2K u^1 \frac{\partial \varphi^1_{[s0]}}{\partial u^2_{[0s]}}, \qquad (6.61)$$

$$\frac{\partial \varphi^2}{\partial u^1_{[s0]}} = K u^1 \frac{\partial \varphi^1_{[s0]}}{\partial u^1_{1[s0]}}, \qquad (6.62)$$

$$\frac{\partial \varphi^3}{\partial u^3_{3[0s]}} K_{[0s]} + K \frac{\partial \varphi^3}{\partial u^3_3} = 2K \frac{\partial \varphi^1_{[s0]}}{\partial u^2_{[0s]}}, \qquad (6.63)$$

$$\frac{\partial \varphi^3}{\partial u^1_{[s0]}} = K \frac{\partial \varphi^1_{[s0]}}{\partial u^1_{1[s0]}}. \qquad (6.64)$$

Let us set $\varphi^4 = \partial \varphi^1 / \partial u^1_1$, $\varphi^5 = 2\partial \varphi^1 / \partial u^2_{[0x]}$. Then the functions φ^4, φ^5 depend on the same variables as φ^1, i.e., on

$$t, \ u^1_{[00]}, \ u^1_{[\infty 0]}, \ u^2_{[0\infty]}, \ u^2_{[\infty\infty]}, \ u^3_{[0\infty]}, \ u^3_{[\infty\infty]}, u^1_{1[00]}, u^1_{1[\infty 0]} \qquad (6.65)$$

and on x, u^1, u^1_1, $u^2_{[0x]}$, $u^3_{[0x]}$, $u^3_{[x\infty]}$ (we take (6.57) into account here).

Multiplying (6.58) by u^1, adding it to (6.59) and cancelling $Ku^1_{[s0]}$, we obtain

$$\frac{\partial \varphi^1}{\partial u^3_{[x\infty]}} = -u^1 \varphi^5. \qquad (6.66)$$

The right-hand side of equation (6.63) is $K\varphi^5_{[s0]}$, and consequently it is independent of the variables listed in (6.65) and of x, s, $u^1_{[s0]}$, $u^1_{1[s0]}$, $u^2_{[0s]}$, $u^3_{[0s]}$, $u^3_{[s\infty]}$. The left-hand side of this equation is independent of $u^1_{1[s0]}$, $u^2_{[0s]}$, $u^3_{[s\infty]}$. Therefore, φ^5 cannot depend on u^1_1, $u^2_{[0x]}$, $u^3_{[x\infty]}$. From (6.64), by similar reasoning, it follows that φ^4 is also independent of u^1_1, $u^2_{[0x]}$, $u^3_{[x\infty]}$.

Due to the equalities

$$\frac{\partial \varphi^1_{[s0]}}{\partial u^1_{1[s0]}} = g_{[s0]} {}^*(\frac{\partial \varphi^1}{\partial u^1_1}) = \varphi^4_{[s0]},$$

equation (6.62) can be rewritten in the form

$$\frac{\partial \varphi^2}{\partial u^1_{[s0]}} = K u^1 \varphi^4_{[s0]}. \qquad (6.67)$$

Let us differentiate (6.60) with respect to $u^1_{[s0]}$. The function $\partial\varphi^1/\partial u^1_1 = \varphi^4$ is independent of $u^1_{[s0]}$, and hence from (6.67) we obtain

$$\frac{\partial^2\varphi^2}{\partial u^1_{[s0]}\partial u^2_1} = 0 \quad \text{and} \quad \frac{\partial^2\varphi^2}{\partial u^1_{[s0]}\partial u^1} = K\varphi^4_{[s0]}.$$

Therefore, after differentiating (6.60) we obtain

$$\frac{\partial\varphi^2}{\partial u^2_1} = \varphi^4 + \varphi^4_{[s0]}, \tag{6.68}$$

and (6.60) can be rewritten in the form

$$\frac{\partial\varphi^2}{\partial u^1} = Ku^1_{[s0]}\varphi^4_{[s0]}. \tag{6.69}$$

Comparing the derivative of (6.68) with respect to u^1 with the derivative of (6.69) with respect to u^2_1, we see that φ^4 is independent of u^1.

From (6.67) it can easily be deduced that the derivatives of φ^2 involved in (6.58) do not depend on $u^1_{[s0]}$. Hence, (6.58) splits into two equations:

$$\frac{\partial\varphi^2}{\partial u^2_3} = \varphi^5, \tag{6.70}$$

$$\frac{\partial\varphi^2}{\partial u^3_{3[0x]}} = 0. \tag{6.71}$$

In a similar way, from (6.69) we see that (6.61) splits into

$$\frac{\partial\varphi^2}{\partial u^2_3} = \varphi^5_{[s0]}, \tag{6.72}$$

$$\frac{\partial\varphi^2}{\partial u^3_{3[0s]}} = 0. \tag{6.73}$$

From equations (6.70) and (6.72) we have the equality $\varphi^5 = \varphi^5_{[s0]}$. Its left-hand side may depend on the variables (6.65) and on x, u^1, $u^3_{[0x]}$ only, while the right-hand side depends on (6.65) and s, $u^1_{[s0]}$, $u^3_{[0s]}$. Therefore, the function φ^5 depends on the variables (6.65) only.

Let us represent the functions φ^1, φ^2, φ^3 in the form

$$\varphi^1 = \varphi^4 u^1_1 + \varphi^5(1/2 u^2_{[0x]} - u^1 u^3_{[x\infty]}) + \varphi^6, \tag{6.74}$$

$$\varphi^2 = \varphi^4_{[s0]} K u^1 u^1_{[s0]} + (\varphi^4 + \varphi^4_{[s0]}) u^2_1 + \varphi^5 u^2_3 + \varphi^7, \tag{6.75}$$

$$\varphi^3 = \varphi^4_{[s0]} K u^1_{[s0]} + \varphi^8, \tag{6.76}$$

where φ^6, φ^7, φ^8 are new functions. Looking through the lists of arguments of the functions φ^4, φ^5, $\varphi^4_{[s0]}$, it can easily be seen that φ^6, φ^7, φ^8 may depend only on the same variables as φ^1, φ^2, and φ^3 respectively. Differentiating (6.74) with respect to $u^2_{[0x]}$, $u^3_{[x\infty]}$, u^1_1 and using (6.66) and the definitions of the functions φ^4 and φ^5, we obtain that φ^6 is independent of $u^2_{[0x]}$, $u^3_{[x\infty]}$, and u^1_1. In a similar way we prove that φ^7 is independent of $u^1_{[s0]}$, u^2_1, u^1, u^2_3, $u^3_{3[0x]}$, $u^3_{3[0s]}$, and φ^8 does not depend on $u^1_{[s0]}$. To this end, it is sufficient to differentiate (6.75) and (6.76) with respect to the corresponding variables and to use equations (6.67)–(6.71), (6.73) and (6.64).

Step 4. Differentiating (6.51) with respect to u_1^2, we obtain the equality

$$g_{[s0]}^*\left(\frac{\partial\varphi^4}{\partial x}\right) - \frac{\partial\varphi^4}{\partial x} + K_{[0s]}u_{[s0]}^1 g_{[s0]}^*\left(\frac{\partial\varphi^4}{\partial u_{[0x]}^3}\right) - K_{[0x]}u^1\frac{\partial\varphi^4}{\partial u_{[0x]}^3} = 0.$$

Since the function φ^4 is independent of u^1 and $u_{[s0]}^1$, we have

$$\frac{\partial\varphi^4}{\partial u_{[0x]}^3} = 0, \qquad g_{[s0]}^*\left(\frac{\partial\varphi^4}{\partial x}\right) = \frac{\partial\varphi^4}{\partial x}.$$

From the last equality it follows that the derivative $\partial\varphi^4/\partial x$ is independent of x. Therefore, $\varphi^4 = \varphi^9 x + \widetilde{\varphi}^9$, where φ^9 are $\widetilde{\varphi}^9$ are functions depending on the variables (6.65).

Substituting our representations for φ^2 and φ^4 into equations (6.75) and (6.54) and taking into account the equalities $u_{1[x,0]}^2 = 0$, $u_{3[x,0]}^2 = 0$ (which are implied by the fifth equation of (6.49)), we rewrite (6.54) in the form

$$\widetilde{\varphi}^9 K_{[x0]} u^1 u_{[00]}^1 + \varphi_{[x0]}^7 = 0.$$

Since the function φ^7 is independent of u^1, the function $\varphi_{[x0]}^7$ is also independent of u^1, and consequently $\widetilde{\varphi}^9 = 0 = \varphi_{[x0]}^7$. Hence, $\varphi^4 = \varphi^9 x$.

Step 5. From now on, when calculating $\varphi_{[x,\infty]}^3$, it is necessary to take into account the equality $g_{[x,\infty]}^*(Ku_{[s0]}^1 s) = 0$, which holds because the improper integral in equation (6.12) converges. Taking the second derivative of (6.51) with respect to the variables u^1 and $u_{[s0]}^1$, we obtain the equation

$$-\frac{\partial\varphi_{[s0]}^6}{\partial u_{[s0]}^1} + \varphi^9 g + \frac{\partial\varphi^7}{\partial u^2} - \frac{\partial\varphi^6}{\partial u^1} = 0, \qquad (6.77)$$

where

$$g = 1 + \frac{1}{K}\left(s\frac{\partial K}{\partial s} + x\frac{\partial K}{\partial x}\right) \qquad (6.78)$$

is a function of x and s. The function φ^6 depends linearly on u^1, since the first three summands in (6.77) are independent of u^1. Substituting the expression $\varphi^6 = u^1\varphi^{10} + \varphi^{11}$ into (6.77), where φ^{10}, φ^{11} are functions depending on the same variables as φ^6, except for u^1, we obtain

$$\varphi^7 = u^2(\varphi^{10} + \varphi_{[s0]}^{10} - \varphi^9 g) + \varphi^{12},$$

where the function φ^{12} depends on the same variables as φ^7, except for u^2.

Step 6. Differentiating (6.50) with respect to u_1^1, $u_{[x\infty]}^3$, $u_{1[x\infty]}^3$, $u_{3[x\infty]}^3$ and twice differentiating (6.51) with respect to u^2 and u^1, we obtain the equalities

$$xD_3(\varphi^9) + u^1\frac{\partial\varphi_{[x\infty]}^8}{\partial u_1^1} = 0, \qquad (6.79)$$

$$u^1 D_3(\varphi^5) = u^1\frac{\partial\varphi_{[x\infty]}^8}{\partial u_{[x\infty]}^3} + \varphi^{11}, \qquad (6.80)$$

$$\frac{\partial\varphi_{[x\infty]}^8}{\partial u_{1[x\infty]}^3} = \varphi^9 x, \qquad \frac{\partial\varphi_{[x\infty]}^8}{\partial u_{3[x\infty]}^3} = \varphi^5, \qquad (6.81)$$

$$\frac{\partial \varphi^{10}}{\partial u^3_{[0x]}} = 0. \tag{6.82}$$

Taking into account all differential consequences of equation (6.51), we transform the latter into

$$\frac{\partial \varphi^{12}}{\partial s} - \frac{\partial \varphi^{12}}{\partial x} = 0. \tag{6.83}$$

Substituting our expression for φ^2 into (6.54), we have $\varphi^{12}_{[x0]} = 0$. Using equation (6.83), we see that φ^{12} is a function of the following form:

$$\varphi^{12} = \widetilde{\varphi}(x+s, u^3_{[0s]}, u^3_{1[0s]}, t, u^1_{[00]}, u^1_{[\infty 0]}, u^2_{[0\infty]}, u^2_{[\infty\infty]}, u^3_{[0\infty]},$$
$$u^3_{[0x]}, u^3_{[\infty\infty]}, u^1_{1[00]}, u^1_{1[\infty 0]}, u^2_{1[0\infty]}, u^2_{1[\infty\infty]}, u^2_{3[0\infty]},$$
$$u^2_{3[\infty\infty]}, u^3_{1[0\infty]}, u^3_{1[0x]}, u^3_{1[\infty\infty]}, u^3_{3[0\infty]}, u^3_{3[\infty\infty]}).$$

The homomorphism $g_{[x0]}{}^*$ changes only the first three arguments of the function φ^{12}:

$$\varphi^{12}_{[x0]} = \widetilde{\varphi}(x, u^3_{[00]}, u^3_{1[00]}, t, u^1_{[00]}, u^1_{[\infty 0]}, u^2_{[0\infty]}, u^2_{[\infty\infty]}, u^3_{[0\infty]},$$
$$u^3_{[0x]}, u^3_{[\infty\infty]}, u^1_{1[00]}, u^1_{1[\infty 0]}, u^2_{1[0\infty]}, u^2_{1[\infty\infty]}, u^2_{3[0\infty]}, \tag{6.84}$$
$$u^2_{3[\infty\infty]}, u^3_{1[0\infty]}, u^3_{1[0x]}, u^3_{1[\infty\infty]}, u^3_{3[0\infty]}, u^3_{3[\infty\infty]}).$$

Since the arguments of the function (6.84) are independent, from the equality $\varphi^{12}_{[x0]} = 0$ we obtain $\varphi^{12} = \widetilde{\varphi} = 0$.

The summands $xD_3(\varphi^9)$ and φ^{11} in equations (6.79) and (6.80) are independent of u^1. Therefore, setting $u^1 = 0$ in these equations, we obtain

$$D_3(\varphi^9) = 0, \tag{6.85}$$
$$\varphi^{11} = 0. \tag{6.86}$$

Equations (6.86) and (6.80) imply the equality

$$\frac{\partial \varphi^8_{[x\infty]}}{\partial u^3_{[x\infty]}} = D_3(\varphi^5). \tag{6.87}$$

Equations (6.87) and (6.81) show that equation (6.50) can be simplified by setting

$$\varphi^8 = u^3 D_3(\varphi^5) - u^3_1 x \varphi^9 + u^3_3 \varphi^5 + \varphi^{13},$$

where φ^{13} is a function of the same variables as φ^8.

Step 7. At this stage, we differentiate (6.85) with respect to the variables of which the function φ^9 is independent. As a result, we obtain equations implying that φ^9 is independent of the variables

$$u^1_{[\infty 0]}, \ u^2_{[0\infty]}, \ u^2_{[\infty\infty]}, \ u^3_{[0\infty]}, \ u^3_{[\infty\infty]}, \ u^1_{1[00]}, \ u^1_{1[\infty 0]}.$$

Hence, this function may depend on $u^1_{[00]}$ and t only. Thus, equation (6.85) acquires the form

$$\frac{\partial \varphi^9}{\partial t} - u^1_{[00]} u^3_{[0\infty]} \frac{\partial \varphi^9}{\partial u^1_{[00]}} = 0,$$

and it follows that φ^9 is constant.

Step 8. Differentiating (6.52) with respect to $u^1_{[s0]}$, we obtain

$$X(\varphi^{13}) = (\varphi^{10}_{[s0]} - \varphi^9 g - D_3(\varphi^5))K, \qquad (6.88)$$

where

$$X = K\frac{\partial}{\partial u^3} + K_1 \frac{\partial}{\partial u^3_1} + K_{[0s]}\frac{\partial}{\partial u^3_{[0s]}} + K_{1[0s]}\frac{\partial}{\partial u^3_{1[0s]}}, \qquad K_1 = \frac{\partial K}{\partial x}.$$

Equations (6.50), (6.52) and (6.55), together with their differential consequences, imply

$$D_3(\varphi^{10}) + \varphi^{13}_{[x\infty]} = 0, \qquad (6.89)$$

$$\frac{\partial \varphi^{13}}{\partial s} = 0,$$

$$\varphi^{13}_{[x0]} = 0. \qquad (6.90)$$

Thus, the function φ^{13} depends on the variables

$$\begin{aligned}
&u^3,\ u^3_1,\ u^3_3,\ u^3_{[0s]},\ u^3_{1[0s]},\ u^3_{3[0s]},\ x,\ t,\ u^1_{[00]},\ u^1_{[\infty 0]}, \\
&u^1,\ u^2_{[0\infty]},\ u^2_{[0x]},\ u^2_{[\infty\infty]},\ u^2_{[x\infty]},\ u^3_{[0\infty]},\ u^3_{[0x]},\ u^3_{[\infty\infty]}, \\
&u^3_{[x\infty]},\ u^1_1,\ u^1_{1[00]},\ u^1_{1[\infty 0]},\ u^2_{1[0\infty]},\ u^2_{1[0x]},\ u^2_{1[\infty\infty]}, \\
&u^2_{1[x\infty]},\ u^2_{3[0\infty]},\ u^2_{3[0x]},\ u^2_{3[\infty\infty]},\ u^2_{3[x\infty]},\ u^3_{1[0\infty]},\ u^3_{1[0x]}, \\
&u^3_{1[\infty\infty]},\ u^3_{1[x\infty]},\ u^3_{3[0\infty]},\ u^3_{3[0x]},\ u^3_{3[\infty\infty]},\ u^3_{3[x\infty]}.
\end{aligned} \qquad (6.91)$$

Equality (6.90) means that the function φ^{13} vanishes on the submanifold $N = \operatorname{im} g_{[x0]}$. In the coordinate space (6.91), the submanifold N is described as the zero set for the functions

$$u^3, u^3_1, u^3_3, u^3_{[0s]}, u^3_{1[0s]}, u^3_{3[0s]} \qquad (6.92)$$

(see the last equation in (6.49)). Therefore, from the first part of Lemma 6.13 we obtain the representation

$$\varphi^{13} = u^3 a^1 + u^3_1 a^2 + u^3_3 a^3 + u^3_{[0s]} a^4 + u^3_{1[0s]} a^5 + u^3_{3[0s]} a^6, \qquad (6.93)$$

where a^i, $i = 1, \ldots, 6$, are smooth functions in the variables (6.91).

Substituting (6.93) into (6.89), we obtain the equality

$$D_3(\varphi^{10}) + u^3_{[x\infty]} a^1_{[x\infty]} + u^3_{1[x\infty]} a^2_{[x\infty]} + u^3_{3[x\infty]} a^3_{[x\infty]}$$
$$+ u^3_{[0\infty]} a^4_{[x\infty]} + u^3_{1[0\infty]} a^5_{[x\infty]} + u^3_{3[0\infty]} a^6_{[x\infty]} = 0.$$

Since the function $D_3(\varphi^{10})$ is independent of the variables

$$u^3_{[x\infty]},\ u^3_{1[x\infty]},\ u^3_{3[x\infty]},\ u^3_{[0\infty]},\ u^3_{1[0\infty]},\ u^3_{3[0\infty]},$$

from the last equality we obtain the equation $D_3(\varphi^{10}) = 0$. Hence (see the solution of equation (6.85) at Step 7 and of equations (6.82) and (6.89)), the function φ^{10} depends on x only, and we have the equality $\varphi^{13}_{[x\infty]} = 0$. The latter means that the function φ^{13} vanishes on the submanifold $N_1 = \operatorname{im} g_{[x\infty]}$. This submanifold is the zero set for the functions

$$\begin{aligned}
&u^3 - u^3_{[x\infty]},\ u^3_1 - u^3_{1[x\infty]},\ u^3_3 - u^3_{3[x\infty]}, \\
&u^3_{[0s]} - u^3_{[0\infty]},\ u^3_{1[0s]} - u^3_{1[0\infty]},\ u^3_{3[0s]} - u^3_{3[0\infty]}.
\end{aligned} \qquad (6.94)$$

Thus, from the second part of Lemma 6.13 we obtain the representation
$$\varphi^{13} = \sum_{i,j} f_i g_j b_{ij},$$
where f_i, $i = 1, \ldots, 6$, denote the functions (6.92), g_j, $j = 1, \ldots, 6$, are the functions (6.94), and b_{ij} are smooth functions in the variables (6.91).

Substituting our representation for φ^{13} into (6.88), we have on the left-hand side of this equation the following expression:
$$\sum_i f_i \left(\sum_j X(g_j b_{ij}) \right) + \sum_j g_j \left(\sum_i X(f_i) b_{ij} \right).$$

The functions on the right-hand side of (6.88) do not depend on the variables on which the functions $f_1, \ldots, f_6, g_1, \ldots, g_6$ depend. Therefore,
$$\varphi^{10}_{[s0]} - \varphi^9 g - D_3(\varphi^5) = 0. \tag{6.95}$$

The functions $\varphi^{10}_{[s0]}$ and φ^5 are independent of x. Hence, from equality (6.7) for $\varphi^9 \neq 0$ it follows that g is also independent of x. But since this function is symmetric in x and s (this is a consequence of its definition (6.78) and the symmetry of the function $K(x,s)$), the function g is also independent of s. Therefore, the product $\varphi^9 g$ is a constant. From (6.7) it follows that $\varphi^{10}_{[s0]}$ does not depend on s, and consequently φ^{10} is a constant.

Introduce the notation
$$\varphi^{14} = \varphi^{10} - \varphi^9 g, \qquad \varphi^{15} = \varphi^5 - \varphi^{14} t.$$

Equation (6.7) is now rewritten in the form $D_3(\varphi^{15}) = 0$. From this and the results of Step 7 it follows that φ^{15} is also a constant. Thus,
$$\varphi^1 = \varphi^9(x u_1^1 + u^1 g) + \varphi^{14}(u^1 + t u_3^1) + \varphi^{15} u_3^1. \tag{6.96}$$

Note that if φ^1, φ^2, φ^3 is a solution of system (6.50)–(6.55), then the functions φ^2 and φ^3 are uniquely determined by φ^1. Indeed, φ^2 and φ^3 determine evolution of the variables u^2 and u^3. The variables u^2 and u^3 are uniquely determined by u^1 (see (6.49)). Therefore, it suffices to find functions φ^2 and φ^3 together with φ^1 satisfying system (6.50)–(6.55). Setting $\varphi^{13} = 0$, we obtain
$$\begin{aligned}\varphi^2 &= \varphi^9(s u_2^2 + x u_1^2 + u^2 g) + \varphi^{14}(2u^2 + t u_3^2) + \varphi^{15} u_3^2, \\ \varphi^3 &= \varphi^9(s u_2^3 + u^3 g + x u_1^3) + \varphi^{14}(u^3 + u_3^3 t) + \varphi^{15} u_3^3. \end{aligned} \tag{6.97}$$

It is easily seen that the functions (6.96)–(6.97) satisfy system (6.50)–(6.55) and determine the classical symmetries corresponding to the liftings of the fields
$$\varphi^9 \left(u^1 g \frac{\partial}{\partial u^1} - x \frac{\partial}{\partial x} \right) + \varphi^{14} \left(u^1 \frac{\partial}{\partial u^1} - t \frac{\partial}{\partial t} \right) + \varphi^{15} \frac{\partial}{\partial t},$$
where φ^9, φ^{14}, φ^{15} are constants. If φ^9 does not vanish, then g is a constant, while from equation (6.78) it follows that $K(x,s)$ is a homogeneous function with degree of homogeneity $\sigma = g - 1$.

EXERCISE 6.14. Reconstruct computations of the above example in all details.

In [18] another system of boundary differential equations was studied, also equivalent to equation (6.12), but possessing a larger semigroup. The symmetry algebra for this system is the same. In [19] the result of symmetry computations for a generalization of equation (6.12) is given. The symmetries searched for were of the closest type to our (6.96)–(6.97). Computations were not too cumbersome in spite of much more complicated equations, and were carried out without a computer.

REMARK 6.5. The choice of a semigroup for a given equation is not unique. One can always extend it, and in some cases make it smaller (compare the semigroups for the coagulation equation (6.12) given in [**18**] and in Example 6.4). Changing the semigroup may lead to changes in the symmetry algebra (see the remark in [**19**, §5]). Thus, the problem of choosing a semigroup so as to obtain the largest symmetry algebra arises. It can be shown that in the differential case, extension of the semigroup consisting of one element (the identity) may lead only to the "trivial" extension of the symmetry algebra. For example, let our equation possess a symmetry $\varphi(c)$ depending on an arbitrary constant c, and assume that there exists an integral (or boundary differential) variable v independent of unknown functions in the equation. For example, v is the value of a dependent variable at some point. Then the change of c to v is a new symmetry $\varphi(v)$. Suppose that the symmetry $\varphi(c)$ depends on an arbitrary function c of some independent variable z. Then, in exactly the same way, we can replace c by a variable v depending on z only and obtain a new symmetry $\varphi(v)$. We call such an extension of the symmetry algebra *extension by a fake constant*. It is trivial in the following sense: in the fiber $v = c$ of our equation, the symmetries $\varphi(v)$ and $\varphi(c)$ coincide.

EXAMPLE 6.8. The Khokhlov–Zabolotskaya equation (see §5.3 of Ch. 3) possesses the symmetry

$$\varphi = q_3 p_{2g} p_1 + 2 p_g p_3 + q_3 p_{22g},$$

where $g\colon (q_1, q_2, q_3, q_4) \mapsto (a, q_2, b, c)$, and a, b, c are arbitrary numbers. This symmetry is obtained from the symmetry $f(A)$ (see (5.16) of Ch. 3) when we rrerplace the function $A(q_2)$ by p_g.

REMARK 6.6. Assume that we are interested not in all solutions of a given equation, but only in the ones satisfying additional conditions (boundary or integral). Then a semigroup extension may lead to extension of the symmetry algebra. An example of integral symmetry can be found in the book [**29**]. The authors are looking for symmetries of the Maxwell equations. They compute classical symmetries of the equations obtained by the Fourier transformation of the initial system. The inverse Fourier transformation delivers integral symmetries of the system consisting of the Maxwell equations and of the existence conditions for the corresponding Fourier transformation.

In conclusion, we consider an example of using the Laplace transformation for obtaining integral symmetries of integro-differential equations.

EXAMPLE 6.9. Consider equation (6.12) in the case $K(x,s) = x + s$. The Laplace transformation with respect to the variable x takes this equation to the boundary differential equation

$$\Phi_2 = \Phi_A \Phi_1 + \Phi_{1A} \Phi - \Phi \Phi_1, \qquad (6.98)$$

where

$$\Phi = \Phi(p,t) = \int_0^\infty u(x,t)e^{-px}\,dx,$$

$$\Phi_A = \int_0^\infty u(x,t)\,dx = \Phi(0,t),$$

$$\Phi_{1A} = -\int_0^\infty xu(x,t)\,dx = \frac{\partial \Phi}{\partial p}(0,t),$$

Φ_1 are Φ_2 being the derivatives of Φ with respect to p and t respectively. The semigroup of this equation consists of the identity mapping and $A\colon (p,t)\mapsto (0,t)$. The equation for the generating function of a symmetry ψ is of the form

$$X(\psi) = \Phi_1 \widehat{A}^*(\psi) + \Phi \widehat{A}^*(D_1(\psi)) + (\Phi_{1A} - \Phi_1)\psi, \qquad (6.99)$$

where $X = D_2 - \Phi_A D_1 + \Phi D_1$. The 0-prolongation of equation (6.98) is the space of variables p, t, Φ_A, Φ_{1A}, Φ, Φ_1. If ψ is a function on the 0-prolongation, then X may be considered as a vector field on the same space, and it is easy to find the first integrals of this field:

$$q_1 = \Phi_{1A}, \qquad q_2 = \Phi_A e^{-q_1 t}, \qquad q_3 = \Phi e^{-q_1 t},$$

$$q_4 = q_1 p + \Phi_A - \Phi, \qquad q_5 = \left(\frac{q_1}{\Phi_1} - 1\right) e^{q_1 t}.$$

Equation (6.99) is easily solved in this case. The general solution, modulo summands corresponding to the trivial extension of the classical symmetry algebra (see Example 6.7), is

$$\psi = (f - \widehat{A}^*(f))\frac{\Phi_1}{q_1} + \frac{\Phi_1}{q_1} e^{q_1 t} \widehat{A}^*\left(\frac{\partial f}{\partial q_5}\right), \qquad (6.100)$$

where f is a function of q_1, q_2, q_3, q_4, q_5 such that $\widehat{A}^*(\partial f/\partial q_3) = 0$. Substituting $f = q_1 q_3 q_4$ into (6.100), we obtain

$$\psi = q_3 q_4 \Phi_1 = \Phi e^{-q_1 t}(q_1 p + \Phi_A - \Phi)\Phi_1.$$

The corresponding symmetry of equation (6.12) for $K(x,s) = x+s$ is of the form

$$\varphi = e^{-q_1 t}\left[\frac{x}{3}\int_0^x \int_0^{s_1} u(x-s_1,t)u(s_1-s_2,t)u(s_2,t)\,ds_2\,ds_1 \right.$$
$$- q_1 \int_0^x u(x-s,t)su_s'(s,t)\,ds - q_1 \int_0^x u(x-s,t)u(s,t)\,ds$$
$$\left. - \frac{x}{2}\int_0^\infty u(s,t)\,ds \int_0^x u(x-s,t)u(s,t)\,ds\right],$$

where $q_1 = -\int_0^\infty su(s,t)\,ds$.

EXERCISE 6.15. Prove formula (6.100). Using this formula, find all symmetries of equation (6.12) for $K(x,s) = x+s$ containing double integrals.

APPENDIX

From Symmetries of Partial Differential Equations Towards Secondary ("Quantized") Calculus

> But science is not yet just a catalogue of ascertained facts about the universe; it is a mode of progress, sometimes tortuous, sometimes uncertain. And our interest in science is not merely a desire to hear the latest facts added to the collection, we like to discuss our hopes and fears, probabilities and expectations.
>
> Sir Arthur Eddington

Introduction

The pre-history of rational mechanics was the study of so-called simple mechanisms. A number of attempts to explain all of Nature as a machine composed of these mechanisms were made back then. The "standard schemes" and "models" of modern quantum field theory (QFT) look much like these simple mechanisms.

This analogy maybe clarifies the reasons for the almost universal feeling that quantum field theory, in its present form, is not yet a "true" well-established theory. Below we attempt to analyze why this is so, and what ingredients should be added to the solution to get the desired crystallization.

Having this in mind, we start with some general observations on the genesis of long-scale theories. These introductory pages furnish our subsequent considerations with the necessary initial impulse. Following it, we eventually arrive at Secondary or, more speculatively, Quantized Differential Calculus, which seems to have some chances to provide the passage from "standard models" to the "true" theory with the necessary mathematical background.

From the very beginning, we would like to stress that Secondary Calculus is only a language with which, hopefully, QFT can be developed smoothly, i.e., without "renormalizations", "anomalies", etc. If this is the case, the fundamental program of translating QFT systematically into Secondary Calculus remains to be carried out separately. Of course, results and experience accumulated in the study of concrete models up to now are indispensable to this purpose.

This appendix is a short version of the article by A. M. Vinogradov, *From symmetries of partial differential equations towards secondary ("quantized") calculus*, first published in J. Geom. Phys. **14** (1994), 146–194, and is included here with permission of Elsevier Science—NL, Sara Burgerhartstraat 25, 1055 KV Amsterdam, The Netherlands.

© 1994 Elsevier Science B.V. All rights reserved

This text is neither a review nor a research account, but a long motivation for this Secondary Calculus. We describe informally some principle ideas and results already obtained in this field, and also indicate some problems and perspectives which seem promising at this moment.

It was not our intention to present here a systematic and rigorous exposition of Secondary Calculus. That would hardly be possible within the limits of such a text. So, we restrict ourselves to a general panorama, which could lead the interested reader to enter the subject by consulting the bibliography. A lot of details and techniques can be found in the main text of this book (see also [**130, 131, 135**]). It should be followed by [**3, 57, 60, 62, 76, 117, 118, 119, 132, 133, 138**].

And, finally, the first "philosophical" pages of this Appendix are to be read half-seriously, keeping an eye on the uncertainty principle: making the meaning of our words more precise will kill the motivating impulses we hope they emit.

1. From symmetries to concepts

It is banal to say that every theory has its origin in rather simple things. But what are they? The word "simple" in common language incorporates many meanings. In linear approximation, they can be displayed by the following diagram:

in which the dots indicate the "intermediate states". In other words, we find enough reasons to interpret "symmetric" as "simple but not banal". Details are just obstructions to symmetry. So, the models manifesting only the essence of the phenomena in question are necessarily symmetric. Recall Euclidean geometry, the Copernican solar system, Newton's laws in mechanics, or special relativity to illustrate this idea. Hence, we accept as a leading principle that the initial stage in the genesis of theories is the study of symmetric models. (Of course, the above remarks are applicable only to rather long-scale situations.)

Symmetry considerations replace conceptual thinking in studying symmetric models quite well. That is why they work well at first, especially for mathematically based theories, owing to the fact that "symmetry" implies "solvability" and "integrability" in this case.

At this point, the theory passes to the next stage of its development when the dominating paradigm states that everything can be composed of simple (symmetric) elements studied earlier and the only thing to be understood is how. Schematically, this period can be characterized as the time when operative concepts of the future "true" theory, not yet discovered, are replaced by their "morphemes" and when more or less mechanical mosaics of the latter replace the calculus of these concepts. This is why we call this stage "morphological".

A serious deficiency of these morphological compositions is that many of them need to be corrected constantly to be in agreement with new experimental data and theoretical demands. This produces numerous perturbation-like schemes which are very characteristic of the morphological era.

Ptolemy's planetary system with its numerous epi- and hypo-cycles and quantum electrodynamics with its renormalizations illustrate this quite well. Also, one can learn from these examples that even an incredibly exact correspondence to experiments is not all what is needed to make a "true" theory. Of course, there is nothing bad in using a perturbation scheme for technical purposes. But it would be hardly reasonable to erect a skyscraper on a perturbed foundation.

Darwin's selection theory seems to be applicable to this selection of concepts as well. For example, one can see many fantastic creations appearing during troubled times (for example, look at the history of QFT in recent decades). This is typical for situations when the expressive powers of the language do not correspond to the subject to be described. Summing up, we represent our idea on genesis of mathematically based theories by the scheme

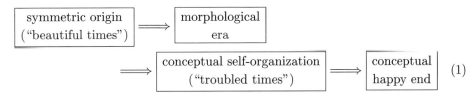 (1)

Of course, in reality, the indicated periods get mixed, and this can happen, sometimes, in a very curious way. For example, nowadays synthetic geometries, typical creations of the morphological era, have almost left the land, being replaced by differential geometry. On the other hand, measure theory, a morphological realization of the idea of integration, coexists peacefully with its future conqueror, the de Rham-like cohomologies.

The passage from attempts to model the scope of new phenomena in terms of the "old", already existing mathematical language to a new one of a higher level, whose expressive potentials are just adequate to the new demands, is the essence of scheme (1). Here we use "mathematical language" in the spirit of "programming language". This enables us to take into account anthropomorphic elements present implicitly in the theories due to the fact that individual brains and scientific communities are something like computers and computer networks, respectively. The history of metric geometry from its Hellenistic symmetric form based on common logic up to its modern Riemannian form based on calculus gives an ideal illustration of the above scheme.

2. "Troubled times" of quantum field theory

Assuming scheme (1) to be true, it becomes quite clear that nowadays QFT is passing through "troubled times". Even some key words of QFT's current vocabulary, such as "renormalizations", "broken symmetries", "anomalies", "ghosts", etc., indicate a deep discrepancy between its physical content and the mathematical equipment being used. Also, one can see too many Lie groups, algebras, etc., up to quantum and quasi-quantum ones, and symmetry considerations based on them, which play a fundamental role in the structure of modern QFT. This shows that the theory is not sufficiently distinct from its symmetric origin. In fact, the strongest and most obvious argument in favor of these "troubled times" comes from the perturbation type structure of the existing theory. However, the absence of real alternatives, and long habits, have reduced the value of this argument almost to zero.

We realize that the sceptic reader, even convinced of these "troubled times", will prefer to follow the current research activity in expectation of times when the aforementioned natural selection mechanisms will have accomplished their work. So, this text is mainly meant for those who would be interested in seeking some possible artificial selection mechanisms, which, as is well known, work much faster.

At this point, we start to look for this "programming language" for QFT, being motivated by the above "evolution theory". Of course, the latter should be expounded in more detail to be perceived correctly. But we do not take the risk of going more in this direction, remembering the attitude toward any philosophy at the end of the "set-theoretic" epoch we are living in. Instead we invite the reader to return to this point once again after having read the whole text. Also, a development of the above general ideas can be found in [**3**, Ch. 1]. In particular, there we touch such topics as: what anthropomorphic factor stands behind the idea of putting set theory into the foundations of all mathematics, and why calculus is the language of classical physics.

3. "Linguization" of the Bohr correspondence principle

We find the initial data in the following two general postulates, which seem to be beyond doubt:

I. Calculus is the language of classical physics.
II. Classical mechanics is the limit case as $h \to 0$ of quantum mechanics (the Bohr correspondence principle).

These are our initial *position* and *momentum*, respectively.

To avoid misunderstanding, we would like to stress that the word "calculus" is used here, and later on, in its direct sense, i.e., as a system of concepts (say, vector fields, differential forms, differential operators, jets, de Rham's, Spencer's, ..., cohomologies, etc., governed by general rules, or formulas like $d^2 = 0$, $L_X = i_X \circ d + d \circ i_X$, etc.). As we showed in [**123**], they all constitute a sort of "logic algebra" due to the fact that differential calculus can be, in fact, developed in a pure algebraic way over an arbitrary (super)commutative algebra A (see also [**60**, Ch. 1]). This algebraically constructed calculus coincides with the standard one for smooth function algebras $A = C^\infty(M)$. Also, one can learn from this algebraic approach—and this is very important to emphasize—that there are many things to discover and to perceive in order to close this logic algebra, i.e., to get the whole calculus. Higher-order analogs of the de Rham complexes [**128**] give such an example.

Thus, the first postulate suggests looking for an extension of calculus, while the second defines more precisely the direction to aim at. Having this in mind, we need to extract the mathematical essence of Bohr's correspondence principle. The following diagram illustrates how it can be done:

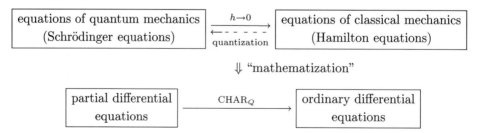

Here CHAR$_\Sigma$ denotes the map which assigns to a given system of partial differential equations $\mathcal{E}$ a system $\mathcal{E}^0_\Sigma$ of ordinary equations describing how Σ-type singularities of solutions of $\mathcal{E}$ propagate. What is meant by Σ-type singularities of solutions and what is, in particular, the above singularity type Q will be discussed later on;

see also [**74, 76, 124, 133**]. But now we shall explain what are the reasons for suspecting CHAR_Q to be behind the Bohr correspondence principle.

First, note that the mathematical background of the passage from wave to geometric optics can be naturally presented in the form $\mathcal{E} \to \mathcal{E}^0_{\text{FOLD}}$, where FOLD stands for the folding type singularity of multi-valued solutions of $\mathcal{E}$ (see §9). On the other hand, multivalence of solutions is related to nonuniqueness of the Cauchy problem and, therefore, to the theory of (bi)characteristics.

REMARK 1. There is a dual way to pass to geometric optics, proposed by Luneburg [**71**] and based on the study of discontinuous solutions. However, the choice of Luneburg's approach instead of ours does not lead to essential changes in our subsequent arguments.

Second, recollecting that Schrödinger discovered his famous equations proceeding from the analogy with wave-geometric optics, one can expect a similar mechanism in the passage from quantum to classical mechanics [**102**]. More precisely, it seems natural to hypothesize that the equations of classical mechanics are the Q-characteristic equations of the corresponding equations of quantum mechanics. These hypothetical Q-characteristic equations should play a similar role with respect to an appropriate "quantum" solution singularity type as the standard characteristic equations do with respect to the singular Cauchy problem. This hypothesis becomes almost evident in the framework of Maslov's approach to semi-classical asymptotics [**83**]. We refer also to the lectures by Levi-Cività [**67**] and the paper by Racah [**96**], one of the first attempts to go this way.

This all motivates us to take the formula

$$\text{QUANTIZATION} = \text{CHAR}_Q^{-1} \qquad (2)$$

as the leading principle, and to seek its consequences.

First of all, the direct attempt to extend (2) to QFT leads immediately to the problem illustrated by the following diagram:

$$\begin{array}{ccc} \boxed{\text{quantum fields}} & \xrightarrow{\hbar \to 0} & \boxed{\text{classical fields}} \\ \Downarrow & \text{"mathematization"} & \Downarrow \\ \boxed{?} & \xrightarrow{\text{CHAR}_Q} & \boxed{\text{partial differential equations}} \end{array} \qquad (3)$$

In other words, we have to discover what kind of mathematical objects are to be placed in the left lower rectangle of (3) or, more precisely, what is the mathematical nature of the equations whose solution singularity propagation is described by means of partial differential equations? The scheme

$$\boxed{\text{ordinary differential equations}} \xleftarrow{\text{CHAR}} \boxed{\text{partial differential equations}} \xleftarrow{\text{CHAR}} \boxed{?}$$

motivates us to call these, yet unknown, mathematical objects *secondary quantized differential equations*.

Thus, the problem to consider next is

What are secondary quantized differential equations? $\qquad (4)$

All the preceding discourses do not furnish us with the necessary impulse to attack it. In searching for such an impulse we consider the simplest situation, when a CHAR-type mapping appears:

$$\sum_i a_i(x)\frac{\partial u}{\partial x_i} = b(x) \xrightarrow{\text{CHAR}} \{\dot{x}_i = a_i(x)\}.$$

In other words, we shall examine the passage from vector fields to ordinary differential equations, making an attempt to understand what secondary ("quantized") vector fields should be.

We can profit from the simplicity of this situation, which, in turn, comes from the symmetry of the context in full accordance with §1. More exactly, infinitesimal symmetries of the system $\dot{x}_i = a_i(x)$ are vector fields $Y = \sum c_i \partial/\partial x_i$ commuting with the field $X = \sum a_i \partial/\partial x_i$; and, as is well known, any vector field admits locally a lot of fields commuting with it. For our purpose it is important to observe that symmetries of the system $\dot{x}_i = a_i(x)$ are objects of the same nature as the differential operator (namely, $X = \sum a_i \partial/\partial x_i$) defining the first-order part of the initial equation $X(u) = b$. For this reason, it seems very likely that secondary quantized vector fields are identical to symmetries of partial differential equations.

So, in checking this hypothesis it is wise to consider the theory of differential equations from the category theory viewpoint, which furnishes a means for drawing an analogy with the category of smooth manifolds.

4. Differential equations are diffieties

A natural generalization of the concept of infinitely prolonged equation motivated by the factorization procedure (see Chapter 3), nonlocal symmetries (see Chapter 6), etc., is given by the following:

DEFINITION 1. A manifold $\mathcal{O}$ supplied with an n-dimensional distribution $\mathcal{C}$ satisfying the Frobenius complete integrability condition is called a *diffiety*[1], if it is locally of the form $\mathcal{E}^\infty$.

The integer n is called the *diffiety dimension* of $\mathcal{O}$ and is denoted by $\text{Dim}\,\mathcal{O}$. Of course, it differs generally from the usual dimension of $\mathcal{O}$, which is equal to infinity as a rule.

EXAMPLE 1 (the projective diffiety). Let $E = E^{n+m}$ be a smooth $(n+m)$-dimensional manifold. The k-jet $[M]_\theta^k$ of an n-dimensional submanifold M at a point $\theta \in E$ is the class of n-dimensional submanifolds of E tangent to M with order k at $\theta \in E$. The totality of k-jets of all possible n-dimensional submanifolds M at all points $\theta \in E$ is denoted by E_n^k. If we recollect the construction of $J^k(\pi)$, $k = 0, \ldots, \infty$, we can easily prove that:

1. The set E_n^k is supplied with a natural structure of smooth manifold.
2. The natural projections $E_n^{k+1} \to E_n^k$ exist.
3. The space $J^\infty(E,n) = E_n^\infty$ defined as the inverse limit of these projections is equipped with the natural completely integrable distribution.

It is also obvious that $J^\infty(E,n)$ has locally the form $J^\infty(\pi)$ for an appropriate m-dimensional bundle π over an n-dimensional manifold and, thus, is a diffiety of dimension n.

---
[1] From "DIFFerential varIETY".

Diffieties are objects of a category called the *category of differential equations*, with the morphisms being the maps that preserve the distribution.

All natural constructions of the theory of differential equations can be extended to arbitrary diffieties with minor or no modifications. In particular, any local isomorphism $\tau\colon \mathcal{O}' \to \mathcal{O}$ is called a *covering* of $\mathcal{O}$, while a symmetry of $\mathcal{O}'$ is said to be a τ-*nonlocal symmetry* of $\mathcal{O}$. Evidently, all τ-symmetries of $\mathcal{O}$ constitute a Lie algebra for a fixed $\tau\colon \mathcal{O}' \to \mathcal{O}$, which coincides with $\operatorname{sym}\mathcal{O}'$. But, at first glance, it seems absurd to look for the commutator of two nonlocal symmetries defined on two different coverings. However, it turns out to be possible to find the desired commutator on a suitable third covering. Therefore, in order to organize all nonlocal symmetries of $\mathcal{O}$ into something like a Lie algebra one must take into consideration all coverings of $\mathcal{O}$ simultaneously.

All coverings of a given diffiety constitute in a natural way a category which we called a cobweb (see [**62**]). This construction implies many important consequences for Secondary Calculus. But at the moment this is only a beautiful perspective to be explored systematically.

We remark also that, while diffieties are analogs of affine varieties of algebraic geometry, cobwebs are analogs of fields of rational functions on them.

There are two different natural inclusions of the category of finite-dimensional manifolds in the category of differential equations:

1. $M^n \Rightarrow J^\infty(M^n, n)$.
2. $M^n \Rightarrow J^\infty(M^n, 0)$.

In the first case an n-dimensional manifold M goes to an n-Dimensional diffiety, i.e., $\dim M = \operatorname{Dim} M$, whereas in the second case $\operatorname{Dim} M = 0$. These two ways are, in a sense, dual to each other, and therefore lead to a kind of duality in the theory of differential equations. Also, one can see that the traditional "differential" mathematics (calculus, geometry, equations, etc.) viewed as a part of diffiety theory becomes conceptually closed only if the underlying manifolds are understood to be zero-dimensional diffieties. In other words, the standard "differential" mathematics forms a zero-dimensional part of diffiety theory. So, from this point of view it would be quite natural to suspect that the relevant mathematics necessary to quantize classical fields smoothly has for the most part not yet been discovered.

Let us now return to the problem posed at the end of the previous section and ask ourselves: what does the notion of symmetry mean when applied to zero-dimensional diffieties, i.e., to usual manifolds. Obviously, $\operatorname{sym} M = \operatorname{D}(M)$, where $\operatorname{D}(M)$ stands for the Lie algebra of all vector fields on M. In an arbitrary coordinate system $u^1, \ldots, u^m$ on M, the standard coordinate expression

$$X = \sum_i \varphi_i(u) \frac{\partial}{\partial u^i}, \qquad u = (u^1, \ldots, u^m),$$

for a vector field X on M is a particular case of formula (2.15) of Ch. 4 for evolutionary derivation,

$$\mathcal{D}_\varphi = \sum_{i,\sigma} D_\sigma(\varphi_i) \frac{\partial}{\partial u^i}.$$

Recall that to each evolutionary derivation $\mathcal{D}_\varphi$ there corresponds the evolution equation (2.19) of Ch. 4,

$$\frac{\partial u^i}{\partial t} = \varphi_i(x, u, \ldots, u^j_\sigma, \ldots), \qquad i = 1, \ldots, m. \tag{5}$$

In the case $\vartheta_\varphi = X$, system (5) takes the form

$$\frac{du^i}{dt} = \varphi_i(u^1,\ldots,u^m), \qquad i = 1,\ldots,m. \tag{6}$$

But in (6) we recognize the ordinary differential equations of characteristics for the first-order partial differential operator $X = \sum_i \varphi_i \partial/\partial u^i$. Now the analogy $X \mapsto \vartheta_\varphi$, (6) $\mapsto$ (5) motivates the following principal statement:

> If $\varphi = (\varphi_1,\ldots,\varphi_m)$ is the generating function of a symmetry $\chi \in \operatorname{sym} \mathcal{E}^\infty$, then the system (5) of partial differential equations can be naturally considered as the characteristic system corresponding to the operator χ.

By virtue of (3), this gives the desired solutions of the main problem (4) for first-order equations:

> Symmetries of partial differential equations are first-order secondary quantized differential operators.

This is our starting point when we look for secondary quantized differential operators of higher order.

5. Secondary ("quantized") functions

It seems natural to define higher-order secondary quantized differential operators as compositions of first-order ones. But in going this way we immediately meet the following difficulty.

Recall that first-order secondary differential operators are elements of the Lie algebra $\operatorname{sym} \mathcal{O}$. On the other hand, they are not proper differential operators but cosets (equivalence classes) of them. So the question arises: how to compose the cosets? We leave to the interested reader the task of verifying that a direct attack to this problem fails.

Another aspect of this problem can be extracted from a similar question: on what kind of objects do secondary differential operators act? No doubt, secondary differential operators should be proper operators, i.e., act on some kind of objects. The usual functions cannot be taken as such. One can see this by trying to define an action of the algebra $\operatorname{sym} \mathcal{O}$ on $C^\infty(\mathcal{O})$. The only natural way to do this is to put $\chi(f) = X(f)$ for $\chi \in \operatorname{sym} \mathcal{O}$, $X \in \mathrm{D}_\mathcal{C}(\mathcal{O})$, $\chi = X \mod \mathcal{C}\mathrm{D}(\mathcal{O})$, and $f \in C^\infty(\mathcal{O})$. But this definition is clearly not correct. Namely, if $X_1, X_2 \in \mathrm{D}_\mathcal{C}(\mathcal{O})$ are two different vector fields and $X_1 \equiv X_2 \mod \mathcal{C}\mathrm{D}(\mathcal{O})$, then, generally, $X_1(f) \neq X_2(f)$.

However, it is clear that, linguistically, secondary operators should act on secondary functions, and we shift this question by asking what they are. The following analogy will help us to answer.

Let $\Lambda^i(M)$ denote the space of differential i-forms on the manifold M. The map

$$C^\infty(M) \xrightarrow{d} \Lambda^1(M) \tag{7}$$

provides extremal problems on smooth functions on M with the "universal solution". Treating smooth manifolds as zero-dimensional diffieties, we see that the analog of (7) should be a map providing variational problems for multiple integrals with the universal solution. But this is the well-known Euler–Lagrange map:

$$\boxed{\text{variational functionals, or "actions"}} \xrightarrow{\mathsf{E}} \boxed{\text{differential operators}} \qquad (8)$$

i.e., E associates with an "action" $\int_\Omega L\, dx_1 \cdots dx_n$ the left-hand side of the corresponding Euler–Lagrange equation. Therefore, this analogy between (7) and (8) suggests that one should adopt "actions" as secondary (or "quantized") functions. This idea is to be corrected because "actions", understood in the standard way, contain a detail that is malignant to our aim and must be eliminated. This detail is the explicit reference to the domain of integration Ω. So, our next problem is to find a meaning for hieroglyphs of the form $\int L\, dx_1 \cdots dx_n$ (without "Ω"!). We shall solve it by interpreting them as some kind of cohomology classes (for more motivations see [**38**]). But before that, we need some preliminaries (cf. Chapter 4).

Let $\mathcal{O}$ be a diffiety, $\operatorname{Dim} \mathcal{O} = n$. Observe that there is the horizontal de Rham complex on $\mathcal{O}$, which is constructed similarly to that on an infinitely prolonged equation. Namely, consider the submodule $\mathcal{C}\Lambda^i(\mathcal{O})$ of $\Lambda^i(\mathcal{O})$ consisting of forms whose restrictions to the contact distribution of $\mathcal{O}$ vanish:

$$\omega \in \mathcal{C}\Lambda^i(\mathcal{O}) \iff (Y_1, \ldots, Y_i) = 0 \qquad \forall Y_1, \ldots, Y_i \in \mathcal{C}\mathrm{D}(\mathcal{O}).$$

We put

$$\Lambda_0^i(\mathcal{O}) = \Lambda^i(\mathcal{O})/\mathcal{C}\Lambda^i(\mathcal{O}).$$

Elements of $\Lambda_0^i(\mathcal{O})$ are called *horizontal differential forms* on $\mathcal{O}$. Evidently, $\Lambda_0^i(\mathcal{O}) = 0$ for $i > n$.

It is easy to see that $d(\mathcal{C}\Lambda^i(\mathcal{O})) \subset \mathcal{C}\Lambda^{i+1}(\mathcal{O})$ and, therefore, the standard differential d induces the *horizontal differential*

$$\widehat{d} \colon \Lambda_0^i(\mathcal{O}) \to \Lambda_0^{i+1}(\mathcal{O}).$$

Of course, $\widehat{d}^2 = 0$, and this enables us to introduce the horizontal de Rham complex of $\mathcal{O}$. Its cohomology is called the *horizontal de Rham cohomology* of $\mathcal{O}$ and is denoted by $\bar{H}^i(\mathcal{O})$, $i = 0, \ldots, n$.

Finally, we accept the following basic interpretation:

> *Secondary (or "quantized") functions on $\mathcal{O}$ are elements of the cohomology group $\bar{H}^n(\mathcal{O})$.*

In other words, we consider the cohomology group $\bar{H}^n(\mathcal{O})$, $\operatorname{Dim} \mathcal{O} = n$, to be an analog of the smooth function algebra in Secondary Calculus.

To justify our choice, we shall describe in coordinates the "horizontal" constructions just given, for $\mathcal{O} = J^\infty(E, n)$ (see Chapters 4 and 5). First of all, we observe that the coset of a differential form $\omega \in \Lambda^i(J^\infty(E, n))$ contains only one element of the form

$$\rho = \sum_{1 \le k_1 < \cdots < k_i \le n} a_{k_1, \ldots, k_i}(x, u, \ldots, u_\sigma^j, \ldots)\, dx_{k_1} \wedge \cdots \wedge dx_{k_i},$$

where $a_{k_1, \ldots, k_i} \in C^\infty(J^\infty(E, n))$. The characteristic feature of such a form is that the differentials du^j, du_σ^j do not enter into its coordinate expression. Therefore, the module $\Lambda_0^i(J^\infty(E, n))$ can be identified locally with the module of forms of this type.

Under this identification, the horizontal de Rham differential $\widehat{d}$ is
$$\widehat{d}\rho = \sum_{s,k_1,\ldots,k_i} D_s(a_{k_1,\ldots,k_i})\, dx_s \wedge dx_{k_1} \wedge \cdots \wedge dx_{k_i},$$
where D_s is the s-th total derivative. In particular, every horizontal $(n-1)$-form can be uniquely represented as
$$\rho = \sum_i (-1)^{i-1} a_i\, dx_1 \wedge \cdots \wedge dx_{i-1} \wedge dx_{i+1} \wedge \cdots \wedge dx_n, \qquad a_i \in C^\infty(J^\infty(E,n)),$$
and
$$\widehat{d}\rho = \operatorname{div}(A)\, dx_1 \wedge \cdots \wedge dx_n,$$
where $A = (a_1, \ldots, a_n)$ and $\operatorname{div}(A) = \sum_i D_i(a_i)$. Also, horizontal n-forms look like
$$L(x, u, \ldots, u_\sigma^j, \ldots)\, dx_1 \wedge \cdots \wedge dx_n, \qquad L \in C^\infty(J^\infty(E,n)),$$
and one can recognize Lagrangian densities in them. So, we see that the horizontal cohomology $\bar{H}^n(J^\infty(E,n))$ can be identified locally with the linear space of equivalence classes of Lagrangian densities on $J^\infty(E,n)$ with respect to the following relation:
$$L_1 \sim L_2 \iff L_1 - L_2 = \operatorname{div}(A) \text{ for some } A.$$
On the other hand, actions $\int_\Omega L_i\, dx_1 \wedge \cdots \wedge dx_n$, $i = 1, 2$, are equivalent in the sense that they lead to identical Euler–Lagrange equations if and only if $L_1 \sim L_2$. This is independent of the choice of Ω. For these reasons, it is natural to identify hieroglyphs $\int L\, dx_1 \wedge \cdots \wedge dx_n$ with n-dimensional horizontal cohomology classes.

We conclude this section by noting that similar reasoning is valid for arbitrary diffieties.

6. Higher-order scalar secondary ("quantized") differential operators

First of all, we must justify our definition of secondary functions by demonstrating that first-order differential operators really act naturally on them. In other words, we must look for a natural action of the algebra $\operatorname{sym} \mathcal{O}$ on the space $\bar{H}^n(\mathcal{O})$. This, however, can be done straightforwardly.

First, note that if $X \in D_\mathcal{C}(\mathcal{O})$, $\omega \in \mathcal{C}\Lambda^i(\mathcal{O})$, and L_X denotes the Lie derivative along X, then $L_X(\omega) \in \mathcal{C}\Lambda^i(\mathcal{O})$, as follows from the definitions and the fact that
$$\mathcal{C}\Lambda^i(\mathcal{O}) = \mathcal{C}\Lambda^1(\mathcal{O}) \wedge \Lambda^{i-1}(\mathcal{O}), \qquad i > 1.$$
This allows us to define the Lie derivative on horizontal forms by passing to quotients:
$$L_X : \Lambda_0^i(\mathcal{O}) \to \Lambda_0^i(\mathcal{O}).$$
Next, let
$$\chi \in \operatorname{sym} \mathcal{O}, \qquad \chi = X \mod \mathcal{CD}(\mathcal{O}) \qquad \text{for } X \in D_\mathcal{C}(\mathcal{O}),$$
$$\theta \in \bar{H}^n(\mathcal{O}), \qquad \theta = \omega \mod \widehat{d}\Lambda_0^{n-1}(\mathcal{O}) \qquad \text{for } \omega \in \Lambda_0^n(\mathcal{O}).$$
We now define the action of χ on θ by putting
$$\chi(\theta) = L_X(\omega) \mod \widehat{d}\Lambda_0^{n-1}(\mathcal{O}) \in \bar{H}^n(\mathcal{O}).$$
The fact that the action is well defined is implied by the following two statements:

1. $L_Y(\omega) \in \widehat{d}\Lambda_0^{n-1}(\mathcal{O})$ if $\omega \in \Lambda_0^n(\mathcal{O})$ and $Y \in \mathcal{CD}(\mathcal{O})$.

2. $L_X \circ \widehat{d} = \widehat{d} \circ L_X$.

They both are direct consequences of the definitions.

Now we see that the above definition of secondary functions correlates nicely with other "secondary" constructions and, therefore, can serve as an example in proceeding to more complicated "secondary" notions. For example, let us observe that we have succeeded in defining a correct action of one quotient (namely, $\mathrm{sym}\,\mathcal{O} = \mathrm{D}_{\mathcal{C}}(\mathcal{O})/\mathcal{CD}(\mathcal{O}))$ on another (namely, $\bar{H}^n(\mathcal{O}) = \Lambda_0^n(\mathcal{O})/\widehat{d}\Lambda_0^{n-1}(\mathcal{O})$), because:

1. The quotient $\mathrm{D}_{\mathcal{C}}(\mathcal{O})$ consists of first-order differential operators which act on the $C^\infty(\mathcal{O})$-module $\Lambda_0^n(\mathcal{O})$ leaving $\widehat{d}\Lambda_0^{n-1}(\mathcal{O})$ invariant.
2. The images of $\Lambda_0^n(\mathcal{O})$ under the action of first-order operators belonging to $\mathcal{CD}(\mathcal{O})$ are contained in $\widehat{d}\Lambda_0^{n-1}(\mathcal{O})$.

We shall get the necessary generalization to higher-order secondary differential operators simply by replacing "first-order" by "k-th order" in 1 and 2 above. More exactly, let $\mathrm{Diff}_k(\Lambda_0^n(\mathcal{O}))$ denote the $C^\infty(\mathcal{O})$-module of all ("usual") differential operators of order $\leq k$ acting on $\Lambda_0^n(\mathcal{O})$, and put

$$\overline{\mathrm{Diff}}_k(\mathcal{O}) = \{\,\Delta \in \mathrm{Diff}_k(\Lambda_0^n(\mathcal{O})) \mid \Delta(\widehat{d}\Lambda_0^{n-1}(\mathcal{O})) \subset \widehat{d}\Lambda_0^{n-1}(\mathcal{O})\,\},$$

$$\underline{\mathrm{Diff}}_k(\mathcal{O}) = \{\,\Delta \in \mathrm{Diff}_k(\Lambda_0^n(\mathcal{O})) \mid \Delta(\Lambda_0^n(\mathcal{O})) \subset \widehat{d}\Lambda_0^{n-1}(\mathcal{O})\,\}.$$

Then the space of all *scalar secondary* (*"quantized"*) *operators* of order $\leq k$ on $\mathcal{O}$ is defined to be the quotient

$$\mathfrak{Diff}_k(\mathcal{O}) = \overline{\mathrm{Diff}}_k(\mathcal{O})/\underline{\mathrm{Diff}}_k(\mathcal{O}). \tag{9}$$

Of course, every secondary operator $\Delta \in \mathfrak{Diff}_k(\mathcal{O})$ can be understood as an operator

$$\Delta \colon \bar{H}^n(\mathcal{O}) \longrightarrow \bar{H}^n(\mathcal{O}),$$

acting on secondary functions. In fact, if

$$\Delta = \delta \mod \underline{\mathrm{Diff}}_k(\mathcal{O}) \qquad \text{for } \delta \in \overline{\mathrm{Diff}}_k(\mathcal{O}),$$

$$\Theta = \omega \mod \widehat{d}\Lambda_0^{n-1}(\mathcal{O}) \qquad \text{for } \omega \in \Lambda_0^n(\mathcal{O}),$$

then the horizontal cohomology class

$$\Delta(\Theta) = \delta(\omega) \mod \widehat{d}\Lambda_0^{n-1}(\mathcal{O})$$

is well defined, i.e., does not depend on the choice of the representatives δ and ω.

For $\mathcal{O} = J^\infty(E, n)$, the secondary differential operators thus defined admit the following coordinate description. Operators of the form

$$\sum_{s=1}^{k} \sum_{\substack{i_1,\ldots,i_s \\ \sigma_1,\ldots,\sigma_s}} a^{i_1,\ldots,i_s}_{\sigma_1,\ldots,\sigma_s} \frac{\partial^s}{\partial u^{i_1}_{\sigma_1} \ldots \partial u^{i_s}_{\sigma_s}} + \mathrm{const},$$

where $\sigma_1, \ldots, \sigma_s$ are multi-indices, are called *vertical* (with respect to the chosen coordinate system). Then it can be proved that every coset

$$\Delta = \delta \mod \underline{\mathrm{Diff}}_k(J^\infty(E,n)) \in \mathfrak{Diff}(J^\infty(E,n)),$$

for $\delta \in \overline{\mathrm{Diff}}_k(J^\infty(E,n))$, contains only one vertical operator. So, the quotient (9) representing secondary differential operators can be identified locally with the set of all *vertical secondary operators*. These operators (of order $\leq k$) can be represented in the form

$$\vartheta_\nabla = \sum_{i=1}^m \sum_\sigma \mathcal{L}(\nabla^i) \circ \frac{\partial}{\partial p^i_\sigma}, \qquad \nabla = (\nabla^1, \ldots, \nabla^m),$$

where $\nabla^i \in \text{Diff}_{k-1}(C^\infty(J^\infty(E,n)))$ are arbitrary $C^\infty(J^\infty(E,n))$-vertical operators and

$$\mathcal{L}_\sigma(\nabla^i) = [D_{i_1}, \ldots, [D_{i_s}, \nabla^i] \ldots],$$

where $\sigma = (i_1, \ldots, i_s)$. The generating operator ∇ is the higher-order analog of the generating functions for evolutionary derivations, but, unlike the latter, it is not defined uniquely if $k > 1$.

Secondary operators of order > 1 are not reduced to compositions of first-order secondary operators. This fact is instructive in connection with the discussion at the beginning of §5.

We note also that an explicit description of secondary differential operators on arbitrary diffieties is a much more difficult problem.

Further details, results, and alternative views concerning secondary differential operators can be found in [**38**].

Finally, turning back to question (4), we can exhibit the simplest k-th order linear secondary (quantized) differential equation as

$$\Delta(H) = 0, \qquad \Delta \in \mathfrak{Diff}_l(\mathcal{O}), \qquad H \in \bar{H}^n(\mathcal{O}).$$

It must be emphasized, however, that these equations form a very special class of secondary differential equations. For instance, differentials $d_1^{p,q} = d_1^{p,q}(\mathcal{O})$ of the $\mathcal{C}$-spectral sequence (see Chapter 5 and §7) give us other examples of secondary quantized differential operators and, therefore, secondary differential equations. One of them looks like

$$\mathsf{E}\left(\int L\,dx\right) = 0,$$

where $\int L\,dx \in \bar{H}^n(\mathcal{O})$ and E is the Euler operator assigning to an action $\int L\,dx$ the corresponding Euler–Lagrange equation. This is due to the fact that $\mathsf{E} = d_1^{0,n}$. Note also that the operators $d_1^{p,q}$ are of finite order, say $\text{ord}(k)$, when restricted to elements of the k-th filtration, but $\text{ord}(k) \to \infty$ as $k \to \infty$. For example, $\text{ord}(k) = 2k$ for the operator E.

7. Secondary ("quantized") differential forms

In this section, we shall consider another aspect of Secondary Calculus, namely, secondary ("quantized") differential forms. What are they? This is a more difficult question than the one about secondary differential operators we have already discussed. For this and other reasons we shall omit here the preliminary motivations, showing, as before, how to arrive at exact definitions. However, some *a posteriori* justifications will be given.

Let $\mathcal{O}$ be a diffiety. Adopting the notation of §6, we consider the algebra

$$\Lambda^*(\mathcal{O}) = \sum_{i \geq 0} \Lambda^i(\mathcal{O})$$

of all differential forms on $\mathcal{O}$ and its ideal
$$\mathcal{C}\Lambda^*(\mathcal{O}) = \sum_{i \geq 0} \mathcal{C}\Lambda^i(\mathcal{O}).$$

Denote by $\mathcal{C}^k\Lambda^*(\mathcal{O})$ the k-th power of this ideal. Then all ideals $\mathcal{C}^k\Lambda^*(\mathcal{O})$ are stable with respect to the exterior differential d. So, we get the following filtration:
$$\Lambda^*(\mathcal{O}) \supset \mathcal{C}\Lambda^*(\mathcal{O}) \supset \mathcal{C}^2\Lambda^*(\mathcal{O}) \supset \cdots \supset \mathcal{C}^k\Lambda^*(\mathcal{O}) \supset \cdots$$
of the de Rham complex of $\mathcal{O}$ by its subcomplexes $\{\mathcal{C}^k\Lambda^*(\mathcal{O}), d\}$. The ideals $\mathcal{C}^k\Lambda^*(\mathcal{O})$ are naturally graded,
$$\mathcal{C}^k\Lambda^*(\mathcal{O}) = \sum_{s \geq 0} \mathcal{C}^k\Lambda^{k+s}(\mathcal{O}),$$
where
$$\mathcal{C}^k\Lambda^{k+s}(\mathcal{O}) = \mathcal{C}^k\Lambda^*(\mathcal{O}) \cap \Lambda^{k+s}(\mathcal{O}).$$

Thus, as in Chapter 5 for infinitely prolonged equations, we can construct the $\mathcal{C}$-spectral sequence
$$E_r(\mathcal{O}) = \sum_{p,q} E_r^{p,q}(\mathcal{O}), \qquad d_r = \sum_{p,q} d_r^{p,q},$$
of the diffiety $\mathcal{O}$.

If $\mathrm{Dim}\,\mathcal{O} = n$, then all nontrivial terms of $E_r^{p,q}(\mathcal{O})$ are situated in the region $p \geq 0$, $0 \leq q \leq n$, i.e., as in the case of infinitely prolonged equations.

DEFINITION 2. Elements of $E_1(\mathcal{O})$ are called *secondary ("quantized") differential forms* on the diffiety $\mathcal{O}$.

Some reasons in favor of this interpretation are as follows. Let M be a finite-dimensional manifold considered as a zero-dimensional diffiety (see §4). Then the term E_1 of its $\mathcal{C}$-spectral sequence degenerates to one line:
$$0 \longrightarrow \Lambda^0(M) \xrightarrow{d_1^{0,0}} \Lambda^1(M) \xrightarrow{d_1^{1,0}} \Lambda^2(M) \longrightarrow \cdots \longrightarrow \Lambda^m(M) \longrightarrow 0.$$
Moreover, $d_1^{p,0} = d$. In other words, we see that the de Rham complex of M coincides with the (generally) nontrivial part of the first term of its $\mathcal{C}$-spectral sequence.

We can observe that standard constructions and formulas connecting "usual" vector fields and differential forms are also valid for their secondary ("quantized") analogs. For instance, the inner product operator of secondary vector fields ("symmetries") with secondary differential forms, and also the corresponding Lie derivatives, are well defined. Moreover, they are related by means of the secondary analog of the infinitesimal Stokes formula
$$L_X = i_X \circ d + d \circ i_X,$$
in which the exterior differential d is to be replaced by its secondary analog, i.e., by $d_1(\mathcal{O})$.

Finally, we remark that secondary differential forms are bigraded objects unlike the "usual" ones, which are only monograded. The reason is clearly seen from the above diagrams. This is an illustration of the fact that secondary objects are richer and more complicated structures than their "primary" analogs. The same idea can be expressed alternatively by saying that the "usual" (or "primary") mathematical objects are degenerate forms of the secondary ones. This statement can be also viewed as the following mathematical paraphrase of the Bohr correspondence principle:

$$\boxed{\text{Secondary Calculus}} \xrightarrow{\text{Dim}\to 0} \boxed{\text{Calculus}}$$

The cobweb theory (see §4) allows us to give an exact meaning to "Dim → 0". This is because the Dimension (not dimension!) is an $\mathbb{R}$-valued function in the framework of this theory.

We saw in Chapter 5 how the $\mathcal{C}$-spectral sequence works. Let us make some further remarks on the term E_2. In the case of infinitely prolonged equations, $E_2(\mathcal{E}^\infty)$ consists of characteristic classes of bordisms composed of solutions of $\mathcal{E}$. The standard "differential characteristic classes" theories can be obtain in this way under a suitable choice of $\mathcal{E}$ (see [**121, 138**]). This approach leads, however, to finer characteristic classes, for instance, *special characteristic classes*. We illustrate this topic for solutions of the (vacuum) Einstein equations, or Einstein manifolds. Let $\mathcal{E}$ be the Einstein system on a manifold M. Then it is possible to show that

$$\mathcal{E}^\infty / \operatorname{Diffeo}(M) = (\mathcal{E}')^\infty,$$

where $\operatorname{Diffeo}(M)$ is the diffeomorphism group of M acting naturally on $\mathcal{E}^\infty$ and $\mathcal{E}'$ is a certain system of partial differential equations. In fact, $\mathcal{E}'$ does not depend on M, and therefore its solutions are just diffeomorphism classes of Einstein manifolds. The corresponding characteristic classes are elements of $E_2((\mathcal{E}')^\infty)$.

8. Quantization or singularity propagation? Heisenberg or Schrödinger?

In the preceding pages, we have reached the coasts of "terra incognita", i.e., diffieties and Secondary Calculus on them, whose existence was predicted by the linguistic version of the Bohr correspondence principle as formulated in §3. Being the exact analog of algebraic geometry for partial differential equations, this branch of pure mathematics deserves to be explored systematically, maybe much more than algebraic geometry itself and independently of the possible physical applications that stimulated the expedition. Later on we shall discuss briefly some other topics related to Secondary Calculus. But now it would be timely to reexamine how close to the solution of the quantization problem for quantum fields we are now that we have Secondary Calculus at our disposal.

It should be stressed from the very beginning that the passage to the "linguistic" version of the Bohr principle inevitably costs us the loss of its original physical context. On the other hand, the accumulated experience in Secondary Calculus convinces us that every *natural* construction in the classical calculus has its secondary analog, which can be found by means of a more or less regular procedure. So, one can expect to deduce fundamental QFT equations by "secondarizing" sample situations in which both the source and the target of the Bohr principle belong to the area of classical calculus.

Evidently, quantum mechanics of particles is exactly such a sample, because the Bohr correspondence principle here starts from differential (Schrödinger) equations and finishes also at differential (Hamilton) equations. However, the Bohr principle in this case is to be reinterpreted exclusively in terms of calculus to become secondarizable. This is the key point.

The desired reinterpretation is not obvious and, in particular, should not be based on the "$h \to 0$" trick, formal series in h, deformations, Hilbert spaces, and similar things. We accept formula (2) to be the first approximation. Then our approach to QFT can be summarized as

$$\text{FIELD QUANTIZATION} = \text{CHAR}_{\mathcal{S}(Q)}^{-1},$$

where $\mathcal{S}$ stands for "secondarization". Hence, the question to be answered first is: what is the solution singularity type (or types) outlined in §3?

The last problem belongs to the theory of solution singularities of partial differential equations, which has not been elaborated enough up to now to provide us with the immediate answer. So, we postpone the direct attack for the future, and limit ourselves here to a quick trip through the theory of some special solution singularities called *geometric*. In addition, the reader can conceive from this model more precise ideas on the general theory as well as more detailed motivation for formula (2). But first we shall make some remarks of a historical nature.

As we know, two different approaches, one by Heisenberg and the other by Schrödinger, were at the origin of quantum mechanics. In modern terms, the first is based on a formal noncommutative deformation of the commutative algebra of classical observables, while the latter proceeds from an analogy with optics. The two of them were proclaimed and even proved equivalent—but this is just the point we would like to doubt now. Namely, it seems that a more exact formulation of this equivalence theorem would be:

The Schrödinger point of view becomes equivalent to the Heisenberg one after being reduced appropriately.

Below some brief general justifications for this assertion are given, and the reader is asked not to confuse "approach" with "picture" in what follows.

First, the Heisenberg approach is "programmed" in the language of operator algebras, while Schrödinger's is in calculus. The former is nonlocalizable in principle, and this is its great disadvantage for applications to fundamental (nontechnical) problems of physics. In particular, the passage from one space-time domain to another cannot be expressed in terms of this language only (see, e.g., [**39**]). But, evidently, fundamental physical theories, at both classical and quantum levels, must be localizable in this sense by their nature. On the other hand, calculus is the only localizable language, due to the fact that localizable operators are just differential ones.

Second, in the Heisenberg approach classical mechanics appears to be a limit case of quantum mechanics or, vice versa, the latter is viewed to be a noncommutative deformation of the former. In particular, this means that they both are treated as things of the same nature differing from each other by a parameter. This is not so in the framework of the Schrödinger approach. In fact, as follows from the general mathematical background of the passage from wave to geometric optics, the latter appears to be a particular aspect of the former. So, applying the analogy between quantum mechanics and optics discovered by Schrödinger, one can conclude that

classical mechanics is a particular aspect of quantum mechanics.

In this connection it would be relevant to note that Planck's constant is a true constant and, therefore, "$\hbar \to 0$" can serve as a trick but not as a foundation stone of the theory.

Thus these are, briefly, the reasons in favor of the Schrödinger alternative. On the other hand, clearly it had no chance to be realized mathematically in the building period of quantum electrodynamics and other quantum field theories. So, the

Heisenberg alternative remained, due to its formality and abstraction, the only possible way for progress in these theories. This was its invaluable historical merit, but it seems almost exhausted now. Finally, we add that this section can be regarded also as an attempt to provide the Schrödinger approach with the mathematical tools which are necessary to extend it to QFT.

9. Geometric singularities of solutions of partial differential equations

In this section we present geometric singularities of solutions of (nonlinear) partial differential equations and some general results on them relevant to our discussion of the quantization problem. Some examples illustrating the general theory and, in particular, the mechanism connecting wave and geometric optics are collected in the next section.

Solution singularities which we called geometric arise naturally in the context of the theory of multi-valued solutions of (nonlinear) partial differential equations. There are different ways to rigorously realize the idea of multivalence, and we choose the one based on the notion of R-manifold. This is as follows.

Recall (see Chapters 3 and 4) that a submanifold $W \subset J^k(E, n)$, $0 \leq k \leq \infty$, is called *integral* if $T_\theta W \subset \mathcal{C}_\theta$ for every $\theta \in J^k(E, n)$ ($\mathcal{C}_\theta$ stands for the Cartan plane at θ). An integral submanifold W is called *locally maximal* if no open part of it belongs to another integral submanifold of greater dimension.

DEFINITION 3. A locally maximal n-dimensional integral submanifold of $J^k(E, n)$ is called an *R-manifold*.

In particular, the graphs $L_{(k)} = M_f^k$ of jets of sections of a bundle $\pi \colon E \to M$ are R-manifolds. They are characterized by the following two properties:
1. $L_{(k)}$ is a locally maximal integral manifold.
2. The restriction of the projection
$$\pi_{k,k-1} \colon J^k(E,n) \to J^{k-1}(E,n)$$
to $L_{(k)}$ is an immersion.

So, omitting property 2, we get the multi-valued analogs of submanifolds $L_{(k)}$, i.e., R-manifolds.

REMARK 2. Recall that there exist different types of locally maximal integral submanifolds of $J^k(E, n)$ which differ from each other by their dimensions. One of these types is formed by fibers of the projection $\pi_{k,k-1}$. These are integral submanifolds of the greatest possible dimension.

Informally, R-manifolds can be treated, generally, as nonsmooth n-dimensional submanifolds of $E = J^0(E, n)$ whose singularities can be resolved by lifting them to a suitable $J^k(E, n)$.

Now we define a *multi-valued solution* of a partial differential equation $\mathcal{E} \subset J^k(E, n)$ to be an R-submanifold, say W, belonging to one of its prolongations $\mathcal{E}^{(s)} \subset J^{k+s}(E, n)$, $0 \leq s < \infty$.

If $W \subset J^k(E, n)$ is an R-manifold, then its *singular* (or *branch*) *points* are defined to be the singular points of the projection $\pi_{k,k-1} \colon J^k(E,n) \to J^{k-1}(E,n)$ restricted to W (see Figure 1).

We stress here that W is a smooth (= nonsingular) submanifold of $J^k(E, n)$, and the adjective "singular" refers to the projection $\pi_{k,k-1}$.

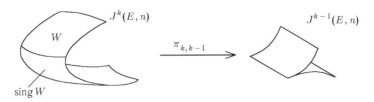

Figure 1

A very rich and interesting structural theory stands behind these simple definitions. This cannot be reduced to the standard singularity (or catastrophe) theory. On the contrary, the latter is a particular degenerate case of the former.

We start with a classification of geometric singularities, which is, of course, the first structural problem to be considered. According to "general principles" we have to classify s-jets of R-manifolds in $J^k(E, \pi)$ for a prescribed integer s under the group of contact transformations of this jet space. The simplest case $s = 1$ is sufficient for our purposes.

Let $W \subset J^k(E, n)$ be an R-manifold and $\theta \in \operatorname{sing} W$. The subspaces of the tangent space $T_\theta J^k(E, n)$ which are of the form $T_\theta W$ are called *singular R-planes* (at θ). So, our problem is to classify singular R-planes.

Let $P = T_\theta W$ be a singular R-plane at θ. The subspace P_0 of P that consists of vectors annihilated by $(\pi_{k,k-1})_*$ is called the *label* of P. It turns out that singular R-planes are equivalent if and only if their labels are equivalent. So, the classification problem in question is reduced to the label classification problem. We define the *type* of a (singular or not) R-plane to be the dimension of its label:

$$\operatorname{type} P = \dim P_0, \quad 0 \leq \operatorname{type} P \leq n.$$

Obviously, $\operatorname{type} P = 0$ if and only if P is nonsingular.

EXAMPLE 2. Branched Riemannian surfaces are identical to multi-valued solutions of the classical Cauchy–Riemann equation. Let W be one of them. Then the set $\operatorname{sing} W$ consists of a number of isolated points, say θ_α. In this case, type $P_\alpha = 2$ for $P_\alpha = T_{\theta_\alpha} W$.

The final result of the label classification is as follows [140]:

THEOREM 1. *Label equivalent classes of geometric singularities are in one-to-one correspondence with isomorphic classes of unitary commutative $\mathbb{R}$-algebras, so that the dimension of a label is equal to that of the algebra corresponding to it.*

Recall that every unitary commutative finite-dimensional algebra splits into a direct sum of algebras $\mathbb{F}_{(k)}$, $k = 1, 2, \ldots$, where $\mathbb{F}_{(k)}$ denotes the unitary $\mathbb{F}$-algebra generated by one element ξ such that $\xi^k = 0$, $\xi^{k-1} \neq 0$ and $\mathbb{F} = \mathbb{R}$ or $\mathbb{C}$. Such a splitting is not unique, but the multiplicity numbers showing how many times a given algebra $\mathbb{F}_{(k)}$ enters it do not depend on the splitting. So, these multiplicity numbers completely determine the isomorphism class of the algebra in question.

Below we speak of A-type geometric singularities, referring to the commutative algebra A corresponding to it by the above theorem.

EXAMPLE 3. Since the only one-dimensional $\mathbb{R}$-algebra is $\mathbb{R}$ itself, there exists only one label type of geometric singularities with one-dimensional label. This type is realized by R-manifolds projected to the manifold of independent variables as foldings. By this reason, it is denoted by FOLD. The standard theory of characteristic covectors plays a natural part in the FOLD-singularity theory.

EXAMPLE 4. There are just three isomorphic classes of two-dimensional unitary commutative algebras, namely, those of $\mathbb{C}$, $\mathbb{R}_{(2)}$, and $\mathbb{R} \oplus \mathbb{R}$, where (see above) $\mathbb{R}_{(2)} = \{\, 1, \xi \mid \xi^2 = 0 \,\}$. For equations with two independent variables, $\mathbb{C}$-type geometric singularities look like ramification points of Riemannian surfaces and like $(n-2)$-dimensional families of such for n independent variables. In four-dimensional space-time, a $\mathbb{C}$-type singularity can be viewed as a vortex around a moving curve. This sets ablaze the suspicion that $\mathbb{C}$-singularities could play an important role in the future of turbulence theory.

The next question that arises immediately when studying concrete equations is:

> *What label type geometric singularities does a given system of partial differential equations admit?*

This is an essentially algebraic problem, which we illustrate with the following examples, omitting the general discussion.

EXAMPLE 5. A system of partial differential equations admits FOLD-type singularities only if it admits nonzero characteristic covectors. For instance, solutions of elliptic equations do not admit FOLD-singularities.

EXAMPLE 6. Let $\mathcal{E}$ be a second-order scalar differential equation with two independent variables. Then it admits only one of the three types of two-dimensional singularities mentioned above. This is the $\mathbb{C}$-type for elliptic equations, the $\mathbb{R}_{(2)}$-type for parabolic equations, and the $\mathbb{R} \oplus \mathbb{R}$-type for hyperbolic equations.

A more delicate problem is to describe submanifolds of the form $\operatorname{sing}_\Sigma W$ for multi-valued solutions W of a given differential equation and a given solution singularity type Σ. Here $\operatorname{sing}_\Sigma W \subset W$ stands for the submanifold of Σ-singular points of W. In other words, we are interested in determining the shape of Σ-singularities admitted by a given equation.

The solution of this problem can be sketched as follows: Let a label solution singularity type Σ be fixed; then it is possible to associate with a given system of partial differential equations $\mathcal{E}$ another system $\mathcal{E}_\Sigma$ such that submanifolds of the form $\operatorname{sing}_\Sigma W$, W being a multi-valued solution of $\mathcal{E}$, satisfy $\mathcal{E}_\Sigma$ and, conversely, every solution of $\mathcal{E}_\Sigma$ is of the form $\operatorname{sing}_\Sigma W$ for (possibly formal) multi-valued solutions of $\mathcal{E}$.

If $\mathcal{E}$ has n independent variables, then $\mathcal{E}_\Sigma$ has $n - s$ independent variables, where s is the dimension of the label Σ. The construction of equations $\mathcal{E}_\Sigma$ is not simple enough to be reproduced here. Instead, in the next section we exhibit some examples from which the reader can conceive an idea of them. Informally speaking, if $\mathcal{E}$ describes a physical substance, say a field or a continuous medium, then $\mathcal{E}_\Sigma$ describes the behavior of a certain kind of singularities of this substance, that can be characterized by the label singularity type Σ. In the case when $\mathcal{E}$ refers to independent space-time variables, the equation $\mathcal{E}_\Sigma$ describes propagation of Σ-type singularities in the substance in question.

Denote by CHAR_Σ the functor that associates the equation $\mathcal{E}_\Sigma$ with a given equation $\mathcal{E}$. The problem:

To what extend does the behavior of singularities of a given type of physical system determine the system itself?

is evidently of a fundamental importance, and the search for the domain of invertibility of the functor CHAR_Σ is maybe the most significant aspect of it. The following result gives an instructive example of this nature.

THE FOLD-RECONSTRUCTION THEOREM. *Every hyperbolic system of partial differential equations $\mathcal{E}$ is determined completely by the associated system $\mathcal{E}_{\mathrm{FOLD}}$.*

In other words, to write down a hyperbolic system $\mathcal{E}$ explicitly, it is sufficient to know the system $\mathcal{E}_{\mathrm{FOLD}}$.

The above theorem can be reformulated by saying that the functor CHAR_Σ is invertible on the class of hyperbolic equations. On the other hand, this functor is not invertible on the class of elliptic equations due to the fact that $\mathcal{E}_{\mathrm{FOLD}}$ is empty for any elliptic $\mathcal{E}$.

The general "singularity reconstruction problem" we are discussing may have various flavors depending on the chosen, not necessary geometric, solution singularity type. For instance, the classical problem of fields and sources can be viewed as a particular case of it. Another remarkable example can be found in the history of electrodynamics. Observing that the elementary laws of electricity and magnetism such as Coulomb's and Faraday's describe the behavior of some kind of singularities of electromagnetic fields, we see that Maxwell's equations deliver a solution to the corresponding singularity reconstruction problem.

The importance of multi-valued solution theory comes in evidence also due to its relations with the Sobolev–Schwartz theory of generalized solutions of linear partial differential equations. These relations are based on the observation that one can get a generalized solution of a given linear differential equation simply by summing up branches of multi-valued solution. As a matter of fact, the procedure assigning the generalized solution to a given multi-valued one is more delicate than a simple summation, and is based on the choice of a de Rham type cohomology theory and a suitable class of test functions. Maslov-type characteristic classes then arise as obstructions to performing this procedure, and their nature depends on the cohomology theory chosen (see [**74, 76, 133**]).

It is worth stressing that generalized solutions assigned to multi-valued ones with no FOLD-singularities are, in fact, smooth, i.e., not properly generalized functions. This correlates nicely with the well-known fact that generalized solutions of elliptic equations are exhausted by smooth solutions, i.e., single-valued ones, while such equations admit nontrivial multi-valued solutions (say, branched Riemannian surfaces for the Cauchy–Riemann equation) with non-FOLD-type singularities. These and other similar facts show multi-valued solutions to be a satisfactory substitution for generalized ones for nonlinear differential equations, where generalized solutions cannot even be defined. Moreover, the former are a finer tool in the framework of the linear theory also.

Further details and results on the topics touched upon in this section can be found in the book [**60**] and in the lecture [**133**]. For a systematic exposition see [**140**]. Many other interesting aspects of solution singularity theory are presented in the recent review by Lychagin [**76**].

Multi-valued solutions were introduced in [**124**], followed by a technically simple but instructive paper [**63**] by Krishchenko. After that, a significant series of papers by Lychagin appeared. Unfortunately, these names almost exhaust the list of contributors in this field. For a full bibliography see [**60, 76, 140**][2].

10. Wave and geometric optics and other examples

In this section we illustrate the generalities of the previous one with some simple examples taken from [**69**]. We enter here neither into technical details nor into interpretations of the equations, referring the reader to [**69**].

10.1. Σ-characteristic equations. Let $\pi\colon E \to M$ be a fiber bundle, $\mathcal{E} \subset J^k(\pi)$ be a system of differential equations, and Σ be a label solution singularity type. The Σ-*characteristic system* of Σ is the system of differential equations whose solutions are of the form $\pi_k(\mathrm{sing}_\Sigma W)$, where $\pi_k\colon J^k(\pi) \to M$ is the natural projection and W is a multi-valued solution of $\mathcal{E}$.

Denote the Σ-characteristic equation of $\mathcal{E}$ by $\mathcal{E}^0_\Sigma$, and observe that the whole system $\mathcal{E}_\Sigma$ is obtained by adding to $\mathcal{E}^0_\Sigma$ some other equations, called *complementary*. If $\mathcal{E}$ refers to independent space-time variables, then $\mathcal{E}^0_\Sigma$ governs motions of the loci of Σ-singularities of the physical system in question, while the complementary equations describe the evolution of the internal structures of Σ-singularities.

Classical characteristic equations, whose theory was initiated by Hugoniot and then developed systematically by Hadamard (see [**40**]), arise naturally in the study of uniqueness of the initial data problem. As we have already mentioned, the uniqueness problem is included in the theory of FOLD-singularities. So, it is not surprising that FOLD-characteristic equations coincide with classical ones. We recommend [**66, 67, 96**], in which the first attempts to apply classical characteristic equations to quantum mechanics and relativity were made.

The coordinate-wise representation of FOLD-characteristic equations looks as follows. Let the basic equation $\mathcal{E}$ be given by

$$F_j(x, u, \ldots, u^i_\sigma, \ldots) = 0, \qquad j = 1, \ldots, l,$$

where $F_j \in \mathcal{F}_k(\pi)$. Introduce the *characteristic matrix* of $\mathcal{E}$,

$$\mathcal{M}_F = \begin{pmatrix} \sum_{|\sigma|=k} \partial F_1/\partial u^1_\sigma p^\sigma & \cdots & \sum_{|\sigma|=k} \partial F_1/\partial u^m_\sigma p^\sigma \\ \cdots\cdots\cdots\cdots\cdots\cdots\cdots\cdots\cdots\cdots\cdots\cdots\cdots\cdots\cdots\cdots \\ \sum_{|\sigma|=k} \partial F_l/\partial u^1_\sigma p^\sigma & \cdots & \sum_{|\sigma|=k} \partial F_l/\partial u^m_\sigma p^\sigma \end{pmatrix},$$

where $p^\sigma = p_1^{i_1} \cdot \ldots \cdot p_n^{i_n}$ for $\sigma = (i_1, \ldots, i_n)$. The FOLD-characteristic equation becomes trivial, i.e., $0 = 0$, for $l < m$. If $l \geq m$, we get the FOLD-characteristic equation $\mathcal{E}^0_{\mathrm{FOLD}}$ of Σ-singular loci representable in the form $x_n = \varphi(x_1, \ldots, x_{n-1})$ by substituting $\partial\varphi/\partial x_i$ for p_i, $i = 1, \ldots, n-1$, and -1 for p_n in $\mathcal{M}_F$, and then equating to zero all m-th order minors of the matrix thus obtained.

REMARK 3. Strictly speaking, the above procedure is valid only for formally integrable $\mathcal{E}$.

10.2. Maxwell's equations and geometric optics. Consider the vacuum Maxwell equations (= "wave optics"):

$$\mathrm{div}\, E = 0, \qquad \mathrm{curl}\, E = -\frac{1}{c}\frac{\partial H}{\partial t},$$

$$\mathrm{div}\, H = 0, \qquad \mathrm{curl}\, H = \frac{1}{c}\frac{\partial E}{\partial t}.$$

---

[2]See also a recent paper by A. Givental [**34**].

In this case, $n = 4$, $m = 6$, $l = 8$. So, the characteristic matrix is 8×6 rectangular and a direct computation shows that all its sixth-order minors are of the form

$$\lambda \left(p_1^2 + p_2^2 + p_3^2 - \frac{1}{c^2} p_4^2 \right)^2,$$

with $\lambda = 0$ or $\pm p_i p_j$. So, putting $x_4 = t$ we see that the equation $\mathcal{E}^0_{\text{FOLD}}$ coincides with the standard eikonal equation

$$\left(\frac{\partial \varphi}{\partial x_1} \right)^2 + \left(\frac{\partial \varphi}{\partial x_2} \right)^2 + \left(\frac{\partial \varphi}{\partial x_3} \right)^2 = \frac{1}{c^2}. \tag{10}$$

In such a way we obtain the interpretation of this well-known fact in terms of the solution singularity theory. However, this gives us something more, namely, the complementary equations that compose, together with the eikonal equation, the whole system $\mathcal{E}_{\text{FOLD}}$. They are

$$\operatorname{div} h_E = \operatorname{grad} \varphi \cdot \operatorname{curl} h_H,$$
$$\operatorname{curl} h_E + \operatorname{div} h_H \cdot \operatorname{grad} \varphi = \operatorname{grad} \varphi \times \operatorname{curl} h_H.$$

Here h_E and h_H are singular values, i.e., values on the singular surface $t = \varphi(x_1, x_2, x_3)$, of the electric and magnetic fields, respectively.

10.3. On the complementary equations. It is obvious from the procedure of §10.1 that very different equations can have the same characteristic equation. For example, the eikonal equation (10) is also the characteristic equation for the Klein–Gordon equation

$$\frac{1}{c^2} \frac{\partial^2 u}{\partial t^2} - \Delta u - m^2 u = 0.$$

So, it is not possible to reconstruct the original equation knowing only its characteristic equation. In view of the reconstruction theorem of §9, the only information one needs for the reconstruction is contained exactly in the complementary equations. Therefore, an independent and direct physical interpretation of quantities in these equations would allow one to make up for missing information needed to solve the corresponding singularity reconstruction problem. One can see now that this *singularity interpretation problem* becomes very important. For example, a solution of this problem for continuous media would provide us with a regular method

> to deduce equations governing a given continuous medium proceeding from observation of how a given type (or types) of singularities propagate in it.

This would be an attractive alternative to the present phenomenological status of continuum mechanics.

It is clear that quantization "à la Schrödinger" can also be treated as such an interpretation problem. In this context the Hamilton–Jacobi equations of classical mechanics, considered as Q-characteristic equations, are to be completed by suitable complementary equations. It is natural to think that the standard formal quantization methods "à la Heisenberg" cover just the remaining gap of these hypothetical complementary equations.

10.4. Alternative singularities via the homogenization trick.
The classical, i.e., FOLD-characteristic, equation for the Schrödinger equation

$$i\hbar \frac{\partial \psi}{\partial t} + \frac{\hbar^2}{2m}\Delta \psi - V\psi = 0, \tag{11}$$

and for singular loci given in the form $t = x_4 = \varphi(x_1, x_2, x_3)$, is

$$\left(\frac{\partial \varphi}{\partial x_1}\right)^2 + \left(\frac{\partial \varphi}{\partial x_2}\right)^2 + \left(\frac{\partial \varphi}{\partial x_3}\right)^2 = 0.$$

This demonstrates that geometric singularities are not adequate for the correspondence between quantum and classical mechanics. For the hypothetical "quantum" singularity type (see §3) the Q-characteristic equation $\mathcal{E}_Q^0$ should be

$$\frac{\partial \varphi}{\partial t} - \frac{1}{2m}\sum_{i=1}^{3}\left(\frac{\partial \varphi}{\partial x_i}\right)^2 - V = 0. \tag{12}$$

It is possible, however, to interpret (12) as the classical characteristic equation for the "homogenized" Schrödinger equation

$$\frac{\partial^2 \widetilde{\psi}}{\partial t \partial s} - \frac{1}{2m}\sum_{i=1}^{3}\frac{\partial^2 \widetilde{\psi}}{\partial x_i^2} + V\frac{\partial^2 \widetilde{\psi}}{\partial s^2} = 0, \tag{13}$$

in five-space with coordinates x_1, x_2, x_3, t, s, if singular loci are given in the form $s - \varphi(x_1, x_2, x_3, t) = 0$. On the other hand, (13) reduces to (11) on the functions

$$\widetilde{\psi} = \psi(x,t)e^{is/\hbar}. \tag{14}$$

This motivates us to define Q-singularities as the reduction of FOLD-singularities on the functions (14). This is not, however, very straightforward, and we refer the reader to [**69**] for some results of this approach.

10.5. $\mathbb{R}_{(k)}$-characteristic equations.
In this subsection, some analogs of the Hamilton–Jacobi equation for extended (i.e., not point-like) singular loci are exhibited. For simplicity, we chose the wave equation

$$\sum_{i=1}^{3}\frac{\partial^2 u}{\partial x_i^2} - \frac{1}{c^2}\frac{\partial^2 u}{\partial t^2} = 0$$

as a basis. Since $\mathbb{R}_{(1)} = \text{FOLD}$, the $\mathbb{R}_{(1)}$-characteristic equation coincides with the standard eikonal equation (10).

For $k = 2$ and the singularity loci given by

$$x_2 = \varphi(s,t), \quad x_3 = \psi(s,t) \quad \text{with } s = x_1,$$

the $\mathbb{R}_{(2)}$-characteristic equation is

$$\left(\frac{\partial \psi}{\partial t}\frac{\partial \varphi}{\partial s} - \frac{\partial \varphi}{\partial t}\frac{\partial \psi}{\partial s}\right)^2 + \left(\frac{\partial \varphi}{\partial t}\right)^2 + \left(\frac{\partial \psi}{\partial t}\right)^2 - c^2\left(\frac{\partial \varphi}{\partial s}\right)^2 - c^2\left(\frac{\partial \psi}{\partial s}\right)^2 - c^2 = 0.$$

Its solutions are two-dimensional surfaces tangent to the light cone.

Finally, the $\mathbb{R}_{(3)}$-characteristic equation for singularity curves of the form $x_i = x_i(t)$, $i = 1, 2, 3$, is

$$\dot{x}_1^2 + \dot{x}_2^2 + \dot{x}_3^2 = c^2.$$

For other examples, results, and discussions, see [**69**].

Bibliography

1. M. J. Ablowitz and H. Segur, *Solitons and the inverse scattering transform*, SIAM Studies in Appl. Math, no. 4, Soc. Ind. Appl. Math., Philadelphia, 1981.
2. R. Abraham and J. E. Marsden, *Foundation of mechanics*, 2nd ed., Benjamin-Cummings, Reading, MA, 1978.
3. D. V. Alekseevsky, A. M. Vinogradov, and V. V. Lychagin, *Basic ideas and concepts of differential geometry*, Geometry I, Encycl. Math. Sci., vol. 28, Springer-Verlag, Berlin, 1991.
4. I. M. Anderson, *Introduction to the variational bicomplex*, Mathematical Aspects of Classical Field Theory (M. Gotay, J. E. Marsden, and V. E. Moncrief, eds.), Contemporary Mathematics, vol. 132, Amer. Math. Soc., Providence, RI, 1992, pp. 51–73.
5. _____, *The variational bicomplex*, to appear.
6. V. I. Arnol'd, *Ordinary differential equations*, Springer-Verlag, Berlin, 1992.
7. V. I. Arnol'd and A. B. Givental, *Symplectic geometry*, Symplectic geometry and its applications, Dynamical Systems IV, Encycl. Math. Sci., vol. 4, Springer-Verlag, Berlin, 1990, pp. 1–136.
8. A. M. Astashov and A. M. Vinogradov, *On the structure of Hamiltonian operators in field theory*, J. Geom. Phys. **3** (1986), 263–287.
9. M. F. Atiyah and I. G. Macdonald, *Introduction to commutative algebra*, Addison-Wesley, Reading, MA, 1969.
10. G. Barnich, F. Brandt, and M. Henneaux, *Local BRST cohomology in Einstein-Yang-Mills theory*, Nuclear Phys. B **445** (1995), 357–408 (E-print hep-th/9505173).
11. _____, *Local BRST cohomology in the antifield formalism*: I. *General theorems*, Comm. Math. Phys. **174** (1995), 57–92 (E-print hep-th/9405109).
12. L. M. Berkovich, *The method of an exact linearization of n-order ordinary differential equations*, Nonlinear Mathematical Physics **3** (1996), 341–350.
13. G. W. Bluman and S. Kumei, *Symmetries and differential equations*, Springer-Verlag, New York, 1989.
14. N. N. Bogolyubov, Jr., A. K. Prikarpatsky, A. M. Kurbatov, and V. G. Samojlenko, *Nonlinear model of Schrödinger type: conservation laws, Hamiltonian structure, and complete integrability*, Theoret. and Math. Phys. **65** (1985), 1054–1164.
15. N. Bourbaki, *Éléments de mathématique. Algèbre. Chapitre X. Algèbre homologique*, Masson, Paris, 1980.
16. R. L. Bryant and P. A. Griffiths, *Characteristic cohomology of differential systems, I: General theory*, J. Amer. Math. Soc. **8** (1995), 507–596 (URL: http://www.math.duke.edu/~bryant).
17. V. N. Chetverikov, *On the structure of integrable C-fields*, Differential Geom. Appl. **1** (1991), 309–325.
18. V. N. Chetverikov and A. G. Kudryavtsev, *A method for computing symmetries and conservation laws of integro-differential equations*, Acta Appl. Math. **41** (1995), 45–56.
19. _____, *Modeling integro-differential equations and a method for computing their symmetries and conservation laws*, in [**78**], pp. 1–22.
20. R. Courant and D. Hilbert, *Methods of mathematical physics. Vol. II: Partial differential equations*, Wiley/Interscience, New York, 1962; reprint, 1989.
21. L. A. Dickey, *Soliton equations and Hamiltonian systems*, World Scientific, Singapore, 1991.
22. R. K. Dodd, J. C. Eilbeck, J. D. Gibbon, and H. C. Morris, *Solitons and nonlinear wave equations*, Academic Press, London, 1982.
23. R. K. Dodd and A. F. Fordy, *The prolongation structures of quasi-polynomial flows*, Proc. Roy. Soc. London Ser. A **385** (1983), 389–429.

24. I. Dorfman, *Dirac structures and integrability of nonlinear evolution equations*, Wiley, Chichester, 1993.
25. B. A. Dubrovin, A. T. Fomenko, and S. P. Novikov, *Modern geometry—methods and applications. Part I. The geometry of surfaces, transformation groups, and fields*, 2nd ed., Springer-Verlag, New York, 1992.
26. S. V. Duzhin and V. V. Lychagin, *Symmetries of distributions and quadrature of ordinary differential equations*, Acta Appl. Math. **24** (1991), 29–57.
27. S. P. Finikov, *A course in differential geometry*, Gostekhizdat Publ., Moscow, 1952 (Russian).
28. A. S. Fokas and V. E. Zakharov (eds.), *Important developments in soliton theory*, Springer-Verlag, Berlin, 1993.
29. V. I. Fushchich and A. G. Nikitin, *Symmetries of Maxwell's equations*, D. Reidel, Dordrecht, 1987.
30. V. I. Fushchich and V. A. Vladimirov, *On an additional invariance of equations for vector fields*, Soviet Physics Dokl. **26** (1981), 396–398.
31. S. I. Gelfand and Yu. I. Manin, *Methods of homological algebra*, Springer-Verlag, Berlin, 1996.
32. D. M. Gessler, *On the Vinogradov C-spectral sequence for determined systems of differential equations*, Differential Geom. Appl. **7** (1997), 303–324 (URL: http://www.botik.ru/~diffiety/).
33. M. Giaquinta and S. Hildebrandt, *Calculus of variations. Vol. 2. The Hamiltonian formalism*, Springer-Verlag, Berlin, 1996.
34. A. B. Givental, *Whitney singularities of solutions of partial differential equations*, J. Geom. Phys. **15** (1995), 353–368.
35. R. Godement, *Topologie algébrique et théorie des faisceaux*, Hermann, Paris, 1958.
36. N. M. Guenther, *Integration of first-order partial differential equations*, ONTI GTTTI Publ., Moscow and Leningrad, 1934 (Russian).
37. V. N. Gusyatnikova, A. V. Samokhin, V. S. Titov, A. M. Vinogradov, and V. A. Yumaguzhin, *Symmetries and conservation laws of Kadomtsev–Pogutse equations (Their computation and first applications)*, in [137], pp. 23–64.
38. V. N. Gusyatnikova, A. M. Vinogradov, and V. A. Yumaguzhin, *Secondary differential operators*, J. Geom. Phys. **2** (1985), 23–65.
39. R. Haag, *Local quantum physics*, Springer-Verlag, 1992.
40. J. Hadamard, *Leçons sur la propagation des ondes*, Hermann, Paris, 1903.
41. J. Hale, *Theory of functional differential equations*, 2nd ed., Springer-Verlag, New York, 1977.
42. M. Henneaux, B. Knaepen, and C. Schomblond, *Characteristic cohomology of p-form gauge theories*, Comm. Math. Phys. **186** (1997), 137–165 (E-print hep-th/9606181).
43. N. H. Ibragimov, *Transformation groups applied to mathematical physics*, Reidel, Dordrecht, 1985.
44. N. H. Ibragimov (ed.), *Symmetries, exact solutions and conservation laws*, CRC Handbook of Lie Group Analysis of Differential Equations, vol. 1, CRC Press, Boca Raton, FL, 1994.
45. N. H. Ibragimov (ed.), *Applications in ingeneering and physical sciences*, CRC Handbook of Lie Group Analysis of Differential Equations, vol. 2, CRC Press, Boca Raton, FL, 1995.
46. N. H. Ibragimov (ed.), *New trends in theoretical developments and computational methods*, CRC Handbook of Lie Group Analysis of Differential Equations, vol. 3, CRC Press, Boca Raton, FL, 1996.
47. C. E. Ince, *University mathematics texts (integration of ordinary differential equations)*, Oliver & Boyd, London, 1939.
48. B. B. Kadomtsev and O. P. Pogutse, *Nonlinear helical plasma perturbation in the tokamak*, Soviet Phys. JETP **38** (1973), 283.
49. E. Kamke, *Differentialgleichungen. Band I: Gewöhnliche Differentialgleichungen*, Akademische Verlag., Geest & Portig, Leipzig, 1962.
50. O. V. Kaptsov, *Extension of the symmetry of evolution equations*, Soviet Math. Dokl. **21** (1982), 173–176.
51. D. J. Kaup, *The Estabrook–Wahlquist method with examples of applications*, Phys. D **1** (1980), 391–411.

52. N. G. Khor'kova, *Conservation laws and nonlocal symmetries*, Math. Notes **44** (1988), 562–568.
53. _____, *Conservation laws and nonlocal symmetries*, Trudy MVTU, no. 512, MVTU Publ., Moscow, 1988, pp. 105–119 (Russian).
54. E. G. Kirnasov, *Wahlquist–Estabrook type coverings over the heat equation*, Math. Notes **42** (1987), 732–739.
55. K. Kiso, *Pseudopotentials and symmetries of evolution equations*, Hokkaido Math. J. **18** (1989), 125–136.
56. N. P. Konopleva and V. N. Popov, *Gauge fields*, Harwood Acad. Publ., Chur, 1981.
57. I. S. Krasil'shchik, *Some new cohomological invariants for nonlinear differential equations*, Differential Geom. Appl. **2** (1992), 307–350.
58. I. S. Krasil'shchik and P. H. M. Kersten, *Deformations and recursion operators for evolution equations*, Geometry in Partial Differential Equations (A. Prastaro and Th. M. Rassias, eds.), World Scientific, Singapore, 1994, pp. 114–154.
59. _____, *Graded differential equations and their deformations: a computational theory for recursion operators*, Acta Appl. Math. **41** (1994), 167–191.
60. I. S. Krasil'shchik, V. V. Lychagin, and A. M. Vinogradov, *Geometry of jet spaces and nonlinear partial differential equations*, Gordon and Breach, New York, 1986.
61. I. S. Krasil'shchik and A. M. Vinogradov, *Nonlocal symmetries and the theory of coverings: An addendum to A. M. Vinogradov's 'Local symmetries and conservation laws'*, Acta Appl. Math. **2** (1984), 79–96.
62. _____, *Nonlocal trends in the geometry of differential equations: Symmetries, conservation laws, and Bäcklund transformations*, Acta Appl. Math. **15** (1989), 161–209.
63. A. P. Krishchenko, *On the bends of R-manifolds*, Moscow Univ. Math. Bull. **32** (1977), no. 1, 13–16.
64. A. B. Kupershmidt, *The variational principles of dynamics*, World Scientific, Singapore, 1992.
65. G. L. Lamb, *Bäcklund transformation at the turn of the centure*, in [**87**], pp. 69–79
66. T. Levi-Città, *Caratteristiche e bicaratteristiche delle equazioni gravitazionali di Einstein*, I, II, Rend. Accad Naz. Lincei (Sci. Fis. Mat. Nat.) **11** (1930), 3–11, 113–121.
67. _____, *Caratteristiche dei sistemi differenziali e propagazione ondosa*, Zanichelli, Bologna, 1988.
68. S. Lie, *Gesammelte Abhandlungen. Bd. 1–4*, Teubner, Leipzig, 1929.
69. F. Lizzi, G. Marmo, G. Sparano, and A. M. Vinogradov, *Eikonal type equations for geometrical singularities of solutions in field theory*, J. Geom. Phys. **14** (1994), 211–235.
70. K. Lonngren and A. Scott (eds.), *Solitons in action*, Academic Press, Boston, 1978.
71. R. K. Luneburg, *Mathematical theory of optics*, University of California Press, Berkeley, CA, 1964.
72. V. V. Lychagin, *Local classification of non-linear first order partial differential equations*, Russian Math. Surveys **30** (1975), no. 1, 105–175.
73. _____, *Contact geometry and non-linear second-order differential equations*, Russian Math. Surveys **34** (1979), no. 1, 149–180.
74. _____, *The geometry and topology of shock waves*, Soviet Math. Dokl. **25** (1982), 685–689.
75. _____, *Singularities of multivalued solutions of nonlinear differential equations, and nonlinear phenomena*, Acta Appl. Math. **3** (1985), 135–173.
76. _____, *Geometric theory of singularities of solutions of non-linear differential equations*, J. Soviet Math. **51** (1990), 2735–2757.
77. _____, *Lectures on geometry of differential equations*, Rome, 1992.
78. _____ (ed.), *The interplay between differential geometry and differential equations*, Amer. Math. Soc. Transl. (2) **167** (1995).
79. S. MacLane, *Homology*, Springer-Verlag, Berlin, 1963.
80. B. Malgrange, *Ideals of differentiable functions*, Oxford University Press, Oxford, 1966.
81. Yu. I. Manin, *Algebraic aspects of nonlinear differential equations*, J. Soviet Math. **11** (1979), 1–122.
82. M. Marvan, *On zero-curvature representations of partial differential equations*, Differential Geometry and Its Applications, Proc. Conf. Opava, 1992, Open Education and Sciences, Opava, 1993, pp. 103–122 (URL: http://www.emis.de/proceedings/).

83. V. P. Maslov, *Asymptotic methods and perturbation theory*, Nauka, Moscow, 1988 (Russian). French transl. of the first Russian edition was published by Dunod, Paris, and Gauthier-Villars, Paris, 1972).
84. J. McCleary, *A user's guide to spectral sequences*, Publish or Perish, Wilmington, DE, 1985.
85. A. V. Mikhajlov, A. B. Shabat, and R. I. Yamilov, *The symmetry approach to the classification of nonlinear equations. Complete lists of integrable systems*, Russian Math. Surveys **42** (1987), no. 4, 1–63.
86. R. M. Miura, *Korteweg–de Vries equation and generalizations.* I: *A remarkable explicit nonlinear transformation*, J. Math. Phys. **9** (1968), 1202–1204.
87. R. M. Miura (ed.), *Bäcklund transformations, the inverse scattering method, solitons, and their applications*, Lecture Notes in Math, vol. 515, Springer-Verlag, Berlin, 1976.
88. R. M. Miura, C. S. Gardner, and M. D. Kruskal, *Korteweg–de Vries equation and generalizations.* II: *Existence of conservation laws and constants of motion*, J. Math. Phys. **9** (1968), 1204–1209.
89. A. C. Newell, *Solitons in mathematics and physics*, CBMS-NSF Reg. Conf. Ser. Appl. Math., vol. 45, SIAM, Philadelphia, PA, 1985.
90. E. Noether, *Invariant variation problems*, Transport Theory Statist. Phys. **1** (1971), 186–207 (German original, *Invariante Variationsprobleme*, Nachr. König. Gesell. Wiss. Göttingen, Math.-Phys. Kl., 1918, pp. 235–257).
91. P. J. Olver, *Applications of Lie groups to differential equations*, 2nd ed., Springer-Verlag, New York, 1993.
92. L.V. Ovsiannikov, *Group analysis of differential equations*, Academic Press, New York, 1982.
93. F. A. E. Pirani, D. C. Robinson, and W. F. Shadwick, *Local jet bundle formulation of Bäcklund transformations*, Reidel, Dordrecht, 1979.
94. M. M. Postnikov, *Lie groups and Lie algebras. Lectures in geometry. Semester V*, Nauka, Moscow, 1982; English transl., Mir, Moscow, 1986.
95. _____, *Smooth manifolds. Lectures in geometry. Semester III. Vol. 1*, URSS Publ., Moscow, 1994.
96. G. Racah, *Characteristiche delle equazioni di Dirac e principio di indeterminazione*, Rend. Accad. Naz. Lincei **13** (1931), 424–427.
97. P. K. Rashevsky, *A course in differential geometry*, GITTL Publ., Moscow and Leningrad, 1950 (Russian).
98. A. V. Samokhin, *Nonlinear MHD-equations*: *Symmetries, solutions, and conservation laws*, Soviet Phys. Dokl. **30** (1985), 1020–1022.
99. _____, *Symmetries of ordinary differential equations*, in [**78**], pp. 193–206.
100. G. Sansone, *Equazioni differenziali nel campo reale.* I, II, 2nd ed., Zanichelli, Bologna, 1948/1949.
101. D. J. Saunders, *The geometry of jet bundles*, Cambrige University Press, Cambrige, 1989.
102. E. Schrödinger, *Abhandlungen zur Wellenmechanik*, Barth, Leipzig, 1927.
103. A. S. Schwarz, *Quantum field theory and topology*, Springer-Verlag, Berlin, 1993.
104. S. I. Senashov and A. M. Vinogradov, *Symmetries and conservation laws of 2-dimensional ideal plasticity*, Proc. Edinburgh Math. Soc. **31** (1988), 415–439.
105. J.-P. Serre, *Lie algebras and Lie groups*, 2nd ed., Springer-Verlag, Berlin, 1992.
106. W. F. Shadwick, *The Bäcklund problem for the equation $z_{xy} = f(z)$*, J. Math. Phys. **19** (1978), 2312–2317.
107. _____, *The KdV prolongation algebra*, J. Math. Phys. **21** (1980), 454–461.
108. I. R. Shafarevich, *Basic algebraic geometry.* I: *Varieties in projective space.* II: *Schemes and complex manifolds*, 2nd rev. exp. ed., Springer-Verlag, Berlin, 1994.
109. N. O. Sharomet, *Symmetries, invariant solutions and conservation laws of the nonlinear acoustics equation*, in [**137**], pp. 83–120.
110. V. E. Shemarulin, *A direct method of inverse problems solution*, to appear.
111. M. Smoluchowski, *Drei Vorträge über Diffusion, Brownische Bewegung und Koagulation von Kolloidteilchen*, Phys. Z. **17** (1916), 557–585.
112. V. V. Sokolov and A. B. Shabat, *Classification of integrable evolution equations*, Soviet Sci. Rev. Section C: Math. Phys. Rev., vol. 4, Harwood Academic Publ., Chur, 1984, pp. 221–280.
113. V. V. Stepanov, *A course in differential equations*, GITTL, Moscow, 1953; Getman transl., VEB Deutscher Verlag. Wiss., Berlin, 1963.

114. H. Stephani, *Differential equations: their solutions using symmetries*, Cambridge University Press, Cambridge, 1989.
115. S. Sternberg, *Lectures on differential geometry*, 2nd ed., Chelsea, New York, 1983.
116. S. I. Svinolupov and V. V. Sokolov, *Factorization of evolution equations*, Russian Math. Surveys **47** (1992), no. 3, 127–162.
117. T. Tsujishita, *On variation bicomplexes associated to differential equations*, Osaka J. Math. **19** (1982), 311–363.
118. _____, *Formal geometry of systems of differential equations*, Sugaku Expositions **3** (1990), 25–73.
119. _____, *Homological method of computing invariants of systems of differential equations*, Differential Geom. Appl. **1** (1991), 3–34.
120. A. M. Verbovetsky, *Notes on the horizontal cohomology*, Secondary Calculus and Cohomological Physics (M. Henneaux, I. S. Krasil'shchik, and A. M. Vinogradov, eds.), Contemporary Mathematics, vol. 219, Amer. Math. Soc., Providence, RI, 1998, pp. 211–231 (E-print math.DG/9803115).
121. A. M. Verbovetsky, A. M. Vinogradov, and D. M. Gessler, *Scalar differential invariants and characteristic classes of homogeneous geometric structures*, Math. Notes **51** (1992), 543–549.
122. E. B. Vinberg and A. L. Onishchik, *A seminar on Lie groups and algebraic groups*, Springer-Verlag, New York, 1990.
123. A. M. Vinogradov, *The logic algebra for the theory of linear differential operators*, Soviet Math. Dokl. **13** (1972), 1058–1062.
124. _____, *Multivalued solutions and a principle of classification of nonlinear differential equations*, Soviet Math. Dokl. **14** (1973), 661–665.
125. _____, *On algebro-geometric foundations of Lagrangian field theory*, Soviet Math. Dokl. **18** (1977), 1200–1204.
126. _____, *Hamilton structures in field theory*, Soviet Math. Dokl. **19** (1978), 790–794.
127. _____, *A spectral sequence associated with a nonlinear differential equation and algebro-geometric foundations of Lagrangian field theory with constraints*, Soviet Math. Dokl. **19** (1978), 144–148.
128. _____, *Some homology systems associated with the differential calculus in commutative algebras*, Russian Math. Surveys **34** (1979), no. 6, 250–255.
129. _____, *The theory of higher infinitesimal symmetries of nonlinear partial differential equations*, Soviet Math. Dokl. **20** (1979), 985–990.
130. _____, *Geometry of nonlinear differential equations*, J. Soviet Math. **17** (1981), 1624–1649.
131. _____, *Local symmetries and conservation laws*, Acta Appl. Math. **2** (1984), 21–78.
132. _____, *The C-spectral sequence, Lagrangian formalism, and conservation laws. I. The linear theory. II. The nonlinear theory*, J. Math. Anal. Appl. **100** (1984), 1–129.
133. _____, *Geometric singularities of solutions of nonlinear partial differential equations*, Differential Geometry and Its Applications, Proc. Conf. Brno, 1986, J. E. Purkyně Univ., Brno, 1987, pp. 359–379.
134. _____, *Integrability and symmetries*, Nonlinear Waves: Structures and Bifurcations (A. V. Gaponov-Grekhov and M. I. Rabinovich, eds.), Nauka, Moscow, 1987, pp. 279–290. (Russian)
135. _____, *An informal introduction to geometry of jet spaces*, Rend. Sem. Fac. Sci. Univ. Cagliari **58** (1988), 301–329.
136. _____, *Symmetries and conservation laws of partial differential equations: Basic notions and results*, in [**137**], pp. 3–21.
137. _____ (ed.), *Symmetries of partial differential equations. Conservation laws - Applications - Algorithms*, Reprinted from Acta Appl. Math. **15** & **16** (1989), Kluwer, Dordrecht, 1989.
138. _____, *Scalar differential invariants, diffieties and characteristic classes*, Mechanics, Analysis and Geometry: 200 Years after Lagrange, Elsevier, 1991, pp. 379–414.
139. _____, *From symmetries of partial differential equations towards secondary ("quantized") calculus*, J. Geom. Phys. **14** (1994), 146–194 (see the Appendix).
140. _____, *On geometric solution singularities*, in preparation.
141. A. M. Vinogradov and I. S. Krasil'shchik, *A method of computing higher symmetries of nonlinear evolution equations and nonlocal symmetries*, Soviet Math. Dokl. **22** (1980), 235–239.
142. _____, *On the theory of nonlocal symmetries of nonlinear partial differential equations*, Soviet Math. Dokl. **29** (1984), 337–341.

143. A. M. Vinogradov and B. A. Kupershmidt, *The structures of Hamiltonian mechanics*, Integrable Systems: Selected Papers, London Math. Soc. Lecture Note Series, no. 60, London, 1981, pp. 173–239 (Reprinted from Russ. Math. Surveys **32** (1977), no. 6, 177–243).
144. V. M. Voloshchuk, *The kinetic coagulation theory*, Gidrometeoizdat Publ., Leningrad, 1984 (Russian).
145. H. D. Wahlquist and F. B. Estabrook, *Prolongation structures of nonlinear evolution equations*, J. Math. Phys. **16** (1975), 1–7.
146. F. W. Warner, *Foundations of differentiable manifolds and Lie groups*, Springer-Verlag, New York, 1983.
147. R. White, D. Monticello, M. N. Rosenbluth, H. Strauss, and B. B. Kadomtsev, *Numerical study of nonlinear evolution of kink and tearing modes in tokamaks*, Proc. Internat. Conf. Plasma Phys. (Tokyo, 1974), vol. 1, IAEA, Vienna, 1975, pp. 495–504.
148. V. V. Zharinov, *Geometrical aspects of partial differential equations*, World Scientific, Singapore, 1992.

Index

action, *see* Lagrangian
adapted coordinates, 73
adjoint operator, 191, 207
algebra of nonlocal symmetries, 251
algebra with filtration, *see* filtered algebra
almost contact manifold, 49

Bäcklund autotransformation, 239
Bäcklund transformation, 239, 249
  for KdV, 240
  for mKdV, 239
  for the sine-Gordon equation, 239
Bianchi–Lie theorem, 26
boundary differential equation, 272, 275
boundary differential operator, 278, 279
branch point, *see* singular point
bundle of infinite jets, 124
bundle of $(k, \mathcal{G})$-jets, 275
Burgers equation, 71, 96, 130, 167, 169, 228
  classical symmetries, 96
  conservation laws, 208
  coverings, 230, 241
  factorization, 114
  higher symmetries, 167, 173
  nonlocal symmetries, 223, 254
  recursion operators, 225

canonical coordinates, 74
  local, 38
Cartan connection, 133
Cartan distribution, 3, 7, 39, 41, 282, 283, 291
  on $\mathcal{E}^\infty$, 159
  on $J^\infty(\pi)$, 139
  on $J^k(\pi)$, 76
  on $J^k(n, m)$, 70
  on a differential equation, 78
Cartan form, 77, 140, 234, 284
Cartan module, 284
Cartan plane, 39, 76
category of differential equations, 227, 311
  dimension of objects, 227
  morphisms, 227
  objects, 227; *see also* diffiety
Cauchy data, 57
$\mathcal{C}$-complete equation, 116
$\mathcal{C}$-differential operator, 189, 207

$\mathcal{C}$-general equation, 116
characteristic, 16, 56
characteristic direction field, 56
characteristic distribution, 17
  of a form, 49
characteristic equations, 324
characteristic field, 10
characteristic matrix, 324
characteristic symmetry of a distribution, 15
characteristic vector field, 49
charge density, *see* conserved density
Clairaut equation, 5
  complete integral, 47
  exceptional integrals, 48
classical finite symmetry of a differential equation, *see* finite symmetry
classical infinitesimal symmetry of a differential equation, 92
classical symmetries, 290
  of the Burgers equation, 96
  of the Kadomtsev–Pogutse equations, 103
  of the KdV equation, 98
  of the Khokhlov–Zabolotskaya equation, 99
coagulation equation, *see* Smoluchowski equation
Cole–Hopf transformation, 224
common equation, 116
commutator relation for symmetries, 164
complementary equations, 324
complete integral, 60, 63
  of the Clairaut equation, 47
completely integrable distribution, 138
composition of differential operators, 128
conservation laws, 186, 188
  of an equation, 168
  of the Burgers equation, 208
  of the filtration equation, 209
  of the heat equation, 209
  of the Kadomtsev–Pogutse equations, 214
  of the KdV equation, 210
  of the Khokhlov–Zabolotskaya equation, 212
  of the Navier–Stokes equations, 213
  of the nonlinear Schrödinger equation, 211
  of the plasticity equations, 212

of the Zakharov equations, 211
conserved current, 186
conserved density, 186
contact element, 37
contact manifold, 49
contact symmetry
  (finite), 46
  (infinitesimal), 54
contact transformation, 42
  (infinitesimal), 50
contact vector field, 32, 50
covering, 226, 227, 311
covering associated to a local symmetry, 228
covering associated to an operator, 227
covering equation, 235
coverings
  over the f-Gordon equation, 248
  over the Burgers equation, 230, 241
  over the equation $u_{xx} + u_{yy} = \varphi(u)$, 249
  over the KdV equation, 236, 244
  over the nonlinear heat equation, 247
  over the potential KdV equation, 225
$\mathcal{C}$-spectral sequence, 187, 317

Darboux transformation, 115
DE category, see category of differential equations
defining equations
  for classical symmetries, 94
  for higher symmetries, 162, 164, 292
density of a Lagrangian, 215
differential equation of order k, 69, 78
differential forms on $J^\infty(\pi)$, 134
differentially closed ideal, 158
diffiety, 310
diffiety dimension, 310
Dimension in the DE category, 227
distribution, 11
distribution on $J^\infty(\pi)$, 137
dressing procedure, 115
Dym equation, see Harry Dym equation

Einstein equations, 199
equivalent coverings, 229
equivalent equations, 46
equivalent predistributions on $J^\infty(\pi)$, 137
Euler operator, 195, 197, 215
Euler transformation, 44
evolutionary derivation, 147, 290
exceptional integral of the Clairaut equation, 48
extrinsic higher symmetry, 161
extrinsic symmetry, 116, 292

f-Gordon equation, 248; see also sine-Gordon equation
  coverings, 248
factorization, 7, 108, 228

of evolution equations in one space variable, 114
of the Burgers equation, 114
of the heat equation, 114, 228
of the Laplace equation, 110, 113
of the wave equation, 114
filtered algebra, 125
filtered module, 130
filtration degree, 125
filtration equation
  conservation laws, 209
finite symmetry
  of a differential equation, 92
  of a distribution, 11
first integral, 59
  of a distribution, 21
first-order differential equation, 41
flux, 186
FOLD-Reconstruction Theorem, 323
formal solution of a boundary differential equation, 291
functional differential equation, see boundary differential equation

Galilean transformation, 84
generalized solution of a differential equation, 41, 71, 78
generating function, 30, 202, 207
  of a contact field, 52
  of a Lie field, 92
generating section, 32
  of a Lie field, 92
  of an evolutionary derivation, 147
generic point of an equation, 180
geometric singularity, 319
G-invariant solution, 95
$\mathfrak{g}$-invariant solution, 95
$\mathcal{G}$-invariant symmetry, 288

Hamilton–Jacobi equation, 58
Hamiltonian, 218
Hamiltonian equation, 205, 206
Hamiltonian evolution equation, 218
Hamiltonian operator, 204, 217
Hamiltonian structure, 204
Hamiltonian vector field, 205
Harry Dym equation
  Hamiltonian structures, 220
heat equation, 167, 170, 224, 228, 256, 260
  conservation laws, 209
  factorization, 114, 228
  higher symmetries, 167, 173
  recursion operators, 175
higher Jacobi bracket, 33, 148
higher KdV equations, 123, 224
higher symmetries
  of a differential equation, 32, 158, 160, 292
  of ordinary differential equations, 180, 184
  of the Burgers equation, 167, 173

of the heat equation, 167, 173
of the KdV equation, 224
of the plasticity equations, 177, 179
transformation under change of variables, 178
hodograph transformation, 85
horizontal de Rham cohomology, 187, 313
horizontal de Rham complex, 136, 187, 188
horizontal differential, 313
horizontal differential form, 136, 234, 313
horizontal module, 189
horizontal operator, see $\mathcal{C}$-differential operator
horizontalization, 30

ideal of a distribution, 137
ideal of an equation, 157
identically conserved current, see trivial conserved current
infinite jet of a section, 124
infinite prolongation of a differential equation, 156, 291
infinitely prolonged equation, see infinite prolongation of a differential equation
infinitesimal automorphism of a distribution, 138
infinitesimal contact symmetry, 54
infinitesimal contact transformation, 50
infinitesimal interior symmetry, 115
infinitesimal symmetry of a distribution, 12, 14
integral manifold of a distribution, 138
integral of motion, see conservation law
integral submanifold, 320
integrating factor, 24
intrinsic coordinates on $\mathcal{E}^\infty$, 164
intrinsic symmetry, 115
invariant manifold of a distribution, 231
invariant solution, 20, 95; see also G-invariant and $\mathfrak{g}$-invariant solution
of the Kadomtsev–Pogutse equations, 104
of the Khokhlov–Zabolotskaya equation, 101
irreducible covering, 231

Jacobi bracket, 53, 93, 148, 290
jet of a section, 75, 275

Kadomtsev–Pogutse equations, 102
classical symmetries, 103
conservation laws, 214
invariant solutions, 104
reproduction of solutions, 106
KdV equation, see Korteweg–de Vries equation
Khokhlov–Zabolotskaya equation, 99
classical symmetries, 99
physically meaningful symmetries, 100
conservation laws, 212

invariant solutions, 101
nonlocal symmetries, 302
Korteweg–de Vries equation, 98, 123
Bäcklund transformations, 240
classical symmetries, 98
conservation laws, 210
coverings, 236, 244
Hamiltonian structures, 219
higher symmetries, 224
master symmetry, 263
nonlocal symmetries, 224, 253, 262
recursion operators, 220, 224, 262

label of a singular R plane, 321
Lagrange–Charpit method, 65
Lagrangian, 215
Lagrangian derivative, 215; see also Euler operator
Lagrangian plane, 56
Laplace equation
factorization, 110, 113
Laplace invariants, 228
Laplace transformation, 227, 303
Legendre transformation, 11, 44, 87
Lenard recursion operator, 220; see also recursion operator
Lie equations, 14
Lie field, 89
Lie fields on $J^\infty(\pi)$, 149
Lie transformation, 86
lifting
of a Lie field, 89
of a linear boundary differential operator, 280
of a linear differential operator, 134
of a point transformation, 85
of a submanifold, 79
of a transformation, 44
of a vector field, 31, 51, 133
linear conservation law, 209
linear Lie equations, 14
linearization, see universal linearization operator
linearization at a section, 151
Liouville equation, 249
ℓ-normal equation, 198, 207
locally maximal integral manifold of a distribution, 138
locally maximal integral submanifold, 320
l-solvable equation, 117

manifold of infinite jets, 124
manifold of $(k, \mathcal{G})$-jets, 275
manifold of 1-jets, 38
master symmetry, 263
matrix nonlinear differential operator, 126
Maxwell equations, 199, 303
Mayer bracket, 53
Miura–Gardner transformation, 236

mKdV equation, *see* modified Korteweg–de Vries equation
modified Korteweg–de Vries equation, 236, 254
  Bäcklund transformations, 239
  recursion operators, 264
module with filtration, *see* filtered module
morphism in the DE category, 227
multi-valued solution of a differential equation, 3, 320

Navier–Stokes equations
  conservation laws, 213
Newton equations
  symmetries and conservation laws, 216
Noether map, 203
Noether theorem, 215
nondegenerate complete integral, 61
nondegenerate subspace, 22
nonlinear differential operator
  matrix, 126
  scalar, 126
nonlinear heat equation
  coverings, 247
nonlinear Lie equations, 14
nonlinear Schrödinger equation
  conservation laws, 211
  recursion operators, 264
nonlocal conservation law, 187, 234
nonlocal coordinates, 229
nonlocal symmetries, 224, 251, 311
  of the Burgers equation, 223, 254
  of the Khokhlov–Zabolotskaya equation, 302
  of the KdV equation, 224, 253, 262
  of the Smoluchowski equation, 294
  of type τ, 251
nonsingular contact element, 37
nontrivial symmetry of a distribution, 15, 16
normal equation, 117

object of the DE category, 227
1-jet, 38
1-jets manifold, 38
order of a covering, 235
ordinary differential equations
  higher symmetries, 180, 184
  point symmetries, 184
oricycle foliation, 18

pedal transformation, 45
pKdV equation, *see* potential Korteweg–de Vries equation
plasticity equations, 175
  conservation laws, 212
  higher symmetries, 177, 179
  recursion operator, 177
point symmetries
  of ordinary differential equations, 184

point transformation, 31, 44, 85
Poisson bracket, 54, 203, 217
potential Korteweg–de Vries equation, 225, 236
  coverings, 225
predistribution on $J^\infty(\pi)$, 137
prolongation
  of a boundary differential equation, 291
  of a boundary differential operator, 280
  of a differential equation, 155
  of a differential operator, 128, 129
  of a submanifold, *see* lifting of a submanifold
prolongation structure, 242, 245
proper conservation law, 202, 208

"quantized" differential form, 317
"quantized" operator, 315
quotient equation, 8, 9

R-manifold, 320
R-plane, 75
ray, *see* ray submanifold
ray submanifold, 79
recursion operator, 175, 177, 220, 224, 261
  for the Burgers equation, 225
  for the heat equation, 175
  for KdV, 224, 262
  for mKdV, 264
  for the nonlinear Schrödinger equation, 264
  for the plasticity equations, 177
  for the sine-Gordon equation, 264
reduced MHD equations, *see* Kadomtsev–Pogutse equations
reducible covering, 231
regular equation, 198
reproduction of solutions, 94
  of the Kadomtsev–Pogutse equations, 106
rigid conservation law, *see* topological conservation law
rigid equation, 115

scalar differential invariants, 114
scalar nonlinear differential operator, 126
scale symmetry, 96
scale transformation, 84
secondary ("quantized") differential form, 317
secondary ("quantized") differential equations, 309
secondary ("quantized") operator, 315
self-similar solution, 96; *see also* invariant solution
semigroup of a boundary differential system, 272
Σ-characteristic system, 324
sine-Gordon equation, 249
  Bäcklund transformations, 239

Lagrangian, 216
  Noether symmetries, 216
  recursion operator, 216, 264
singular point, 3, 320
  of an equation, 42
singular R-plane, 321
singularity interpretation problem, 325
Smoluchowski equation, 274
  symmetries, 294
smooth distribution, 11
smooth functions on $J^\infty(\pi)$, 124
smooth mapping of jet manifolds, 126
solution of a boundary differential equation, 275
solution of a differential equation, 3
solvable Lie algebra, 26
space of k-jets of a bundle, 74
special characteristic classes, 318
special coordinates, 74
  local, 38
symmetry of a differential equation, 92
symmetry of a distribution, 11, 12, 14
  on $J^\infty(\pi)$, 138
symmetry of a dynamical system, *see* symmetry of a vector field
symmetry of a vector field, 18
symplectic operator, 206
system in involution, 67

topological conservation law, 202, 207
total derivative operator, 29, 86, 134, 282
total Jacobian, 86
translation, 84
trivial conserved current, 186
trivial covering, 230
trivial symmetry of a distribution, *see* characteristic symmetry of a distribution
type of a singular R plane, 321

universal Abelian covering, 260
universal algebra of a covering, 242
universal element, 40
universal evolutionary differential, 148
universal linearization operator, 31, 152, 290

variational complex, 197
variational derivative, *see* Lagrangian derivative
variational functional, *see* Lagrangian
vector field depending on time, 49
vector field on $J^\infty(\pi)$, 131
vertical operator, 315
vertical secondary operator, 316
vertical vector field, 132, 161, 234

wave equation, 249
  factorization, 114
weights, 174
Whitney product of coverings, 231

Yang–Mills equations, 199

Zakharov equations
  conservation laws, 211
  Hamiltonian structure, 218
  symmetries, 218

Selected Titles in This Series

(Continued from the front of this publication)

148 **Vladimir I. Piterbarg,** Asymptotic methods in the theory of Gaussian processes and fields, 1996

147 **S. G. Gindikin and L. R. Volevich,** Mixed problem for partial differential equations with quasihomogeneous principal part, 1996

146 **L. Ya. Adrianova,** Introduction to linear systems of differential equations, 1995

145 **A. N. Andrianov and V. G. Zhuravlev,** Modular forms and Hecke operators, 1995

144 **O. V. Troshkin,** Nontraditional methods in mathematical hydrodynamics, 1995

143 **V. A. Malyshev and R. A. Minlos,** Linear infinite-particle operators, 1995

142 **N. V. Krylov,** Introduction to the theory of diffusion processes, 1995

141 **A. A. Davydov,** Qualitative theory of control systems, 1994

140 **Aizik I. Volpert, Vitaly A. Volpert, and Vladimir A. Volpert,** Traveling wave solutions of parabolic systems, 1994

139 **I. V. Skrypnik,** Methods for analysis of nonlinear elliptic boundary value problems, 1994

138 **Yu. P. Razmyslov,** Identities of algebras and their representations, 1994

137 **F. I. Karpelevich and A. Ya. Kreinin,** Heavy traffic limits for multiphase queues, 1994

136 **Masayoshi Miyanishi,** Algebraic geometry, 1994

135 **Masaru Takeuchi,** Modern spherical functions, 1994

134 **V. V. Prasolov,** Problems and theorems in linear algebra, 1994

133 **P. I. Naumkin and I. A. Shishmarev,** Nonlinear nonlocal equations in the theory of waves, 1994

132 **Hajime Urakawa,** Calculus of variations and harmonic maps, 1993

131 **V. V. Sharko,** Functions on manifolds: Algebraic and topological aspects, 1993

130 **V. V. Vershinin,** Cobordisms and spectral sequences, 1993

129 **Mitsuo Morimoto,** An introduction to Sato's hyperfunctions, 1993

128 **V. P. Orevkov,** Complexity of proofs and their transformations in axiomatic theories, 1993

127 **F. L. Zak,** Tangents and secants of algebraic varieties, 1993

126 **M. L. Agranovskiĭ,** Invariant function spaces on homogeneous manifolds of Lie groups and applications, 1993

125 **Masayoshi Nagata,** Theory of commutative fields, 1993

124 **Masahisa Adachi,** Embeddings and immersions, 1993

123 **M. A. Akivis and B. A. Rosenfeld,** Élie Cartan (1869–1951), 1993

122 **Zhang Guan-Hou,** Theory of entire and meromorphic functions: deficient and asymptotic values and singular directions, 1993

121 **I. B. Fesenko and S. V. Vostokov,** Local fields and their extensions: A constructive approach, 1993

120 **Takeyuki Hida and Masuyuki Hitsuda,** Gaussian processes, 1993

119 **M. V. Karasev and V. P. Maslov,** Nonlinear Poisson brackets. Geometry and quantization, 1993

118 **Kenkichi Iwasawa,** Algebraic functions, 1993

117 **Boris Zilber,** Uncountably categorical theories, 1993

116 **G. M. Fel'dman,** Arithmetic of probability distributions, and characterization problems on abelian groups, 1993

115 **Nikolai V. Ivanov,** Subgroups of Teichmüller modular groups, 1992

114 **Seizô Itô,** Diffusion equations, 1992

113 **Michail Zhitomirskiĭ,** Typical singularities of differential 1-forms and Pfaffian equations, 1992

112 **S. A. Lomov,** Introduction to the general theory of singular perturbations, 1992

(See the AMS catalog for earlier titles)

Copying and reprinting. Individual readers of this publication, and nonprofit libraries acting for them, are permitted to make fair use of the material, such as to copy a chapter for use in teaching or research. Permission is granted to quote brief passages from this publication in reviews, provided the customary acknowledgment of the source is given.

Republication, systematic copying, or multiple reproduction of any material in this publication (including abstracts) is permitted only under license from the American Mathematical Society. Requests for such permission should be addressed to the Assistant to the Publisher, American Mathematical Society, P. O. Box 6248, Providence, Rhode Island 02940-6248. Requests can also be made by e-mail to reprint-permission@ams.org.